Geometry of Banach Spaces and Related Fields

Proceedings of Symposia in
PURE MATHEMATICS

Volume 106

Geometry of Banach Spaces and Related Fields

Gilles Godefroy
Mohammad Sal Moslehian
Juan Benigno Seoane-Sepúlveda
Editors

AMS
AMERICAN MATHEMATICAL SOCIETY
Providence, Rhode Island

2020 *Mathematics Subject Classification.* Primary 43A22, 46B04, 46B15, 46B20, 46E15, 46E30, 46E40, 46H25, 46L52, 47B49.

For additional information and updates on this book, visit
www.ams.org/bookpages/pspum-106

Library of Congress Cataloging-in-Publication Data

Names: Godefroy, Gilles, editor. | Moslehian, Mohammad Sal, 1966– editor. | Seoane Sepúlveda, Juan B., editor.
Title: Geometry of Banach spaces and related fields / Gilles Godefroy, Mohammad Sal Moslehian, Juan Benigno Seoane-Sepúlveda, editors.
Description: Providence, Rhode Island : American Mathematical Society, [2024] | Series: Proceedings of symposia in pure mathematics, 0082-0717 ; volume 106 | Includes bibliographical references.
Identifiers: LCCN 2023053394 | ISBN 9781470475703 (paperback) | ISBN 9781470476748 (ebook)
Subjects: LCSH: Banach spaces. | AMS: Abstract harmonic analysis – Abstract harmonic analysis – Homomorphisms and multipliers of function spaces on groups, semigroups, etc.. | Functional analysis – Normed linear spaces and Banach spaces; Banach lattices – Isometric theory of Banach spaces. | Functional analysis – Normed linear spaces and Banach spaces; Banach lattices – Summability and bases. | Functional analysis – Normed linear spaces and Banach spaces; Banach lattices – Geometry and structure of normed linear spaces. | Functional analysis – Linear function spaces and their duals – Banach spaces of continuous, differentiable or analytic functions. | Functional analysis – Linear function spaces and their duals – Spaces of measurable functions spaces, Orlicz spaces, Köthe function spaces, Lorentz spaces, rearrangement invariant spaces, ide | Functional analysis – Linear function spaces and their duals – Spaces of vector- and operator-valued functions. | Functional analysis – Topological algebras, normed rings and algebras, Banach algebras – Normed modules and Banach modules, topological modules. | Functional analysis – Self-adjoint operator algebras (algebras, von Neumann algebras, etc.) – Noncommutative function spaces. | Operator theory – Special classes of linear operators – Transformers, preservers (operators on spaces of operators).
Classification: LCC QA322.2 .G495 2024 | DDC 515/.732–dc23/eng/20231220
LC record available at https://lccn.loc.gov/2023053394

10 9 8 7 6 5 4 3 2 1   29 28 27 26 25 24

Contents

Preface

This book presents a comprehensive examination of contemporary theories in the field of Banach spaces, uniformly convex spaces, function spaces, and Banach algebras. This scholarly work delves into the intriguing realm of geometric and analytic properties associated with Banach spaces and Banach algebras, encompassing current subjects and presenting open problems. Renowned mathematicians have contributed to each chapter, ensuring high-quality content that is both accessible and self-contained. Extensive bibliographies accompany each chapter, enhancing the scholarly value of the text.

The book consists of twelve chapters. The chapters are written in a reader-friendly style and can be read independently.

Chapter 1 gives a concise introduction to the significant mathematical contributions of Per Enflo.

Chapter 2 is devoted to exploring the quasi-greedy basic sequence problem, an evolutionary concept that has roots in significant issues from classical Banach space theory, including the basis problem and the unconditional basic sequence problem.

Chapter 3 delves into the examination of the finiteness principle, the structure of balls of geometric preduals, simultaneous extension from trace spaces, and approximation and biduality properties of geometric preduals.

Chapter 4 provides an overview of recent findings regarding the denseness of the set of mappings attaining their norms or numerical radii. It also explores the applications of these results to geometric concepts of Banach spaces, particularly the Daugavet property.

Chapter 5 offers a comprehensive review of different aspects related to translation-invariant Banach function spaces over an infinite compact abelian group, along with a focus on order-bounded operators.

Chapter 6 explores the minimal projections onto cycle spaces in the respective ℓ_1-spaces. This investigation is closely tied to invariant projections on the cycle space, which are determined by the properties of representations of the automorphism group of the associated graph.

Chapter 7 introduces a novel approach to extending classical convexity to modules through the concept of unit segments, facilitating the definition of locally convex module topologies.

Chapter 8 is a survey of the latest advancements in the topic of derivations, specifically those with values in an order ideal of the $*$-algebra of all measurable operators affiliated with a semifinite von Neumann algebra.

Chapter 9 deals with the action of finitely truncated singular integral operators on functions taking values in a Banach space. Additionally, this chapter addresses a query posed by Naor and Hytönen.

Chapter 10 comprises a collection of intriguing problems within the realms of analysis and geometry of Banach spaces.

Chapter 11 provides an up-to-date survey of the results related to the geometry of low-dimensional polynomial Banach spaces.

Chapter 12 assembles notable contributions concerning metric invariants for complex Banach algebras and Jordan-Banach algebras, including comprehensive arguments and presenting numerous novel findings.

The book encompasses cutting-edge progressions in the theories of Banach spaces, uniformly convex spaces, function spaces, and Banach algebras, thereby ensuring that readers remain abreast of the latest advancements in the field. With its comprehensive coverage of Banach space operators, the book serves as an indispensable resource for graduate students, mathematicians, and researchers engaged in functional analysis, operator theory, and allied disciplines.

The editors extend their heartfelt gratitude to both the authors for their outstanding chapter contributions and the numerous mathematicians who dedicated their time to meticulously reviewing the chapters featured in this book. The invaluable feedback and insightful comments offered by these reviewers have significantly enhanced the quality of the chapters.

G. Godefroy
Institut de Mathématiques de Jussieu, France

M. S. Moslehian
Ferdowsi University of Mashhad, Iran

J. B. Seoane-Sepúlveda
Universidad Complutense de Madrid, Spain

Proceedings of Symposia in Pure Mathematics
Volume **106**, 2024
https://doi.org/10.1090/pspum/106/01924

A Few Lines on Per Enflo's Works

Gilles Godefroy, Mohammad Sal Moslehian,
and Juan Benigno Seoane-Sepúlveda

Abstract. A brief historical overview of the notable mathematical contributions of Per Enflo is presented.

Saying that Per Enflo (born on the 20th May 1944 in Sweden; Figure 1) is a multi-talented man is an understatement. He was a child prodigy in mathematics and music as well, and his first public piano performance with the Royal Opera Orchestra of Sweden took place when he was about twelve years old. His versatility within science is impressive as well, since besides mathematics he contributed significantly to population genetics and paleoanthropology. However, we will focus on his utterly important contribution to mathematics. Let us try to outline, with as few technicalities as possible, some of the outstanding theorems and original ideas which we owe to Per Enflo, and display their lasting influence on fundamental or applied mathematics.

Figure 1. Per H. Enflo, Rome (2017). Courtesy of Lena Enflo.

Per Enflo's early works, when he was still a student working under the supervision of Hans Radström at Stockholm University, primarily concerned non-linear geometry of Banach spaces, and Hilbert's fifth problem in the infinite-dimensional

2020 *Mathematics Subject Classification.* Primary 01A60; Secondary 47A15.
Key words and phrases. Per H. Enflo, history of mathematics, invariant subspace problem, approximation problem, basis problem.

case: if a topological group is locally homeomorphic to a Banach space, does it carry a differential structure? The synergy between technical power and original thinking which is Per Enflo's trademark led him to spectacular progress on this topic. It should be stressed that at the time, non-linear questions were sometimes regarded as features of intellectual curiosity since the world of Banach spaces consisted mainly of linear concepts and results. Today's point of view is quite different since ideas and results from the non-linear geometry of Banach spaces turned out to be well-suited for applications to theoretical computer science and algorithmic complexity, as shown e.g. by Assaf Naor and some of his co-authors. Enflo's metric type was an early bird in this field of research, a crucial definition, and a starting point for many of his followers.

In 1972, the world of functional analysis was shaken by the news that a relatively unknown young Swedish mathematician had solved negatively the approximation problem [**2**], and thus the basis problem for Banach spaces, and Per Enflo immediately became a star of first magnitude. Banach spaces that have the approximation property, and among them those that have a Schauder basis, are spaces in which computation is possible and elementary linear algebra is applicable. In rough terms, linear operators can be represented by matrices and this is extremely useful. Before Enflo's work, experts were hesitant about the statements themselves, and some conjectured that all reflexive spaces have the approximation property. The problem was already present in the Scottish book, and Grothendieck's work had demonstrated its relevance to giving a well-defined value to the trace of nuclear operators. Enflo's deep and somewhat unexpected result opened a new era in functional analysis, which is lasting to this day since the approximation property in its various guises remains a very active field of research. It should be noted however that more than half a century after Enflo's discovery, all proofs that some Banach spaces fail the approximation property remain quite involved. With the peculiar sense of humor that mathematicians can have, Per Enflo received from Stanislaw Mazur's hands a live goose, Figure 2, as was promised in the Scottish book: a modest reward maybe, but a proof that the whole community acknowledged and confirmed Per Enflo's major achievement.

The local theory of Banach spaces focuses on properties of an infinite-dimensional Banach space X which can be checked on the finite-dimensional subspaces of X. For instance, although reflexivity is a topological property, the reflexivity of some Banach spaces X can sometimes be checked on their finite-dimensional subspaces. When it is so, X is said to be super-reflexive. Per Enflo [**1**] proved that super-reflexivity is characterized by the existence of a uniformly convex renorming, and equivalently of a uniformly smooth renorming, which shows in particular that super-reflexivity is a self-dual property. A quantitative version of this theorem was established later on by Gilles Pisier. The Hilbert space l_2 is itself characterized by its finite-dimensional subspaces and all its subspaces are complemented. However, Per Enflo solved in 1975 (with Joram Lindenstrauss and Gilles Pisier [**3**]) the following three-space problem: there exist two nested Banach spaces $Y \subseteq X$ such that Y and X/Y are both isomorphic to l_2, but X is not isomorphic to l_2. Such three-space problems have been the centers of attention since then, in relation to interpolation theory, and some of Nigel Kalton's articles stress their importance.

During the seventies, Per Enflo solved by himself another major problem: there exists a separable Banach space X and a linear continuous operator T from X to

FIGURE 2. Stanisław Mazur and Per H. Enflo. (The original uploader was Stako at Polish Wikipedia. –Kazimierz Kuratowski "Pol wieku matematyki polskiej 1920–1970" (Ksiazka i Wiedza, 1973 r, bez zastrzezonych praw autoskich). Transferred from pl.wikipedia to Commons by Pjahr using CommonsHelper. Public Domain, `https://commons.wikimedia.org/w/index.php?curid=5847552`)

itself with no closed non-trivial invariant subspace. In other words, the inclusion $T(Y) \subseteq Y$ cannot take place with $Y \subseteq X$ a closed linear subspace, except of course if $Y = \{0\}$ or $Y = X$. Once again, the finite-dimensional linear algebra fails to be even weakly applicable to this operator T, and Per Enflo has provided the mathematical world with a fundamental new example. Although Per delivered lectures on his work and did not try to conceal his proofs, this major progress was not immediately recognized: indeed some analysts were focusing on the Hilbert case and did not fully grasp the importance of the Banach space case. Moreover, the complexity of Enflo's proof was such that the peer-review process took several years: in fact, the paper [**4**] was published in a full issue of Acta Matematica in 1987, six years after being submitted there. The invariant subspace problems are now a major field of research. It should be stressed that more than forty years after Enflo's breakthrough, the complexity of any known example remains high. The reflexive case remains a mystery.

Per Enflo gave us an outstanding example of what a profoundly original mind can achieve; see [**5**]. His work is a brilliant achievement. More importantly, it is a bridge to the future.

References

[1] P. Enflo, *Banach spaces which can be given an equivalent uniformly convex norm*, Israel J. Math. **13** (1972), 281–288 (1973), DOI 10.1007/BF02762802. MR336297

[2] P. Enflo, *A counterexample to the approximation problem in Banach spaces*, Acta Math. **130** (1973), 309–317, DOI 10.1007/BF02392270. MR402468

[3] P. Enflo, J. Lindenstrauss, and G. Pisier, *On the "three space problem"*, Math. Scand. **36** (1975), no. 2, 199–210, DOI 10.7146/math.scand.a-11571. MR383047

[4] P. Enflo, *On the invariant subspace problem for Banach spaces*, Acta Math. **158** (1987), no. 3-4, 213–313, DOI 10.1007/BF02392260. MR892591

[5] P. H. Enflo, M. S. Moslehian, and J. B. Seoane-Sepúlveda, *A history of solving some famous problems in mathematical analysis*, Br. J. Hist. Math. **37** (2022), no. 1, 64–80, DOI 10.1080/26375451.2022.2037358. MR4425919

Institut de Mathématiques de Jussieu, Université Pierre et Marie Curie (Paris VI), 4 Place Jussieu, 75252 Paris cedex 05, France

Email address: godefroy@imj-prg.fr

Department of Pure Mathematics, Center of Excellence in Analysis on Algebraic Structures (CEAAS), Ferdowsi University of Mashhad, P.O. Box 1159, Mashhad 91775, Iran

Email address: moslehian@um.ac.ir; moslehian@yahoo.com

Instituto de Matemática Interdisciplinar (IMI), Departamento de Análisis Matemático y Matemática Aplicada, Facultad de Ciencias Matemáticas, Plaza de Ciencias 3, Universidad Complutense de Madrid, Madrid, 28040, Spain

Email address: jseoane@mat.ucm.es

Proceedings of Symposia in Pure Mathematics
Volume **106**, 2024
https://doi.org/10.1090/pspum/106/01925

On the Quasi-greedy Basic Sequence Problem

Fernando Albiac and José L. Ansorena

This paper is dedicated to Professor Per H. Enflo

Abstract. The quasi-greedy basic sequence problem is a successor of some of the main problems in classical Banach space theory such as the basis problem or the unconditional basic sequence problem. In this article we shed light onto it and walk the reader to the most recent stage in its development.

1. Introduction and preliminaries

One of the key ideas in the isomorphic theory of Banach spaces is to use the properties of bases and basic sequences as a tool to recognizing the differences and similarities between spaces. Banach himself was already aware of the fact that every infinite-dimensional Banach space contains a basic sequence, i.e., a sequence that is a Schauder basis of its closed linear span, and the most elemental procedure for constructing such sequences perhaps goes back to Mazur. In contrast, solving the *basis problem*, formulated by Banach in his book [**9**, p. 111] at the dawn of the theory, would end up being an important challenge.

Problem 1.1. Does every infinite-dimensional separable Banach space have a Schauder basis?

Chances are that Mazur might have known the connection between the famous 153rd problem that he raised in 1936 in the *Scottish Book* (an informal collection of problems kept in Lviv) [**28**] with the basis problem and the *approximation problem*, which was first formulated by Hildebrandt in 1931 [**22**]. However, the first systematic study of the variants of the approximation problem was initiated by Grothendieck, who, without reference to the Scottish Book, asked the same question as Mazur and showed that it was equivalent to the approximation problem [**21**].

These questions motivated a great deal of research over the next forty years, which led to the advancement of a theory that had gone into retreat after the Second World War. From the mid 1950's until well into the 1970's, Pełczyński together

2020 *Mathematics Subject Classification.* Primary 46B15, 41A65; Secondary 46A35.

Key words and phrases. Quasi-greedy basis, partial unconditionality, extraction principles for sequences in Banach spaces.

The first author acknowledges the support of the Spanish Ministry for Science and Innovation under Grants PID2019-107701GB-I00 and PID2022-138342NB-I00.

The second author acknowledges the support of the Spanish Ministry for Science and Innovation under Grant PID2022-138342NB-I00.

with his collaborators developed the theory of bases and basic sequences into a subtle and effective technique in the understanding of the isomorphic structure of Banach spaces. Eventually, in this atmosphere, the time was ripe and both the basis problem and the approximation problem were solved in the negative by Enflo [**19**], who gave an example of a reflexive separable Banach space failing to have the Approximation Property and *a fortiori* also failing to have a Schauder basis.

Enflo's results changed the theoretical framework of the subject dramatically. After the solution of the basis problem the research focus turned to the study of the existence of unconditional bases and unconditional basic sequences. Certainly, having an unconditional basis is very special and, although most of the spaces which are useful in harmonic analysis or partial differential equations such as the spaces L_p for $1 < p < \infty$ do have unconditional bases, one cannot a priori rely on the existence of such bases in a given separable space. So, the single most important problem left in the area was the *unconditional basic sequence problem.*

PROBLEM 1.2. Does every Banach space contain at least an unconditional basic sequence?

In 1993, twenty years after the solution of the basis problem, Gowers and Maurey found a counterexample to the unconditional basic sequence problem [**20**]. Thus from the point of view of abstract Banach space theory, and in connection with the unconditional basic sequence problem, it becomes natural to ask about the existence of basic sequences that satisfy weakened forms of unconditionality.

Additional motivation to investigate unconditional bases and partial forms of unconditionality comes from approximation theory, where new types of bases of interest in functional analysis came into play at the turn of the 21st century. *Greedy* and *quasi-greedy* bases were defined by Konyagin and Temlyakov [**25**] in their study of the efficiency of the thresholding greedy algorithm, and surprisingly characterized intrinsically in terms of well-known properties for the Banach space specialist. Specifically, Konyagin and Temlyakov showed that greedy bases can be characterized as those bases that are unconditional with the supplementary feature of being *democratic.* Further, Dilworth et al. [**14**] introduced a third type of greedy-like basis that they called *almost greedy* which can be characterized as those bases that are quasi-greedy and democratic.

Of course, a Banach space cannot have a greedy basis unless it has an unconditional basis and this rules out many spaces such as L_1 and $\mathcal{C}([0,1])$. One can also give examples of spaces failing to have a greedy basis like $\ell_1 \oplus \ell_2$; here it is a classical result of Edelstein and Wojtasczcyk [**17**] that any normalized unconditional basis of $\ell_1 \oplus \ell_2$ is equivalent to a permutation of the canonical basis of the space, which is plainly not democratic. However, it is much easier to have an almost greedy basis, and a very general construction was given in [**14**]. From this method one gets immediately that L_1 has an almost greedy basis.

At this point we should observe the connection between quasi-greediness and unconditionality. Unconditional bases are quasi-greedy, and although the converse is not true in general, quasi-greedy bases always retain some vestiges of unconditionality. For example, a quasi-greedy basis $(x_n)_{n=1}^\infty$ is *unconditional for constant coefficients*, i.e., there is a constant $C \geq 1$ so that

$$\left\| \sum_{j \in B} a_n x_n \right\| \leq C \left\| \sum_{j \in A} a_n x_n \right\|$$

whenever $B \subseteq A$ and $|a_j| = 1$ for all $n \in A$. It is known **[14]** that there exist separable Banach spaces such as $\mathcal{C}([0,1])$ which despite having a Schauder basis do not possess a quasi-greedy basis. Still, it should be *a priori* easier for a Banach space to contain quasi-greedy basic sequences than unconditional basic sequences, a question that remains unanswered as of today.

PROBLEM 1.3. Does every infinite-dimensional Banach space contain a quasi-greedy basic sequence?

A related question is the *quasi-greedy subsequence extraction problem.*

PROBLEM 1.4. Does every weakly null sequence in a Banach space have a quasi-greedy subsequence?

If the answer to Problem 1.4 is positive then so is the answer to Problem 1.3 by Rosenthal's theorem **[30]**, which states that any bounded sequence in a Banach space either is equivalent to the canonical ℓ_1-basis or has a weakly Cauchy subsequence.

The quasi-greedy subsequence extraction problem was first examined in **[14]**, where a partial solution is obtained.

THEOREM 1.5. *Let $\mathcal{X}$ be a Banach space which does not have c_0 as a spreading model (e.g., suppose $\mathcal{X}$ has nontrivial cotype). Then every semi-normalized weakly null sequence has a quasi-greedy subsequence.*

Problem 1.4 in general seems rather hard, and curiously intersects with ideas that have already been considered in Banach space theory. In 1978, Elton **[18]** examined a similar phenomenon. His research culminated in the following theorem.

THEOREM 1.6. *For each $0 < d < 1$ there is a constant $C = C(d) \geq 1$ such that every normalized weakly null sequence in a Banach space has a d-near unconditional subsequence $(y_n)_{n=1}^{\infty}$ with constant C, i.e.,*

$$\left\| \sum_{j \in A} a_n y_n \right\| \leq C \left\| \sum_{n=1}^{\infty} a_n y_n \right\|$$

for all $(a_n)_{n=1}^{\infty} \in c_{00}$ with $\max_n |a_n| \leq 1$ and all $A \subseteq \mathbb{N}$ with $\min_{n \in A} |a_n| \geq d$.

Thus, Elton's theorem gives a restricted form of unconditionality for subsequences of weakly null sequences. It is worth noting that at the time of Elton's work, examples were already known of weakly null sequences which contain no unconditional basic sequences **[27]** and that a more recent result of Johnson et al. **[24]** gives such an example in L_1.

Our aim in this paper is to survey the development of the quasi-greedy basic sequence problem up to the recent important results by Dilworth and Causey **[13]**, who have been able to obtain almost isometric constants for the basic sequence they extract. We tackle this in Section 3. In Section 4 we study what could be called the *almost greedy basic sequence problem.* Here we improve the constants that can be found in the literature and show that the dual of the basic sequence whose existence we guarantee is almost greedy too. We close in Section 5 solving a debilitated form of the quasi-greedy basic sequence problem, namely that every Banach space possesses a truncation quasi-greedy basic sequence. To do this we take advantage of the recently proved equivalence between truncation quasi-greedy bases and a form of partial unconditionality called bounded oscillation unconditionality. Dilworth et al. showed in [**16**] that every *real* Banach space has a bounded oscillation unconditional basic sequence. Our modest contribution to the theory in this direction permits to extend this result to Banach spaces over the complex scalar field. For the sake of self reference, in the preparatory Section 2 we review the main subsequence extraction principles which we will apply in the upcoming sections. Sometimes proofs or indications of proofs are also given to convey the flavour of the techniques that are employed.

2. Partial unconditionality and partial symmetry

Ramsey's theorem is a powerful tool for extracting subsequences that fulfil certain weak forms of unconditionality such as Schreier unconditionality, a variant that we will discuss after introducing some specific terminology.

A *minimal system* in a Banach space $\mathcal{X}$ is a sequence $\mathbf{x} = (x_n)_{n=1}^\infty$ in $\mathcal{X}$ for which there is a sequence $(x_n^*)_{n=1}^\infty$ in the dual space $\mathcal{X}^*$ biorthogonal to $\mathbf{x}$, that is, $x_n^*(x_k) = \delta_{n,k}$ for all $(n,k) \in \mathbb{N}^2$. If $\mathbf{x}$ is *complete*, i.e., its closed linear span is the entire space $\mathcal{X}$, then $(x_n^*)_{n=1}^\infty$ is unique. Given a set N and $m \in \mathbb{N}$, let $\mathcal{P}_{\le m}(N)$ denote the set consisting of all subsets of N of cardinality at most m. Put

$$\mathcal{P}_{<\infty}(N) = \cup_{m=0}^\infty \mathcal{P}_{\le m}(N).$$

For $A \in \mathcal{P}_{<\infty}(\mathbb{N})$ we define the *coordinate projection on* A relative to a complete minimal system $\mathbf{x}$ by

$$S_A \colon \mathcal{X} \to \mathcal{X}, \quad x \mapsto S_A(x) = \sum_{n\in A} x_n^*(x) x_n.$$

Given $\mathcal{U} \subseteq \mathcal{P}_{<\infty}(\mathbb{N})$ and $C \ge 1$, we say that $\mathbf{x}$ is a *$\mathcal{U}$-unconditional basis of $\mathcal{X}$ with constant C* if $\|S_A\| \le C$ for all $A \in \mathcal{U}$. If $\mathbf{x}$ is a $\mathcal{U}$-unconditional basis of its closed linear span with constant C we say that it is a *$\mathcal{U}$-unconditional basic sequence with constant C*. In this language, the complete minimal system $\mathbf{x}$ is *C-suppression unconditional* if and only if it is $\mathcal{P}_{<\infty}(\mathbb{N})$-unconditional with constant C. Note also that a sequence $(y_n)_{n=1}^\infty$ in $\mathcal{X}$ is a basic sequence if and only if there is a constant C such that $(y_n)_{n=1}^\infty$ is $\mathcal{F}$-unconditional with constant C, where

$$\mathcal{F} = \{\mathbb{N}_{\le m} := \{1, \dots, m\} \colon m \in \mathbb{N}\},$$

in which case C is an upper bound for the basis constant of $\mathbf{x}$. The following result guarantees the existence of subsequences that are basic sequences inside weakly null sequences.

THEOREM 2.1 (see, e.g., [**10**, Théorème 1, Appendice A]). *Let $\mathbf{x} = (x_n)_{n=1}^\infty$ be a weakly null sequence in a Banach space $\mathcal{X}$. If $\liminf_n \|x_n\| > 0$ then $\mathbf{x}$ has for every $C > 1$ a subsequence that is a basic sequence with basis constant not larger than C.*

In turn, Odell [**29**] proved the following extraction principle, also for weakly null sequences.

THEOREM 2.2. *Let $m \in \mathbb{N}$ and $C > 1$. Every semi-normalized weakly null sequence in a Banach space has a subsequence that is $\mathcal{P}_{\le m}(\mathbb{N})$-unconditional with constant C.*

Recall that a sequence is *Schreier unconditional with constant C* if and only if it is $\mathcal{C}$-unconditional with constant C, where

$$\mathcal{C} = \{A \subseteq \mathbb{N} \colon |A| \le \min A\}.$$

Given an increasing sequence $\mathbf{q} = (q_m)_{m=0}^\infty$ in $\mathbb{N}$ we will also consider the set

$$\mathcal{C}_\mathbf{q} = \cup_{m=0}^\infty \{A \subseteq \mathbb{N} \colon m < \min A,\ |A| \le q_m\}.$$

Note that a suitable subsequence of a Schreier unconditional sequence with constant C is $\mathcal{C}_\mathbf{q}$-unconditional with constant C; in particular, it is $\mathcal{P}_{\le q_0}(\mathbb{N})$-unconditional. Conversely, Odell [**29**] combined Theorem 2.2 with Theorem 2.1 to prove that any semi-normalized weakly null sequence of any Banach space has, for any $C > 2$, a subsequence that is Schreier unconditional with constant C. However, a stronger result had been formerly achieved.

THEOREM 2.3 ([**10**, Chapitre 3, Théorème 1]). *Let $\mathbf{x}$ be a weakly null semi-normalized sequence in a Banach space $\mathbb{X}$. For every $C > 1$ and every increasing sequence $\mathbf{q} = (q_m)_{m=0}^\infty$ in $\mathbb{N}$, $\mathbf{x}$ has a subsequence that is $\mathcal{C}_\mathbf{q}$-unconditional with constant C.*

Combining Theorem 2.3 with Theorem 2.1 gives that every semi-normalized weakly null sequence of a Banach space has a subsequence that is $\mathcal{D}_\mathbf{q}$-unconditional with constant arbitrarily close to 2 for any increasing sequence $\mathbf{q} = (q_m)_{m=0}^\infty$ in $\mathbb{N}$, where

$$\mathcal{D}_\mathbf{q} = \cup_{m=0}^\infty \{\mathbb{N}_{\le m} \cup A \colon m < \min A,\ |A| \le q_m\}.$$

Causey and Dilworth [**13**] improved this result more recently.

THEOREM 2.4 ([**13**, Theorem 1.1]). *Let $\mathbf{q} = (q_m)_{m=0}^\infty$ be an increasing sequence in $\mathbb{N}$ and let $C > 1$. Then every semi-normalized weakly null sequence in a Banach space has a subsequence that is $\mathcal{D}_\mathbf{q}$-unconditional with constant C.*

Ramsey's theorem has been used as well in extraction principles which permit to obtain subsequences satisfying weakened forms of symmetry that can be described in terms of spreading models, an important tool that we briefly recall for ease of exposition.

Let us denote by c_{00} the vector space of all finitely nonzero sequences with values in the real or complex field $\mathbb{F}$, that is, the vector space spanned by the unit vectors $(e_j)_{j=1}^\infty$. Given an increasing map $\psi \colon \mathbb{N} \to \mathbb{N}$, consider the linear map $L_\psi \colon c_{00} \to c_{00}$ defined by $L_\psi(e_j) = e_{\psi(j)}$ for all $j \in \mathbb{N}$. If α and β are such that $\beta = L_\psi(\alpha)$ for some ψ we say that *β is a right shift of α*. Given $m \in \mathbb{N}$, we endow the set $\mathbb{N}^{(m)}$ of all increasing m-tuples in $\mathbb{N}$ with the partial order given by $(k_j)_{j=1}^m \prec (n_j)_{j=1}^m$ if $k_1 \le n_1$. Suppose that a sequence $\mathbf{y} = (y_n)_{n=1}^\infty$ in a Banach space $\mathbb{X}$ has the property that for all $m \in \mathbb{N}$ and all $(a_j)_{j=1}^m$ in $\mathbb{F}^\mathbb{N}$ there exists

$$F\left((a_j)_{j=1}^m\right) := \lim_{(n_j)_{j=1}^m \in \mathbb{N}^{(m)}} \left\| \sum_{j=1}^m a_j\, y_{n_j} \right\| \in [0, \infty),$$

where the limit is taken in the sense of the directed set $\mathbb{N}^{(m)}$. In this case we say that $\mathbf{y}$ is a *good sequence* and we can define a map $\|\cdot\|_\mathbf{y} \colon c_{00} \to [0, \infty)$ by

$\|(a_j)_{j=1}^\infty\|_{\mathbf{y}} = F((a_j)_{j=1}^m)$, where $m \in \mathbb{N}$ is chosen so that $a_j = 0$ whenever $j > m$. Clearly, $\|\cdot\|_{\mathbf{y}}$ is a semi-norm. Moreover,

(S) $\|\beta\|_{\mathbf{y}} = \|\alpha\|_{\mathbf{y}}$ whenever β is a right shift of α, and

- $\|e_1\|_{\mathbf{y}} = \nu_1 := \lim_n \|y_n\|$, $\|e_1 - e_2\|_{\mathbf{y}} = \nu_2 := \lim_{(n,m)\in\mathbb{N}^{(2)}} \|y_n - y_m\|$.

We infer that, given $\alpha = (a_j)_{j=1}^\infty \in c_{00}$ and $j \in \mathbb{N}$,

$$|a_j|\nu_2 = \|a_j e_j - a_j e_{j+1}\| = \|\alpha_1 - \alpha_2\| \le 2\|\alpha\|,$$

where α_1 and α_2 are suitable right shifts of α. Hence, assuming that $\nu_2 \neq 0$, we have that $\|\cdot\|_{\mathbf{y}}$ is a norm over c_{00}. In this case we denote by $S[\mathbf{y}]$ the completion of c_{00} with respect to $\|\cdot\|_{\mathbf{y}}$.

Two sequences $\mathbf{x} = (x_n)_{n=1}^\infty$ and $\mathbf{y} = (y_n)_{n=1}^\infty$ in respective Banach spaces $\mathcal{X}$ and $\mathcal{Y}$ are said to be *equivalent* if there is an isomorphism T from the closed linear span of $\mathbf{x}$ in $\mathcal{X}$ onto the closed linear span of $\mathbf{y}$ in $\mathcal{Y}$ with $T(x_n) = y_n$ for all $n \in \mathbb{N}$. If T is an isometry, we say that $\mathbf{x}$ and $\mathbf{y}$ are *isometrically equivalent*. A sequence $(s_j)_{j=1}^\infty$ in a Banach space $\mathcal{S}$ is said to be the *spreading model generated by the good sequence* $\mathbf{y}$ if it is isometrically equivalent to the unit vector system of $S[\mathbf{y}]$. If $(s_j)_{j=1}^\infty$ is the spreading model generated by some good subsequence of a Banach space $\mathcal{X}$ we say that $\mathcal{X}$ *has* $(s_j)_{j=1}^\infty$ *as a spreading model.*

Suppose now that a good sequence $\mathbf{y} = (y_n)_{n=1}^\infty$ is weakly null. Fix $\alpha = (a_j)_{j=1}^\infty \in c_{00}$ and $k \in \mathbb{N}$ with $a_k = 0$. Set $b_k = 1$ and $b_j = a_j$ if $j \neq k$. Pick $m \in \mathbb{N}$ such that $b_j = 0$ if $j > m$. By definition, given $\varepsilon > 0$ there is $q \in \mathbb{N}$ such that

$$\left| \|\alpha\|_{\mathbf{y}} - \left\| \sum_{j=1}^m a_j y_{n_j} \right\| \right| < \varepsilon, \quad \left| \|\alpha + e_k\|_{\mathbf{y}} - \left\| \sum_{j=1}^m b_j y_{n_j} \right\| \right| < \varepsilon$$

for all $(n_j)_{j=1}^m \in \mathbb{N}^{(m)}$ with $n_1 > q$. By Mazur's lemma (see, e.g., [**7**, Appendix G, Corollary G4]) there are $r \ge q + k$ and $(c_n)_{n=q+k}^r$ in $[0, \infty)$ such that

$$\sum_{n=q+k}^r c_n = 1, \quad \left\| \sum_{n=q+k}^r c_n y_n \right\| < \varepsilon.$$

Combining the above inequalities and using convexity we get

$$\begin{aligned}
\|\alpha\|_{\mathbf{y}} &< \varepsilon + \left\| \sum_{j=1}^{k-1} a_j y_{j+q} + \sum_{j=k+1}^m a_j y_{j+r-k} \right\| \\
&< 2\varepsilon + \left\| \sum_{j=1}^{k-1} a_j y_{j+q} + \sum_{n=q+k}^r c_n y_n + \sum_{j=k+1}^m a_j y_{j+r-k} \right\| \\
&\le 2\varepsilon + \sum_{n=q+k}^r c_n \left\| \sum_{j=1}^{k-1} a_j y_{j+q} + y_n + \sum_{j=k+1}^m a_j y_{j+r-k} \right\| < 3\varepsilon + \|\alpha + e_k\|_{\mathbf{y}}.
\end{aligned}$$

Since $\varepsilon > 0$ is arbitrary, $\|\alpha\|_{\mathbf{y}} \le \|\alpha + e_k\|_{\mathbf{y}}$. In particular, $\nu_1 \le \nu_2$. In the language of bases we have showed that if $\nu_1 > 0$ and $\mathbf{y}$ is weakly null then the spreading model generated by $\mathbf{y}$ is a 1-suppression unconditional basis; besides, by (S), it is isometrically equivalent to all its subsequences. In the literature, the 1-suppression unconditional bases that are isometrically equivalent to all their subsequences are called 1-subsymmetric. The dual basis of a 1-subsymmetric basis is 1-subsymmetric (see, e.g., [**8**]). Moreover, if φ denotes the *fundamental function* of a 1-subsymmetric basis $(s_j)_{j=1}^\infty$, given by

$$\varphi(m) = \left\| \sum_{j=1}^m s_j \right\|, \quad m \in \mathbb{N},$$

and φ^* denotes the fundamental function of its dual basis, we have

$$m \le \varphi(m)\varphi^*(m) \le 2m, \quad m \in \mathbb{N} \tag{2.1}$$

(see [**26**, Proposition 3.a.6]). A 1-subsymmetric basic sequence $(s_j)_{j=1}^\infty$ is equivalent to the standard c_0-basis if and only if φ is bounded, so a 1-subsymmetric basic sequence is equivalent to the canonical ℓ_1-basis if and only if $\varphi(m) \approx m$ for $m \in \mathbb{N}$.

An application of Ramsey's theorem shows that every bounded sequence has a subsequence that is good in the sense defined above (see [**10**–**12**]). For further reference we record the consequence of this result for weakly null sequences.

THEOREM 2.5. *Let* $\mathbf{x} = (x_n)_{n=1}^\infty$ *be a weakly null sequence in a Banach space* $\mathcal{X}$. *Suppose that* $\liminf_n \|x_n\| > 0$. *Then* $\mathbf{x}$ *generates a* 1*-suppression unconditional (hence* 1*-subsymmetric) spreading model.*

From now on we will use $\mathbb{B}$ for the set of all scalars of modulus at most one. Given a basic sequence $\mathbf{x} = (x_n)_{n=1}^\infty$ of a Banach space $\mathcal{X}$ we denote by $\mathbf{x}^*$ its dual basis $(x_n^*)_{n=1}^\infty$, defined on the Banach space spanned by $\mathbf{x}$, say $\mathcal{Y}$. Note that $\mathbf{x}^*$ is a basic sequence in $\mathcal{Y}^*$. Combining Theorems 2.5, 2.3, and 2.1 yields the following corollary. To properly understand its significance we must be aware that in the lack of unconditionality, the dual basis $\mathbf{z}^*$ of a subbasis $\mathbf{z}$ of $\mathbf{y}$ could not be equivalent to the corresponding subsequence of $\mathbf{y}^*$. So, it might be convenient to guarantee that the subsequence extraction principles dualize well.

COROLLARY 2.6 (cf. [**10**, Chapitre 3, Corollaire 5]). *Let* $\mathbf{x} = (x_n)_{n=1}^\infty$ *be a weakly null seminormalized sequence in a Banach space* $\mathcal{X}$. *Let* $(q_m)_{m=1}^\infty$ *be an increasing sequence in* $\mathbb{N}$, *and* $(\varepsilon_m)_{m=1}^\infty$ *be a sequence in* $(0,\infty)$ *with* $\lim_m \varepsilon_m = 0$. *Then there is a basic subsequence* $\mathbf{y} = (y_n)_{n=1}^\infty$ *of* $\mathbf{x}$ *and a* 1*-suppression unconditional spreading model* $(s_j)_{j=1}^\infty$ *satisfying the following properties.*

(1) *For all* $m \in \mathbb{N}$, *all* $(a_j)_{j=1}^{q_m} \in \mathbb{F}^{q_m}$, *and all increasing sequences* $(n_j)_{j=1}^{q_m}$ *in* $\mathbb{N}$ *with* $n_1 > m$ *we have*

$$\frac{1}{1+\varepsilon_m}\left\| \sum_{j=1}^{q_m} a_j\, s_j \right\| \le \left\| \sum_{j=1}^{q_m} a_j\, y_{n_j} \right\| \le (1+\varepsilon_m) \left\| \sum_{j=1}^{q_m} a_j\, s_j \right\|.$$

(2) *Let* $(s_j^*)_{j=1}^\infty$ *denote the dual spreading model of* $(s_j)_{j=1}^\infty$. *Let* $(z_k^*)_{k=1}^\infty$ *be the dual system of a subsequence* $\mathbf{z} = (y_{n_k})_{n=1}^\infty$ *of* $\mathbf{y}$. *For all* $m \in \mathbb{N}$, *all* $(a_j)_{j=1}^{q_m} \in \mathbb{F}^{q_m}$, *and all increasing sequences* $(k_j)_{j=1}^{q_m}$ *in* $\mathbb{N}$ *with* $n_{k_1} > m$ *we have*

$$\frac{1}{1+\varepsilon_m}\left\| \sum_{j=1}^{q_m} a_j\, s_j^* \right\| \le \left\| \sum_{j=1}^{q_m} a_j\, z_{k_j}^* \right\| \le (1+\varepsilon_m) \left\| \sum_{j=1}^{q_m} a_j\, s_j^* \right\|.$$

In particular, $\mathbf{y}$ *generates* $(s_j)_{j=1}^\infty$ *and* $\mathbf{z}^*$ *generates* $(s_j^*)_{j=1}^\infty$.

PROOF. Without loss of generality we assume that $\lim_n \|x_n\| = 1$. Let $(\delta_m)_{m=1}^\infty$ be the sequence in $(0,\infty)$ defined for $m \in \mathbb{N}$ by $(1+2\delta_m^2)(1+\delta_m)/(1-\delta_m) = 1+\varepsilon_m$.

A standard argument based on the compactness of $\mathbb{B}^m$ yields that if $(s_j)_{j=1}^\infty$ is the spreading model generated by a subsequence $\mathbf{y} = (y_n)_{n=1}^\infty$ of $\mathbf{x}$, then for every choice of $\delta > 0$ and $m \in \mathbb{N}$ there is $q = q(m, \varepsilon) \in \mathbb{N}$ such that

$$\left| \left\| \sum_{j=1}^m a_j\, y_{n_j} \right\| - \left\| \sum_{j=1}^m a_j\, s_j \right\| \right| \le \delta$$

for all $(a_n)_{n=1}^m \in \mathbb{B}^m$ and all increasing sequences $(n_j)_{j=1}^m$ in $\mathbb{N}$ with $n_1 \ge q$. Consequently, passing to a furher subsequence,

$$\left| \left\| \sum_{j=1}^{q_m} a_j\, y_{n_j} \right\| - \left\| \sum_{j=1}^{q_m} a_j\, s_j \right\| \right| \le \delta_m \tag{2.2}$$

for all $(a_j)_{j=1}^{q_m} \in \mathbb{B}^{q_m}$ and all increasing sequences $(n_j)_{j=1}^{q_m}$ in $\mathbb{N}$ with $n_1 > m$. Besides, by Theorem 2.1 we can take $\mathbf{y}$ so that $\|y_n\| > 1 - \delta_m$ whenever $n > m$ and $\|S_A\| \le 1 + \delta_m$ if $m < \min A$ and $|A| \le q_m$, where $\mathcal{Y}$ is the Banach space spanned by $\mathbf{y}$ and $S_A \colon \mathcal{Y} \to \mathcal{Y}$ is the coordinate projection on A relative to $\mathbf{y}$. We then have,

$$\max_{1\le j\le q_m} |a_j| \le \frac{1+\delta_m}{1-\delta_m} \left\| \sum_{j=1}^{q_m} a_j\, y_{n_j} \right\|, \quad \max_{1\le j\le q_m} |a_j| \le \left\| \sum_{j=1}^{q_m} a_j\, s_j \right\| \tag{2.3}$$

for $(n_j)_{j=1}^{q_m}$as above and for all $(a_j)_{j=1}^{q_m} \in \mathbb{F}^{q_m}$. Combining (2.2) and (2.3) gives

$$\frac{1-\delta_m}{1+2\delta_m^2} \left\| \sum_{j=1}^{q_m} a_j\, s_j \right\| \le \left\| \sum_{j=1}^{q_m} a_j\, y_{n_j} \right\| \le (1+\delta_m) \left\| \sum_{j=1}^{q_m} a_j\, s_j \right\|,$$

and

$$\frac{1}{1+\varepsilon_m} \left\| \sum_{j=1}^{q_m} a_j\, s_j^* \right\| \le \left\| \sum_{j=1}^{q_m} a_j\, z_{k_j}^* \right\| \le \frac{1+\delta_m^2}{1-\delta_m} \left\| \sum_{j=1}^{q_m} a_j\, s_j^* \right\|.$$

Let $\mathbf{z}$ be a subsequence of $\mathbf{y}$ and let $\eta := (k_j)_{j=1}^{q_m}$ be as in (2). Let $\mathcal{Z}$ be the Banach space generated by $\mathbf{z}$. The linear operators

$$S_\eta \colon \operatorname{span}(y_{n_{k_j}} \colon 1 \le j \le q_m) \to \mathcal{S}, \quad T_\eta \colon \operatorname{span}(s_j \colon 1 \le j \le q_m) \to \mathcal{Z}$$

given by $y_{n_{k_j}} \mapsto s_j$ and $s_j \mapsto y_{n_{k_j}}$, respectively, satisfy $\|S_\eta\| \le (1+2\delta_m^2)/(1-\delta_m)$ and $\|T_\eta\| \le 1 + \delta_m$. By duality, the linear operators

$$S'_\eta \colon \operatorname{span}(z_{k_j}^* \colon 1 \le j \le m) \to \mathcal{S}^*, \quad T'_\eta \colon \operatorname{span}(s_j^* \colon 1 \le j \le m) \to \mathcal{Z}^*,$$

given respectively by $z_{k_j}^* \mapsto s_j^*$ and $s_j^* \mapsto y_{k_j}^*$ satisfy $\|S'_\eta\| \le (1+\delta_m)^2$ and $\|T'_\eta\| \le (1+2\delta_m^2)(1+\delta_m)/(1-\delta_m)$. □

Theorems 2.5, 2.1, and 2.4 highlight the importance of knowing whether a given Banach space $\mathcal{X}$ has a weakly null sequence that does not converge to zero in norm. If such sequences do not exist, $\mathcal{X}$ is called a *Schur space*. The following consequences of Rosenthal's ℓ_1-theorem reveal the role played by ℓ_1 is this matter.

THEOREM 2.7. *Any infinite-dimensional Schur space contains a subspace isomorphic to ℓ_1.*

PROOF. Let $\mathcal{X}$ be an infinite-dimensional Banach space which does not contain a copy of ℓ_1. Pick $\delta > 0$ and a δ-separated sequence $\mathbf{x}$ in the unit ball of $\mathcal{X}$. There is a subsequence $(x_n)_{n=1}^\infty$ of $\mathbf{x}$ that is weakly Cauchy. Then the sequence defined for $n \in \mathbb{N}$ by $y_n = x_{2n-1} - x_{2n}$ is weakly null and $\inf_n \|y_n\| \geq \delta$. □

LEMMA 2.8. *Suppose that a Banach space $\mathcal{X}$ that does not contain any copy of ℓ_1 has $(s_j)_{j=1}^\infty$ as an unconditional spreading model. Then there is a weakly null sequence of $\mathcal{X}$ that generates a spreading model equivalent to $(s_j)_{j=1}^\infty$.*

PROOF. There is a weakly Cauchy sequence $(x_n)_{n=1}^\infty$ that defines $(s_j)_{j=1}^\infty$. Then the weakly null sequence $(x_{2n-1} - x_{2n})_{n=1}^\infty$ defines the spreading model $(s_{2j-1} - s_{2j})_{j=1}^\infty$, which is equivalent to $(s_j)_{j=1}^\infty$. □

James' ℓ_1 distortion theorem is an important result in the almost isometric theory of bases.

THEOREM 2.9 ([**23**]). *Any Banach space containing ℓ_1 contains an almost isometric copy of ℓ_1.*

Note that combining Theorems 2.1, 2.7, and 2.9 we obtain that any Banach space contains a basic sequence with basis constant as close to one as desired. This observation gives for free an alternative proof of Mazur's solution of the basic sequence problem.

3. Extraction principles for quasi-greedy bases

Let $\mathbf{x} = (x_n)_{n=1}^\infty$ be a semi-normalized, complete, minimal system in a Banach space $\mathcal{X}$ with biorthogonal functionals $(x_n^*)_{n=1}^\infty$. A finite set $A \subseteq \mathbb{N}$ is said to be a *greedy set* of $x \in \mathcal{X}$ with respect to $\mathbf{x}$ if $|x_n^*(x)| \geq |x_k^*(x)|$ whenever $n \in A$ and $k \in \mathbb{N} \setminus A$. If $\sup_n \|x_n^*\| < \infty$ then $\lim_n x_n^*(x) = 0$ and so there is an increasing sequence $(A_m(x))_{m=1}^\infty$ in $\mathcal{P}(\mathbb{N})$ such that $A_m(x)$ is a greedy set of x of cardinality m for all $m \in \mathbb{N}$. If we choose a criterion to make this sequence unique we obtain for each $m \in \mathbb{N}$ a nonlinear operator

$$\mathcal{G}_m \colon \mathcal{X} \to \mathcal{X}, \quad x \mapsto S_{A_m(x)}(x).$$

The *thresholding greedy algorithm* (TGA for short) relative to $\mathbf{x}$ is the sequence of maps $(\mathcal{G}_m)_{m=1}^\infty$. Wojtaszczyk [**31**] (see also [**6**, Theorem 4.1]) proved that the TGA relative to $\mathbf{x}$ is well-defined and converges, i.e., $(x_n^*)_{n=1}^\infty$ is bounded and $\lim_m \mathcal{G}_m(x) = x$ for all $x \in \mathcal{X}$ if and only if $\mathbf{x}$ is quasi-greedy. Konyagin and Temlyakov [**25**] had coined the term quasi-greedy bases to name those semi-normalized complete minimal systems for which the greedy projections are uniformly bounded, i.e.,

$$\|S_A(x)\| \leq C\|x\|$$

for all $x \in \mathcal{X}$ and all greedy sets A of x, for some $C \geq 1$. If this inequality holds for a particular constant C we say that $\mathbf{x}$ is quasi-greedy with constant C. A quasi-greedy basic sequence will be a sequence that is a quasi-greedy basis of its closed linear span.

Given a sequence $\mathbf{x} = (x_n)_{n=1}^\infty$ of a Banach space $\mathcal{X}$ and a $A \subseteq \mathbb{N}$ finite, we put

$$\mathbf{1}_A = \mathbf{1}_A[\mathbf{x}] = \sum_{n \in A} x_n.$$

We extend this notation as follows. We let $\mathbb{E} = \{a \in \mathbb{F} \colon |a| = 1\}$, and if $\gamma = (\gamma_n)_{n\in A} \in \mathbb{E}^A$ then we put

$$\mathbf{1}_{\gamma,A} = \mathbf{1}_{\gamma,A}[\mathbf{x}] = \sum_{n\in A} \gamma_n \, x_n.$$

A sequence $\mathbf{x} \in \mathcal{X}$ is said to be *democratic* if there is a non-decreasing sequence $(\sigma_m)_{m=1}^\infty$ in $(0,\infty)$ such that $\|\mathbf{1}_A\| \approx \sigma_{|A|}$ for $A \subseteq \mathbb{N}$ finite. Such a sequence $(\sigma_m)_{m=1}^\infty$ must be equivalent to the *upper democracy function* of $\mathbf{x}$, given by

$$\varphi_u[\mathbf{x}](m) = \sup\{\|\mathbf{1}_{\gamma,A}[\mathbf{x}]\| \colon |A| \le m, \gamma \in \mathbb{E}^A\}, \quad m \in \mathbb{N}.$$

We will also consider the *lower fundamental function* of $\mathbf{x}$, defined for $m \in \mathbb{N}$ as

$$\varphi_l[\mathbf{x}](m) = \inf\{\|\mathbf{1}_{\gamma,A}[\mathbf{x}]\| \colon |A| \ge m, \gamma \in \mathbb{E}^A\}.$$

Sequences $\mathbf{x}$ such that $\varphi_u[\mathbf{x}] \le D\varphi_l[\mathbf{x}]$ for some constant $D \ge 1$ are called *super-democratic*. A sequence is superdemocratic if and only if it is democratic and unconditional for constant coefficients.

A semi-normalized complete minimal system $\mathbf{x}$ is said to be *greedy* if for all $m \in \mathbb{N}$ and all $x \in \mathcal{X}$, $\mathcal{G}_m(x)$ provides essentially optimal approximations to x amongst all arbitrary m-term linear combinations of vectors from the system $\mathbf{x}$. If $\mathcal{G}_m$ provides essentially optimal approximations amongst the coordinate projections on sets of cardinality m then $\mathbf{x}$ is said to be *almost greedy*. That is, $\mathbf{x}$ is almost greedy if there exists a constant $C \ge 1$ such that for all $x \in \mathcal{X}$, all greedy sets A of x, and all $B \subseteq \mathbb{N}$ with $|A| = |B|$,

$$\|x - S_A(x)\| \le C\|x - S_B(x)\|.$$

We next record the estimates we will use for the almost greedy basis constant.

THEOREM 3.1 ([**15**, Theorem 3.3] and [**6**, Proposition 6.6]). *Let $\mathbf{x}$ be a semi-normalized complete minimal system. Then $\mathbf{x}$ is almost greedy if and only if it is quasi-greedy and democratic. Morever, if $\mathbf{x}$ is quasi-greedy with constant C and super-democratic with constant D, then it is almost greedy with constant $(1+C)+CD$.*

Here we would like to point out that all known proofs of the characterization of almost greedy bases provided by Theorem 3.1 rely on showing first that quasi-greedy bases $\mathbf{x}$ are *truncation quasi-greedy*, i.e., there is a constant $C \ge 1$ such that for all $x \in \mathcal{X}$ and all greedy sets A of x,

$$\min_{n\in A} |x_n^*(x)| \|\mathbf{1}_{\gamma(x),A}\| \le C\|x\|,$$

where $(x_n^*)_{n=1}^\infty$ are the biorthogonal functionals of $\mathbf{x}$, and $\gamma(x) = (\gamma_n)_{n=1}^\infty$ is the *sign* of x with respect to $\mathbf{x}$, i.e., $\gamma_n = x_n^*(x)/|x_n^*(x)|$ if $x_n^*(x) \ne 0$ and $\gamma_n = 1$ otherwise. Although quasi-greedy bases are truncation quasi-greedy in general quasi-Banach spaces [**6**], local convexity makes a quantitative difference: if $\mathbf{x}$ is quasi-greedy with constant C in a Banach space then $\mathbf{x}$ is truncation quasi-greedy with the same constant [**15**].

Our next extraction principle guarantees the existence of subsequences with some democracy features. Put $\Upsilon_{\mathbb{F}} = 2$ if $\mathbb{F} = \mathbb{R}$ and $\Upsilon_{\mathbb{F}} = 4$ if $\mathbb{F} = \mathbb{C}$.

THEOREM 3.2. *Let $\mathbf{x} = (x_n)_{n=1}^\infty$ be a semi-normalized weakly null sequence in a Banach space $\mathcal{X}$. Suppose that $\mathbf{x}$ generates a spreading model $(s_j)_{j=1}^\infty$ nonequivalent to the canonical c_0-basis. Let φ denote the fundamental function of $(s_j)_{j=1}^\infty$. Then,* *given $\varepsilon > 0$ there is a basic subsequence $\mathbf{y}$ of $\mathbf{x}$ that is superdemocratic with constant $\Upsilon_{\mathbb{F}} + \varepsilon$. Further, the democracy functions φ_l and φ_u of $\mathbf{y}$ satisfy $\varphi_u(m) \le (\Upsilon_{\mathbb{F}} + \varepsilon)\varphi(m) \le \varphi_l(m)$ and $\varphi(m) \le (\Upsilon_{\mathbb{F}} + \varepsilon)\varphi_l(m)$ for all $m \in \mathbb{N}$. Moreover, if $(s_j)_{j=1}^\infty$ is also nonequivalent to the canonical ℓ_1-basis, we can choose $\mathbf{y}$ so that the dual basis $\mathbf{z}^*$ of any subbasis $\mathbf{z}$ of $\mathbf{y}$ is super-democratic with constant $\Upsilon_{\mathbb{F}} + \varepsilon$, and the lower and upper democracy function φ_l^* and φ_u^* of $\mathbf{z}^*$ satisfy $\varphi_u^*(m) \le 2(\Upsilon_{\mathbb{F}} + \varepsilon)m/\varphi(m)$ and $m/\varphi(m) \le 2(\Upsilon_{\mathbb{F}} + \varepsilon)\varphi_l^*(m)$ for all $m \in \mathbb{N}$.*

PROOF. Bearing in mind Theorem 2.1 we can assume that $\mathbf{x}$ is a Schauder basis of $\mathcal{X}$. Let $C \in [1, \infty)$ be such that $\max\{\|x_n\|, \|x_n^*\|\} \le C$ for all $n \in \mathbb{N}$. Note that $(s_j)_{j=1}^\infty$ is 1-suppression unconditional by Theorem 2.5, and that φ is unbounded. By convexity,

$$\frac{1}{\Upsilon_{\mathbb{F}}}\varphi(m) \le \left\| \sum_{j\in A} a_j s_j \right\| \le \Upsilon_{\mathbb{F}}\varphi(m), \quad |a_j| = 1,\ |A| = m.$$

Pick $0 < \delta < 1/\Upsilon_{\mathbb{F}}$. There are $(\varepsilon_m)_{m=1}^\infty$ in $(0, \delta/\Upsilon_{\mathbb{F}})$ and $(p_m)_{m=1}^\infty$ in $\mathbb{N}$ with $\lim_m p_m = \infty$ such that

$$2Cp_m \le (\delta - \varepsilon_m \Upsilon_{\mathbb{F}})\varphi(m), \quad m \in \mathbb{N}. \tag{3.1}$$

Let $(q_m)_{m=1}^\infty$ be an increasing sequence in $\mathbb{N}$ with $q_{p_m} \ge m$ for all $m \in \mathbb{N}$. Let $\mathbf{y} = (y_n)_{n=1}^\infty$ be the subsequence of $\mathbf{x}$ supplied by Corollary 2.6. For all $m \in \mathbb{N}$, all $(a_j)_{j=1}^m \in \mathbb{F}^m$, and all $(n_j)_{j=1}^m$ increasing with $p_m < n_1$, we have

$$\frac{1}{1+\varepsilon_m}\left\| \sum_{j=1}^m a_j\, s_j \right\| \le \left\| \sum_{j=1}^m a_j\, y_{n_j} \right\| \le (1+\varepsilon_m)\left\| \sum_{j=1}^m a_j\, s_j \right\|.$$

Fix $A \subseteq \mathbb{N}$ with $|A| = m$ and $\gamma \in \mathbb{E}^A$. In order to estimate the norm of $\mathbf{1}_{\gamma,A} = \mathbf{1}_{\gamma,A}[\mathbf{y}]$ we set $A_1 = A \cap [1, p_m]$, $A_2 = A \cap (p_m, \infty)$, $m_1 = |A_1|$, and $m_2 = |A_2|$. Choose a bijection $\sigma\colon \mathbb{N}_{\le m} \to A$ whose restriction to $\mathbb{N}_{\le m_2}$ is an increasing bijection onto A_2. Set $Q = \| \sum_{j=1}^m \gamma_{\sigma(j)} s_j \|$, $Q_1 = \| \sum_{j=1+m_2}^m \gamma_{\sigma(j)} s_j \|$ and $Q_2 = \| \sum_{j=1}^{m_2} \gamma_{\sigma(j)} s_j \|$. Since $m_1 \le p_m$, applying the triangle law gives

$$\begin{aligned} \big|\|\mathbf{1}_{\gamma,A}\| - Q\big| &\le \big|\|\mathbf{1}_{\gamma,A}\| - \|\mathbf{1}_{\gamma,A_2}\|\big| + \big|\|\mathbf{1}_{\gamma,A_2}\| - Q_2\big| + |Q_2 - Q| \\ &\le \|\mathbf{1}_{\gamma,A_1}\| + \varepsilon_m Q_2 + Q_1 \le 2Cp_m + \varepsilon_m \Upsilon_{\mathbb{F}}\varphi(m) \le \delta\varphi(m). \end{aligned}$$

On one hand, $Q \le \Upsilon_{\mathbb{F}}\varphi(m)$ so that $\|\mathbf{1}_{\gamma,A}\| \le (\Upsilon_{\mathbb{F}} + \delta)\varphi(m)$. On the other hand, $\varphi(m) \le \Upsilon_{\mathbb{F}} Q$, whence $(1 - \delta\Upsilon_{\mathbb{F}})Q \le \|\mathbf{1}_{\gamma,A}\|$. Let $B \subseteq \mathbb{N}$ with $m_3 := |B| \ge m$, and let $\beta \in \mathbb{E}^B$. If $\beta\colon \mathbb{N}_{\le m_3} \to B$ is a bijection,

$$\|\mathbf{1}_{\gamma,A}\| \le \delta\varphi(m) + Q \le (1+\delta)\Upsilon_{\mathbb{F}} \left\| \sum_{j=1}^{m_3} \gamma_{\beta(j)} t_j \right\| \le \frac{1+\delta}{1-\delta\Upsilon_{\mathbb{F}}}\Upsilon_{\mathbb{F}}\|\mathbf{1}_{\beta,B}\|.$$

If $(s_j)_{j=1}^m$ is not equivalent to the canonical ℓ_1-basis, then the fundamental function φ^* of the dual basis of $(s_j)_{j=1}^\infty$ is also unbounded. So, we can choose $(p_m)_{m=1}^\infty$ so that, besides (3.1),

$$2Cp_m + \varepsilon_m\varphi^*(m) \le \delta\varphi^*(m), \quad m \in \mathbb{N}.$$

Let $\mathbf{z}$ be a subsequence of $\mathbf{y}$. Let m, A, B, γ and β be as before. Replacing $(s_j)_{j=1}^\infty$ with $(s_j^*)_{j=1}^\infty$, the argument carried on above gives

$$\frac{1-\delta\Upsilon_{\mathbb{F}}}{\Upsilon_{\mathbb{F}}}\varphi^*(m) \le \|\mathbf{1}_{\gamma,A}[\mathbf{z}^*]\| \le (\Upsilon_{\mathbb{F}}+\delta)\varphi^*(m),\ \|\mathbf{1}_{\gamma,A}[\mathbf{z}^*]\| \le \frac{1+\delta}{1-\delta\Upsilon_{\mathbb{F}}}\Upsilon_{\mathbb{F}}\|\mathbf{1}_{\beta,B}[\mathbf{z}^*]\|.$$

Using (2.1) and choosing δ small enough we obtain the desired estimates. □

The following lemma shows that Schreier unconditionality serves as a tool to obtain quasi-greediness. The pattern to observe here is that despite the fact that unconditionality and democracy are properties of rather different nature, democracy-related estimates can play a role to get nonlinear unconditionality properties from linear ones.

LEMMA 3.3 (cf. [**2**, Proposition 2.9]). *Let $\mathbf{x} = (x_n)_{n=1}^\infty$ be a sequence of a Banach space $\mathcal{X}$. Suppose that there is an increasing sequence $(q_m)_{m=1}^\infty$ in $\mathbb{N}$ and some constants C_1, C_2, and C_3 such that:*
- *$\mathbf{x}$ is a Schauder basis with constant C_1 of $\mathcal{X}$,*
- *$\varphi_u(m) \le C_2\varphi_l(q_m - m)$ for all $m \in \mathbb{N}$, and*
- *$\mathbf{x}$ is $\mathcal{D}_{\mathbf{q}}$-unconditional with constant C_3, where $\mathbf{q} = (q_{m+1})_{m=0}^\infty$.*

Then $\mathbf{x}$ is quasi-greedy with constant $C := C_3 + C_1C_2 + C_2C_3$.

PROOF. Let $(p_m)_{m=0}^\infty$ in $\mathbb{N}\cup\{0\}$ be the left inverse of $(q_m)_{m=1}^\infty$ defined by $p_m = p$ if $q_p \le m < q_{p+1}$, with the convention that $q_0 = 0$. Since $q_{p_m} \le m$, $\varphi_u(p_m) \le C_2\varphi_l(m-p_m)$ for all $m \in \mathbb{N}\cup\{0\}$. Since $m < q_{1+p_m}$, $\mathbf{x}$ is $\mathcal{E}_0$-unconditional with constant C_3, where

$$\mathcal{E}_0 = \cup_{m=1}^\infty\{\mathbb{N}_{\le p_m} \cup A\colon p_m < \min A,\ |A| \le m\}.$$

Consequently, $\mathbf{x}$ is $\mathcal{E}_1$-unconditional with constant $C_1 + C_3$, where $\mathcal{E}_1$ consists of all nonempty subsets $A \subseteq \mathbb{N}$ with $p_{|A|} < \min A$. By convexity,

$$\left\|\sum_{n\in A}\lambda_n\, a_n\, y_n\right\| \le (C_1 + C_3)\left\|\sum_{n=1}^\infty a_n\, x_n\right\|$$

for all $A \in \mathcal{E}_1$, $(a_n)_{n\in A} \in \mathbb{F}^A$, and $(\lambda_n)_{n\in A} \in [0,1]^A$. Convexity also yields

$$\left\|\sum_{n\in B} a_n\, x_n\right\| \le \varphi_u(p), \quad |B| \le p,\ |a_n| \le 1.$$

Let A be a greedy set of $x \in \mathcal{X}$. Set $m = |A|$ and $t = \max_{n\in A}|x_n^*(x)|$. Let $E = ([1,p_m]\cap\mathbb{Z}) \cup A$, $F = ([1,p_m]\cap\mathbb{Z}) \setminus A$, and $G = A \setminus [1,p_m]$. Note that $S_A(x) = S_E(x) - S_F(x)$, $|F| \le p_m$ and $|G| \ge m - p_m$. We have

$$\|S_F(x)\| \le t\varphi_u(p_m) \le C_2 t\varphi_l(m-p_m) \le C_2 t\|\mathbf{1}_{\gamma(x),G}\| \le C_2(C_1 + C_3)\|x\|$$

and $\|S_E(x)\| \le C_3\|x\|$. Therefore, $\|S_A(x)\| \le C\|x\|$. □

Since quasi-greedy bases with constant 1 are 1-suppression unconditional [**1**], the following result is sharp.

THEOREM 3.4 (cf. [**13**, Theorem 3.1]). *Let $\mathbf{x} = (x_n)_{n=1}^\infty$ be a semi-normalized weakly null sequence in a Banach space. Suppose that $(x_n)_{n=1}^\infty$ generates a spreading model $(s_j)_{j=1}^\infty$ nonequivalent to the canonical c_0-basis. Then, for every $\varepsilon > 0$, $\mathbf{x}$ has a basic subsequence $\mathbf{y}$ that is quasi-greedy with constant $1+\varepsilon$. Moreover, if $(s_j)_{j=1}^\infty$*

is also nonequivalent to the canonical ℓ_1*-basis, then we can choose* $\mathbf{y}$ *so that for any further subsequence* $\mathbf{z}$ *of* $\mathbf{y}$*,* $\mathbf{z}^*$ *is quasi-greedy with constant* $1+\varepsilon$*.*

PROOF. Pick $\delta > 0$ such that $(1+\delta)(1+2\delta) \le 1+\varepsilon$. By Theorem 2.1 we can assume that $\mathbf{x}$ is a basic sequence with basis constant not larger that $1+\delta$. By Theorem 3.2, passing to a subsequence we can assume that there is an increasing sequence $(q_m)_{m=1}^\infty$ in $\mathbb{N}$ such that the democracy functions of $\mathbf{x}$ satisfy $\varphi_u(m) \le \delta\varphi_l(q_m - m)$. By Theorem 2.4, passing to a further subsequence we can assume that $\mathbf{x}$ is $\mathcal{D}_{\mathbf{q}}$-unconditional with constant $1+\delta$, where $\mathbf{q} = (q_{m+1})_{m=1}^\infty$. By Lemma 3.3, $\mathbf{z}$ is quasi-greedy with constant $(1+\delta)(1+2\delta)$. In the case when $\mathbf{x}$ is also nonequivalent to the canonical ℓ_1-basis we can further assume that the democracy functions φ_u^* and φ_l^* of the dual basis of any subbasis $\mathbf{z}$ of $\mathbf{x}$ satisfy $\varphi_u^*(m) \le \delta\varphi_l^*(q_m - m)$ for all $m \in \mathbb{N}$. Since $\mathcal{D}_{\mathbf{q}}$-unconditionality passes to subsequences and to dual bases without increasing the constant, $\mathbf{z}^*$ is also quasi-greedy with constant $(1+\delta)(1+2\delta)$. $\square$

We observe that the conclusion of Theorem 3.4 does not hold for all weakly null sequences in general Banach spaces. Indeed, there is a renorming of c_0 such that the quasi-greedy constant of any subsequence of the canonical basis is at least 8/7 ([**16**, p. 59]).

THEOREM 3.5. *Let* $\mathcal{X}$ *be a Banach space that has an unconditional spreading model* $(s_j)_{j=1}^\infty$ *nonequivalent to the canonical* c_0*-basis. Then for any* $\varepsilon > 0$*,* $\mathcal{X}$ *has a basic sequence* $\mathbf{x}$ *that is quasi-greedy with constant* $1+\varepsilon$*. Moreover, if* ℓ_1 *embeds into* $\mathcal{X}$ *or* $(s_j)_{j=1}^\infty$ *is nonequivalent to the canonical* ℓ_1*-basis, then we can choose* $\mathbf{x}$ *so that* $\mathbf{y}^*$ *is quasi-greedy with constant* $1+\varepsilon$ *for any subbasis* $\mathbf{y}$ *of* $\mathbf{x}$*.*

PROOF. If ℓ_1 does not embed in $\mathcal{X}$ the result follows from combining Lemma 2.8 with Theorem 3.4. Otherwise, it is a consequence of Theorem 2.9. $\square$

4. Extraction principles for almost greedy bases

In this section we apply the results from Sections 2 and 3 to the study of subsequence extraction principles for almost greedy bases.

THEOREM 4.1 (cf. [**13**, Remark 3.2]). *Let* $\varepsilon > 0$ *and* $\mathbf{x} = (x_n)_{n=1}^\infty$ *be a semi-normalized weakly null sequence in a Banach space. Suppose that* $(x_n)_{n=1}^\infty$ *generates a spreading model* $(s_j)_{j=1}^\infty$ *nonequivalent to the canonical* c_0*-basis. Then* $\mathbf{x}$ *has a basic subsequence* $\mathbf{y}$ *that is almost greedy with constant* $2+\Upsilon_{\mathbb{F}}+\varepsilon$*. Moreover, if* $(s_j)_{j=1}^\infty$ *is also nonequivalent to the canonical* ℓ_1*-basis, then* $\mathbf{y}$ *can be choosen so that* $\mathbf{z}^*$ *is almost greedy with constant* $2+\Upsilon_{\mathbb{F}}+\varepsilon$ *for any subsequence* $\mathbf{z}$ *of* $\mathbf{y}$*.*

PROOF. Fix $\delta > 0$. Combining Theorem 3.4 with Theorem 3.2 gives a subsequence $\mathbf{y}$ of $\mathbf{x}$ that is quasi-greedy with constant $1+\delta$ and super-democratic with constant $\Upsilon_{\mathbb{F}}+\delta$. In the case when $(s_j)_{j=1}^\infty$ is also nonequivalent to the canonical ℓ_1-basis, we choose $\mathbf{y}$ so that the same holds for $\mathbf{z}^*$. By Theorem 3.1, $\mathbf{y}$ and $\mathbf{z}^*$ are almost greedy with constant $2+\delta+(1+\delta)(\Upsilon_{\mathbb{F}}+\delta)$. Choosing δ small enough we obtain the desired inequality. $\square$

THEOREM 4.2. *Let* $\mathcal{X}$ *be a Banach space that generates a* 1*-suppression unconditional spreading model nonequivalent to the canonical* c_0*-basis. Then for any* $\varepsilon > 0$*,* $\mathcal{X}$ *has a basic sequence* $\mathbf{x}$ *that is almost greedy with constant* $2+\Upsilon_{\mathbb{F}}+\varepsilon$*. Moreover, if no subspace of* $\mathcal{X}$ *is isomorphic to* ℓ_1 *or* $(s_j)_{j=1}^\infty$ *is nonequivalent to the* *canonical ℓ_1-basis, then we can choose* $\mathbf{x}$ *so that* $\mathbf{y}^*$ *is almost greedy with constant* $2+\Upsilon_{\mathbb{F}}+\varepsilon$ *for any subbasis* $\mathbf{y}$ *of* $\mathbf{x}$.

PROOF. If ℓ_1 embeds into $\mathcal{X}$ the result follows from combining Lemma 2.8 with Theorem 4.1. Otherwise, it is a consequence of Theorem 2.9. □

A complete minimal system $\mathbf{x}$ of a Banach space $\mathcal{X}$ is said to be *bidemocratic* if there is a constant C such that $\varphi_u(m)\varphi_u^*(m) \le Cm$ for all $m \in \mathbb{N}$, where φ_u and φ_u^* are the upper democracy functions of $\mathbf{x}$ and its biorthogonal functionals, respectively. It is known [**15**] that $\mathbf{x}$ is quasi-greedy and bidemocratic if and only if both $\mathbf{x}$ and its biorthogonal functionals are almost greedy.

THEOREM 4.3 (cf. [**14**, Corollary 5.7]). *Given a Banach space* $\mathcal{X}$, *the following are equivalent.*

- $\mathcal{X}$ *has a bidemocratic basic sequence.*
- $\mathcal{X}$ *has a quasi-greedy bidemocratic basic sequence.*
- *Either* $\mathcal{X}$ *contains* c_0 *or* ℓ_1, *or it has a suppression unconditional spreading model which is neither equivalent to the canonical basis of* c_0 *nor the canonical basis of* ℓ_1.

PROOF. It suffices to prove that if $\mathcal{X}$ has a bidemocratic basic sequence $\mathbf{x} = (x_n)_{n=1}^\infty$ and contains neither c_0 nor ℓ_1 then $\mathbf{x}$ has a spreading model which is neither equivalent to the canonical c_0-basis nor the canonical ℓ_1-basis. Passing to a subsequence and replacing $(x_n)_{n=1}^\infty$ with $(x_{2n-1}-x_{2n})_{n=1}^\infty$ we can assume that $\mathbf{x}$ is weakly null. Then, passing to a further subsequence we can assume that $\mathbf{x}$ generates a 1-suppression unconditional spreading model $(s_j)_{j=1}^\infty$ whose fundamental function φ is equivalent to the upper democracy function of $\mathbf{x}$, and the fundamental function φ^* of $(s_j^*)_{j=1}^\infty$ is equivalent to that of $\mathbf{x}^*$. Since neither $\mathbf{x}$ nor $\mathbf{x}^*$ are equivalent to the canonical c_0-basis, both φ and φ^* are unbounded. Hence, $(s_j)_{j=1}^\infty$ is neither equivalent to the canonical c_0-basis nor to canonical ℓ_1-basis. □

5. Existence of truncation quasi-greedy basic sequences

In the scale of greedy-like bases, truncation quasi-greedy bases are just one step below the level of quasi-greedy bases. Moreover, many features relative to the performance of the TGA using quasi-greedy bases pass down to truncation quasi-greedy bases (see [**5**]). We start by summarizing several characterizations of this relatively new class of bases.

THEOREM 5.1. *Let* $\mathbf{x}$ *be a complete minimal system of a Banach space* $\mathcal{X}$, *and let* $(x_n^*)_{n=1}^\infty$ *be the sequence of its biorthogonal functionals. The following are equivalent.*

(1) $\mathbf{x}$ *is truncation quasi-greedy.*
(2) *There is a constant* C *such that* $\|\sum_{n\in A} a_n\, x_n\| \le C\|x\|$ *for all* $x \in \mathcal{X}$, *all* $A \subseteq \mathbb{N}$ *with* $|x_n^*(x)| \ge 1$ *for all* $n \in A$, *and all* $(a_n)_{n\in A} \in \mathbb{B}^A$.
(3) *For every* $0 < d < 1$ *there is a constant* $C = C(d)$ *such that* $\|S_A(x)\| \le C\|x\|$ *for all* $x \in \mathcal{X}$ *and all* $A \subseteq \mathbb{N}$ *with* $d \le |x_n^*(x)| \le 1$ *for all* $n \in A$.
(4) *There are* $0 < d < 1$ *and a constant* C *such that* $\|S_A(x)\| \le C\|x\|$ *for all* $x \in \mathcal{X}$ *and all* $A \subseteq \mathbb{N}$ *with* $d \le |x_n^*(x)| \le 1$ *for all* $n \in A$.
(5) *There is a constant* C *such that* $\|S_A(x)\| \le C\|x\|$ *for all* $x \in \mathcal{X}$ *and all* $A \subseteq \mathbb{N}$ *with* $|x_n^*(x)| = 1$ *for all* $n \in A$.

(6) *There is a constant C such that $\|S_A(x)\| \le C\|x\|$ for all $x \in \mathcal{X}$ and all $A \subseteq \mathbb{N}$ such that $x_n^*(x) = 1$ for all $n \in A$.*
(7) *There is a constant C such that $\|\sum_{n\in A} a_n\, x_n\| \le C\|x\|$ for all $x \in \mathcal{X}$, all $A \subseteq \mathbb{N}$ with $|x_n^*(x)| = 1$ for $n \in A$, and all $(a_n)_{n\in A} \in \mathbb{B}^A$.*

PROOF. If $\mathbf{x}$ is truncation quasi-greedy with constant C then it is unconditional for constant coefficients with constant C. Let A, $(a_n)_{n\in A}$, and x be as in (2). Set $B = \{n \in \mathbb{N} \colon |x_n^*(x)| \ge 1\}$. By convexity,

$$\left\|\sum_{n\in A} a_n\, x_n\right\| \le \Upsilon_{\mathbb{F}} C\|\mathbf{1}_{\gamma(x),B}\| \le \Upsilon_{\mathbb{F}} C^2\|x\|.$$

If (2) holds and x and A are as in (3) then $|x_n^*(x)| \le 1 \le |x_n^*(x/d)|$ for all $n \in \mathbb{N}$. Therefore

$$\|S_A(x)\| = \left\|\sum_{n\in A} x_n^*(x)\, x_n\right\| \le C\|x/d\| = \frac{C}{d}\|x\|.$$

(3) is formally stronger than (4), (4) is formally stronger than (5), and (5) is formally stronger than (6). Assume that (6) holds. Then, in particular, $\|\mathbf{1}_B\| \le C\|\mathbf{1}_A\|$ whenever $B \subseteq A$, whence, by convexity,

$$\left\|\sum_{n\in A} a_n\, x_n\right\| \le \Upsilon_{\mathbb{F}} C\|\mathbf{1}_A\|, \quad A \subseteq \mathbb{N},\ |A| < \infty,\ |a_n| \le 1. \tag{5.1}$$

Set $\delta = 1/(2\Upsilon_{\mathbb{F}} C^2)$. Fix $A \subseteq \mathbb{N}$ finite, $x \in \mathcal{X}$ with $\operatorname{supp}(x) \cap A = \emptyset$, $\omega \in \mathbb{E}$, and $\gamma = (\gamma_n)_{n\in A}$ with $|\varepsilon_n - \omega| \le \delta$ for all $n \in A$. Then,

$$\begin{aligned}\|\mathbf{1}_A\| &\le C\|\mathbf{1}_A + x/\omega\| = C\|\omega\mathbf{1}_A + x\|\\ &\le C\|\omega\mathbf{1}_A - \mathbf{1}_{\gamma,A}\| + C\|\mathbf{1}_{\gamma,A} + x\| \le \Upsilon_{\mathbb{F}} C^2\delta\|\mathbf{1}_A\| + C\|\mathbf{1}_{\varepsilon,A} + x\|,\end{aligned}$$

so that $\|\mathbf{1}_A\| \le 2C\|\mathbf{1}_{\gamma,A} + x\|$. To obtain a similar estimate without assuming that the scalars in γ are close enough to a suitable scalar $\omega \in \mathbb{E}$, we pick a finite partition $(\mathbb{E}_k)_{k=1}^N$ of $\mathbb{E}$ for which there is $(\omega_k)_{k=1}^N \in \mathbb{E}^N$ such that

$$\sup_{\omega\in\mathbb{E}_k} |\omega - \omega_k| \le \delta, \quad k = 1, \dots, N.$$

Fix $A \subseteq \mathbb{N}$ finite, $\gamma = (\gamma_n)_{n\in A} \in \mathbb{E}^A$, and $u \in \mathcal{X}$ with $\operatorname{supp}(x) \cap A = \emptyset$. Given $B \subseteq A$, put

$$B_k = \{n \in B \colon \gamma_n \in \mathbb{E}_k\} \quad \text{and} \quad u_k = \mathbf{1}_{\gamma, A\setminus B_k} + u, \quad \text{for } k = 1, \dots, N.$$

Since $\operatorname{supp}(u_k) \cap B_k = \emptyset$ for all $k = 1, \dots, N$,

$$\|\mathbf{1}_B\| \le \sum_{k=1}^N \|\mathbf{1}_{B_k}\| \le 2C\sum_{k=1}^N \|\mathbf{1}_{\gamma,B_k} + u_k\| = 2NC\|\mathbf{1}_{\gamma,A} + u\|.$$

By convexity, if $(a_n)_{n\in A}$ and x are as in (7), $\|\sum_{n\in A} a_n\, x_n\| \le 2NC\Upsilon_{\mathbb{F}}\|x\|$.

(7) is formally stronger than (5). Assume that (5) holds. Choose $1 - 1/C < d < 1$. Given A and x as in (4), choose $B \subseteq A$ such that

$$\|\mathbf{1}_{\gamma(x),E}\| \le \|\mathbf{1}_{\varepsilon(x),B}\|, \quad E \subseteq A.$$

By convexity,

$$\|S_A(x)\| \le \|\mathbf{1}_{\gamma(x),B}\| \quad \text{and} \quad \|\mathbf{1}_{\gamma(x),B} - S_B(x)\| \le (1-d)\|\mathbf{1}_{\varepsilon(x),B}\|. \tag{5.2}$$

The second inequality in (5.2) yields

$$\|\mathbf{1}_{\gamma(x),B}\| \le C\|\mathbf{1}_{\gamma(x),B} + x - S_B(x)\| \le C\|x\| + C(1-d)\|\mathbf{1}_{\gamma(x),B}\|,$$

therefore $(1 - C(1-d))\|\mathbf{1}_{\gamma(x),B}\| \le \|x\|$. Combining with the first inequality in (5.2) we obtain $\|S_A(x)\| \le 1/(1-C(1-d))\|x\|$.

Suppose that (4) holds, and let A be a greedy set of $x \in \mathcal{X}$. Assume that $b := \min_{n\in A} |x_n^*(x)| > 0$; otherwise there is nothing to prove. Put

$$A_k = \{n \in A \colon bd^{1-k} \le |x_n^*(x)| < bd^{-k}\}, \quad k \in \mathbb{N},$$

so that $(A_k)_{k=1}^\infty$ is a partition of A. For each $k \in \mathbb{N}$, choose $B_k \subseteq A_k$ such that

$$\|S_B(x)\| \le \|S_{B_k}(x)\|, \quad B \subseteq A_k.$$

By convexity,

$$\left\| \sum_{n\in A_k} a_n\, x_n^*(x)\, \mathbf{x}_n \right\| \le \|S_{B_k}(x)\|, \quad 0 \le a_n \le 1.$$

Since $d \le |x_n^*(d^k x/b)| \le 1$ for all $k \in \mathbb{N}$ and all $n \in B_k$,

$$b\|\mathbf{1}_{\gamma(x),A}(x)\| \le \sum_{k=1}^\infty b\|\mathbf{1}_{\gamma(x),A_k}(x)\| \le \sum_{k=1}^\infty d^{k-1}\|S_{B_k}(x)\| \le C\|x\| \sum_{k=1}^\infty d^{k-1}.$$

Hence, $\mathbf{x}$ is truncation quasi-greedy with constant $C/(1-d)$. □

All the characterizations of truncation quasi-greedy bases in Theorem 5.1 other than (6) were obtained in [**4**]. Precisely this characterization, which is especially relevant when the scalar field is $\mathbb{C}$, is better fitted for the combinatorial techniques needed to obtain the following extraction principle, which we will prove later.

THEOREM 5.2. *There is a constant C such that any semi-normalized weakly null basic sequence in a Banach space has a truncation quasi-greedy basic subsequence with constant C.*

Bases that satisfy (2) in Theorem 5.1 were introduced in [**16**], where they were named *bounded-oscillation unconditional.* Note that d-near unconditional bases are those bases that satisfy property (4) in Theorem 5.1 under the following extra assumption:

(A) $|x_n^*(x)| \le 1$ for all $n \in \mathbb{N}$.

In turn, bases that satisfy property (5) in Theorem 5.1 under the restriction (A) are called *quasi-greedy for largest coefficients* (QGLC for short) (see [**6**]). It is known [**4**] that quasi-greediness for largest coefficients is a strictly weaker property than truncation quasi-greediness, but it is equivalent to d-near unconditionality for some (all) $0 < d < 1$ (see [**3**]). So, Theorem 5.2 improves Theorem 1.6. Nonetheless, the study of Elton's near unconditionality continues to hold interest and, in fact, the existence of a bound independent of d for the constant $K(d)$ in Theorem 1.6 would imply a positive answer to Problem 1.4 (see [**16**, Proposition 3.2]).

To prove Theorem 5.2 we will require a few combinatorial lemmas that rely on Ramsey's theory. Given $N \subseteq \mathbb{N}$, let $\mathcal{P}_\infty(N)$ denote the set consisting of all infinite subsets of N. We will assume that a given subset of $\mathcal{P}(\mathbb{N})$ is endowed with the topology it inherits from the identification of $\mathcal{P}(\mathbb{N})$ with $2^\mathbb{N}$, where the power set $2^\mathbb{N}$ is equipped with the product topology. Note that if $\mathcal{A} \subseteq \mathcal{P}_{<\infty}(\mathbb{N})$, the topology we obtain this way is the discrete topology. Given A, $B \in \mathcal{P}_{<\infty}(\mathbb{N}) \setminus \{\emptyset\}$ we say that A is an *initial segment* of B if $A \subseteq B$ and $\max(A) < \min(B \setminus A)$. A set $\mathcal{A} \subseteq \mathcal{P}_{<\infty}(\mathbb{N}) \setminus \{\emptyset\}$ is said to be *thin* if no set in $\mathcal{A}$ is an initial segment of another set $B \in \mathcal{A}$ unless $A = B$.

LEMMA 5.3 (cf. [**16**, Theorem 2.2]). *Suppose that for each $M \in \mathcal{P}_\infty(\mathbb{N})$, some $F_M \in \mathcal{P}_{<\infty}(\mathbb{N}) \setminus \{0\}$ is given. Suppose also that the map $M \mapsto F_M$ is continuous. Then for every $N \in \mathcal{P}_\infty(\mathbb{N})$ there exist L and $M \in \mathcal{P}_\infty(N)$ such that F_L is an initial segment of F_M and $L \cap M = F_L$.*

LEMMA 5.4 ([**16**, Proposition 2.3]). *Let $\mathcal{C}$ be a set of 'colours'. Let $(\mathcal{C}_k)_{k=1}^\infty$ be a sequence of finite sets such that $\mathcal{C} = \cup_{k=1}^\infty \mathcal{C}_k$. Let $\mathcal{A} \subseteq \mathcal{P}_{<\infty}(\mathbb{N}) \setminus \{\emptyset\}$ be thin. Let $c\colon \mathcal{A} \to \mathcal{C}$ be a 'colouring' such that $c(A) \in \mathcal{C}_{\min A}$ for all $A \in \mathcal{A}$. Then for all $L \in \mathcal{P}_\infty(\mathbb{N})$ there is $M \in \mathcal{P}_\infty(L)$ such that $c(A) = c(B)$ whenever A, $B \in \mathcal{A}$ are such that A, $B \subseteq M$ and $\min A = \min B$.*

LEMMA 5.5 (cf. [**16**, Lemma 2.5]). *Let Ω be a well-ordered set equipped with the discrete topology. Suppose that for each $M \in \mathcal{P}_\infty(\mathbb{N})$ a nonempty subset $\Phi(M)$ of Ω is given. Let $(\Omega_k)_{k=1}^\infty$ be such that $\Omega = \cup_{k=1}^\infty \Omega_k$ and $\Phi(L) \cap \Omega_k = \Phi(M) \cap \Omega_k$ whenever*

$$M \cap \mathbb{N}_{\le k} = L \cap \mathbb{N}_{\le k}.$$

Then for each $M \in \mathcal{P}_\infty(\mathbb{N})$ we can choose $\phi(M) \in \Phi(M)$ in such a way that the map ϕ is continuous. Moreover, for each $M \in \mathcal{P}_\infty(\mathbb{N})$ there is $k \in \mathbb{N}$ such that $\phi(M)$ is the least element in $\Phi(M) \cap \Omega_k$ and $\Phi(M) \cap \Omega_j = \emptyset$ if $j < k$.

LEMMA 5.6 ([**16**, Lemma 2.6]). *Assume that $c_0 \subseteq \mathbb{F}^\mathbb{N}$ is endowed with the product topology. Suppose that for each $M \in \mathcal{P}_\infty(\mathbb{N})$ a sequence $f(M) = (a_n^M)_{n=1}^\infty \in c_0$ is given in such a way that the map $M \mapsto f(M)$ is continuous, and for every sequence $(M_n)_{n=1}^\infty$, $(f(M_n))_{n=1}^\infty$ has a cluster point. Then for each $\varepsilon > 0$ and every $M \in \mathcal{P}_\infty(\mathbb{N})$ there is $N \in \mathcal{P}_\infty(M)$ so that for all $P \in \mathcal{P}_\infty(N)$ we have*

$$\sum_{n \in N \setminus P} |a_n^M| \le \varepsilon,$$

i.e., the support of $f(P)$ relative to N is almost contained in P.

We will obtain Theorem 5.2 as a consequence of the (new) almost isometric extraction principle we write below. Its proof is an adaptation to our setting of the proof of [**16**, Theorem 2.1]. Note that, in light of the equivalence between conditions (1) and (3) in Theorem 5.1, Theorem 5.2 could be derived from [**16**, Theorem 2.1], although we should point out that its proof only works for real scalars.

THEOREM 5.7. *Let $\mathbf{x} = (x_n)_{n=1}^\infty$ be a weakly null seminormalized sequence in a Banach space $\mathbb{X}$. For every $C > 1$ there is a subsequence $\mathbf{y} = (y_n)_{n=1}^\infty$ of $\mathbf{x}$ such that*

$$\left\| \sum_{n \in A} a_n\, y_n \right\| \le C \left\| \sum_{n=1}^\infty a_n\, y_n \right\|$$

for all $(a_n)_{n=1}^\infty \in c_{00}$ and all $A \subseteq \mathbb{N}$ with $a_n = 1$ for $n \in A$.

PROOF. Assume that $\lim_n \|x_n\| = 1$ and that $\mathbf{x}$ is a Schauder basis of $\mathbb{X}$ whose dual basis $(x_n^*)_{n=1}^\infty$ satisfies $\|x_n^*\| \le 2$ for all $n \in \mathbb{N}$. In what follows, $\operatorname{supp}(y)$ denotes the support of $y \in \mathbb{X}$ with respect to $\mathbf{x}$.

Suppose, by contradiction, that there is $C>1$ such that no subsequence of $\mathbf{x}$ satisfies the conclusion of Theorem 5.7. Fix $\varepsilon>0$ and choose a function $\rho\colon \mathbb{N}\to(0,\infty)$ with $\lim_k \rho(k)=\infty$ and

$$\rho(k)\le(1+\varepsilon)(\rho(k-1)-k-\varepsilon), \quad k\ge 2.$$

We claim that for every $M\in\mathcal{P}_\infty(\mathbb{N})$ there are $F\in\mathcal{P}_{<\infty}(M)\setminus\{\emptyset\}$, $y\in\operatorname{span}(x_n\colon n\in M)$, and $y^*\in B_{\mathcal{X}^*}$ such that $x_n^*(y)=1$ and $\operatorname{Re}(y^*(x_n))>0$ for all $n\in F$, $F\subseteq\operatorname{supp}(y)$, and

$$\frac{C}{1+\varepsilon}\|y\|<\sum_{n\in F}\operatorname{Re}(y^*(x_n))<\rho(\min F).$$

We call such a triple (F,y,y^*) a *witness for* M. To see this, choose $M_0\in\mathcal{P}_\infty(M)$ so that $\|x_n\|\le 1+\varepsilon$ for all $n\in M_0$. Next we pick $E\subseteq M_0$ and $y\in\operatorname{span}(x_n\colon n\in M_0)$ such that $x_n^*(y)=1$ for all $n\in E$, and $\|\mathbf{1}_E\|>C\|y\|$. Then, for a suitable $y^*\in B_{\mathcal{X}^*}$,

$$\sum_{n\in E}\operatorname{Re}(y^*(x_n))>C\|y\|.$$

Passing to a subset of E if necessary we can suppose that $Re(y^*(x_n))>0$ for all $n\in E$. Set $j=\min E$. Since, unlike ρ, the function $h\colon\mathbb{N}\to[0,\infty)$ given by

$$h(k)=\sum_{\substack{n\in E\\ n>k}}\operatorname{Re}(y^*(x_n))=\sum_{\substack{n\in E\\ n\ge k+1}}\operatorname{Re}(y^*(x_n))$$

is bounded, there is $k\in\mathbb{N}\cap[j,\infty)$ minimal with the property that $h(k)<\rho(k)-1-\varepsilon$. Let us prove that (F,y,y^*) is a witness for M, where $F=E\cap[k,\infty)$. We have

$$\sum_{n\in F}\operatorname{Re}(y^*(x_n))=\operatorname{Re}(y^*(x_k))+h(k)\le 1+\varepsilon+h(k)<\rho(k).$$

The minimality of k implies that $k\in E$. If $k>j$ then $\rho(k-1)\le h(k-1)+1+\varepsilon$ and

$$\sum_{n\in E}\operatorname{Re}(y^*(x_n))\le(1+\varepsilon)k+h(k)<(1+\varepsilon)(k-1)+\rho(k)\le(1+\varepsilon)(\rho(k-1)-1-\varepsilon).$$

Consequently, regardless of whether $k>j$ or $k=j$,

$$\sum_{n\in E}\operatorname{Re}(y^*(x_n))\le(1+\varepsilon)\sum_{\substack{n\in E\\ n\ge k}}\operatorname{Re}(y^*(x_n)).$$

Consider a well-ordering on the set Ω of all witnesses of some $M\in\mathcal{P}_\infty(\mathbb{N})$. Assume that Ω is equipped with the discrete topology. For each $k\in\mathbb{N}$, let Ω_k be the set of all $(F,y,y^*)\in\Omega$ with $\max\operatorname{supp}(y)\le k$. An application of Lemma 5.5 gives for each $M\in\mathcal{P}_\infty(\mathbb{N})$ a witness $\phi(M)=(F_M,y_M,y_M^*)$ so that the map $M\mapsto\phi(M)$ is continuous, and $\phi(M)=\phi(L)$ whenever $\operatorname{supp}(y_L)$ is an initial segment of $\operatorname{supp}(y_M)$. The last condition implies that

$$\mathcal{A}:=\{\operatorname{supp}(y_M)\colon M\in\mathcal{P}_\infty(\mathbb{N})\}$$

is thin. Note that $\operatorname{supp}(y_L)=\operatorname{supp}(y_M)$ implies $\phi(L)=\phi(M)$. For convenience, set

$$S_M=\sum_{n\in F_M}y_M^*(x_n), \quad a_k^M=x_k^*(y_M), \quad M\in\mathcal{P}_\infty(\mathbb{N}),\ k\in\mathbb{N}.$$

Since $\mathbf{x}$ is weakly null, $f(M) := (y_M^*(x_n))_{n=1}^\infty \in c_0$ for all $M \in \mathcal{P}_\infty(\mathbb{N})$. By Banach-Alaoglu's theorem, the range of the map $f\colon \mathcal{P}_\infty(\mathbb{N}) \to c_0$ is relatively compact (with respect to the topology of pointwise convergence). Since ϕ is continuous, so is f. Hence, by Lemma 5.6, there is $N_0 \in \mathcal{P}_\infty(\mathbb{N})$ such that

$$\sum_{n \in N_0 \setminus P} |y_P^*(x_n)| \le \varepsilon, \quad P \in \mathcal{P}_\infty(N_0).$$

Let us colour Ω. Set $T = \varepsilon(\mathbb{N} \cup \{0\})$ and, for each $k \in \mathbb{N}$, put

$$\mathcal{C}_k = \{(k, \omega) \in \mathbb{N} \times T \colon \omega \le \rho(k)\}.$$

To any witness (F, y, y^*) of an $M \in \mathcal{P}_\infty(\mathbb{N})$ we assign the colour (k, ω), where $k = \min F$ and ω is the largest number in T with $\omega \le \sum_{n\in F} y^*(x_n)$. Via the choice of witnesses we have made, this colouring induces a colouring of $\mathcal{P}_\infty(\mathbb{N})$ which, in turn, induces a colouring $c\colon \mathcal{A} \to \cup_{k=1}^\infty \mathcal{C}_k$. Observe that this last colouring satisfies $c(A) \in \mathcal{C}_{\min A}$ for all $A \in \mathcal{A}$. Hence, by Lemma 5.4, there is $N \in \mathcal{P}_\infty(N_0)$ such that two sets M and L in $\mathcal{P}_\infty(N)$ have the same colour whenever $\min F_M = \min F_L$. Note that if (k, ω) is the colour of (F, y, y^*), then

$$\omega \le \sum_{n \in F} y^*(x_n) < \omega + \varepsilon.$$

To finish the proof we use Lemma 5.3 to pick L and $M \in \mathcal{P}_\infty(N)$ such that F_L is an initial segment of F_M and $L \cap M = F_L$. Since $\min F_L = \min F_M$, L and M have the same colour, say (k, ω). Consequently,

$$S_M < \omega + \varepsilon \le S_L + \varepsilon.$$

Since $|a_n^M| \le 2\|y_M\|$ for all $n \in \mathbb{N}$, $\|y_M\| \ge 1/2$, and the number

$$R := \sum_{n \in M \setminus L} a_n^M y_L^*(x_n)$$

satisfies $|R| \le 2\varepsilon \|y_M\|$. Summing up,

$$\begin{aligned} \frac{C\|y_M\|}{1+\varepsilon} &< S_L + \varepsilon = \operatorname{Re}(y_L^*(y_M) - R) + \varepsilon \\ &\le \operatorname{Re}(y_L^*(y_M)) + 2\varepsilon\|y_M\| + \varepsilon \le (1+2\varepsilon)\|y_M\| + \varepsilon \le (1+4\varepsilon)\|y_M\|. \end{aligned}$$

Dropping the factor $\|y_M\|$ and letting ε tend to zero we get $C \le 1$. □

PROOF OF THEOREM 5.2. The estimate for the truncation quasi-greedy constant implicit in the implication (6) ⇒ (1) in Theorem 5.1 only depends on C. So, the result follows from combining this implication with Theorem 5.7. □

COROLLARY 5.8. *There is a constant $C \ge 1$ such that any Banach space has a truncation quasi-greedy basic subsequence with constant C.*

PROOF. Just combine Theorem 5.2, Theorem 2.7, and Theorem 2.9. □

References

[1] F. Albiac and J. L. Ansorena, *Characterization of 1-quasi-greedy bases*, J. Approx. Theory **201** (2016), 7–12, DOI 10.1016/j.jat.2015.08.006. MR3424006

[2] F. Albiac, J. L. Ansorena, G. Bello, and P. Wojtaszczyk, *Existence of almost greedy bases in mixed-norm sequence and matrix spaces, including Besov spaces*, Constr. Approx. 2023. `https://doi.org/10.1007/s00365-023-09662-0`.

[3] F. Albiac, J. L. Ansorena, and M. Berasategui, *Elton's near unconditionality of bases as a threshold-free form of greediness*, J. Funct. Anal. **285** (2023), no. 7, Paper No. 110060, 24, DOI 10.1016/j.jfa.2023.110060. MR4605406

[4] F. Albiac, J. L. Ansorena, and M. Berasategui, *Linear versus nonlinear forms of partial unconditionality of bases*, `arXiv:2305.12253`, 2023.

[5] F. Albiac, J. L. Ansorena, M. Berasategui, P. M. Berná, and S. Lassalle, *Weak forms of unconditionality of bases in greedy approximation*, Studia Math. **267** (2022), no. 1, 1–17, DOI 10.4064/sm210601-2-2. MR4460220

[6] F. Albiac, J. L. Ansorena, P. M. Berná, and P. Wojtaszczyk, *Greedy approximation for biorthogonal systems in quasi-Banach spaces*, Dissertationes Math. **560** (2021), 88, DOI 10.4064/dm817-11-2020. MR4262900

[7] F. Albiac and N. J. Kalton, *Topics in Banach space theory*, 2nd ed., Graduate Texts in Mathematics, vol. 233, Springer, [Cham], 2016. With a foreword by Gilles Godefroy, DOI 10.1007/978-3-319-31557-7. MR3526021

[8] J. L. Ansorena, *A note on subsymmetric renormings of Banach spaces*, Quaest. Math. **41** (2018), no. 5, 615–628, DOI 10.2989/16073606.2017.1393704. MR3836410

[9] S. Banach, *Théorie des opérations linéaires* (French), Éditions Jacques Gabay, Sceaux, 1993. Reprint of the 1932 original. MR1357166

[10] B. Beauzamy and J.-T. Lapresté, *Modèles étalés des espaces de Banach* (French), Publ. Dép. Math. (Lyon) (N.S.) **4/A** (1983), ii+199+11. MR721416

[11] A. Brunel, *Espaces associés à une suite bornée dans un espace de Banach* (French), Séminaire Maurey-Schwartz 1973–1974: Espaces L^p, applications radonifiantes et géométrie des espaces de Banach, École Polytech., Paris, 1974, pp. Exp. Nos. 15, 16 et 18, 23. MR405070

[12] A. Brunel and L. Sucheston, *On B-convex Banach spaces*, Math. Systems Theory **7** (1974), no. 4, 294–299, DOI 10.1007/BF01795947. MR438085

[13] R. M. Causey and S. J. Dilworth, *Almost isometric constants for partial unconditionality*, Proc. Amer. Math. Soc. **144** (2016), no. 8, 3397–3404, DOI 10.1090/proc/13004. MR3503707

[14] S. J. Dilworth, N. J. Kalton, and D. Kutzarova, *On the existence of almost greedy bases in Banach spaces*, Studia Math. **159** (2003), no. 1, 67–101, DOI 10.4064/sm159-1-4. Dedicated to Professor Aleksander Pełczyński on the occasion of his 70th birthday. MR2030904

[15] S. J. Dilworth, N. J. Kalton, D. Kutzarova, and V. N. Temlyakov, *The thresholding greedy algorithm, greedy bases, and duality*, Constr. Approx. **19** (2003), no. 4, 575–597, DOI 10.1007/s00365-002-0525-y. MR1998906

[16] S. J. Dilworth, E. Odell, Th. Schlumprecht, and A. Zsák, *Partial unconditionality*, Houston J. Math. **35** (2009), no. 4, 1251–1311. MR2577152

[17] I. S. Èdel′šteĭn and P. Wojtaszczyk, *On projections and unconditional bases in direct sums of Banach spaces*, Studia Math. **56** (1976), no. 3, 263–276, DOI 10.4064/sm-56-3-263-276. MR425585

[18] J. H. Elton, *Weakly null normalized sequences in Banach spaces*, ProQuest LLC, Ann Arbor, MI, 1978. Thesis (Ph.D.)–Yale University. MR2628434

[19] P. Enflo, *A counterexample to the approximation problem in Banach spaces*, Acta Math. **130** (1973), 309–317, DOI 10.1007/BF02392270. MR402468

[20] W. T. Gowers and B. Maurey, *The unconditional basic sequence problem*, J. Amer. Math. Soc. **6** (1993), no. 4, 851–874, DOI 10.2307/2152743. MR1201238

[21] A. Grothendieck, *Produits tensoriels topologiques et espaces nucléaires* (French), Mem. Amer. Math. Soc. **16** (1955), Chapter 1: 196 pp.; Chapter 2: 140. MR75539

[22] T. H. Hildebrandt, *Linear functional transformations in general spaces*, Bull. Amer. Math. Soc. **37** (1931), no. 4, 185–212, DOI 10.1090/S0002-9904-1931-05129-6. MR1562128

[23] R. C. James, *Uniformly non-square Banach spaces*, Ann. of Math. (2) **80** (1964), 542–550, DOI 10.2307/1970663. MR173932

[24] W. B. Johnson, B. Maurey, and G. Schechtman, *Weakly null sequences in L_1*, J. Amer. Math. Soc. **20** (2007), no. 1, 25–36, DOI 10.1090/S0894-0347-06-00548-0. MR2257395

[25] S. V. Konyagin and V. N. Temlyakov, *A remark on greedy approximation in Banach spaces*, East J. Approx. **5** (1999), no. 3, 365–379. MR1716087

[26] J. Lindenstrauss and L. Tzafriri, *Classical Banach spaces. I: Sequence spaces*, Ergebnisse der Mathematik und ihrer Grenzgebiete [Results in Mathematics and Related Areas], vol. 97, Springer-Verlag, Berlin-New York, 1977.

[27] B. Maurey and H. P. Rosenthal, *Normalized weakly null sequence with no unconditional subsequence*, Studia Math. **61** (1977), no. 1, 77–98, DOI 10.4064/sm-61-1-77-98. MR438091

[28] R. D. Mauldin (ed.), *The Scottish Book*, 2nd ed., Birkhäuser/Springer, Cham, 2015. Mathematics from the Scottish Café with selected problems from the new Scottish Book; Including selected papers presented at the Scottish Book Conference held at North Texas University, Denton, TX, May 1979, DOI 10.1007/978-3-319-22897-6. MR3242261

[29] E. Odell, *On Schreier unconditional sequences*, Banach spaces (Mérida, 1992), Contemp. Math., vol. 144, Amer. Math. Soc., Providence, RI, 1993, pp. 197–201, DOI 10.1090/conm/144/1209461. MR1209461

[30] H. P. Rosenthal, *A characterization of Banach spaces containing l^1*, Proc. Nat. Acad. Sci. U.S.A. **71** (1974), 2411–2413, DOI 10.1073/pnas.71.6.2411. MR358307

[31] P. Wojtaszczyk, *Greedy algorithm for general biorthogonal systems*, J. Approx. Theory **107** (2000), no. 2, 293–314, DOI 10.1006/jath.2000.3512. MR1806955

Department of Mathematics, Statistics, and Computer Sciencies–InaMat2, Universidad Pública de Navarra, Campus de Arrosadía, Pamplona, 31006 Spain

Email address: fernando.albiac@unavarra.es

Department of Mathematics and Computer Sciences, Universidad de La Rioja, Logroño, 26004 Spain

Email address: joseluis.ansorena@unirioja.es

Proceedings of Symposia in Pure Mathematics
Volume **106**, 2024
https://doi.org/10.1090/pspum/106/01926

On Properties of Geometric Preduals of Classical Smoothness Spaces

Alexander Brudnyi

This paper is dedicated to Professor Per H. Enflo.

ABSTRACT. A closed subspace Z of the dual Y^* of a Banach space $Y \subset C(S)$, $S \subset \mathbb{R}^n$, is said to be the geometric predual of Y if Z is generated by evaluation functionals at points of S and Z^* is isomorphic to Y. In this paper we survey some recent results on geometric preduals of trace spaces of the classical smoothness spaces $C_b^{k,\omega}(\mathbb{R}^n)$, $J_b^{k,\omega}(\mathbb{R}^n)$ and $\Lambda^{k,\varphi}(\mathbb{R}^n)$. We focus on the following issues: the finiteness principle and the structure of balls of geometric preduals, simultaneous extension from trace spaces, approximation and biduality properties of geometric preduals.

1. Introduction

Let $Y \subset C(S)$ be a Banach space of real-valued continuous functions on a subset S of $\mathbb{R}^n$ such that each evaluation functional δ_x at $x \in S$ belongs to the dual space Y^*. Let $Z \subset Y^*$ be the closure of the linear hull of the set of evaluations $\delta_S := \{\delta_x : x \in S\}$ and $\Delta_S : S \to Z$, $x \mapsto \delta_x$, be the corresponding inclusion.

The space Z is said to be the *geometric predual* of Y if the pullback $\Delta_S^* : Z^* \to C(S)$, $\Delta_S^*(f) := f \circ \Delta_S$, determines an isomorphism of Z^* onto Y.

In a sense, the geometric predual of Y is the natural extension of the domain S of functions in Y containing important information about their functional properties.

It is worth noting that if $Y \subset C(\mathbb{R}^n)$ has the geometric predual Z and if $S \subset \mathbb{R}^n$ is a closed subset, then the minimal closed subspace $Z(S)$ of Z containing the set of evaluations $\delta_S := \{\delta_x : x \in S\}$ is the geometric predual of the trace space $Y|_S$ consisting of restrictions to S of functions from Y equipped with the trace norm

$$\|f\|_{Y|_S} := \inf\{\|F\|_Y : F|_S = f, \quad F \in Y\}. \tag{1.1}$$

(Here and below for a Banach space Y its norm is denoted by $\|\cdot\|_Y$.)

In this review, we present some basic properties of geometric preduals of classical smoothness spaces and their relation to some deep problems of analysis. Specifically, we will discuss properties of geometric preduals of the trace spaces

---

2020 *Mathematics Subject Classification.* Primary 46E15; Secondary 46B20.

Key words and phrases. Geometric predual, smoothness space, finiteness principle, simultaneous extension, approximation property, biduality property.

The author was supported in part by NSERC Grant #10010444.

$C_b^{k,\omega}(S) := C_b^{k,\omega}(\mathbb{R}^n)|_S$, $S \subset \mathbb{R}^n$, of the function space $C_b^{k,\omega}(\mathbb{R}^n)$ consisting of all C^k functions whose derivatives up to order k are bounded on $\mathbb{R}^n$, and whose k-th derivatives have moduli of continuity $O(\omega)$. Here, we will focus on the properties of the corresponding geometric preduals $G_b^{k,\omega}(S)$ related to the classical Whitney problems, see [**38**], [**39**], asking about the characterization of trace spaces of C^k functions on $\mathbb{R}^n$ (see survey [**17**] and book [**6**] and references therein for recent developments in the area). Also, we will discuss similar properties of geometric preduals $G_J^{k,\omega}(S) \subset G_b^{k,\omega}(\mathbb{R}^n)$ of jet spaces $J_b^{k,\omega}(S)$ related to $C_b^{k,\omega}(S)$. Finally, along these lines we will discuss properties of geometric preduals of trace spaces of the space of Lipschitz functions of higher order $\Lambda^{k,\varphi}(\mathbb{R}^n)$ and formulate some open problems in this area.

2. Basic Definitions

To formulate the results of the paper we recall some definitions and notation. We use the standard notation of differential analysis. In particular, $\alpha = (\alpha_1, \dots, \alpha_n) \in \mathbb{Z}_+^n$ denotes a multi-index and $|\alpha| := \sum_{i=1}^n \alpha_i$. Also, for $x = (x_1, \dots, x_n) \in \mathbb{R}^n$,

$$x^\alpha := \prod_{i=1}^n x_i^{\alpha_i} \quad \text{and} \quad D^\alpha := \prod_{i=1}^n D_i^{\alpha_i}, \quad \text{where} \quad D_i := \frac{\partial}{\partial x_i}. \tag{2.1}$$

Let ω be a nonnegative function on $(0, \infty)$ (referred to as *modulus of continuity*) satisfying the following conditions:

(i) $\omega(t)$ and $\dfrac{t}{\omega(t)}$ are nondecreasing functions on $(0, \infty)$;

(ii) $\lim_{t \to 0+} \omega(t) = 0$.

DEFINITION 2.1. $C_b^{k,\omega}(\mathbb{R}^n)$ is the Banach subspace of functions $f \in C^k(\mathbb{R}^n)$ with norm

$$\|f\|_{C_b^{k,\omega}(\mathbb{R}^n)} := \max\left(\|f\|_{C_b^k(\mathbb{R}^n)}, |f|_{C_b^{k,\omega}(\mathbb{R}^n)}\right), \tag{2.2}$$

where

$$\|f\|_{C_b^k(\mathbb{R}^n)} := \max_{|\alpha| \le k}\left\{\sup_{x \in \mathbb{R}^n} |D^\alpha f(x)|\right\} \tag{2.3}$$

and

$$|f|_{C_b^{k,\omega}(\mathbb{R}^n)} := \max_{|\alpha| = k}\left\{\sup_{x,y \in \mathbb{R}^n,\, x \ne y} \frac{|D^\alpha f(x) - D^\alpha f(y)|}{\omega(\|x - y\|)}\right\}. \tag{2.4}$$

Here $\|\cdot\|$ is the Euclidean norm of $\mathbb{R}^n$.

If $S \subset \mathbb{R}^n$ is a closed subset, then by $C_b^{k,\omega}(S)$ we denote the trace space of functions $g \in C_b^{k,\omega}(\mathbb{R}^n)|_S$ equipped with the quotient norm

$$\|g\|_{C_b^{k,\omega}(S)} := \inf\{\|G\|_{C_b^{k,\omega}(\mathbb{R}^n)} : G \in C_b^{k,\omega}(\mathbb{R}^n),\ G|_S = g\}. \tag{2.5}$$

DEFINITION 2.2. The jet space $J_b^{k,\omega}(S)$ consists of vector functions $\vec{f} = (f_\alpha)_{|\alpha| \le k}$ on S such that $(D^\alpha F)|_S = f_\alpha$ for some $F \in C_b^{k,\omega}(\mathbb{R}^n)$ and all α, equipped with the norm $\|\vec{f}\|_{J_b^{k,\omega}(S)}$ which is equal to the infimum of norms $\|F\|_{C_b^{k,\omega}(\mathbb{R}^n)}$ over the set of all such F.

In turn, the space of Lipschitz functions of higher order $\Lambda^{k,\varphi}(\mathbb{R}^n)$ is defined as follows.

DEFINITION 2.3. $\Lambda^{k,\varphi}(\mathbb{R}^n)$ is the Banach space of bounded functions f on $\mathbb{R}^n$ equipped with the norm

$$\|f\|_{\Lambda^{k,\varphi}(\mathbb{R}^n)} := \max\left\{\|f\|_{L_\infty(\mathbb{R}^n)}, \sup_{t>0}\frac{\omega_k(f,t)}{\varphi(t)}\right\},$$

where $\omega_k(f,\cdot)$ is the kth modulus of continuity of f, and φ is a k-majorant, i.e., $\varphi\in C(\mathbb{R}_+)$ is nondecreasing with $\varphi(0)=0$ and $\frac{\varphi(t)}{t^k}$, $t>0$, is nonincreasing.

We denote be $\Lambda^{k,\varphi}(S)$ the trace space $\Lambda^{k,\varphi}(\mathbb{R}^n)|_S$.

3. Geometric Preduals

3.1. Space $\mathbf{G_b^{k,\omega}(S)}$. Let $\left(C_b^{k,\omega}(\mathbb{R}^n)\right)^*$ be the dual of $C_b^{k,\omega}(\mathbb{R}^n)$. Clearly, each evaluation functional δ_x at $x\in\mathbb{R}^n$ (i.e., $\delta_x(f):=f(x)$, $f\in C_b^{k,\omega}(\mathbb{R}^n)$) belongs to $\left(C_b^{k,\omega}(\mathbb{R}^n)\right)^*$ and has norm one. Let $G_b^{k,\omega}(S)\subset\left(C_b^{k,\omega}(\mathbb{R}^n)\right)^*$ be the minimal closed subspace containing all δ_x, $x\in S$. Then we have (see [**3**, Thm. 1.2]):

PROPOSITION 3.1. *The map $\Delta_S: S\to G_b^{k,\omega}(S)$, $x\mapsto\delta_x$, is continuous and the pullback map Δ_S^* determines an isometric isomorphism between the dual of $G_b^{k,\omega}(S)$ and $C_b^{k,\omega}(S)$.*

Thus, $G_b^{k,\omega}(S)$ is the geometric predual of the trace space $C_b^{k,\omega}(S)$.

By δ_x^α, $x\in\mathbb{R}^n$, $\alpha\in\mathbb{Z}_+^n$, we denote the evaluation functional $D^\alpha|_{\{x\}}$ (so that $\delta_x^0:=\delta_x$). By definition each δ_x^α, $|\alpha|\le k$, belongs to $\left(C_b^{k,\omega}(\mathbb{R}^n)\right)^*$ and has norm ≤ 1. Similarly, functionals $\frac{\delta_x^\alpha-\delta_y^\alpha}{\omega(\|x-y\|)}$, $|\alpha|=k$, $x,y\in\mathbb{R}^n$, $x\ne y$, belong to $\left(C_b^{k,\omega}(\mathbb{R}^n)\right)^*$ and have norm ≤ 1. Then Corollary 2.3 of [**3**] states the following:

PROPOSITION 3.2. *The closed unit ball $B_{G_b^{k,\omega}(\mathbb{R}^n)}$ of $G_b^{k,\omega}(\mathbb{R}^n)$ is the balanced closed convex hull of the set of all functionals δ_x^α, $|\alpha|\le k$, and $\frac{\delta_x^\alpha-\delta_y^\alpha}{\omega(\|x-y\|)}$, $|\alpha|=k$, $x,y\in\mathbb{R}^n$, $x\ne y$.*

Although formally the closed unit ball $B_{G_b^{k,\omega}(S)}$ of a general space $G_b^{k,\omega}(S)$ is the intersection of $B_{G_b^{k,\omega}(\mathbb{R}^n)}$ with $G_b^{k,\omega}(\mathbb{R}^n)$ its constructive description similar to that of Proposition 3.2 is much more complicated and will be given in Section 4.

3.2. Space $\mathbf{J_b^{k,\omega}(S)}$. Let $\delta_{(s,\alpha)}\in(J_b^{k,\omega}(S))^*$ be the evaluation functional at the point (s,α), $s\in S$, $|\alpha|\le k$, i.e., for $\vec{f}=(f_\alpha)_{|\alpha|\le k}\in J_b^{k,\omega}(S)$,

$$\delta_{(s,\alpha)}(\vec{f}) := f_\alpha(s).$$

By $G_J^{k,\omega}(S)\subset(J_b^{k,\omega}(S))^*$ we denote the minimal closed subspace containing all functionals $\delta_{(s,\alpha)}$, $s\in S$, $|\alpha|\le k$, equipped with the induced norm. Consider a bounded linear map $I_S: J_b^{k,\omega}(S)\to(G_J^{k,\omega}(S))^*$,

$$I_S(\vec{f})(v) := v(\vec{f}),\quad v\in G_J^{k,\omega}(S). \tag{3.1}$$

According to [**9**, Prop. 3.1] we have:

PROPOSITION 3.3. *The map I_S is an isometric isomorphism between $J_b^{k,\omega}(S)$ and $(G_J^{k,\omega}(S))^*$.*

Thus, $G_J^{k,\omega}(S)$ is the geometric predual of $J_b^{k,\omega}(S)$. Such spaces naturally appear in the context of Whitney extension problems for smooth functions (see [**40**]).

Next, let us consider the map $D_S : C_b^{k,\omega}(\mathbb{R}^n) \to J_b^{k,\omega}(S)$,

$$D_S(f)(s) := ((D^\alpha f)(s))_{|\alpha|\le k}, \quad s \in S,\ f \in C_b^{k,\omega}(\mathbb{R}^n). \tag{3.2}$$

Then D_S is a linear surjection of norm one such that the adjoint map $D_S^* : (J_b^{k,\omega}(S))^* \to (C_b^{k,\omega}(\mathbb{R}^n))^*$ is an isometry. It is easy to check that for all α and s

$$D_S^*(\delta_{(s,\alpha)}) = \delta_s^\alpha. \tag{3.3}$$

Therefore, D_S^* maps $G_J^{k,\omega}(S)$ isometrically onto the minimal subspace of $(C_b^{k,\omega}(\mathbb{R}^n))^*$ containing all functionals δ_s^α, $s \in S$, $|\alpha| \le k$. In particular,

$$G_b^{k,\omega}(S) \subset D_S^*(G_J^{k,\omega}(S)) \subset G_b^{k,\omega}(\mathbb{R}^n).$$

The constructive description of the closed ball of $G_J^{k,\omega}(S)$ is given in Proposition 4.3 and is based on the classical Whitney-Glaeser theorem, see, e.g., [**6**, Ch. 2].

3.3. Space $\mathbf{\Lambda^{k,\varphi}(S)}$. Let $G_\Lambda^{k,\varphi}(S)$ be the closure in $(\Lambda^{k,\varphi}(S))^*$ of the linear hull of the set of evaluations $\delta_S := \{\delta_x : x \in S\}$ (which clearly is a subset of the dual). It is easy to check that the pullback $\Delta_S^* : (G_\Lambda^{k,\varphi}(S))^* \to C(S)$ determines an isometric isomorphism onto $\Lambda^{k,\varphi}(S)$, so that $G_\Lambda^{k,\varphi}(S)$ is the geometric predual of $\Lambda^{k,\omega}(S)$, cf. [**6**, Thm. 4.93]. For $x, h \in \mathbb{R}^n$ we set

$$\Delta_h^k(\delta_x) := \sum_{i=0}^{k} (-1)^{k-i} \binom{k}{i} \delta_{x+ih}. \tag{3.4}$$

Then we have

PROPOSITION 3.4. *The closed unit ball $B_{G_\Lambda^{k,\varphi}(\mathbb{R}^n)}$ of $G_\Lambda^{k,\varphi}(\mathbb{R}^n)$ is the balanced closed convex hull of the set of all functionals δ_x and $\frac{\Delta_h^k(\delta_x)}{\varphi(\|h\|)}$, $x, h \in \mathbb{R}^n$, $h \ne 0$.*

The proof is straightforward, see, e.g., [**22**, Sect. 3]. The constructive description of the closed ball of a general space of $G_\Lambda^{k,\varphi}(S)$ will be discussed in the next section.

4. Finiteness Principle

4.1. Formulation. In the general setting of a Banach space $Y \subset C(\mathbb{R}^n)$ of real-valued continuous functions on $\mathbb{R}^n$, the finiteness principle reads as follows.

Finiteness Principle. To decide whether a given $f : S \to \mathbb{R}$, $S \subset \mathbb{R}^n$, extends to a function $F \in Y$, it is enough to look at all restrictions $f|_{S'}$, where $S' \subset S$ is an arbitrary d-element subset. Here d is an integer constant, independent of S.

More precisely, if $f|_{S'}$ extends to a function $F^{S'} \in Y$ of norm at most 1 (for each $S' \subset S$ with at most d elements), then f extends to a function $F \in Y$, whose norm is bounded by a constant independent of S.

Suppose that Y has a geometric predual Z. In this case the finiteness principle can be reformulated in terms of the structure of the unit closed ball B_Z of Z.

Specifically, let $Z(S) \subset Z$ be the geometric predual of the trace space $Y|_S$, and let $B(S;m) \subset B_{Z(S)}$, $m \in \mathbb{N}$, be the balanced closed convex hull of the union of all finite-dimensional balls $B_{Z(S')} \subset Z(S')$, $S' \subset S$, $\operatorname{card} S' \le m$. At the level of geometric preduals, the finiteness principle is equivalent to the following statement (for the proof of the equivalence, see, e.g., the proof of [**3**, Thm. 1.3]):

Geometric Finiteness Principle. There exist constants $d \in \mathbb{N}$ and $c \in (1, \infty)$ independent of S such that

$$B(S;d) \subset B_{Z(S)} \subset cB(S;d). \tag{4.1}$$

Since in most cases balls $B_{Z(S')}$ with $\operatorname{card} S' \le d$ can be described explicitly, we call (4.1) a constructive description of the ball $B_{Z(S)}$.

4.2. Finiteness principle for $\mathbf{C}_b^{k,\omega}(S)$ spaces. For the trace spaces $C_b^{k,\omega}(S)$, $S \subset \mathbb{R}^n$, the finiteness principle was conjectured by Yu. Brudnyi and P. Shvartsman in the 1980s, cf. [**17**, p. 210]. Their formulation requires that *the corresponding constants d and c depend only on k and n.* The problem goes back to the classical 1934 Whitney papers [**38**], [**39**]. Let us briefly describe the known results in this direction.

For $k = 0$ (the Lipschitz case) the McShane extension theorem [**29**] implies the finiteness principle with the optimal constants $d = 2$ and $c = 1$. Also, for $n = 1$ the finiteness principle is valid with the optimal constant $d = k + 2$ (with $c = c(k)$ growing polynomially in k). The result is essentially due to Merrien [**30**].

In the multidimensional case the finiteness principle was proved by Yu. Brudnyi and Shvartsman for $k = 1$ with the optimal constant $d = 3 \cdot 2^{n-1}$ (with $c = c(n)$), see [**10**]. In the early 2000s the finiteness principle was proved by C. Fefferman for all k and n with $d = d(k,n)$ and $c = \frac{c(k,n)}{\omega(1)}$, see [**15**]. The upper bound for the constant d in the Fefferman proof was reduced later to $d = 2^{\binom{k+n}{k}}$ by Bierstone and P. Milman [**1**] and independently and by a different method by Shvartsman [**36**]. The obtained results admit the following geometric reformulation, see (4.1). Let $B_b^{k,\omega}(S;m) \subset B_{G_b^{k,\omega}(S)}$, $m \in \mathbb{N}$, be the balanced closed convex hull of the union of all finite-dimensional balls $B_{G_b^{k,\omega}}(S') \subset G_b^{k,\omega}(S')$, $S' \subset S$, $\operatorname{card} S' \le m$.

THEOREM 4.1. *There exist constants $d \in \mathbb{N}$ and $c \in (1, \infty)$ such that*

$$B_b^{k,\omega}(S;d) \subset B_{G_b^{k,\omega}(S)} \subset cB_b^{k,\omega}(S;d).$$

Here for $k = 0$, $d = 2$ (- optimal) and $c = 1$, for $n = 1$, $d = k + 2$ (- optimal) and c depends only on k, for $k = 1$, $d = 3 \cdot 2^{n-1}$ (- optimal) and c depends only on k and n, and for $k \ge 2$, $d = 2^{\binom{k+n}{k}}$ and $c = \frac{\tilde{c}}{\omega(1)}$, where $\tilde{c}$ depends only on k and n.

4.3. Finiteness principle for $\mathbf{J}_b^{k,\omega}(S)$ spaces. Recall that the classical Whitney-Glaeser theorem (see, e.g., [**6**, Ch. 2]) gives necessary and sufficient conditions for a vector function $\vec{f} = (f_\alpha)_{|\alpha| \le k}$ on $S \subset \mathbb{R}^n$ to lie in $J_b^{k,\omega}(S)$. Specifically, one associates with $\vec{f}$ the family $\{T_x^k \vec{f}\}_{x \in S}$ of polynomials on $\mathbb{R}^n$ by the following formula

$$T_x^k \vec{f}(z) := \sum_{\alpha \in \mathbb{Z}_+^n : |\alpha| \le k} f_\alpha(x) \frac{(z - x)^\alpha}{\alpha!}, \qquad x \in S, \ z \in \mathbb{R}^n. \tag{4.2}$$

THEOREM 4.2 (Whitney-Glaeser). *The vector function $\vec{f}$ lies in $J_b^{k,\omega}(S)$ if and only if there exists a constant c such that for all $\alpha \in \mathbb{Z}_+$, $|\alpha| \le k$, and all $x, y \in S$, $z \in \{x, y\}$*

$$|f_\alpha(x)| \le c \quad \textit{and} \quad |D^\alpha(T_x^k\vec{f} - T_y^k\vec{f})(z)| \le c\|x-y\|^{k-|\alpha|}\omega(\|x-y\|). \tag{4.3}$$

The smallest such c determines a norm on $J_b^{k,\omega}(S)$ equivalent to the norm $\|\cdot\|_{J_b^{k,\omega}(S)}$ with the constants of equivalence depending only on k and n.

As a corollary of the Whitney-Glaeser theorem we obtain the following constructive description of balls in $G_J^{k,\omega}(S)$ and the finiteness principle for $J_b^{k,\omega}(S)$. Let $B_{G_J^{k,\omega}(S)}$ be the closed unit ball in $G_J^{k,\omega}(S)$ and $B'_{G_J^{k,\omega}(S)}$ be the balanced closed convex hull of the set of vectors $\delta_{(s,\alpha)}$, $s \in S$ (see Section 3.2), and

$$\frac{\delta_{(x,\alpha)} - \sum_{|\beta|\le k-|\alpha|}(x-y)^\beta/\beta! \cdot \delta_{(y,\alpha+\beta)}}{\|x-y\|^{k-\alpha}\omega(\|x-y\|)}, \quad x, y \in S, \ |\alpha| \le k.$$

Let $B_J^{k,\omega}(S,m)$, $m \in \mathbb{N}$, be the balanced closed convex hull of the union of all finite-dimensional balls $B_{G_J^{k,\omega}(S')} \subset G_J^{k,\omega}(S')$, $S' \subset S$, $\operatorname{card} S' \le m$.

PROPOSITION 4.3 ([**9**, Prop. 3.2]). *There exist $c_1, c_2 \in \mathbb{R}_+$ depending only on k, n such that*

(a)
$$c_1 B'_{G_J^{k,\omega}(S)} \subset B_{G_J^{k,\omega}(S)} \subset c_2 B'_{G_J^{k,\omega}(S)};$$

(b)
$$B_J^{k,\omega}(S,2) \subset B_{G_J^{k,\omega}(S)} \subset \frac{c_2}{c_1} B_J^{k,\omega}(S,2).$$

4.4. Finiteness principle for $\Lambda^{k,\varphi}(S)$ spaces. For the trace spaces $\Lambda^{k,\varphi}(S)$, $S \subset \mathbb{R}^n$, the finiteness principle was conjectured by Yu. Brudnyi in the early 1980s. In the strongest form, this requires that *the corresponding constants d and c depend only on k and n.*

The finiteness principle has been proven in only a few cases. For instance, for $\Lambda^{k,\varphi}(S)$ with $S \subset \mathbb{R}$, the finiteness principle was established by Shevchuk [**34**] with $d = k+1$ and $c = c(\omega, k)$. Also, it is valid for the spaces $\Lambda^{k,\varphi}(S)$, where $\varphi(t) = t^k\omega(t)$, $t > 0$, and ω is a quasipower majorant, i.e., such that

$$\sup_{t>0}\left\{\frac{1}{\omega(t)}\int_0^t \frac{\omega(u)}{u}\,du\right\} < \infty, \tag{4.4}$$

as in this case $\Lambda^{k,\varphi}(\mathbb{R}^n) = C_b^{k,\omega}(\mathbb{R}^n)$ and the corresponding norms are equivalent with the constants of equivalence depending on k, n and ω only, see, e.g., [**6**, Ch. 9] for the references. This implies that each $\Lambda^{k,\varphi}(S) = C_b^{k,\omega}(S)$ with the same equivalence of norms and so according to Theorem 4.1, the finiteness principle for such spaces $\Lambda^{k,\varphi}(S)$ is valid with $d = 2^{\binom{k+n}{k}}$ and $c = c(k,n,\omega)$. This gives also the constructive description of unit closed balls in $G_\Lambda^{k,\omega}(S)$, see (4.1).

Finally, the selection theorem of P. Shvartsman, see [**35**] or [**6**, Thm. 5.56], implies that the finiteness principle is valid for the spaces $\Lambda^{2,\varphi}(S)$ (i.e., the trace spaces of the generalized Zygmund space $\Lambda^{2,\varphi}(\mathbb{R}^n)$) with the optimal constant $d = 3 \cdot 2^{n-1}$ and $c = c(n)$, see, e.g., [**6**, Rem. 10.120].

5. Simultaneous Extension

5.1. Problem statement. For a Banach space $Y \subset C(\mathbb{R}^n)$ of real-valued continuous functions on $\mathbb{R}^n$ the problem requires the existence of a bounded linear extension operator from the trace space $Y|_S$ into Y. The set of such operators is denoted by $\mathrm{Ext}(Y|_S, Y)$. Some of the simplest such operators are finite depth extension operators defined as follows:

DEFINITION 5.1. An operator $T \in Ext(Y|_S, Y)$ has depth $d \in \mathbb{N}$ if for all $x \in \mathbb{R}^n$ and $f \in Y|_S$,

$$(Tf)(x) = \sum_{i=1}^{d} \lambda_i^x \cdot f(y_i^x), \tag{5.1}$$

where $y_i^x \in S$ and λ_i^x depend only on x.

Suppose that Y has a geometric predual Z. Then each finite depth operator T is weak* continuous with respect to the weak* topologies on $Y|_S$ and Y relative to $Z(S)$ and Z, respectively, and in particular there exists a bounded linear projection $P : Z \to Z(S)$ such that $T = P^*$ (the proof of this fact is similar to that of [**3**, Thm. 1.6]). Hence, the subspace $Z(S)$ is complemented in Z. It is easy to see that if T has depth d and is defined by (5.1), then

$$p(x) := P(\delta_x) = \sum_{i=1}^{d} \lambda_i^x \cdot \delta_{y_i^x} \quad \text{for all} \quad x \in \mathbb{R}^n, \tag{5.2}$$

and the $Z(S)$-valued function p on $\mathbb{R}^n$ is such that $f \circ p \in Y$ for all $f \in Y|_S$ $(= Z(S)^*)$ and

$$\|T\| = \sup_{\|f\|_{Y|_S} \le 1} \|f \circ p\|_Y.$$

5.2. Simultaneous extensions for $\mathbf{C}_b^{k,\omega}(S)$ spaces. The problem of the existence of a simultaneous (i.e., linear continuous) extension from $C_b^{k,\omega}(S)$ into $C_b^{k,\omega}(\mathbb{R}^n)$ was posed by Yu. Brudnyi and P. Shvartsman in the 1980s and is known as *second Whitney problem*, see, e.g., [**6**, Ch. 10]. (In the strongest form, this requires that the norm of the existing extension operator depends only on k and n.) The problem is now completely solved and simultaneous extensions are given by operators of finite depth. Specifically, for $k = 0$ (the Lipschitz case) the Whitney-Glaeser linear extension operators $C_b^{0,\omega}(S) \to C_b^{0,\omega}(\mathbb{R}^n)$, see [**20**], have depth d depending only on n and norms bounded by a constant depending only on n. Further, in the 1990s bounded linear extension operators $C_b^{1,\omega}(S) \to C_b^{1,\omega}(\mathbb{R}^n)$ of depth d depending only on n with norms bounded by a constant depending only on n were constructed by Yu. Brudnyi and Shvartsman [**11**]. Finally, Luli [**28**] recently constructed bounded linear extensions operators of depth d that depends only on k and n for all spaces $C_b^{k,\omega}(S)$; their norms are bounded by $\frac{c}{\omega(1)}$, where $c \in (1, \infty)$ is a constant that depends only on k and n. (Earlier such extension operators were constructed for finite sets S by C. Fefferman [**16**, Thm. 8].) In particular, these results imply that all geometric predual spaces $G_b^{k,\omega}(S)$ are complemented in $G_b^{k,\omega}(\mathbb{R}^n)$.

5.3. Simultaneous extensions for $\mathbf{J}_b^{k,\omega}(S)$ spaces. In this case, the simultaneous extension is given by the Whitney-Glaeser extension operator T of depth and norm bounded by constants depending only on k and n, see [**20**] or [**6**, Ch. 2]. The operator determines a bounded linear projection of the geometric predual spaces $P : G_J^{k,\omega}(\mathbb{R}^n) \to G_J^{k,\omega}(S)$ such that $T = P^*$. Specifically, given a closed subset $S \subset \mathbb{R}^n$ let $\mathcal{W}_S := \{Q\}$ be the Whitney cover by cubes of $S^c := \mathbb{R}^n \setminus S$, $\{\varphi_Q\}_Q \subset C^\infty(\mathbb{R}^n)$ be a smooth partition of unity subordinate to $\mathcal{W}_S$ and $s_Q \in S$ be such that $\operatorname{dist}(s_Q, Q) = \operatorname{dist}(S, Q)$.

THEOREM 5.2 ([**9**, Thm. 3.3]). *The bounded linear projection $P : G_J^{k,\omega}(\mathbb{R}^n) \to G_J^{k,\omega}(S)$ is given by the formula:*

(5.3)
$$P(\delta_{(x,\alpha)}) := \begin{cases} \delta_{(x,\alpha)} & \text{if } \ x \in S,\ |\alpha| \le k \\ \displaystyle\sum_{|Q|\le 1} \sum_{|\beta|\le k} \frac{D^\alpha\big((x-s_Q)^\beta \varphi_Q(x)\big)}{\beta!} \delta_{(s_Q,\beta)} & \text{if } \ x \in S^c,\ |\alpha| \le k; \end{cases}$$

here $|Q|$ stands for the Lebesgue measure of Q.

5.4. Simultaneous extensions for $\Lambda^{k,\varphi}(S)$ spaces. For the trace spaces $\Lambda^{k,\varphi}(S)$, $S \subset \mathbb{R}^n$, the existence of simultaneous extensions was conjectured by Yu. Brudnyi in the early 1980s. (In its strongest form, this requires that the norm of the extension operator be bounded by a constant that depends only on k and n.)

With the exception of a few known cases, the problem is still open. In particular, the existence of simultaneous extensions (given by operators of finite depth) was proved for the following trace spaces: $\Lambda^{k,\varphi}(S)$, where $\varphi(t) = t^k\omega(t)$, $t > 0$, and ω is a quasipower majorant (in this case the norm of the existed extension operator is bounded by a constant that depends on k, n and ω, see the explanation in Section 4.4), the generalized Zygmund space $\Lambda^{2,\varphi}(S)$, see [**11**], with the norm of the extension operator bounded by a constant that depends only on n, and $\Lambda^{k,\varphi}(S)$ with a quasipower majorant φ for a Markov set $S \subset \mathbb{R}^n$, see [**6**, Thm. 9.30] and references therein. In the latter case, the norm of the extension operator is bounded by a constant depending on k, n, φ and the Markov constant $c(n, S)$.

Let us recall, see [**24**], that a closed subset $S \subset \mathbb{R}^n$ is said to be *Markov* if for some constant $c = c(n, S) \ge 1$ and every cube Q with center in S of radius $0 < r_Q \le 2 \operatorname{diam} S$ and every polynomial p of degree one

$$\sup_{x\in Q} \|\nabla p(x)\| \le \frac{c}{r_Q} \max_{Q\cap S} |p|. \tag{5.4}$$

6. Approximation Property

6.1. Basic definitions. A Banach space X is said to have the *approximation property*, if, for every compact set $K \subset X$ and every $\varepsilon > 0$, there exists an operator $T : X \to X$ of finite rank so that $\|Tx - x\|_X \le \varepsilon$ for every $x \in K$.

Although it is strongly believed that the class of spaces with the approximation property includes practically all spaces which appear naturally in analysis, it is not known yet even for the space H^∞ of bounded holomorphic functions on the open unit disk. The first example of a space which fails to have the approximation property was constructed by Enflo [**14**]. Since Enflo's work several other examples of such spaces were constructed, for the references see, e.g., [**26**].

A Banach space has the λ-*approximation property*, $1 \le \lambda < \infty$, if it has the approximation property with the approximating finite rank operators of norm $\le \lambda$. A Banach space is said to have the *bounded approximation property*, if it has the λ-approximation property for some λ. If $\lambda = 1$, then the space is said to have the *metric approximation property*.

Every Banach space with a basis has the bounded approximation property. Also, it is known that the approximation property does not imply the bounded approximation property, see [**18**]. It was established by Pełczyński [**33**] that a separable Banach space has the bounded approximation property if and only if it is isomorphic to a complemented subspace of a separable Banach space with a basis.

Next, for Banach spaces X, Y we denote by $\mathcal{L}(X, Y)$ the space of bounded linear operators from X into Y and by $\mathcal{F}(X, Y) \subset \mathcal{L}(X, Y)$ the subspace of operators of finite rank. Let us consider the trace mapping V from the projective tensor product $Y^*\hat{\otimes}_\pi X \to \mathcal{F}(X, Y)^*$ defined by

$$(Vu)(T) = \mathrm{trace}(Tu), \quad \text{where} \quad u \in Y^*\hat{\otimes}_\pi X,\ T \in \mathcal{F}(X, Y),$$

that is, if $u = \sum_{n=1}^{\infty} y_n^* \otimes x_n$, then $(Vu)(T) = \sum_{n=1}^{\infty} y_n^*(Tx_n)$.

It is easy to see that $\|Vu\| \le \|u\|_\pi$. The λ-bounded approximation property of X is equivalent to the fact that $\|u\|_\pi \le \lambda\|Vu\|$ for all Banach spaces Y. This well-known result (see, e.g., [**13**, page 193]) is essentially due to Grothendieck [**19**].

The main open problem is whether the classical smoothness spaces have the approximation property. While this problem is very complicated and inaccessible by modern methods (for some particular results, see, e.g., [**25**]), the problem about the approximation property for geometric preduals is significantly simpler. In the next section, we describe some known results in this direction.

6.2. Approximation property for geometric preduals. All $G_b^{k,\omega}(S)$ spaces have the bounded approximation property. In fact it suffices to prove that only for the space $G_b^{k,\omega}(\mathbb{R}^n)$ and then use the fact that all other spaces $G_b^{k,\omega}(S)$ are complemented in it, see Section 5.2. Thus, the following result is obtained.

THEOREM 6.1 ([**3**, Thm. 1.8]).

(1) *Geometric predual spaces $G_b^{k,\omega}(\mathbb{R}^n)$ have the λ-approximation property with* $\lambda = \lambda(k, n, \omega) := 1 + c \cdot \lim_{t\to\infty} \dfrac{1}{\omega(t)}$, *where c depends only on k and n.*

(2) *All other geometric preduals $G_b^{k,\omega}(S)$ have the λ-approximation property with $\lambda = c' \cdot \lambda(1, n, \omega)$, where c' is a constant depending only on n, if $k = 0, 1$, and with $\lambda = \frac{c''\cdot\lambda(k,n,\omega)}{\omega(1)}$, where c'' is a constant depending only on k and n, if $k \ge 2$.*

The proof of (1) is based on the construction from the proof of the classical Jackson approximation theorem, see, e.g., [**37**, Ch. V].

If $\lim_{t\to\infty} \omega(t) = \infty$, then (1) implies that the corresponding space $G_b^{k,\omega}(\mathbb{R}^n)$ has the metric approximation property. In turn, if $\lim_{t\to\infty} \omega(t) < \infty$, then one can introduce the new modulus of continuity $\widetilde{\omega}$ (cf. properties (i) and (ii) in its definition in Section 2) by the formula

$$\widetilde{\omega}(t) = \max\{\omega(t), t\}, \quad t \in (0, \infty).$$

It is easy to see that the spaces $C_b^{k,\omega}(\mathbb{R}^n)$ and $C_b^{k,\widetilde{\omega}}(\mathbb{R}^n)$ are isomorphic. Thus $G_b^{k,\omega}(\mathbb{R}^n)$ is isomorphic to the space $G_b^{k,\widetilde{\omega}}(\mathbb{R}^n)$ having the metric approximation property. However, the distortion of this isomorphism depends on ω. Therefore, in general, it is not clear whether $G_b^{k,\omega}(\mathbb{R}^n)$ itself has the metric approximation property. In fact, in some cases the spaces $G_b^{k,\omega}(S)$ do have the metric approximation property even if $\lim_{t\to\infty}\omega(t) < \infty$. E.g., by the classical result of Grothendieck [**19**, Ch. I], separable dual spaces with the approximation property have the metric approximation property. The class of such spaces $G_b^{k,\omega}(S)$ will be discussed in Theorem 7.3.

Next, all geometric preduals $G_J^{k,\omega}(S)$ of jet spaces $J_b^{k,\omega}(S)$ have the bounded approximation property. This follows from the fact that the spaces $G_J^{k,\omega}(\mathbb{R}^n)$ and $G_b^{k,\omega}(\mathbb{R}^n)$ are isometrically isomorphic, see Section 3.2, and that all other spaces $G_J^{k,\omega}(S)$ are complemented in $G_J^{k,\omega}(\mathbb{R}^n)$, see Theorem 5.2. In particular, using Theorem 6.1 (1) one obtains:

THEOREM 6.2 ([**9**, Thm. 3.5]). *Geometric preduals $G_J^{k,\omega}(S)$ of jet spaces $J_b^{k,\omega}(S)$ have the λ-approximation property with λ depending only on k, n and $\lim\limits_{t\to\infty}\dfrac{1}{\omega(t)}$.*

Whether general geometric preduals $G_\Lambda^{k,\varphi}(S)$ of spaces $\Lambda^{k,\varphi}(S)$ possess the bounded approximation property remains an open question. We only note that this is always true if $\varphi(t) = t^k\omega(t)$, $t > 0$, where ω is a quasipower majorant, since in this case the question reduces to the question for the corresponding $G_b^{k,\omega}$ spaces, see Section 4.4 for an explanation.

7. Biduality Property

7.1. Problem statement. A function space X defined in terms of a big-O condition is said to have the *biduality property* with respect to its subspace x defined by the corresponding little-o condition if the bidual x** of x is isomorphic to X.

The property is known for pairs of many classical function spaces such as (ℓ_∞, c_0), (BMO, VMO), (Lip, lip), etc. (see, e.g., [**27**], [**12**], [**23**], [**21**], [**22**], [**31**], [**32**], [**7**], [**8**] and references therein) and plays an important role in the study of their geometric structure.

Let $(X, \|\cdot\|_X)$ be a function space of continuous functions on $\mathbb{R}^n$ and let $X(S) := X|_S$ be the trace space of X to a closed subset $S \subset \mathbb{R}^n$. The main problem is the question of the biduality property of trace spaces. In particular, we are interested in the solution of the following general problem:

Define those pairs (X, x) for which the biduality property is hereditary, i.e., such that if X has the biduality property with respect to x, then $X(S)$ has this property with respect to x(S) for every closed subset $S \subset \mathbb{R}^n$.

For instance, this holds for the pair $(\mathrm{Lip}_d(\mathbb{R}^n), \mathrm{lip}_d(\mathbb{R}^n))$ of bounded Lipschitz functions on $\mathbb{R}^n$ equipped with a metric d compatible with topology because $\mathrm{Lip}_d(S)$ and $\mathrm{lip}_d(S)$ are analogous spaces on the metric space $(S, d|_S)$. Next, we describe other results extending the previous example to other classes of classical smoothness spaces.

7.2. Preduals of $G_b^{k,\omega}(S)$ spaces. Let $C_0^{k,\omega}(\mathbb{R}^n)$ be the subspace of functions $f \in C_b^{k,\omega}(\mathbb{R}^n)$ such that

(i) for all $\alpha \in \mathbb{Z}_+^n$, $0 \le |\alpha| \le k$,

$$\lim_{\|x\|\to\infty} D^\alpha f(x) = 0;$$

(ii) for all $\alpha \in \mathbb{Z}_+^n$, $|\alpha| = k$,

$$\lim_{\|x-y\|\to 0} \frac{D^\alpha f(x) - D^\alpha f(y)}{\omega(\|x-y\|)} = 0.$$

It is easy to see that $C_0^{k,\omega}(\mathbb{R}^n)$ equipped with the norm induced from $C_b^{k,\omega}(\mathbb{R}^n)$ is a Banach space. By $C_0^{k,\omega}(S)$ we denote the trace of $C_0^{k,\omega}(\mathbb{R}^n)$ to a closed subset $S \subset \mathbb{R}^n$ equipped with the trace norm.

If $\lim_{t\to 0+} \frac{t}{\omega(t)} > 0$ (see condition (i) for ω in Section 2), then clearly, the corresponding space $C_0^{k,\omega}(\mathbb{R}^n)$ is trivial. Thus, we can naturally assume that ω satisfies the condition

$$\lim_{t\to 0+} \frac{t}{\omega(t)} = 0. \tag{7.1}$$

In the following result, the weak* topology of $C_b^{k,\omega}(S)$ is defined using functionals in $G_b^{k,\omega}(S) \subset \left(C_b^{k,\omega}(\mathbb{R}^n)\right)^*$.

THEOREM 7.1 ([**3**, Thm. 1.12]). *Suppose that ω satisfies condition* (7.1).

(1) *The dual space $\left(C_0^{k,\omega}(\mathbb{R}^n)\right)^*$ is isomorphic to the geometric predual $G_b^{k,\omega}(\mathbb{R}^n)$, isometrically if $\lim_{t\to\infty} \omega(t) = \infty$.*

(2) *If there exists a weak* continuous operator $T \in Ext(C_b^{k,\omega}(S), C_b^{k,\omega}(\mathbb{R}^n))$ such that $T\left(C_0^{k,\omega}(S)\right) \subset C_0^{k,\omega}(\mathbb{R}^n)$, then $\left(C_0^{k,\omega}(S)\right)^*$ is isomorphic to $G_b^{k,\omega}(S)$.*

Since under the conditions of Theorem 7.1 the space $G_b^{k,\omega}(S)$ with ω satisfying (7.1) is separable dual and has the bounded approximation property by Theorem 6.1, it follows from Grothendieck's result [**19**, Ch. I] that the spaces $G_b^{k,\omega}(S)$ and $C_0^{k,\omega}(S)$ possess the metric approximation property.

It is not clear whether the condition of the second part of the theorem is true for all spaces $C_b^{k,\omega}(S)$ with ω satisfying (7.1). Let us describe the class of sets S satisfying this condition. In the following definition, $\mathcal{P}_{k,n}$ denotes the space of real polynomials on $\mathbb{R}^n$ of degree k, and $Q_r(x) \subset \mathbb{R}^n$ is a closed cube with center at x of sidelength $2r$.

DEFINITION 7.2. A point x of a subset $S \subset \mathbb{R}^n$ is said to be weak k-Markov if

$$\varliminf_{r\to 0} \left\{ \sup_{p\in\mathcal{P}_{k,n}\setminus 0} \left(\frac{\sup_{Q_r(x)} |p|}{\sup_{Q_r(x)\cap S} |p|} \right) \right\} < \infty.$$

A closed set $S \subset \mathbb{R}^n$ is said to be weak k-Markov if it contains a dense subset of weak k-Markov points.

The class of weak k-Markov sets, denoted by $\mathrm{Mar}_k^*(\mathbb{R}^n)$, was introduced and studied by Yu. Brudnyi and the author, see [**5**], [**2**]. It contains, in particular, the closure of any open set, the Ahlfors p-regular compact subsets of $\mathbb{R}^n$ with $p > n-1$, a wide class of fractals of arbitrary positive Hausdorff measure, direct products $\prod_{j=1}^{l} S_j$, where $S_j \in \mathrm{Mar}_k^*(\mathbb{R}^{n_j})$, $1 \le j \le l$, $n = \sum_{j=1}^{l} n_j$, and closures of unions of any combination of such sets. Solutions of Whitney problems (see Sections 4.2 and 5.2 above) for sets S in $\mathrm{Mar}_k^*(\mathbb{R}^n)$ are relatively simple, see [**5**].

In [**3**] we proved the following result.

THEOREM 7.3 ([**3**, Thm. 1.15]). *Let $S' \in \mathrm{Mar}_k^*(\mathbb{R}^n)$ and ω satisfy (7.1). Suppose $H : \mathbb{R}^n \to \mathbb{R}^n$ is a differentiable map such that*

(a) *the entries of its Jacobian matrix belong to $C_b^{k-1,\omega_o}(\mathbb{R}^n)$, where ω_o satisfies*

$$\lim_{t\to 0^+} \frac{\omega_o(t)}{\omega(t)} = 0; \tag{7.2}$$

(b) *the map $H|_{S'} : S' \to S =: H(S')$ is a proper retraction.*[1]

Then condition (2) of Theorem 7.1 holds for $C_b^{k,\omega}(S)$. Hence, $G_b^{k,\omega}(S)$ is isomorphic to $\big(C_0^{k,\omega}(S)\big)^*$ and therefore $G_b^{k,\omega}(S)$ and $C_0^{k,\omega}(S)$ have the metric approximation property.

In addition to weak k-Markov sets $S \subset \mathbb{R}^n$, Theorem 7.3 holds, e.g., for a compact subset S of a C^{k+1}-manifold $M \subset \mathbb{R}^n$ such that the base for the topology of S consists of relatively open subsets of Hausdorff dimension $> \dim M-1$. Indeed, in this case there exist tubular open neighbourhoods $U_M \subset V_M \subset \mathbb{R}^n$ of M such that $\mathrm{cl}(U_M) \subset V_M$ together with a C^{k+1} retraction $r : U_M \to M$. Then, by virtue of the hypothesis for S, the base for the topology of $S' := r^{-1}(S) \cap \mathrm{cl}(U_M)$ consists of relatively open subsets of Hausdorff dimension $> n-1$ and so $S' \in \mathrm{Mar}_p^*(\mathbb{R}^n)$ for all $p \in \mathbb{N}$, see, e.g., [**2**, page 536]. Moreover, it is easy to see that $r|_{S'}$ is the restriction to S' of a map $H \in C_b^{k+1}(\mathbb{R}^n, \mathbb{R}^n)$. Reducing V_M, if necessary, we can assume that S' is compact, and then the triple (H, S', S) satisfies the hypothesis of the theorem.

Note that under the conditions of Theorem 7.3, the pair $(C_b^{k,\omega}(S), C_0^{k,\omega}(S))$ possesses the biduality property.

7.3. Preduals of $\mathbf{G}_J^{k,\omega}(S)$ spaces. We define $J_0^{k,\omega}(S)$ as the subspace of $J_b^{k,\omega}(S)$ of vector functions $\vec{f} = (f_\alpha)_{|\alpha|\le k}$ on S such that $D^\alpha F|_S = f_\alpha$ for some $F \in C_0^{k,\omega}(\mathbb{R}^n)$ and all $|\alpha| \le k$ equipped with the norm $\|\vec{f}\|_{J_0^{k,\omega}(S)}$ that equal to the infimum of norms $\|F\|_{C_b^{k,\omega}(\mathbb{R}^n)}$ over the set of all such F. Then we have

THEOREM 7.4 ([**9**, Thm. 3.7]). *Suppose that ω satisfies (7.1). Then $(J_0^{k,\omega}(S))^*$ is isomorphic to $G_J^{k,\omega}(S)$, isometrically if $\lim_{t\to\infty} \omega(t) = \infty$. Hence, the spaces $G_J^{k,\omega}(S)$ and $J_0^{k,\omega}(S)$ have the metric approximation property.*

The proof of the first statement of the theorem is analogous to that of Theorem 7.1 (2) since the Whitney-Glaeser extension operator $T \in \mathit{Ext}(J_b^{k,\omega}(S), J_b^{k,\omega}(\mathbb{R}^n))$ is weak* continuous and $T\big(J_0^{k,\omega}(S)\big) \subset J_0^{k,\omega}(\mathbb{R}^n)$. The second statement of the theorem follows from Grothendieck's result [**19**, Ch. I] since the space $G_J^{k,\omega}(S)$ with ω satisfying (7.1) is separable dual and has the bounded approximation property by Theorem 6.2.

[1] I.e., $S \subset S'$ and $H|_{S'}(x) = x$ for all $x \in S$, and for each compact $K \subset S$ its preimage $(H|_{S'})^{-1}(K)$ is compact.

As a direct corollary of Theorem 7.4 we obtain the biduality property for the pair $(J_b^{k,\omega}(S), J_0^{k,\omega}(S))$:

Corollary 7.5. *If ω satisfies* (7.1), *then the second dual of $J_0^{k,\omega}(S)$ is isomorphic to $J_b^{k,\omega}(S)$, isometrically if $\lim_{t\to\infty}\omega(t) = \infty$.*

7.4. Preduals of $\mathbf{G}_\Lambda^{k,\varphi}(S)$ spaces. In this section, we briefly discuss the biduality property for $\Lambda^{k,\varphi}(S)$ spaces.

We define the closed subspace $\lambda^{k,\varphi}(\mathbb{R}^n) \subset \Lambda^{k,\varphi}(\mathbb{R}^n)$ of functions f satisfying the condition

$$\lim_{t\to 0+} \frac{\omega_k(f,t)}{\varphi(t)} = 0. \tag{7.3}$$

For a closed subset $S \subset \mathbb{R}^n$, let $\lambda^{k,\varphi}(S)$ denote the corresponding trace space.

It was proved by Hanin in [**22**, Thm. 7.2] that under condition

$$\lim_{t\to 0+} \frac{\varphi(t)}{t^k} = \infty \tag{7.4}$$

the dual $(\lambda^{k,\varphi}(\mathbb{R}^n))^*$ is isomorphic to $G_\Lambda^{k,\varphi}(\mathbb{R}^n)$ and, hence, the pair of Banach spaces $(\Lambda^{k,\varphi}(\mathbb{R}^n), \lambda^{k,\varphi}(\mathbb{R}^n))$ has the biduality property.

Earlier Jonsson [**23**] proved that the dual $(\lambda^{k,\varphi}(S)^*$ is isomorphic to $G_\Lambda^{k,\varphi}(S)$ for $\varphi(t) := t^\alpha$, $0 < \alpha < k$, and $S \subset \mathbb{R}^n$ is a Markov set, see (5.4). Hence, such pairs $(\Lambda^{k,\varphi}(S), \lambda^{k,\varphi}(S))$ possess the biduality property.

If $S \subset \mathbb{R}^n$ is a weak Markov set, see Section 7.2, and $\varphi(t) = t^{k-1}\omega(t)$, $t > 0$, with ω a quasipower majorant satisfying $\lim_{t\to 0+} \frac{\omega(t)}{t} = \infty$, then the space $\Lambda^{k,\varphi}(\mathbb{R}^n)$ is isomorphic to the space $C_b^{k-1,\omega}(\mathbb{R}^n)$, see Section 4.4, and this isomorphism maps $\lambda^{k,\varphi}(\mathbb{R}^n)$ isomorphically to $C_0^{k-1,\omega}(\mathbb{R}^n)$. Since the pair $(C_b^{k-1,\omega}(S), C_0^{k-1,\omega}(S))$ has the biduality property, see Theorem 7.3, the pair $(\Lambda^{k,\varphi}(S), \lambda^{k,\varphi}(S))$ also has this property.

Finally, in [**4**] using Shvartsman's finiteness theorem [**36**] the author proved that $(\lambda^{2,\varphi}(S))^*$ is isomorphic to $G_\Lambda^{2,\varphi}(S)$ for all closed $S \subset \mathbb{R}^n$ and φ satisfying (7.4). This implies the biduality property for all such pairs $(\Lambda^{2,\varphi}(S), \lambda^{2,\varphi}(S))$.

In general, it is not known whether the biduality property is true for all pairs of trace spaces $(\Lambda^{k,\varphi}(S), \lambda^{k,\varphi}(S))$ with φ satisfying (7.4).

References

[1] E. Bierstone and P. D. Milman, *$\mathcal{C}^m$-norms on finite sets and $\mathcal{C}^m$ extension criteria*, Duke Math. J. **137** (2007), no. 1, 1–18, DOI 10.1215/S0012-7094-07-13711-6. MR2309142

[2] A. Brudnyi, *Differential calculus on topological spaces with weak Markov structure I*, Rev. Mat. Iberoam. **31** (2015), no. 2, 531–574, DOI 10.4171/RMI/844. MR3375862

[3] A. Brudnyi, *On properties of geometric preduals of $C^{k,\omega}$ spaces*, Rev. Mat. Iberoam. **35** (2019), no. 6, 1715–1744, DOI 10.4171/rmi/1099. MR4029781

[4] A. Brudnyi, *Two stars theorems for traces of the Zygmund space*, St. Petersburg Math. J. **34** (2023), no. 1, 25–44. Reprinted from Algebra i Analiz **34** (2022), No. 1. MR4528762

[5] A. Brudnyĭ and Yu. Brudnyĭ, *Traces of functions of the class C^k on weakly Markov subsets of $\mathbb{R}^n$* (Russian, with Russian summary), Algebra i Analiz **23** (2011), no. 1, 61–86, DOI 10.1090/S1061-0022-2011-01185-9; English transl., St. Petersburg Math. J. **23** (2012), no. 1, 39–56. MR2760147

[6] A. Brudnyi and Y. Brudnyi, *Methods of geometric analysis in extension and trace problems. Volume 2*, Monographs in Mathematics, vol. 103, Birkhäuser/Springer Basel AG, Basel, 2012. MR2868143

[7] A. Brudnyi and Y. Brudnyi, *On the Banach structure of multivariate BV spaces*, Dissertationes Math. **548** (2020), 52, DOI 10.4064/dm801-7-2019. MR4078209

[8] A. Brudnyi and Y. Brudnyi, *Multivariate bounded variation functions of Jordan-Wiener type*, J. Approx. Theory **251** (2020), 105346, 70, DOI 10.1016/j.jat.2019.105346. MR4040115

[9] A. Brudnyi and A. Butaev, *On geometric preduals of jet spaces on closed subsets of $\mathbb{R}^n$* (English, with English and French summaries), C. R. Math. Acad. Sci. Soc. R. Can. **42** (2020), no. 1, 10–20. MR4119677

[10] Y. Brudnyi and P. Shvartsman, *Generalizations of Whitney's extension theorem*, Internat. Math. Res. Notices **3** (1994), 129 ff., approx. 11, DOI 10.1155/S1073792894000140. MR1266108

[11] Y. Brudnyi and P. Shvartsman, *The Whitney problem of existence of a linear extension operator*, J. Geom. Anal. **7** (1997), no. 4, 515–574, DOI 10.1007/BF02921632. MR1669235

[12] R. R. Coifman and G. Weiss, *Extensions of Hardy spaces and their use in analysis*, Bull. Amer. Math. Soc. **83** (1977), no. 4, 569–645, DOI 10.1090/S0002-9904-1977-14325-5. MR447954

[13] A. Defant and K. Floret, *Tensor norms and operator ideals*, North-Holland Mathematics Studies, vol. 176, North-Holland Publishing Co., Amsterdam, 1993. MR1209438

[14] P. Enflo, *A counterexample to the approximation problem in Banach spaces*, Acta Math. **130** (1973), 309–317, DOI 10.1007/BF02392270. MR402468

[15] C. Fefferman, *A generalized sharp Whitney theorem for jets*, Rev. Mat. Iberoamericana **21** (2005), no. 2, 577–688, DOI 10.4171/RMI/430. MR2174917

[16] C. Fefferman, *Extension of $\mathrm{C}^{m,\omega}$-smooth functions by linear operators*, Rev. Mat. Iberoam. **25** (2009), no. 1, 1–48, DOI 10.4171/RMI/568. MR2514337

[17] C. Fefferman, *Whitney's extension problems and interpolation of data*, Bull. Amer. Math. Soc. (N.S.) **46** (2009), no. 2, 207–220, DOI 10.1090/S0273-0979-08-01240-8. MR2476412

[18] T. Figiel and W. B. Johnson, *The approximation property does not imply the bounded approximation property*, Proc. Amer. Math. Soc. **41** (1973), 197–200, DOI 10.2307/2038840. MR341032

[19] A. Grothendieck, *Produits tensoriels topologiques et espaces nucléaires* (French), Mem. Amer. Math. Soc. **16** (1955), Chapter 1: 196 pp.; Chapter 2: 140. MR75539

[20] G. Glaeser, *Étude de quelques algèbres tayloriennes* (French), J. Analyse Math. **6** (1958), 1–124; erratum, insert to 6 (1958), no. 2, DOI 10.1007/BF02790231. MR101294

[21] L. G. Hanin, *Kantorovich-Rubinstein norm and its application in the theory of Lipschitz spaces*, Proc. Amer. Math. Soc. **115** (1992), no. 2, 345–352, DOI 10.2307/2159251. MR1097344

[22] L. G. Hanin, *Duality for general Lipschitz classes and applications*, Proc. London Math. Soc. (3) **75** (1997), no. 1, 134–156, DOI 10.1112/S0024611597000294. MR1444316

[23] A. Jonsson, *The duals of Lipschitz spaces defined on closed sets*, Indiana Univ. Math. J. **39** (1990), no. 2, 467–476, DOI 10.1512/iumj.1990.39.39025. MR1089048

[24] A. Jonsson and H. Wallin, *Function spaces on subsets of $\mathbf{R}^n$*, Math. Rep. **2** (1984), no. 1, xiv+221. MR820626

[25] N. J. Kalton, *Spaces of Lipschitz and Hölder functions and their applications*, Collect. Math. **55** (2004), no. 2, 171–217. MR2068975

[26] J. Lindenstrauss, *Some open problems in Banach space theory*, Séminaire Choquet **18** (1975), 1–9.

[27] K. de Leeuw, *Banach spaces of Lipschitz functions*, Studia Math. **21** (1961/62), 55–66, DOI 10.4064/sm-21-1-55-66. MR140927

[28] G. K. Luli, *$C^{m,\omega}$ extension by bounded-depth linear operators*, Adv. Math. **224** (2010), no. 5, 1927–2021, DOI 10.1016/j.aim.2010.01.027. MR2646115

[29] E. J. McShane, *Extension of range of functions*, Bull. Amer. Math. Soc. **40** (1934), no. 12, 837–842, DOI 10.1090/S0002-9904-1934-05978-0. MR1562984

[30] J. Merrien, *Prolongateurs de fonctions différentiables d'une variable réelle* (French), J. Math. Pures Appl. (9) **45** (1966), 291–309. MR207937

[31] K.-M. Perfekt, *Duality and distance formulas in spaces defined by means of oscillation*, Ark. Mat. **51** (2013), no. 2, 345–361, DOI 10.1007/s11512-012-0175-7. MR3090201

[32] K.-M. Perfekt, *On M-ideals and o-O type spaces*, Math. Scand. **121** (2017), no. 1, 151–160, DOI 10.7146/math.scand.a-96626. MR3708971

[33] A. Pełczyński, *Any separable Banach space with the bounded approximation property is a complemented subspace of a Banach space with a basis*, Studia Math. **40** (1971), 239–243, DOI 10.4064/sm-40-3-239-243. MR308753

[34] I. A. Shevchuk, *Approximation by polynomials and traces of functions continuous on closed intervals*, Kiev, "Naukova Dumka", 1992 (in Russian).

[35] P. A. Shvartsman, *Lipschitz sections of multivalued mappings* (Russian), Studies in the theory of functions of several real variables (Russian), Yaroslav. Gos. Univ., Yaroslavl′, 1986, pp. 121–132, 149. MR878806

[36] P. Shvartsman, *The Whitney extension problem and Lipschitz selections of set-valued mappings in jet-spaces*, Trans. Amer. Math. Soc. **360** (2008), no. 10, 5529–5550, DOI 10.1090/S0002-9947-08-04469-3. MR2415084

[37] A. F. Timan, *Theory of approximation of functions of a real variable*, Dover Publications, Inc., New York, 1994. Translated from the Russian by J. Berry; Translation edited and with a preface by J. Cossar; Reprint of the 1963 English translation. MR1262128

[38] H. Whitney, *Analytic extensions of differentiable functions defined in closed sets*, Trans. Amer. Math. Soc. **36** (1934), no. 1, 63–89, DOI 10.2307/1989708. MR1501735

[39] H. Whitney, *Differentiable functions defined in closed sets. I*, Trans. Amer. Math. Soc. **36** (1934), no. 2, 369–387, DOI 10.2307/1989844. MR1501749

[40] A. Zygmund, *Smooth functions*, Duke Math. J. **12** (1945), 47–76. MR12691

Department of Mathematics and Statistics, University of Calgary, Calgary, Alberta, Canada T2N 1N4

Email address: abrudnyi@ucalgary.ca

Proceedings of Symposia in Pure Mathematics
Volume **106**, 2024
https://doi.org/10.1090/pspum/106/01927

On Norm-attaining Mappings on Banach Spaces

Yun Sung Choi, Sun Kwang Kim, and Han Ju Lee

This paper is dedicated to Professor Per H. Enflo

ABSTRACT. The main aim of the present paper is to survey recent results on the denseness of the set of mappings attaining their norms or numerical radii, and also to show their applications for geometric concepts of Banach spaces. Lots of classical results of bounded linear operators are also presented, which enable us to understand the motivation of extending those problems to nonlinear mappings such as multilinear mappings, homogeneous polynomials and holomorphic mappings. We also deal with a stronger property called the Bishop-Phelps-Bollobás property, and its stronger variation called the pointwise Bishop-Phelps-Bollobás property. We review these topics from the classical bounded linear operators to the nonlinear mapping, and also geometric properties related with them. We demonstrate their applications to the Daugavet property and the polynomial Daugavet property of a function space with a base algebra, which contains some new results. Finally, we focus on the numerical radius-attaining mappings and also on the geometric properties such as the numerical index, homogeneous polynomial numerical index and analytic numerical index. We present interesting results to characterize a Banach space whose polynomial (analytic) numerical index is 1.

1. Introduction

Lots of attention has been paid to the question if the set of norm-attaining operators is dense in the Banach space of such operators, since the celebrated the Bishop-Phelps theorem [**19**] came out in 1961. Actually, it was firstly motivated by V. Klee [**66**] who asked if a bounded closed convex subset of a Banach space has a support point. In parallel to that the same questions has been studied for the numerical radius of a bounded linear operator. We present in Section 2 the most recent results on the denseness of the set of mappings attaining their norms and also on the geometric properties of a Banach space to insure the denseness. Here the mappings can be nonlinear like multilinear mappings, homogeneous polynomials and holomorphic mappings. Lots of classical results of bounded linear operators are

---

2020 *Mathematics Subject Classification.* Primary 46B04; Secondary 46B07, 46B20.

Key words and phrases. Banach space; norm-attaining operators; Bishop-Phelps-Bollobás property.

The first author was supported in part by POSTECH Basic Science Research Institute Grant [NRF-2021R1A6A1A10042944].

The third author was supported in part by Basic Science Research Program through the National Research Foundation of Korea (NRF) funded by the Ministry of Education, Science and Technology [NRF-2020R1A2C1A01010377].

also presented, because they enable us to understand the motivation of extending those problems to the nonlinear mappings.

In Section 3, a stronger property called the Bishop-Phelps-Bollobás property is introduced, and we study this property from the classical bounded linear operators to the nonlinear mappings, and also some geometric properties related with them.

In Section 4 we deal with a stronger variation than the Bishop-Phelps-Bollobás property, called the pointwise Bishop-Phelps-Bollobás property. Indeed, the pointwise Bishop-Phelps-Bollobás property implies the Bishop-Phelps-Bollobás property, and it can characterize the uniform smoothness of a Banach space. We present interesting results on this property from the spaces of bounded linear operators to nonlinear cases including a function space with a base algebra.

In Section 5, we demonstrate applications of norm-attaining mappings to the Daugavet property and the polynomial Daugavet property concerning from a Banach space to a function space with a base algebra, and some new results are also provided.

In Section 6, we survey the results on the denseness of numerical radius-attaining mappings and also on the geometric properties such as the numerical index, homogeneous polynomial numerical index and analytic numerical index. We also show interesting results to characterize a Banach space whose polynomial (analytic) numerical index is 1.

2. Norm Attaining Mappings

In the present survey, $\mathcal{X}$ and $\mathcal{Y}$ are reserved for Banach spaces based on the scalar field $\mathbb{K}$ which can be the real field $\mathbb{R}$ or the complex field $\mathbb{C}$. We denote the closed unit ball and the unit sphere of $\mathcal{X}$ by $B_{\mathcal{X}}$ and $S_{\mathcal{X}}$ respectively. The set of all natural numbers is denoted by $\mathbb{N}$.

For $n \in \mathbb{N}$ and Banach spaces $\mathcal{X}_1, \ldots, \mathcal{X}_n$ and $\mathcal{Y}$, let $\mathbb{L}^n(\mathcal{X}_1 \times \cdots \times \mathcal{X}_n; \mathcal{Y})$ be the space of bounded n-linear mappings from $\mathcal{X}_1 \times \cdots \times \mathcal{X}_n$ into $\mathcal{Y}$. It is well known that it is a Banach space equipped with the norm

$$\|A\| := \sup\{\|A(x_1, \ldots, x_n)\| : (x_1, \ldots, x_n) \in B_{\mathcal{X}_1} \times \cdots \times B_{\mathcal{X}_n}\}$$

for $A \in \mathbb{L}^n(\mathcal{X}_1 \times \cdots \times \mathcal{X}_n; \mathcal{Y})$.

If $\mathcal{X}_1 = \cdots = \mathcal{X}_n = \mathcal{X}$, then we simply write $\mathbb{L}(^n\mathcal{X}; \mathcal{Y})$ for $\mathbb{L}^n(\mathcal{X}_1 \times \cdots \times \mathcal{X}_n; \mathcal{Y})$, and for the case $n = 1$ we just write $\mathbb{B}(\mathcal{X}, \mathcal{Y})$ for the space of 1-linear mappings. Especially, the dual $\mathbb{B}(\mathcal{X}, \mathbb{K})$ of $\mathcal{X}$ will be denoted by $\mathcal{X}^*$ shortly.

For the diagonal mapping $\Delta_k : \mathcal{X} \to \mathcal{X}^n$ defined by

$$\Delta_n(x) = \underbrace{(x, x, \ldots, x)}_{n \text{ times}},$$

a mapping $P : \mathcal{X} \to \mathcal{Y}$ is called a bounded *n-homogeneous polynomial* from $\mathcal{X}$ to $\mathcal{Y}$ if it is the composition of Δ_n with an element $A \in \mathbb{L}(^n\mathcal{X}; \mathcal{Y})$, that is, $P = A \circ \Delta_n$. We denote by $\mathbb{P}(^n\mathcal{X}; \mathcal{Y})$ the set of all bounded n-homogenous polynomials from $\mathcal{X}$ into $\mathcal{Y}$. A *bounded polynomial* is a finite sum of bounded homogenous polynomials from $\mathcal{X}$ to $\mathcal{Y}$. The notation $\mathbb{P}(\mathcal{X}; \mathcal{Y})$ denotes the set of all bounded polynomials from $\mathcal{X}$ into $\mathcal{Y}$. The vector space $\mathbb{P}(\mathcal{X}; \mathcal{Y})$ is a normed space under the norm

$$\|P\| = \{\|P(x)\| : x \in B_{\mathcal{X}}\}, \qquad (P \in \mathbb{P}(\mathcal{X}; \mathcal{Y})).$$

Note that $\mathbb{P}(\mathcal{X}; \mathcal{Y})$ is not a Banach space in general. However, $\mathbb{P}(^n\mathcal{X}; \mathcal{Y})$ is a Banach space under the same norm for every $n \in \mathbb{N}$. If $\mathcal{Y}$ is the scalar field, we usually omit $\mathcal{Y}$ in the notation if it does not cause any confusion. For example, we simply write $\mathbb{P}(^n\mathcal{X})$ for $\mathbb{P}(^n\mathcal{X};\mathcal{Y})$.

A bounded n-linear mapping $A \in \mathbb{L}^n(\mathcal{X}_1 \times \cdots \times \mathcal{X}_n;\mathcal{Y})$ is said to be *norm-attaining* if there exists an element $(x_1,\ldots,x_n) \in B_{\mathcal{X}_1} \times \cdots \times B_{\mathcal{X}_n}$ such that $\|A\| = \|A(x_1,\ldots,x_n)\|$. A polynomial $P \in \mathbb{P}(^n\mathcal{X},\mathcal{Y})$ is said to be *norm-attaining* if there exists $x_0 \in B_{\mathcal{X}}$ such that $\|P\| = \|P(x_0)\|$. To denote the set of norm-attaining mappings we write $\mathbb{NA}$ in front of the notation for each space of mappings. For example, the set of norm-attaining n-linear mappings and n-homogeneous polynomials are $\mathbb{NAL}^n(\mathcal{X}_1 \times \cdots \times \mathcal{X}_n;\mathcal{Y})$ and $\mathbb{NAP}(^n\mathcal{X};\mathcal{Y})$, respectively.

The theory of norm-attaining mappings, which is one of the most fruitful topics in Banach space theory, is motivated by the question of V. Klee [**66**] whether a bounded closed convex subset of a Banach space has a support point or not. This question can be rewritten in the way that for each bounded closed convex subset of a Banach space is it possible to find a non-zero bounded linear functional whose modulus attains its maximum on it? For a real Banach space, the answer was given in 1961 by E. Bishop and R. Phelps [**19**]. They proved a more stronger statement that the set of bounded linear functionals attaining their maximum modulus on a bounded closed convex subset is dense in its dual space. Unlike the case of a real Banach space, the problem for a complex Banach space had not been solved for a very long time. In 2001, V. Lomonosov [**74**] constructed a closed bounded convex subset in a complex Banach space without a support point.

According to [**19**], for a particular case that the bounded closed convex set is the closed unit ball, the set of maximum modulus-attaining functionals is dense.

THEOREM 2.1 ([**19**]). *For an arbitrary Banach space $\mathcal{X}$, $\overline{\mathbb{NAB}(\mathcal{X},\mathbb{K})} = \mathbb{B}(\mathcal{X},\mathbb{K})$.*

It is natural to ask if this theorem can be generalized to arbitrary mappings, and it was posed for bounded linear operators in the same paper. The first result for bounded linear operators was deduced by J. Lindenstrauss [**73**]. He proved that there is no Bishop-Phelps theorem for bounded linear operators by observing that every norm-attaining operator from c_0 into a strictly convex space has finite rank.

THEOREM 2.2 ([**73**]). *For a strictly convex Banach space $\mathcal{Y}$, $\overline{\mathbb{NAB}(c_0,\mathcal{Y})} \subset \mathbb{K}(c_0,\mathcal{Y})$ where $\mathbb{K}(c_0,\mathcal{Y})$ is the space of compact operators from c_0 to $\mathcal{Y}$.*

It is well-known that every separable space can be renormed to be strictly convex and c_0 is separable. Hence, there is a strictly convex Banach space $\mathcal{Y}$ which is a renorming of c_0, and the identity operator from c_0 into $\mathcal{Y}$ is not compact. Therefore, we see that the denseness does not hold in general.

Nevertheless, he also found some geometric conditions on the domain space $\mathcal{X}$ or the range space $\mathcal{Y}$ for the denseness of norm-attaining linear operators. For example, the reflexivity of $\mathcal{X}$ or the condition that $B_{\mathcal{X}}$ is the closed convex hull of its uniformly strongly exposed points are some of them.

THEOREM 2.3 ([**73**]). *For Banach spaces $\mathcal{X}$ and $\mathcal{Y}$, if $\mathcal{X}$ is reflexive or the closed unit ball $B_{\mathcal{X}}$ is the closed convex hull of its uniformly strongly exposed points, then $\overline{\mathbb{NAB}(\mathcal{X},\mathcal{Y})} = \mathbb{B}(\mathcal{X},\mathcal{Y})$*

It is worth to comment that the result on the reflexivity was generalized by J. Bourgain [**21**] for the domain space $B_{\mathcal{X}}$ with the Radon-Nikodým property.

THEOREM 2.4 ([**21**]). *For Banach spaces $\mathcal{X}$ and $\mathcal{Y}$, if $\mathcal{X}$ has the Radon-Nikodým property, then $\overline{\mathbb{NAB}(\mathcal{X},\mathcal{Y})} = \mathbb{B}(\mathcal{X},\mathcal{Y})$*

Motivated by the paper of Lindenstrauss, the following property α was introduced by W. Schachermayer [**83**].

DEFINITION 2.5. A Banach space $\mathcal{X}$ is said to have *property* α with a constant $0 \leq \rho < 1$ if there exists a set $\{(x_i, x_i^*) \in \mathcal{X} \times \mathcal{X}^* : \|x_i\| = \|x_i^*\| = 1, i \in \Lambda\}$ with an index set Λ satisfying

(i) $x_i^*(x_i) = 1$ for every $i \in \Lambda$,
(ii) $|x_i^*(x_j)| \leq \rho < 1$ for every $i, j \in \Lambda$ with $i \neq j$,
(iii) $B_{\mathcal{X}} = \overline{aconv}\{x_i \ : \ i \in \Lambda\}$,

where $\overline{aconv}$ is the closed convex hull of $\{\pm x : x \in A\}$ for $A \subset \mathcal{X}$.

Since $\{x_i \ : \ i \in \Lambda\}$ in the definition is a set of uniformly strongly exposed points, if $\mathcal{X}$ has the property α, then the the set of norm-attaining operators is dense. Moreover, it is known that every weakly compactly generated space can be renormed to have the property α [**83**]. Motivated by these results and [**2**], a general property was introduced.

DEFINITION 2.6 ([**39**]). A Banach space $\mathcal{X}$ is said to have *property quasi-*α if there exist a set $\{(x_i, x_i^*) \in \mathcal{X} \times \mathcal{X}^* : \|x_i\| = \|x_i^*\| = 1, \ i \in \Lambda\}$ with an index set Λ and a function $\lambda\colon \{x_i\}_{i\in\Lambda} \to \mathbb{R}$ satisfying that

(i) $x_i^*(x_i) = 1$ for all $i \in \Lambda$,
(ii) $|x_i^*(x_j)| \leq \lambda(x_i) < 1$ for all $i, j \in \Lambda$ with $i \neq j$,
(iii) For every $e \in \mathrm{Ext}(B_{\mathcal{X}^{**}})$, there exists $\Lambda_e \subset \Lambda$ such that $te \in \overline{\{x_i\}_{i\in\Lambda_e}}^{w^*}$ for some $t \in S_{\mathbb{K}}$ and $r_e = \sup\{\lambda(x_i)\colon i \in \Lambda_e\} < 1$.

We comment that similar results on range space $B_{\mathcal{Y}}$ can be also found in [**2**, **73**, **78**]. Afterwards, this topic has been developed in various directions by many mathematicians. Many interesting geometric properties of a Banach space which insure denseness have been found, and it is also studied for specific spaces. We now present some well-known examples, specifically focusing on Lebesgue spaces and spaces of continuous functions, as follows.

EXAMPLE 2.7.

(1) [**55**] $\overline{\mathbb{NAB}(L_1(\mu), L_1(\nu))} = \mathbb{B}(L_1(\mu), L_1(\nu))$ for finite positive Borel measures μ and ν on $[0,1]$.
(2) [**88**] $\overline{\mathbb{NAB}(L_1(\mu), L_q(\nu))} = \mathbb{B}(L_1(\mu), L_q(\nu))$ for finite positive measures μ and ν, and every $q \in (1, \infty)$.
(3) [**79**] $\overline{\mathbb{NAB}(L_1(\mu), L_\infty(\nu))} = \mathbb{B}(L_1(\mu), L_\infty(\nu))$ for an arbitrary measure μ and a localizable measure ν.
(4) [**21**] $\overline{\mathbb{NAB}(L_p(\mu), L_q(\nu))} = \mathbb{B}(L_p(\mu), L_q(\nu))$ for a finite positive measure μ and an arbitrary measure ν, and every $(p, q) \in (1, \infty) \times [1, \infty]$.
(5) [**57**] $\overline{\mathbb{NAB}(\mathcal{C}(K), \mathcal{C}(S))} = \mathbb{B}(\mathcal{C}(K), \mathcal{C}(S))$ for compact Hausdorff spaces K and S when the scalar field is $\mathbb{R}$.

As a final remark for linear operators, we comment that there are some classical Banach spaces such that denseness does not hold. For example, it is known in [**84**] that $\overline{\mathbb{NAB}(L_1([0,1]), \mathcal{C}([0,1]))} \neq \mathbb{B}(L_1([0,1]), \mathcal{C}([0,1]))$.

It is quite natural to ask the same question for nonlinear mappings. The first try was for the space of n-linear mappings and n-homogeneous polynomials. Similarly to the case of linear operators, some geometric conditions ensure denseness. Aron,

Finet and Werner proved [**15**] that if a Banach space $\mathcal{X}$ has the Radon-Nikodým property or the property α, then for $n \in \mathbb{N}$ the set of norm-attaining elements is dense in $\mathbb{L}(^n\mathcal{X}, \mathcal{Y})$ for every Banach space $\mathcal{Y}$. The first named author and Kim [**33**] also showed the positive result for n-homogeneous polynomials, that is, the set of norm-attaining elements is dense in $\mathbb{P}(^n\mathcal{X}, \mathcal{Y})$ for every Banach space $\mathcal{Y}$ if $\mathcal{X}$ has the Radon-Nikodým property.

THEOREM 2.8. *For Banach spaces $\mathcal{X}$ and $\mathcal{Y}$ and for $n \in \mathbb{N}$*

(1) [**15**] $\overline{\mathbb{NAL}(^n\mathcal{X}, \mathcal{Y})} = \mathbb{L}(^n\mathcal{X}, \mathcal{Y})$ *whenever $\mathcal{X}$ has the Radon-Nikodým property or property α.*
(2) [**33**] $\overline{\mathbb{NAP}(^n\mathcal{X}, \mathcal{Y})} = \mathbb{P}(^n\mathcal{X}, \mathcal{Y})$ *whenever $\mathcal{X}$ has the Radon-Nikodým property.*

Moreover, there are many particular examples such that denseness holds. It is shown in [**11**] that the set of norm-attaining bilinear forms in $\mathbb{L}(^2\mathcal{C}(K))$ is dense, where K is a compact Hausdorff space. In [**56**], the authors proved that for every $N \geq 2$ the set of norm-attaining elements is dense in $\mathbb{L}(^N d_*(w,1))$ if and only if $w \notin \ell_N$, where $w \in c_0 \backslash \ell_1$ is a decreasing sequence of positive numbers with $w(1) = 1$ and $d_*(w,1)$ is the canonical predual of Lorentz sequence space $d(w,1)$. However, unlike what we observed regarding the denseness of norm-attaining operators on classical spaces, the denseness does not hold on L_1-space even for the simplest case $N = 2$. More precisely, the first named author showed the following.

THEOREM 2.9 ([**26**]). *For the Lebesgue measure μ on $[0,1]$,*

$$\overline{\mathbb{NAL}(^2L_1(\mu))} \neq \mathbb{L}(^2L_1(\mu)).$$

For holomorphic functions, we introduce a stronger concept of a norm-attaining function, which is called strongly peaking. Since this is the main ingredient of the present survey, let us recall some necessary definitions about separating function algebras.

Given a topological space Ω and a complex Banach space $\mathcal{Y}$, let

$$\mathcal{C}_b(\Omega, \mathcal{Y}) := \{f : \Omega \to \mathcal{Y} \ : \ f \text{ is a bounded continuous function}\}.$$

Then $\mathcal{C}_b(\Omega, \mathcal{Y})$ is a Banach space under the supremum norm

$$\|f\| = \sup\{\|f(t)\| : t \in \Omega\} \qquad (f \in \mathcal{C}_b(\Omega, \mathcal{Y})).$$

If $\mathcal{Y}$ is the complex field, then $\mathcal{C}_b(\Omega, \mathcal{Y})$ is abbreviated to $\mathcal{C}_b(\Omega)$. Note that $\mathcal{C}_b(\Omega)$ is a unital Banach algebra. A closed subalgebra $\mathcal{A}$ of $\mathcal{C}_b(\Omega)$ is called a *function algebra.* $\mathcal{A}$ is said to be *separating* if $\mathcal{A}$ separates the points of Ω. That is, given two distinct points $s, t \in \Omega$, there is $f \in \mathcal{A}$ such that $f(s) \neq f(t)$. A function algebra is said to be *unital* if it contains all constants.

If Ω is a compact Hausdorff space, a separating function algebra $\mathcal{A}$ of $C(\Omega)$ is called a *uniform algebra* on Ω. In this paper we are dealing with only unital uniform algebras.

For a unital uniform algebra $\mathcal{A}$ on Ω and given $t \in \Omega$, the evaluation δ_t at t is defined by $\delta_t(f) = f(t)$ for each $f \in \mathcal{A}$ and let $S = \{x^* \in \mathcal{A}^* : \|x^*\| = x^*(\mathbf{1}) = 1\}$, where $\mathbf{1}$ is unit in $\mathcal{A}$. It is known that the set $\text{Ext}(S)$ of all extreme points of S is contained in $\{\delta_t : t \in \Omega\}$ and the set $\{t \in \Omega : \delta_t \in \text{Ext}(S)\}$ is called the *Choquet boundary* of $\mathcal{A}$. For more on uniform algebras, refer to [**40**].

Given a function algebra $\mathcal{A}$ on a Hausdorff space Ω, $f \in \mathcal{A}$ is said to be a *strong peak function* at $s \in \Omega$ if $\|f\| = |f(s)| > \sup\{|f(t)| : t \in \Omega \setminus V\}$ for every open neighborhood V of s, and the point $s \in \Omega$ is called a *strong peak point* for $\mathcal{A}$. It is clear that strong peak function is norm-attaining. The set of all strong peak points for $\mathcal{A}$ is denoted by $\rho\mathcal{A}$. A subset $N \subset \Omega$ is called a *norming set* for $\mathcal{A}$ if

$$\|f\| = \sup\{|f(t)| : t \in N\} \quad (f \in \mathcal{A}).$$

It is clear that the closed norming subset contains the closure of $\rho\mathcal{A}$.

The smallest closed norming subset for a separating function algebra is called the *Shilov boundary*, which always uniquely exists for uniform algebras on a compact Hausdorff space ([**85**]). A separating function algebra on a noncompact set need not to have the Shilov boundary ([**14, 31, 53, 80**]). If the Shilov boundary of a unital separating function algebra $\mathcal{A}$ exists, its Shilov boundary is denoted by $\partial\mathcal{A}$. Bishop [**18**] showed that if $\mathcal{A}$ is a unital uniform algebra on compact metric space, every element in $\mathcal{A}$ attains its maximum modulus on $\rho\mathcal{A}$ and so $\partial\mathcal{A} = \overline{\rho\mathcal{A}}$.

Note that the Shilov boundary of a function algebra exists if either Ω is a compact Hausdorff space or the set of strong peak points is norming. Note also that the Shilov boundary of a unital uniform algebra is the closure of the Choquet boundary [**40**]. For a compact metrizable space Ω and for a unital uniform algebra $\mathcal{A} \subset \mathcal{C}(\Omega)$, $\rho\mathcal{A}$ is the same as the Choquet boundary of $\mathcal{A}$ [**40**].

For two complex Banach spaces $\mathcal{X}$, $\mathcal{Y}$, let

$$\mathcal{A}_b(B_{\mathcal{X}}, \mathcal{Y}) := \{f \in \mathcal{C}_b(B_{\mathcal{X}}, \mathcal{Y}) : f \text{ is holomorphic on } B_{\mathcal{X}}^\circ\},$$
$$\mathcal{A}_u(B_{\mathcal{X}}, \mathcal{Y}) := \{f \in \mathcal{A}_b(B_{\mathcal{X}}, \mathcal{Y}) : f \text{ is uniformly continuous on } B_{\mathcal{X}}\}$$

where $B_{\mathcal{X}}^\circ$ is the interior of $B_{\mathcal{X}}$. Both $\mathcal{A}_b(B_{\mathcal{X}}, \mathcal{Y})$ and $\mathcal{A}_u(B_{\mathcal{X}}, \mathcal{Y})$ are closed subspaces of $\mathcal{C}_b(B_{\mathcal{X}}, \mathcal{Y})$. In case that $\mathcal{Y}$ is the complex scalar field $\mathbb{C}$, we omit the scalar field as usual and simply write $\mathcal{C}_b(B_{\mathcal{X}})$, $\mathcal{A}_b(B_{\mathcal{X}})$ and $\mathcal{A}_u(B_{\mathcal{X}})$, respectively. Note that they are also separating function algebras.

In [**3**], the denseness of norm-attaining holomorphic functions is studied.

THEOREM 2.10 ([**3**]).
(1) *If $\mathcal{X}$ has the Radon-Nikodým property, the set of norm-attaining elements is dense in $\mathcal{A}_u(B_{\mathcal{X}})$.*
(2) *If $\mathcal{X}$ is $\mathcal{C}(K)$ for a scattered compact Hausdorff space K, the set of norm-attaining elements is dense in $\mathcal{A}_u(B_{\mathcal{X}})$.*

Motivated by these results, the following is obtained. From now on, $\mathcal{A}(B_{\mathcal{X}}, \mathcal{Y})$ denotes either $\mathcal{A}_u(B_{\mathcal{X}}, \mathcal{Y})$ or $\mathcal{A}_b(B_{\mathcal{X}}, \mathcal{Y})$.

THEOREM 2.11 ([**37**, Theorem 4.4]). *Assume that a complex Banach space $\mathcal{X}$ has the Radon-Nikodým property.*
(1) *Then the set of all strong peak functions in $\mathcal{A}(B_{\mathcal{X}}, \mathcal{Y})$ is dense in $\mathcal{A}(B_{\mathcal{X}}, \mathcal{Y})$. In particular, the set of all smooth points of $B_{\mathcal{A}(B_{\mathcal{X}}, \mathcal{Y})}$ is dense in $S_{\mathcal{A}(B_{\mathcal{X}}, \mathcal{Y})}$*
(2) *$\rho\mathcal{A}(B_{\mathcal{X}})$ is a norming subset of $\mathcal{A}(B_{\mathcal{X}}, \mathcal{Y})$ and $\partial\mathcal{A}(B_{\mathcal{X}}) = \overline{\rho\mathcal{A}(B_{\mathcal{X}})}$. In particular, the set of complex extreme points of $B_{\mathcal{X}}$ is a norming subset for $\mathcal{A}(B_{\mathcal{X}})$.*

On the other hand, recall that, for a given complex Banach space $\mathcal{X}$, $x \in S_{\mathcal{X}}$ is called a *complex extreme point* if $\max\{\|x + e^{i\theta}y\| : \theta \in [0, 2\pi]\} > 1$ for every nonzero $y \in \mathcal{X}$. Given $\epsilon > 0$, and $x \in S_{\mathcal{X}}$, let

$$H(\epsilon, x) = \inf\left\{ \max_{\theta \in [0,2\pi]} \|x + e^{i\theta}y\| - 1 : \|y\| \geq \epsilon, \ y \in \mathcal{X} \right\}$$

and for each $\epsilon > 0$, put

$$H(\epsilon) = \inf\{H(\epsilon, x) : x \in S_{\mathcal{X}}\}.$$

If $H(\epsilon, x) > 0$ for all $x \in S_{\mathcal{X}}$ and for all $\epsilon > 0$, $\mathcal{X}$ is said to be *locally uniformly c-convex.* If $H(\epsilon) > 0$ for all $\epsilon > 0$, $\mathcal{X}$ is said to be *uniformly complex convex.* To see interesting aspects of the complex convexity, we refer the reader to [**47, 49, 50, 52, 86**].

The complex extreme points of the unit ball of a complex Banach space are closely related to norming sets for $\mathcal{A}(B_{\mathcal{X}})$ ([**12, 32, 53**]). It is easy to observe that every strong peak point for $\mathcal{A}(B_{\mathcal{X}})$ is a complex extreme point of $B_{\mathcal{X}}$ and every (real) extreme point of $B_{\mathcal{X}}$ is a complex extreme point of $B_{\mathcal{X}}$.

The complex convexity of some sequence spaces had been applied to show that the set of norm-attaining operators is not dense in $\mathbb{B}(\mathcal{X}, \mathcal{Y})$ for some complex sequence spaces $\mathcal{X}$, $\mathcal{Y}$ ([**25**]). The uniform complex convexity is equivalent to the uniform monotonicity in a complex Banach lattice ([**68, 69**]).

It is observed in [**37**] that the differentiability of a norm at f in a closed subspace of $\mathcal{C}_b(\Omega, \mathcal{X})$ is closely related to the convergence of a norming sequence for f. Using this observation, it is shown in [**37**] that the norm of $\mathcal{A}(B_{\mathcal{X}})$ is never Fréchet differentiable for every nontrivial complex Banach space $\mathcal{X}$.

For the other spaces without Radon-Nikodým property, the variational technique in [**48**] is used to get the denseness of strong peak functions.

THEOREM 2.12 ([**60**]). *Let K be a complete metric space and $\mathcal{X}$ be a Banach space. Suppose that $\mathcal{A}$ is a closed subspace of $\mathcal{C}_b(K, \mathcal{X})$ such that $\mathcal{A}$ contains all elements of the form $t \mapsto (x^*f(t))^m x$ for every $(x, x^*) \in \mathcal{X} \times \mathcal{X}^*$ and for every $m \in \mathbb{N}$. Then $\rho\mathcal{A}$ is norming if and only if the set of all strong peak functions in $\mathcal{A}$ is dense in $\mathcal{A}$.*

It is easy to check that if a complex Banach space $\mathcal{X}$ is locally uniformly convex, then every point $S_{\mathcal{X}}$ is strongly exposed and $S_{\mathcal{X}} \subset \rho\mathcal{A}(B_{\mathcal{X}})$. So $\rho\mathcal{A}(B_{\mathcal{X}})$ is norming for $\mathcal{A}(B_{\mathcal{X}})$ ([**70**]). It is also observed in [**32**] that if $\mathcal{X}$ is an order continuous, locally uniformly c-convex sequence space, then $\rho\mathcal{A}(B_{\mathcal{X}})$ is dense in $S_{\mathcal{X}}$. So, for such spaces $\mathcal{X}$, the set of strong peak functions in $\mathcal{A}(B_{\mathcal{X}})$ is dense.

3. Bishop-Phelps-Bollobás property

In 1970, B. Bollobás [**20**] strengthened the Bishop-Phelps theorem by proving that if a functional almost attains its norm at a point in the closed unit ball, then both functional and point can be approximated by norm-attaining functionals and their norm-attaining points simultaneously.

THEOREM 3.1. *For a Banach space $\mathcal{X}$ and a positive number $\varepsilon \in (0, 2)$, if a pair $(x, x^*) \in B_{\mathcal{X}} \times B_{\mathcal{X}^*}$ satisfies*

$$Re\, x^*(x) > 1 - \frac{\varepsilon^2}{2},$$

then there is a pair $(y, y^*) \in S_{\mathcal{X}} \times S_{\mathcal{X}^*}$ such that

$$|y^*(y)| = 1, \quad \|y - x\| < \varepsilon \quad \text{and} \quad \|y^* - x^*\| < \varepsilon.$$

It is clear that this result implies the Bishop-Phelps theorem, and we call this the Bishop-Phelps-Bollobás theorem. Afterwards, similarly to the case of the Bishop-Phelps theorem, lots of authors started to investigate the extension of the

Bishop-Phelps-Bollobás theorem for other mappings. The extension for operators was firstly studied in [**4**], and the authors introduced the Bishop-Phelps-Bollobás property. Later, the version for n-linear mappings and polynomials was introduced as follows.

DEFINITION 3.2 ([**6**]).

(1) For Banach spaces $\mathcal{X}_1, \dots, \mathcal{X}_n$ and $\mathcal{Y}$, the space $\mathbb{L}^n(\mathcal{X}_1 \times \cdots \times \mathcal{X}_n, \mathcal{Y})$ is said to have the Bishop-Phelps-Bollobás property (BPBp) if for each $\epsilon > 0$, there exists $\eta > 0$ such that, whenever $A \in \mathbb{L}^n(\mathcal{X}_1 \times \cdots \times \mathcal{X}_n, \mathcal{Y})$ with $\|A\| = 1$ and $(x_1, \dots, x_n) \in \prod_{k=1}^n S_{\mathcal{X}_k}$ satisfy $\|A(x_1, \dots, x_n)\| > 1 - \eta$, there exist $(y_1, \dots, y_n) \in \prod_{k=1}^n S_{\mathcal{X}_k}$ and $B \in \mathbb{L}^n(\mathcal{X}_1 \times \cdots \times \mathcal{X}_n, \mathcal{Y})$ such that $\|B\| = \|B(y_1, \dots, y_n)\| = 1$, $\max\{\|x_k - y_k\| : 1 \leq k \leq n\}\| < \epsilon$, and $\|A - B\| < \epsilon$.

(2) For Banach spaces $\mathcal{X}$ and $\mathcal{Y}$, the space $\mathbb{P}(^n\mathcal{X}, \mathcal{Y})$ is said to have the Bishop-Phelps-Bollobás property (BPBp) if for each $\epsilon > 0$, there exists $\eta > 0$ such that, whenever $A \in \mathbb{P}(^n\mathcal{X}, \mathcal{Y})$ with $\|A\| = 1$ and $x \in S_{\mathcal{X}}$ satisfy $\|Ax\| > 1 - \eta$, there exist $y \in S_{\mathcal{X}}$ and $B \in \mathbb{P}(^n\mathcal{X}, \mathcal{Y})$ such that $\|B\| = \|By\| = 1$, $\|x - y\| < \epsilon$, and $\|A - B\| < \epsilon$.

In [**4**], the authors showed that the space $\mathbb{B}(\mathcal{X}, \mathcal{Y})$ has the the BPBp for all finite dimensional Banach spaces $\mathcal{X}$ and $\mathcal{Y}$. Their remarkable result is that $\mathbb{B}(\ell_1, \mathcal{Y})$ fails to have BPBp for some $\mathcal{Y}$. Since ℓ_1 is a typical example having both the Radon-Nikodým property and the property α, this shows the big difference between the classical theory of norm-attaining operators and that of the BPBp. Moreover, it enables us to consider new interesting geometric properties on Banach spaces. The authors of [**4**] introduced the Approximate Hyperplane Series property as follows to characterize the Banach space $\mathcal{Y}$ such that the couple $\mathbb{B}(\ell_1, \mathcal{Y})$ has the BPBp.

DEFINITION 3.3 ([**4**]). A Banach space $\mathcal{X}$ is said to have the Approximate Hyperplane Series property (AHSp) if for each $\varepsilon > 0$ there exists $\eta > 0$ satisfying the following. If a sequence $(x_k) \subset S_{\mathcal{X}}$ and a convex series $\sum_{n=1}^{\infty} \alpha_k$ satisfy $\|\sum_{n=1}^{\infty} \alpha_k x_k\| > 1 - \eta$, then there exist a subset $\Gamma \subset \mathbb{N}$ and a subset $\{z_k : k \in \Gamma\} \subset S_{\mathcal{X}}$ such that

(i) $\sum_{k \in \Gamma} \alpha_k > 1 - \varepsilon$
(ii) $\|z_k - x_k\| < \varepsilon$ for all $k \in \Gamma$ and
(iii) for a certain $x^* \in S_{\mathcal{X}^*}$ $x^*(z_k) = 1$ for all $k \in \Gamma$.

The authors proved that $\mathbb{B}(\ell_1, \mathcal{X})$ has the BPBp if and only if $\mathcal{X}$ has the AHSp, and every strictly convex space is uniformly convex if it has the AHSp. They also found many classes of spaces having the AHSp.

EXAMPLE 3.4 ([**4**]). The following Banach space $\mathcal{X}$ has the AHSp.

(1) $\mathcal{X}$ is a finite dimensional space.
(2) $\mathcal{X} = L_1(\mu)$ for a σ-finite measure μ.
(3) $\mathcal{X} = \mathcal{C}(K)$ for a compact Hausdorff space K
(4) $\mathcal{X}$ is uniformly convex.

Later, it is shown that every lush space has the AHSp ([**34**]), and it is well known that L_1 and $\mathcal{C}(K)$ spaces are lush spaces. Similarly to the denseness of norm attaining operators, many authors paid attention to find examples having BPBp. We introduce some of them.

EXAMPLE 3.5.

(1) [**36**] $\mathbb{B}(L_1(\mu), L_1(\nu))$ has the BPBp for arbitrary measures μ and ν.
(2) [**35, 36**] $\mathbb{B}(L_1(\mu), L_q(\nu))$ has the BPBp for arbitrary measures μ and ν and for every $q \in (1, \infty)$.
(3) [**13, 36**] $\mathbb{B}(L_1(\mu), L_\infty(\nu))$ has the BPBp for an arbitrary measure μ and a localizable measure ν.
(4) [**7, 63**] $\mathbb{B}(L_p(\mu), L_q(\nu))$ has the BPBp for arbitrary measures μ and ν and for every $(p, q) \in (1, \infty) \times [1, \infty]$.
(5) [**5**] $\mathbb{B}(\mathcal{C}(K), \mathcal{C}(S))$ has the BPBp for compact Hausdorff spaces K and S when the scalar field is $\mathbb{R}$.

It is worth to remark that the results (2) and (3) are firstly proved for specific measures in [**13, 35**], respectively and generalized later for general measures in [**36**]. As a result, ones in Example 2.7 are true for these measures since BPBp is stronger than the denseness of norm-attaining operators. For the case of (5), the complex version is still not known even for the denseness of norm-attaining operators.

For n-linear mappings, the denseness of norm-attaining elements does not hold in general as mentioned in Theorem 2.9, but we have also seen that there are many examples with such denseness. However, the case of the BPBp is more restrictive.

THEOREM 3.6 ([**38**]). *$\mathbb{L}(^2\ell_1)$ does not have the BPBp.*

Since ℓ_1 is a space having the Radon-Nikodým property and the property α, the norm-attaining elements are dense. Moreover, it is interesting that $\mathbb{L}(^2\ell_1)$ is isometrically isomorphic to $\mathbb{B}(\ell_1, \ell_\infty)$ which has the BPBp (see e.g. [**4**]). However, if the domain Banach spaces $\mathcal{X}_i$'s are uniformly convex, a positive result is obtained.

THEOREM 3.7 ([**7**]). *If $\mathcal{X}_i$ $(1 \leq i \leq n)$ are uniformly convex, then, for every Banach space $\mathcal{Y}$, $\mathbb{L}^n(\mathcal{X}_1 \times \cdots \times \mathcal{X}_n, \mathcal{Y})$ has the BPBp.*

In [**7**], the problem that $\mathbb{L}^2(c_0 \times c_0)$ has the BPBp has been raised, that is still open for real c_0 space. However, the problem was solved positively for the complex case. In the proof, the uniform complex convexity of L_1-space [**52**] played an essential role.

In [**1**], it is proved that $\mathbb{B}(\mathcal{C}_0(L), \mathcal{X})$ has the BPBp if $\mathcal{X}$ is complex uniformly convex and L is a locally compact Hausdorff space where $\mathcal{C}_0(L)$ is the Banach space of complex-valued continuous functions on L vanishing at infinity. Motivated by the argument there, it is shown that $\mathbb{L}^2(\mathcal{C}_0(L_1) \times \mathcal{C}_0(L_2))$ has the BPBp for locally compact Hausdorff spaces L_1 and L_2 since the dual of $\mathcal{C}_0(L_2)$ is complex uniformly convex.

THEOREM 3.8.

(1) [**1**] *$\mathbb{B}(\mathcal{C}_0(L), \mathcal{X})$ has the BPBp if $\mathcal{X}$ is complex uniformly convex and L is a locally compact Hausdorff space.*
(2) [**65**] *The space $\mathbb{L}^2(\mathcal{C}_0(L_1) \times \mathcal{C}_0(L_2))$ has the BPBp if L_1 and L_2 are locally compact Hausdorff spaces.*

There are not many known results for polynomials, but we have a positive case when the domain space is uniformly convex.

THEOREM 3.9 ([**6**]). *If $\mathcal{X}$ is uniformly convex, then $\mathbb{P}(^n\mathcal{X}, \mathcal{Y})$ has the BPBp for every Banach space $\mathcal{Y}$.*

Before we discuss results on holomorphic functions, we introduce a function space with a base algebra which is formalized in [**71**] as follows.

DEFINITION 3.10. A closed subspace $\mathcal{F}(\Omega, \mathcal{X})$ of $\mathcal{C}_b(\Omega, \mathcal{X})$ is said to be a *function space with a base algebra* $\mathcal{F}$ if it satisfies the following three conditions:

(i) $\mathcal{F} = \{x^* \circ f : x^* \in \mathcal{X}^*, f \in \mathcal{F}(\Omega, \mathcal{X})\}$ is a function algebra on Ω,
(ii) $\mathcal{F} \otimes \mathcal{X} \subset \mathcal{F}(\Omega, \mathcal{X})$ and
(iii) $fg \in \mathcal{F}(\Omega, \mathcal{X})$ whenever $f \in \mathcal{F}$ and $g \in \mathcal{F}(\Omega, \mathcal{X})$.

DEFINITION 3.11. Let (Ω, d) be a metric space with distance d, $\mathcal{F}(\Omega, \mathcal{X})$ be a function space with a base algebra $\mathcal{F}$ on a Hausdorff space Ω and $\mathcal{G}$ be a subset of $\mathcal{F}(\Omega, \mathcal{X})$. $\mathcal{F}(\Omega, \mathcal{X})$ is said to have the *BPBp* (resp. *BPBp with* $\mathcal{G}$) if, given $\epsilon > 0$, there is $\eta > 0$ such that whenever $\omega_0 \in \Omega$ and $f \in \mathcal{F}(\Omega, \mathcal{X})$ satisfy $|f(\omega_0)| > 1 - \eta$, there exist $\omega \in \Omega$ and $g \in \mathcal{F}(\Omega, \mathcal{X})$ (resp. $g \in \mathcal{G}$) such that

$$\|g\| = \|g(\omega)\| = 1, \quad d(\omega, \omega_0) < \epsilon, \quad \|f - g\| < \epsilon.$$

We recall a Urysohn type lemma given by Cascales, Guirao and Kadets [**24**] for unital uniform algebras.

LEMMA 3.12 ([**24**]). *Let K be a compact Hausdorff space and $\mathcal{A}$ be a unital uniform algebra of $\mathcal{C}(K)$. Suppose that Γ_0 is the Choquet boundary of $\mathcal{A}$ and U is an open subset of K such that $U \cap \Gamma_0 \neq \emptyset$. Then, given $0 < \varepsilon < 1$, there is $\phi \in S_{\mathcal{A}}$ and $\omega_0 \in U \cap \Gamma_0$ such that*

$$\phi(\omega_0) = \|\phi\|_\infty = 1, \quad \sup_{\omega \in K \setminus U} |\phi(\omega)| < \varepsilon$$

and for every $\omega \in K$

$$|\phi(\omega)| + (1 - \varepsilon)|1 - \phi(\omega)| \leq 1.$$

This lemma was firstly obtained to study the BPBp for Asplund operators from a Banach space to a uniform algebra in [**24**]. Later, it was also used to get a result on the following unital uniform algebra.

$$\mathcal{A}_{w^*u}(B_{\mathcal{X}^*}) := \{f \in \mathcal{A}_b(B_{\mathcal{X}^*}, \mathbb{C}) : f \text{ is } w^*\text{-uniformly continuous}\}$$

THEOREM 3.13 ([**42**]). *For a Banach space $\mathcal{X}$, if $\rho\mathcal{A}_{w^*u}(B_{\mathcal{X}^*})$ is dense in $S_{\mathcal{X}^*}$, then $\mathcal{A}_{w^*u}(B_{\mathcal{X}^*})$ has the BPBp.*

In the same paper, it is commented that if $\mathcal{X}^*$ is isometric to ℓ_p $(1 \leq p < \infty)$, then $\rho\mathcal{A}_{w^*u}(B_{\mathcal{X}^*})$ is dense in $B_{\mathcal{X}^*}$. In its proof, the Urysohn type lemma plays a crucial role, and the Urysohn type lemma needs to be generalized for considering other function spaces like $\mathcal{A}_b$ and $\mathcal{A}_u$, because $B_{\mathcal{X}}$ is not compact in general.

THEOREM 3.14 ([**64**]). *For a Hausdorff space Ω, let $\mathcal{A}$ be a closed subalgebra of $\mathcal{C}_b(\Omega)$. If $U \subset \Omega$ is an open set having $\omega_0 \in U \cap \rho\mathcal{A}$, then, for $0 < \epsilon < 1$, there is a strong peak function $\phi \in \mathcal{A}$ such that $\|\phi\| = |\phi(\omega_0)| = 1$, $\sup_{\omega \in \Omega \setminus U} |\phi(\omega)| < \epsilon$ and $|\phi(\omega)| + (1 - \epsilon)|1 - \phi(\omega)| \leq 1$ for all $\omega \in \Omega$.*

This is applicable to get the BPBp of $\mathcal{A}(B_{\mathcal{X}})$ ([**64**]).

THEOREM 3.15 ([**64**]). *Let $\mathcal{X}$ and $\mathcal{Y}$ be Banach spaces. If $\rho\mathcal{A}(B_{\mathcal{X}})$ is norming for $\mathcal{A}(B_{\mathcal{X}})$, then $\mathcal{A}(B_{\mathcal{X}}, \mathcal{Y})$ has the BPBp.*

4. Pointwise Bishop-Phelps-Bollobás property

Very recently, several interesting variations of the BPBp were considered. In this section, we introduce the pointwise Bishop-Phelps-Bollobás property. For other variations and their applications to geometric properties of a Banach space, we refer the readers to [**41**, **45**, **81**, **82**].

DEFINITION 4.1 ([**44**]). For Banach spaces $\mathcal{X}$ and $\mathcal{Y}$, the space $\mathbb{B}(\mathcal{X},\mathcal{Y})$ is said to have the pointwise Bishop-Phelps-Bollobás property (pointwise BPBp) if for each $\epsilon > 0$, there exists $\eta > 0$ such that, whenever $A \in \mathbb{B}(\mathcal{X},\mathcal{Y})$ with $\|A\| = 1$ and $x \in S_{\mathcal{X}}$ satisfy $\|Ax\| > 1 - \eta$, there exists $B \in \mathbb{B}(\mathcal{X},\mathcal{Y})$ such that

$$\|B\| = \|Bx\| = 1 \text{ and } \|A - B\| < \epsilon.$$

The difference between the BPBp and the pointwise BPBp is that the pointwise BPBp insures to approximate the operator by ones which attain their norm at the same point where the given operator almost attains its norm. Clearly the pointwise BPBp implies the BPBp. In [**44**], it is found that the uniform smoothness can be characterized by this property.

THEOREM 4.2 ([**44**]). *For a Banach space $\mathcal{X}$, $\mathcal{X}$ is uniformly smooth if and only if $\mathbb{B}(\mathcal{X},\mathbb{K})$ has the pointwise BPBp.*

Moreover, it is also proved in the same paper that if $\mathbb{B}(\mathcal{X},\mathcal{Y})$ has the pointwise BPBp, then $\mathcal{X}$ is uniformly smooth. This means that the pointwise BPBp is more restrictive than the BPBp, but there are many examples having this property as follows.

EXAMPLE 4.3 ([**44**]).

(1) $\mathbb{B}(\mathcal{H},\mathcal{Y})$ has the pointwise BPBp for a Hilbert space $\mathcal{H}$ and a Banach space $\mathcal{Y}$.
(2) $\mathbb{B}(\mathcal{X},\mathcal{A})$ has the pointwise BPBp for a uniformly smooth space $\mathcal{X}$ and a unital uniform algebra $\mathcal{A}$.
(3) $\mathbb{B}(\mathcal{X},\mathcal{Y})$ has the pointwise BPBp for a uniformly smooth space $\mathcal{X}$ and $\mathcal{Y} = c_0$ or ℓ_∞.

It is remarkable that for $\mathcal{X} = L_p$ for $p \in (2,\infty)$ there is a space $\mathcal{Y}$ such that $\mathbb{B}(\mathcal{X},\mathcal{Y})$ does not have the pointwise BPBp ([**43**]).

The pointwise BPBp for n-linear mappings is also discussed, but there are not many known results. For example, $\mathbb{L}^2(\mathcal{H}_1 \times \mathcal{H}_2, \mathbb{K})$ has the pointwise BPBp for Hilbert spaces $\mathcal{H}_1$ and $\mathcal{H}_2$ [**44**]. On the other hand, we have interesting results in the setting of holomorphic functions.

DEFINITION 4.4. A function space $\mathcal{F}(\Omega,\mathcal{X})$ is said to have the *pointwise BPBp* (resp. *pointwise BPBp with* $\mathcal{G} \subset \mathcal{F}(\Omega,\mathcal{X})$) if, given $\epsilon > 0$, there is $\eta > 0$ such that whenever $\omega_0 \in \Omega$ and $f \in \mathcal{F}(\Omega,\mathcal{X})$ satisfy $|f(\omega_0)| > 1 - \eta$, there exists $g \in \mathcal{F}(\Omega,\mathcal{X})$ (resp. $g \in \mathcal{G}$) such that

$$\|g\| = \|g(\omega_0)\| = 1, \quad \|f - g\| < \epsilon.$$

THEOREM 4.5 ([**67**]). *Let $\mathcal{X}$ be a complex Banach space and let $\mathcal{F}(\Omega,\mathcal{X})$ be a function space with a base algebra $\mathcal{F}$. Then, given $\epsilon > 0$, whenever $f \in S_{\mathcal{F}(\Omega,\mathcal{X})}$ and $\omega_0 \in \rho\mathcal{F}$ satisfy $\|f(\omega_0)\| > 1 - \epsilon/5$, there is a norm-one peak function $g \in \mathcal{F}(\Omega,\mathcal{X})$ at ω_0 such that $\|f - g\| < \epsilon$. Moreover, if $\rho\mathcal{F}$ is norming for $\mathcal{F}$, then $\mathcal{F}(\Omega,\mathcal{X})$ has* *the pointwise BPBp with the set of strong peak functions and the set of strong peak functions is dense.*

It is easy to check that $\rho\mathcal{F}$ is a norming subset for $\mathcal{F}$ if and only if $\rho\mathcal{F}(\Omega, \mathcal{X})$ is norming for $\mathcal{F}(\Omega, \mathcal{X})$ for every complex Banach space $\mathcal{X}$. By the remark after Theorem 2.12, we get the following:

THEOREM 4.6 ([**64**]). *Assume that $\rho\mathcal{A}(B_{\mathcal{X}})$ is norming for $\mathcal{A}(B_{\mathcal{X}})$ for a complex Banach space $\mathcal{X}$. Then $\mathcal{A}(B_{\mathcal{X}}, \mathcal{Y})$ has the pointwise BPBp for every complex Banach space $\mathcal{Y}$. In particular, the assumption is satisfied if $\mathcal{X}$ is one of the following:*

(i) *$\mathcal{X}$ has the Radon-Nikodým property.*
(ii) *$\mathcal{X}$ is locally uniformly convex.*
(iii) *$\mathcal{X}$ is an order continuous, locally uniformly c-convex sequence space.*

It is easy to check that if $f \in \mathcal{F}(\Omega, \mathcal{X})$ is a norm-one strong peak function, then $x^* \circ f$ is a norm-one strong peak function in the base algebra $\mathcal{F}$ for some $x^* \in \mathcal{X}^*$. It is also easy to check that if the set of strong peak functions is dense in $\mathcal{F}$, then $\rho\mathcal{F}$ is norming [**67**]. So we get the following.

THEOREM 4.7. *Let $\mathcal{X}$ be a complex Banach space and let $\mathcal{F}(\Omega, \mathcal{X})$ be a function space with base algebra $\mathcal{F}$ on a Hausdorff space Ω. Then the following are equivalent:*

(i) *$\rho\mathcal{F}$ is norming for $\mathcal{F}$.*
(ii) *$\mathcal{F}(\Omega, \mathcal{X})$ has the pointwise BPBp with the set of strong peak functions.*
(iii) *$\mathcal{F}(\Omega, \mathcal{X})$ has the BPBp with the set of strong peak functions.*
(iv) *The set of strong peak functions in $\mathcal{F}(\Omega, \mathcal{X})$ is dense.*

5. Applications to the Daugavet property

Now we extend the previous results and demonstrate their applications to the Daugavet property of a function space with a base algebra. To generalize the Gelfand transform of a function algebra, we need the notion of compatibility.

A locally convex Hausdorff topology (LCH) τ on a Banach space $\mathcal{X}$ is said to be *compatible to the dual pair* $(\mathcal{X}, \mathcal{X}^*)$ if $(\mathcal{X}, \tau)^* = \mathcal{X}^*$ and a LCH topology τ on $\mathcal{X}^*$ is said to be *compatible to a dual pair* $(\mathcal{X}^*, \mathcal{X})$ if $(\mathcal{X}^*, \tau)^* = \mathcal{X}$. In this paper, we say that a LCH topology τ on $\mathcal{X}$ *is compatible to a dual pair* if $(\mathcal{X}, \tau)^* = \mathcal{X}^*$ or $(\mathcal{X}, \tau)^* = \mathcal{X}_*$, where $\mathcal{X}_*$ is a predual of $\mathcal{X}$.

By the Mackey-Arens theorem [**77**, Theorem 8.7.4], for any LCH topology τ compatible to a dual pair $(\mathcal{X}, \mathcal{X}^*)$ (resp. $(\mathcal{X}^*, \mathcal{X})$), we have $(\mathcal{X}, w) \subset \tau \subset (\mathcal{X}, \|\cdot\|)$ (resp. $(\mathcal{X}^*, w^*) \subset \tau \subset (\mathcal{X}^*, \|\cdot\|)$). Also for the topology τ compatible to a dual pair, τ-bounded sets are norm-bounded. We refer the reader to [**71**] for details.

Let $\mathcal{C}_b(\Omega, (\mathcal{X}, \tau))$ be the set of all τ-bounded continuous functions from a Hausdorff space Ω to $(\mathcal{X}, \tau)$, where $\mathcal{X}$ is a Banach space and τ is a locally convex Hausdorff topology compatible to a dual pair. Then it is a Banach space equipped with the supremum norm. Its proof is omitted, because it is very similar to that of [**71**, Proposition 2.1]. Then a closed subspace $\mathcal{F}(\Omega, (\mathcal{X}, \tau))$ is said to be a $(\mathcal{X}, \tau)$-*valued function space with a base algebra $\mathcal{F}$* if it satisfies the conditions in Definition 3.10. The Gelfand transform shows the following. Its proof is also omitted, because it is an easy extension of [**71**, Theorem 4.5].

THEOREM 5.1. *Let $\mathcal{X}$ be a Banach space and τ be a LCH topology compatible to a dual pair. Suppose that $\mathcal{F}(\Omega, (\mathcal{X}, \tau))$ is an $(\mathcal{X}, \tau)$-valued function space with a base* *algebra $\mathcal{F}$ on a Hausdorff space Ω. Then $\mathcal{F}(\Omega,(\mathcal{X},\tau))$ is isometrically isomorphic to an $((\mathcal{X},\tau)^{**},w^*)$-valued function space with the base algebra $\hat{\mathcal{F}}$, which is a uniform algebra on the compact Hausdorff (maximal ideal) space of $\mathcal{F}$ with the Gelfand topology. In particular, $\hat{\mathcal{F}}$ is the Gelfand transform of $\mathcal{F}$.*

THEOREM 5.2. *Let $\mathcal{X}$ be a Banach space and τ be a LCH topology compatible to a dual pair. If $\mathcal{F}(\Omega,(\mathcal{X},\tau))$ is an $\mathcal{X}$-valued function space with an infinite dimensional base algebra $\mathcal{F}$ on a Hausdorff space Ω, then it has the diameter 2 property (D2P). That is, every nonempty weak neighborhood of the unit ball of $\mathcal{F}(\Omega,(\mathcal{X},\tau))$ has diameter 2.*

It is worth to remark that in both Theorem 5.1 and 5.2, a function space with a base algebra only requires conditions (i) and (iii) in Definition 3.10. Note also that if $\mathcal{X}_*$ is a predual of a Banach space $\mathcal{X}$ and τ is a LCH topology such that $(\mathcal{X},\tau)^*=\mathcal{X}_*$, then $\mathcal{F}(\Omega,(\mathcal{X},\tau))$ is isometrically isomorphic to an $(\mathcal{X},w^*)$-valued function space with a uniform algebra over the maximal ideal space as its base algebra. Theorem 5.1 implies the diameter 2 property by [**71**, Theorem 4.2].

Recall that a Banach space $\mathcal{X}$ is said to have the *Daugavet property* if every weakly compact operator $T:\mathcal{X}\to\mathcal{X}$ satisfy the *Daugavet equation*

$$\|T+Id\|=\|T\|+1.$$

Due to a famous result of Daugavet [**46**], it is known that a compact operator $T:\mathcal{C}[0,1]\to\mathcal{C}[0,1]$ satisfies the Daugavet equation. Afterwards, this norm identity has been studied by many authors on various Banach spaces and various classes of operators, since it is related to important geometric properties such as the dentability of the closed unit ball of a Banach space (see [**58**,**59**] for instance). Motivated by this, recently, the polynomial Daugavet property was introduced in [**30**] by saying that a Banach space $\mathcal{X}$ has the *polynomial Daugavet property* if the Daugavet equation $\|P+Id\|=\|P\|+1$ holds for every weakly compact polynomial $P:\mathcal{X}\to\mathcal{X}$.

Let us comment that the Daugavet property on a uniform algebra has been studied by Wojtaszczyk [**89**]. In this section, we study these properties on a vector-valued function space using some new Urysohn type lemma. To do so, we introduce the following facts [**30**, Proposition 1.3 and Corollary 2.2].

LEMMA 5.3. *For a Banach space $\mathcal{X}$, the following are equivalent.*

(i) *$\mathcal{X}$ has the polynomial Daugavet property.*

(ii) *For every $P\in\mathbb{P}(\mathcal{X})$ with $\|P\|=1$, every $x_0\in S_{\mathcal{X}}$ and every $\epsilon>0$, there exist $z_0\in B_{\mathbb{C}}$ and $y\in B_{\mathcal{X}}$ such that*

$$\operatorname{Re} z_0P(y)>1-\epsilon,\quad \text{and}\quad \|x_0+z_0y\|>2-\epsilon.$$

(iii) *Every weakly compact $P\in\mathbb{P}(\mathcal{X},\mathcal{X})$ satisfies $\sup\operatorname{Re} W(P)=\|P\|$, where*

$$\operatorname{Re} W(P)=\{\operatorname{Re} x^*(P(x)):(x,x^*)\in S_{\mathcal{X}}\times S_{\mathcal{X}^*},\ x^*(x)=1\}.$$

It is proved that a uniform algebra has the Daugavet property if its Shilov boundary does not contain an isolated point [**89**]. In fact, such a uniform algebra has the polynomial Daugavet property ([**30**]). Recently, it was shown that given a compact Hausdorff space K and a complex Banach space $\mathcal{X}$ having the uniform convexity, a function space $\mathcal{F}(K,\mathcal{X})$ has the Daugavet property if and only if the Shilov boundary of its base algebra $\mathcal{F}$ does not have an isolated point ([**71**]). Moreover, for a compact Hausdorff space K and for a complex Banach space $\mathcal{X}$ the function space $\mathcal{F}(K, \mathcal{X})$ has the Daugavet property if and only if either $\mathcal{F}$ has the Daugavet property or $\mathcal{X}$ has the Daugavet property [**72**].

The following theorem is a direct application of the proof in [**29**] or [**72**] by applying the Urysohn type lemma for closed subalgebra and we omit their proofs.

THEOREM 5.4. *Let $\mathcal{X}$ be a complex Banach space. Suppose that $\mathcal{F}(\Omega, \mathcal{X})$ is a function space with a base algebra $\mathcal{F}$ on a Hausdorff space Ω and that either Ω is a compact Hausdorff space or $\rho\mathcal{F}$ is norming for $\mathcal{F}$. If the Shilov boundary of $\mathcal{F}$ has no isolated points, then $\mathcal{F}(\Omega, \mathcal{X})$ has the polynomial Daugavet property.*

If we apply the argument given in [**72**, Corollary 4.10], we get the following.

THEOREM 5.5. *Suppose that $\mathcal{F}(\Omega)$ is a function algebra on a Hausdorff space Ω and that either Ω is a compact Hausdorff space or $\rho\mathcal{F}$ is norming for $\mathcal{F}$. Then $\mathcal{F}(\Omega)$ has the polynomial Daugavet property if and only if its Shilov boundary has no isolated points.*

The connectedness of the spectrum can be used to distinguish the spaces with the Daugavet property. Indeed, as shown in the following theorem, if a unital uniform algebra does not have the Daugavet property, it contains an element whose spectrum is $\{0, 1\}$, which is disconnected.

THEOREM 5.6. *$\mathcal{A}(B_{\mathcal{X}})$ has the Daugavet property for every nontrivial complex Banach space $\mathcal{X}$.*

PROOF. Given an element ξ in a unital Banach algebra, the spectrum of ξ is denoted by $\sigma(\xi)$. Let $\mathcal{A} = \mathcal{A}(B_{\mathcal{X}})$ and $\Psi : \mathcal{A} \to \mathcal{C}(M)$ be the Gelfand transform, where M is the maximal ideal space of $\mathcal{A}$. For simplicity, we write $\hat{f} = \Psi(f)$ for each $f \in \mathcal{A}$ and let $\hat{\mathcal{A}} = \Psi(\mathcal{A})$. Note that $\sigma(f) = \sigma(\hat{f})$ for each $f \in \mathcal{A}$, since Ψ is an algebraic isometric isomorphism. Then $\sigma(f) = \overline{f(B_{\mathcal{X}})}$. Indeed, it is clear that $f(B_{\mathcal{X}}) \subset \sigma(f)$, and thus $\overline{f(B_{\mathcal{X}})} \subset \sigma(f)$ since $\sigma(f)$ is compact. On the other hand, if $\lambda \in \mathbb{C} \setminus \overline{f(B_{\mathcal{X}})}$, then

$$\delta = \inf\{|f(x) - \lambda| : x \in B_{\mathcal{X}}\} > 0.$$

Hence $g(x) = \frac{1}{f(x)-\lambda}$ is bounded, continuous on $B_{\mathcal{X}}$ and holomorphic on $B_{\mathcal{X}}^{\circ}$. If f is uniformly continuous on $B_{\mathcal{X}}$, g is also uniformly continuous. So $\lambda \notin \sigma(f)$ and hence $\sigma(f) = \overline{f(B_{\mathcal{X}})}$.

For each $f \in \mathcal{A}$, we get $\sigma(\hat{f}) = \sigma(f) = \overline{f(B_{\mathcal{X}})}$ and $\sigma(f)$ is connected.

Suppose that $\mathcal{A}$ does not have the Daugavet property. Then neither does $\hat{\mathcal{A}}$. So its Shilov boundary Γ has an isolated point t_0 by [**71**, Corollary 5.7] and there is a function $\hat{g} \in \hat{\mathcal{A}}$ such that $\hat{g}(t_0) = 1$ and $\hat{g}(\Gamma \setminus \{t_0\}) = \{0\}$ by [**71**, Lemma 2.5]. Since $\mathcal{A}$ is not one-dimensional, $\Gamma \setminus \{t_0\}$ is not empty.

Note that $\hat{\mathcal{A}}$ is algebraically isometrically isomorphic to

$$\hat{\mathcal{A}}|_\Gamma = \{\hat{f}|_\Gamma : \hat{f} \in \hat{\mathcal{A}}\},$$

where $\hat{f}|_\Gamma$ is the restriction of $\hat{f}$ to Γ. Hence $\sigma(\hat{g}|_\Gamma) = \sigma(\hat{g}) = \sigma(g)$ is connected. It is clear that $\sigma(\hat{g}|_\Gamma)$ contains $\hat{g}(\Gamma) = \{0, 1\}$. Since $(\hat{g}|_\Gamma)^2 = \hat{g}|_\Gamma$, $\sigma(\hat{g}|_\Gamma) = \{0, 1\}$ by the spectral mapping theorem. This is a contradiction. □

A similar proof shows that if Ω is a connected Hausdorff space containing more than one point, then $\mathcal{C}_b(\Omega)$ has the Daugavet property, and also that the space $H^\infty(V)$ of bounded holomorphic functions on V, equipped with the supremum norm, has the Daugavet property if V is a nonempty open connected subset of a complex Banach space. Since the Shilov boundary of $\mathcal{C}([0,1]\cup[2,3])$ is $[0,1]\cup[2,3]$, it has the Daugavet property [**89**], so the domain is not necessarily connected in order to have the Daugavet property.

For the vector-valued case, we get the following by Theorem 5.4, 5.5 and 5.6.

THEOREM 5.7. *Suppose that a complex Banach space $\mathcal{X}$ satisfies one of the following:*

(i) *$\mathcal{X}$ has the Radon-Nikodým property.*
(ii) *$\mathcal{X}$ is locally uniformly convex.*
(iii) *$\mathcal{X}$ is an order continuous, locally uniformly c-convex sequence space.*

Then $\mathcal{A}(B_{\mathcal{X}},\mathcal{Y})$ has the Daugavet property for every nontrivial complex Banach space $\mathcal{Y}$.

THEOREM 5.8. *Let $\mathcal{F}(\Omega,\mathcal{X})$ be an $\mathcal{X}$-valued function space on a Hausdorff space Ω and suppose that each $f\in\mathcal{F}(\Omega,\mathcal{X})$ has a precompact range in $\mathcal{X}$. Then $\mathcal{F}(\Omega,\mathcal{X})$ is isometrically isomorphic to an $\mathcal{X}$-valued function space $\hat{\mathcal{F}}(\Delta,\mathcal{X})$ on a compact Hausdorff space Δ such that the base algebra $\{x^*\circ f : f\in\hat{\mathcal{F}}, x^*\in\mathcal{X}^*\}$ is a uniform algebra on Δ. In particular, Δ can be chosen to be the maximal ideal space of the base algebra of $\mathcal{F}$.*

PROOF. Let $\mathcal{F}$ be a function algebra and Δ be the set of all nonzero algebra homomorphisms from $\mathcal{F}$ to $\mathbb{C}$. The maximal ideal space Δ is a compact Hausdorff space with the Gelfand topology. The Gelfand transform $\hat{f}$ of $f\in\mathcal{F}$ is defined by $\hat{f}(\phi)=\phi(f)$ for $\phi\in\Delta$. Given $f\in\mathcal{F}(\Omega,\mathcal{X})$, define $\hat{f}(\phi)(x^*)=\phi(x^*\circ f)$ for $x^*\in\mathcal{X}^*$ and $\phi\in\Delta$. Then $|\hat{f}(\phi)(x^*)|\le\|f\|\|x^*\|$ shows that $\hat{f}(\phi)$ is in $\mathcal{X}^{**}$. So $\hat{f}:\Delta\to\mathcal{X}^{**}$ is a well-defined continuous function if we consider the weak-$*$ topology in $\mathcal{X}^{**}$. Let $\hat{\mathcal{F}}$ be the image of this linear and isometric mapping, which is a subspace of $\mathcal{C}_b(\Delta,(\mathcal{X}^{**},w^*))$. Let $\tilde{\mathcal{F}}=\{x^*\circ\hat{f} : f\in\mathcal{F},\ x^*\in\mathcal{X}^*\}$. From the definition, we have that $\widehat{x^*\circ f}=x^*\circ\hat{f}$ and

$$\{\widehat{x^*\circ f} : f\in\mathcal{F},\ x^*\in\mathcal{X}^*\}=\tilde{\mathcal{F}}.$$

Let $\tilde{\Omega}=\{z\delta_t : t\in\Omega, z\in B_{\mathbb{C}}\}$ in $\mathcal{F}^*$ and we claim that the weak-$*$ closure of the convex hull of $\tilde{\Omega}$ contains Δ. Indeed, if not, there exist $\phi\in\Delta$, $f\in\mathcal{F}$ and real numbers $r<s$ such that

$$\text{Re } z\delta_t(f)=\text{Re } zf(t)\le r<s\le\text{Re }\phi(f)$$

for all $t\in\Omega$ and for all $z\in B_{\mathbb{C}}$. So $\|f\|\le r<s\le\phi(f)\le\|f\|$ and this is a contradiction.

We may assume that the range of f is contained in a balanced compact convex subset K_f of $\mathcal{X}$. For each ϕ, there is a net $(\phi_\alpha)_\alpha$ convergent to ϕ in Gelfand topology such that each $\phi_\alpha\in\text{conv}(\tilde{\Omega})$ has the expression $\phi_\alpha=\sum_{j=1}^n\lambda_j z_j\delta_{t_j}$, where $\lambda_j\ge 0$, $z_j\in B_{\mathbb{C}}$ $(j=1,\dots,n)$ and $\sum_j\lambda_j=1$. Since $\hat{f}(\phi_\alpha)\in K_f$ for each α, we have $\hat{f}(\phi)\in K_f$. Hence $\hat{f}(\Delta)\subset K_f$ and the relative weak-$*$ and the relative norm topologies are the same in K_f. So $\hat{f}:\Delta\to K_f$ is continuous with respect to the relative norm topology on K_f in $\mathcal{X}$.

Let $\hat{\mathcal{F}}(\Delta,\mathcal{X})=\{\hat{f} : f\in\mathcal{F}(\Omega,\mathcal{X})\}$. We claim that $\hat{\mathcal{F}}(\Delta,\mathcal{X})$ is a function space with a base algebra $\hat{\mathcal{F}}$ in Definition 3.10: (i) is already shown. If $f\in\mathcal{F}$ and $x\in\mathcal{X}$, then $\widehat{f \otimes x} = \hat{f} \otimes x$. Hence $\hat{\mathcal{F}} \otimes \mathcal{X} \subset \mathcal{F}(\Delta, \mathcal{X})$. So (ii) is proved. For (iii), note that $\widehat{fg} = \hat{f}\hat{g}$ for every $f \in \mathcal{A}(\Omega, \mathcal{X})$ and for every $g \in \mathcal{A}$. The proof is complete. □

THEOREM 5.9. *Let $\mathcal{X}, \mathcal{Y}$ be nontrivial complex Banach spaces. Let $\mathcal{A}_K(B_{\mathcal{X}}, \mathcal{Y})$ be the closed subspace of $\mathcal{A}(B_{\mathcal{X}}, \mathcal{Y})$ consisting of all $f \in \mathcal{A}(B_{\mathcal{X}}, \mathcal{Y})$ whose image is precompact. Then $\mathcal{A}_K(B_{\mathcal{X}}, \mathcal{Y})$ has the polynomial Daugavet property.*

PROOF. By Theorem 5.8, $\mathcal{A}_K(B_{\mathcal{X}}, \mathcal{Y})$ is isometrically isomorphic to a function space $\hat{\mathcal{A}}(\Delta, \mathcal{Y})$ with the base algebra $\hat{\mathcal{A}}$, which is the Gelfand transform of $\mathcal{A}(B_{\mathcal{X}})$. So $\hat{\mathcal{A}}$ has the polynomial Daugavet property by Theorem 5.6. So its Shilov boundary does not contain an isolated point. Hence, by Theorem 5.5, $\hat{\mathcal{A}}(\Delta, \mathcal{Y})$ has the polynomial Daugavet property. This means that $\mathcal{A}_K(B_{\mathcal{X}}, \mathcal{Y})$ has the polynomial Daugavet property. □

It is natural to ask if $\mathcal{A}(B_{\mathcal{X}}, \mathcal{Y})$ has the Daugavet property for every complex Banach space $\mathcal{Y}$, which remains unknown to us.

PROBLEM 1. Let $\mathcal{X}$ be a complex Banach space. Then does $\mathcal{A}(B_{\mathcal{X}}, \mathcal{Y})$ have the Daugavet property for every complex Banach space $\mathcal{Y}$?

In the proof of Theorem 5.6, the Gelfand transform plays an important role. So Theorem 5.1 and Theorem 5.4 suggests the following interesting question.

PROBLEM 2. Let $\mathcal{X}$ be a complex Banach space and τ be a LCH topology compatible to a dual pair. Suppose that $\mathcal{F}(\Omega, (\mathcal{X}, \tau))$ be a function space with a base algebra $\mathcal{F}$ on a Hausdorff space Ω and that either Ω is a compact Hausdorff space or $\rho\mathcal{F}$ is norming for $\mathcal{F}$. If the Shilov boundary of $\mathcal{F}$ has no isolated points, then does $\mathcal{F}(\Omega, (\mathcal{X}, \tau))$ have the polynomial Daugavet property?

Note that if Problem 2 has an positive answer, then Theorem 5.1 shows that we may assume Ω is compact. Since $\mathcal{F}$ has the Daugavet property, its Shilov boundary does not have any isolated points, so we get also a positive answer to the following problem and Problem 1.

PROBLEM 3. Let $\mathcal{X}$ be a complex Banach space and τ be a LCH topology compatible to a dual pair. Suppose that $\mathcal{F}(\Omega, (\mathcal{X}, \tau))$ is a function space with a base algebra $\mathcal{F}$ on a Hausdorff space Ω. If $\mathcal{F}$ has the Daugavet property, then does $\mathcal{F}(\Omega, (\mathcal{X}, \tau))$ have the Daugavet property?

6. The numerical radius attainment and its applications to the numerical index

In this section, we focus on the numerical radius-attaining mappings and their related topics. The numerical radius for a linear operator is the supremum modulus of the numerical range and it is a seminorm on the space of linear operators. We give their definitions in a more general setting for k-homogeneous polynomials of [**27**].

For Banach space $\mathcal{X}$, define

$$\Pi(\mathcal{X}) = \{(x, x^*) \in \mathcal{X} \times \mathcal{X}^* \ : \ \|x\| = \|x^*\| = 1 = x^*(x)\}.$$

The *spatial numerical range* of $P \in \mathbb{P}({}^k\mathcal{X}, \mathcal{X})$ is defined by

$$W(P) = \{x^*(Px) : (x, x^*) \in \Pi(\mathcal{X})\}$$

The *numerical radius* of P is defined by

$$v(P) = \sup\{|\lambda| : \lambda \in W(P)\}.$$

The *k-homogeneous polynomial numerical index* $n^{(k)}(\mathcal{X})$ is defined in [**27**] by

$$n^{(k)}(\mathcal{X}) = \inf\{v(P) : P \in \mathbb{P}(^k\mathcal{X}, \mathcal{X}), \|P\| = 1\}.$$

The numerical range was first considered by O. Toeplitz [**87**] for matrices, and generalized later to the Banach space setting by G. Lumer [**75**] and F. Bauer [**16**]. It is worth to note that the numerical range of an operator is closely related to its spectrum, and the numerical index is related to the geometry of the Banach space. These concepts had been studied intensively by many researchers, see [**58**] for the excellent survey about this topic.

Similarly to norm-attaining mappings, a numerical radius-attaining mapping is a fruitful research topic. A polynomial $P \in \mathbb{P}(^k\mathcal{X}, \mathcal{X})$ is said to be *numerical radius-attaining* if there exists $(x, x^*) \in \Pi(\mathcal{X})$ such that $v(P) = |x^*(Px)|$. To denote the set of numerical radius-attaining mappings we put $\mathbb{NRA}$ in front of the space of mappings. For example, $\mathbb{NRAP}(^k\mathcal{X}, \mathcal{X})$ is the set of numerical radius-attaining k-homogeneous polynomials.

As it is shown for norm-attaining mappings, some geometric conditions of a Banach space imply the denseness of numerical radius-attaining mappings. We introduce some notable results. This problem was firstly considered by Berg and Sims in [**17**] and they found that the uniform convexity implies the denseness of numerical radius-attaining operators. This was generalized by Acosta and Payá for reflexive spaces in [**9**] and Radon-Nikodým property in [**10**]. More generally, the same statement for polynomials is also true.

THEOREM 6.1. [**8**] *Let $\mathcal{X}$ be a Banach space. If $\mathcal{X}$ has the Radon-Nikodým property, then* $\overline{\mathbb{NRAP}(^k\mathcal{X}, \mathcal{X})} = \mathbb{P}(^k\mathcal{X}, \mathcal{X})$.

It is also shown for the Lebesgue spaces and the Banach space of continuous functions on a compact Hausdorff space.

EXAMPLE 6.2.

(1) [**22**] $\overline{\mathbb{NRAB}(L_1(\mu), L_1(\mu))} = \mathbb{B}(L_1(\mu), L_1(\mu))$ for a finite positive measure μ.
(2) [**17**] $\overline{\mathbb{NRAB}(L_p(\mu), L_p(\mu))} = \mathbb{B}(L_p(\mu), L_p(\mu))$ for an arbitrary measure μ and $1 < p < \infty$.
(3) [**23**] $\overline{\mathbb{NRAB}(\mathcal{C}(K), \mathcal{C}(K))} = \mathbb{B}(\mathcal{C}(K), \mathcal{C}(K))$ for a compact Hausdorff space K when the scalar field is $\mathbb{R}$.

In [**8**], it is shown that the set of numerical radius-attaining elements in $\mathbb{P}(^2d_*(w, 1), d_*(w, 1))$ is not dense. For more results on the denseness of numerical radius-attaining polynomials (or n-linear mappings) we refer to [**28, 33**] for the readers.

Harris [**54**] introduced the notion of the numerical radius for holomorphic functions. Similarly to k-homogeneous polynomials we can define the following notions for holomorphic functions.

The *spatial numerical range* of $f \in \mathcal{C}_b(B_{\mathcal{X}}, \mathcal{X})$ is defined by

$$W(f) = \{x^*(f(x)) : (x, x^*) \in \Pi(\mathcal{X})\},$$

and the *numerical radius* of f is defined by

$$v(f) = \sup\{|\lambda| : \lambda \in W(f)\}.$$

For a complex Banach space $\mathcal{X}$, *b-analytic numerical index* $n_{ba}(\mathcal{X})$ and *u-analytic index* $n_{ua}(\mathcal{X})$ are defined by, respectively,

$$n_{ba}(\mathcal{X}) = \inf\{v(f) : f \in \mathcal{A}_b(B_{\mathcal{X}} : \mathcal{X}), \|f\| = 1\},$$

$$n_{ua}(\mathcal{X}) = \inf\{v(f) : f \in \mathcal{A}_u(B_{\mathcal{X}} : \mathcal{X}), \|f\| = 1\}.$$

It is easy to check that $0 \leq n^{(k+1)}(\mathcal{X}) \leq n^{(k)}(\mathcal{X}) \leq 1$ for every $k \in \mathbb{N}$ and for every Banach space $\mathcal{X}$. If $n^{(k)}(\mathcal{X}) > 0$, then $v(\cdot)$ is an equivalent norm in $\mathbb{P}(^k\mathcal{X}, \mathcal{X})$. It is also clear from the definitions that $0 \leq n_{ba}(\mathcal{X}) \leq n_{ua}(\mathcal{X}) \leq n^{(k)}(\mathcal{X}) \leq 1$ for all $k \geq 1$ and for every complex Banach space $\mathcal{X}$.

Harris [**54**] showed that, for a complex Banach space, the numerical radius is an equivalent norm in the space of k-homogeneous polynomials. More precisely, if $\mathcal{X}$ is a complex Banach space and $k \geq 2$, then

$$n^{(k)}(\mathcal{X}) \geq \exp\left(\frac{k \log k}{k-1}\right).$$

It was also proved in [**54**] that the above inequalities are sharp. In the real case, Harris' result above is false since the polynomial numerical indices of a real space may vanish as in the Example 8.b [**58**]. For more information about polynomial numerical indices, we refer the reader to the survey paper [**58**],

Analogously, a numerical radius-attaining holomorphic function can be defined as follows.

DEFINITION 6.3. Let $\mathcal{X}$ be a Banach space. A function $f \in \mathcal{C}_b(B_{\mathcal{X}}, \mathcal{X})$ is said to be *numerical radius-attaining* if $v(f) = |x^* f(x)|$ for some $(x, x^*) \in \Pi(\mathcal{X})$.

Similarly, we attach $\mathbb{NRA}$ in front of the space of mappings to denote the set of numerical radius-attaining functions. Acosta and Kim [**8**] also investigated the denseness of them, which influenced further studies in [**37, 60, 67**]. As is shown for polynomials, it is proved that the Radon-Nikodým property implies the denseness for holomorphic functions.

THEOREM 6.4 ([**8**]). *For a complex Banach space* $\mathcal{X}$,

(1) *if* $\mathcal{X}$ *has the Radon-Nikodým property, then* $\overline{\mathbb{NRA}\mathcal{A}(B_{\mathcal{X}}, \mathcal{X})} = \mathcal{A}(B_{\mathcal{X}}, \mathcal{X})$.
(2) *if* $\mathcal{X} = c_0$, *then* $\overline{\mathbb{NRA}\mathcal{A}(B_{\mathcal{X}}, \mathcal{X})} = \mathcal{A}(B_{\mathcal{X}}, \mathcal{X})$.
(3) *if* $\mathcal{X} = \mathcal{C}(K)$ *where* K *is a compact Hausdorff space, then* $\overline{\mathbb{NRA}\mathcal{A}_b(B_{\mathcal{X}}, \mathcal{X})} = \mathcal{A}_b(B_{\mathcal{X}}, \mathcal{X})$.

The main topic of this section is to present the applications of the norm and numerical radius-attainment to the numerical index. We recall some known facts for that. It is shown in [**29**] that $n^{(k)}(\mathcal{A}) = 1$ and $n_{ua}(\mathcal{A}) = 1$ for a unital uniform algebra $\mathcal{A}$. In fact, their proof can be easily modified to show that $n_{ba}(\mathcal{A}) = 1$. In general, $n_{ba}(\mathcal{A}) = 1$ holds for every function algebra $\mathcal{A}$. Indeed, as observed in [**67**], it is not difficult to see that if $\mathcal{A}$ is a unital function algebra on a Hausdorff space, then, by the Gelfand transform, $\mathcal{A}$ is isometric to a unital uniform algebra on a compact Hausdorff space Δ where Δ can be chosen to be the maximal ideal space of $\mathcal{A}$. This gives a positive answer to the third question in [**8**], where it is asked if $n_{ba}(\mathcal{C}_0(\Omega)) = 1$ for a noncompact space Ω.

A characterization of a Banach space with its polynomial (analytic) numerical index 1 was obtained [**70**] when a Banach space has the Radon-Nikodým property. Its proof is closely related to the denseness of strongly norm-attaining polynomials on Banach spaces. We generalize the notion introduced by Ferrera [**51**] for the k-homogeneous polynomials.

DEFINITION 6.5. Let $\mathcal{X}$ and $\mathcal{Y}$ be Banach spaces over a scalar field $\mathbb{K}$. A nontrivial function $f \in \mathcal{C}_b(B_\mathcal{X}, \mathcal{Y})$ is said to *strongly attain its norm* at x_0 if whenever $\lim_n \|f(x_n)\| = \|f\|$ for a sequence $\{x_n\}_n$ in $B_\mathcal{X}$, it has a subsequence $\{x_{n_k}\}$ which is converging to αx_0 for some $|\alpha| = 1$, $\alpha \in \mathbb{K}$. Such a point x_0 is said to be *strongly norm-attained* by f. Let H be a subspace of $\mathcal{C}_b(B_\mathcal{X}, \mathcal{Y})$. We denote by $\tilde{\rho}H$ the set of all strongly norm-attained points in $B_\mathcal{X}$ by an element in H. That is,

$$\tilde{\rho}H = \{x_0 \in B_\mathcal{X} : \exists f \in H \ \text{ strongly attains its norm at } \ x_0\}.$$

It was shown in [**70**] that if H is a closed subspace of $\mathcal{A}(B_\mathcal{X}, \mathcal{Y})$ for nontrivial complex Banach spaces $\mathcal{X}$ and $\mathcal{Y}$, then

$$\tilde{\rho}H \subset \tilde{\rho}\mathcal{A}(B_\mathcal{X}) = \rho\mathcal{A}(B_\mathcal{X}) \subset \operatorname{Ext}_\mathbb{C}(B_\mathcal{X}),$$

where $\operatorname{Ext}_\mathbb{C}(B_\mathcal{X})$ is the set of all complex extreme points of $B_\mathcal{X}$. Note also that $\rho\mathcal{A}(B_\mathcal{X}) = \operatorname{Ext}_\mathbb{C}(B_\mathcal{X})$ for every finite dimensional complex Banach space $\mathcal{X}$ ([**32**]).

For a real or complex Banach space $\mathcal{X}$ and for a nontrivial Banach space $\mathcal{Y}$, it is easy to check that

$$\operatorname{Sep}(B_\mathcal{X}) \subset \tilde{\rho}\mathbb{P}(^k\mathcal{X}) = \tilde{\rho}\mathbb{P}(^k\mathcal{X}, \mathcal{Y}),$$

where $\operatorname{Sep}(B_\mathcal{X})$ is the set of all strongly exposed points of $B_\mathcal{X}$. It was shown in [**37, 51**] that if $\mathcal{X}$ is a nontrivial Banach space with the Radon-Nikodým property, the set of strongly norm-attaining polynomials in $P(^k\mathcal{X}, \mathcal{Y})$ is dense in $P(^k\mathcal{X}, \mathcal{Y})$ for every nontrivial Banach space $\mathcal{Y}$ and for every natural number k.

THEOREM 6.6 ([**70**]). *Suppose that k is a positive integer and $\mathcal{X}$ is a nontrivial real or complex Banach space with the Radon-Nikodým property. The following are equivalent:*

(i) $n^{(k)}(\mathcal{X}) = 1$;

(ii) $|x^*(x)| = 1$ *for each* $x \in \tilde{\rho}\mathbb{P}(^k\mathcal{X})$ *and* $x^* \in \operatorname{Ext}(B_{\mathcal{X}^*})$,

where $\operatorname{Ext}(B_{\mathcal{X}^*})$ *is the set of all extreme point of* $B_{\mathcal{X}^*}$. *In the case where $\mathcal{X}$ is a finite dimensional real Banach space, we have* $\tilde{\rho}\mathbb{P}(^2\mathcal{X}) = S_\mathcal{X}$, *and* $n^{(2)}(\mathcal{X}) = 1$ *if and only if $\mathcal{X}$ is one-dimensional.*

Theorem 6.6 is an extension of the result of McGregor [**76**] who characterized the finite dimensional Banach spaces with numerical index 1.

THEOREM 6.7 ([**70**]). *Suppose that $\mathcal{X}$ is a nontrivial complex Banach space with the Radon-Nikodým property. The following are equivalent:*

(i) $n_{ua}(\mathcal{X}) = 1$;

(ii) *For each* $x \in \rho A_u(B_\mathcal{X})$ *and* $x^* \in \operatorname{Ext}(B_{\mathcal{X}^*})$, *we have* $|x^*(x)| = 1$.

As it is done for norms, the notions of a numerical (strong) peak function and a numerical Shilov boundary were introduced in [**61, 62**].

DEFINITION 6.8. Let $\mathcal{X}$ be a Banach space and $f \in \mathcal{C}_b(B_\mathcal{X}, \mathcal{X})$ is said to be a *numerical peak function* at $(x_0, x_0^*) \in \Pi(\mathcal{X})$ if $v(f) = |x_0^* f(x_0)|$ and $v(f) > |y^* f(y)|$ for every $(y, y^*) \in \Pi(\mathcal{X}) \setminus \{(x_0, x_0^*)\}$. A numerical peak function f at (x_0, x_0^*) is said to be *strong numerical peak function* if whenever $(x_n, x_n^*) \in \Pi(\mathcal{X})$ is a sequence satisfying $v(f) = \lim_n |x_n^* f(x_n)|$, we have $\lim_n \|x_n - x_0\| = 0$ and $\{x_n^*\}$ converges weak-$*$ to x_0^*.

THEOREM 6.9 ([**62**]). *Let $\mathcal{X}$ be a complex Banach space. Then the set of all numerical strong peak functions in $\mathcal{A}(B_{\mathcal{X}}, \mathcal{X})$ is dense if $\mathcal{X}$ satisfies one of the following:*

(i) *$\mathcal{X}$ is finite dimensional;*
(ii) *$\mathcal{X}$ is smooth and has the Radon-Nikodým property;*
(iii) *$\mathcal{X} = \ell_1$.*

Consider the product topology τ in $S_{\mathcal{X}} \times (S_{\mathcal{X}^*}, w^*)$. The τ-closed subset F is called a *numerical boundary* of a closed subspace H of $\mathcal{C}_b(B_{\mathcal{X}}, \mathcal{X})$ if

$$v(f) = \sup\{|x^* f(x)| : (x, x^*) \in F\}$$

for all $f \in H$. If the intersection of all numerical boundaries of H is the numerical boundaries of H, the intersection is called the *numerical Shilov boundary* of H. It is shown in [**62**] that if the set of numerical strong peak functions in H is dense, then the τ-closure of the numerical strong peak points is the numerical Shilov boundary of H.

There are some negative results about the denseness. If K is an infinite compact Hausdorff space, then there is no numerical strong peak function in $\mathcal{A}(B_{\mathcal{C}(K)}, \mathcal{C}(K))$ ([**61**]). It was also shown in [**62**] that if Ω is a locally compact Hausdorff space with more than two elements, there is no numerical strong peak function in $\mathcal{A}(B_{\mathcal{C}_0(\Omega)}, \mathcal{C}_0(\Omega))$ ([**62**]).

Acknowledgment

The authors express their gratitude to the anonymous referee for a thorough review of the paper.

References

[1] M. D. Acosta, *The Bishop-Phelps-Bollobás property for operators on $\mathcal{C}(K)$*, Banach J. Math. Anal. **10** (2016), no. 2, 307–319, DOI 10.1215/17358787-3492875. MR3474841

[2] M. D. Acosta, F. J. Aguirre, and R. Payá, *A new sufficient condition for the denseness of norm attaining operators*, Rocky Mountain J. Math. **26** (1996), no. 2, 407–418, DOI 10.1216/rmjm/1181072066. MR1406488

[3] M. D. Acosta, J. Alaminos, D. García, and M. Maestre, *On holomorphic functions attaining their norms*, J. Math. Anal. Appl. **297** (2004), no. 2, 625–644, DOI 10.1016/j.jmaa.2004.04.010. Special issue dedicated to John Horváth. MR2088685

[4] M. D. Acosta, R. M. Aron, D. García, and M. Maestre, *The Bishop-Phelps-Bollobás theorem for operators*, J. Funct. Anal. **254** (2008), no. 11, 2780–2799, DOI 10.1016/j.jfa.2008.02.014. MR2414220

[5] M. D. Acosta, J. Becerra-Guerrero, Y. S. Choi, M. Ciesielski, S. K. Kim, H. J. Lee, M. L. Lourenço, and M. Martín, *The Bishop-Phelps-Bollobás property for operators between spaces of continuous functions*, Nonlinear Anal. **95** (2014), 323–332, DOI 10.1016/j.na.2013.09.011. MR3130526

[6] M. D. Acosta, J. Becerra-Guerrero, Y. S. Choi, D. García, S. K. Kim, H. J. Lee, and M. Maestre, *The Bishop-Phelps-Bollobás property for bilinear forms and polynomials*, J. Math. Soc. Japan **66** (2014), no. 3, 957–979, DOI 10.2969/jmsj/06630957. MR3238324

[7] M. D. Acosta, J. Becerra-Guerrero, D. García, and M. Maestre, *The Bishop-Phelps-Bollobás theorem for bilinear forms*, Trans. Amer. Math. Soc. **365** (2013), no. 11, 5911–5932, DOI 10.1090/S0002-9947-2013-05881-3. MR3091270

[8] M. D. Acosta and S. G. Kim, *Denseness of holomorphic functions attaining their numerical radii*, Israel J. Math. **161** (2007), 373–386, DOI 10.1007/s11856-007-0083-x. MR2350167

[9] M. D. Acosta and R. Paya, *Denseness of operators whose second adjoints attain their numerical radii*, Proc. Amer. Math. Soc. **105** (1989), no. 1, 97–101, DOI 10.2307/2046740. MR937841

[10] M. D. Acosta and R. Payá, *Numerical radius attaining operators and the Radon-Nikodým property*, Bull. London Math. Soc. **25** (1993), no. 1, 67–73, DOI 10.1112/blms/25.1.67. MR1190367

[11] J. Alaminos, Y. S. Choi, S. G. Kim, and R. Payá, *Norm attaining bilinear forms on spaces of continuous functions*, Glasgow Math. J. **40** (1998), no. 3, 359–365, DOI 10.1017/S0017089500032717. MR1660038

[12] E. L. Arenson, *Gleason parts and the Choquet boundary of the algebra of functions on a convex compactum* (Russian, with English summary), Zap. Nauchn. Sem. Leningrad. Otdel. Mat. Inst. Steklov. (LOMI) **113** (1981), 204–207, 268. Investigations on linear operators and the theory of functions, XI. MR629841

[13] R. M. Aron, Y. S. Choi, D. García, and M. Maestre, *The Bishop-Phelps-Bollobás theorem for* $\mathcal{L}(L_1(\mu), L_\infty[0,1])$, Adv. Math. **228** (2011), no. 1, 617–628, DOI 10.1016/j.aim.2011.05.023. MR2822241

[14] R. M. Aron, Y. S. Choi, M. L. Lourenço, and O. W. Paques, *Boundaries for algebras of analytic functions on infinite-dimensional Banach spaces*, Banach spaces (Mérida, 1992), Contemp. Math., vol. 144, Amer. Math. Soc., Providence, RI, 1993, pp. 15–22, DOI 10.1090/conm/144/1209443. MR1209443

[15] R. M. Aron, C. Finet, and E. Werner, *Some remarks on norm-attaining n-linear forms*, Function spaces (Edwardsville, IL, 1994), Lecture Notes in Pure and Appl. Math., vol. 172, Dekker, New York, 1995, pp. 19–28. MR1352216

[16] F. L. Bauer, *On the field of values subordinate to a norm*, Numer. Math. **4** (1962), 103–113, DOI 10.1007/BF01386300. MR145329

[17] I. D. Berg and B. Sims, *Denseness of operators which attain their numerical radius*, J. Austral. Math. Soc. Ser. A **36** (1984), no. 1, 130–133. MR720006

[18] E. Bishop, *A minimal boundary for function algebras*, Pacific J. Math. **9** (1959), 629–642. MR109305

[19] E. Bishop and R. R. Phelps, *A proof that every Banach space is subreflexive*, Bull. Amer. Math. Soc. **67** (1961), 97–98, DOI 10.1090/S0002-9904-1961-10514-4. MR123174

[20] B. Bollobás, *An extension to the theorem of Bishop and Phelps*, Bull. London Math. Soc. **2** (1970), 181–182, DOI 10.1112/blms/2.2.181. MR267380

[21] J. Bourgain, *On dentability and the Bishop-Phelps property*, Israel J. Math. **28** (1977), no. 4, 265–271, DOI 10.1007/BF02760634. MR482076

[22] C. S. Cardassi, *Numerical radius attaining operators*, Banach spaces (Columbia, Mo., 1984), Lecture Notes in Math., vol. 1166, Springer, Berlin, 1985, pp. 11–14, DOI 10.1007/BFb0074687. MR827753

[23] C. S. Cardassi, *Numerical radius-attaining operators on* $C(K)$, Proc. Amer. Math. Soc. **95** (1985), no. 4, 537–543, DOI 10.2307/2045839. MR810159

[24] B. Cascales, A. J. Guirao, and V. Kadets, *A Bishop-Phelps-Bollobás type theorem for uniform algebras*, Adv. Math. **240** (2013), 370–382, DOI 10.1016/j.aim.2013.03.005. MR3046314

[25] C. Choi, A. Kamińska, and H. J. Lee, *Complex convexity of Orlicz-Lorentz spaces and its applications*, Bull. Pol. Acad. Sci. Math. **52** (2004), no. 1, 19–38, DOI 10.4064/ba52-1-3. MR2070025

[26] Y. S. Choi, *Norm attaining bilinear forms on* $L^1[0,1]$, J. Math. Anal. Appl. **211** (1997), no. 1, 295–300, DOI 10.1006/jmaa.1997.5461. MR1460172

[27] Y. S. Choi, D. Garcia, S. G. Kim, and M. Maestre, *The polynomial numerical index of a Banach space*, Proc. Edinb. Math. Soc. (2) **49** (2006), no. 1, 39–52, DOI 10.1017/S0013091502000810. MR2202141

[28] Y. S. Choi, D. Garcia, S. G. Kim, and M. Maestre, *Norm or numerical radius attaining polynomials on* $C(K)$, J. Math. Anal. Appl. **295** (2004), no. 1, 80–96, DOI 10.1016/j.jmaa.2004.03.005. MR2064409

[29] Y. S. Choi, D. García, S. K. Kim, and M. Maestre, *Some geometric properties of disk algebras*, J. Math. Anal. Appl. **409** (2014), no. 1, 147–157, DOI 10.1016/j.jmaa.2013.07.002. MR3095025

[30] Y. S. Choi, D. García, M. Maestre, and M. Martín, *The Daugavet equation for polynomials*, Studia Math. **178** (2007), no. 1, 63–82, DOI 10.4064/sm178-1-4. MR2282490
[31] Y. S. Choi and K. H. Han, *Boundaries for algebras of holomorphic functions on Marcinkiewicz sequence spaces*, J. Math. Anal. Appl. **323** (2006), no. 2, 1116–1133, DOI 10.1016/j.jmaa.2005.11.028. MR2260168
[32] Y. S. Choi, K. H. Han, and H. J. Lee, *Boundaries for algebras of holomorphic functions on Banach spaces*, Illinois J. Math. **51** (2007), no. 3, 883–896. MR2379728
[33] Y. S. Choi and S. G. Kim, *Norm or numerical radius attaining multilinear mappings and polynomials*, J. London Math. Soc. (2) **54** (1996), no. 1, 135–147, DOI 10.1112/jlms/54.1.135. MR1395073
[34] Y. S. Choi and S. K. Kim, *The Bishop-Phelps-Bollobás property and lush spaces*, J. Math. Anal. Appl. **390** (2012), no. 2, 549–555, DOI 10.1016/j.jmaa.2012.01.053. MR2890536
[35] Y. S. Choi and S. K. Kim, *The Bishop-Phelps-Bollobás theorem for operators from $L_1(\mu)$ to Banach spaces with the Radon-Nikodým property*, J. Funct. Anal. **261** (2011), no. 6, 1446–1456, DOI 10.1016/j.jfa.2011.05.007. MR2813477
[36] Y. S. Choi, S. K. Kim, H. J. Lee, and M. Martín, *The Bishop-Phelps-Bollobás theorem for operators on $L_1(\mu)$*, J. Funct. Anal. **267** (2014), no. 1, 214–242, DOI 10.1016/j.jfa.2014.04.008. MR3206513
[37] Y. S. Choi, H. J. Lee, and H. G. Song, *Bishop's theorem and differentiability of a subspace of $C_b(K)$*, Israel J. Math. **180** (2010), 93–118, DOI 10.1007/s11856-010-0095-9. MR2735057
[38] Y. S. Choi and H. G. Song, *The Bishop-Phelps-Bollobás theorem fails for bilinear forms on $l_1 \times l_1$*, J. Math. Anal. Appl. **360** (2009), no. 2, 752–753, DOI 10.1016/j.jmaa.2009.07.008. MR2561271
[39] Y. S. Choi and H. G. Song, *Property (quasi-α) and the denseness of norm attaining mappings*, Math. Nachr. **281** (2008), no. 9, 1264–1272, DOI 10.1002/mana.200510676. MR2442704
[40] H. G. Dales, *Banach algebras and automatic continuity*, London Mathematical Society Monographs. New Series, vol. 24, The Clarendon Press, Oxford University Press, New York, 2000. Oxford Science Publications. MR1816726
[41] S. Dantas, *Some kind of Bishop-Phelps-Bollobás property*, Math. Nachr. **290** (2017), no. 5-6, 774–784, DOI 10.1002/mana.201500487. MR3636377
[42] S. Dantas, D. García, S. K. Kim, U. Y. Kim, H. J. Lee, and M. Maestre, *A nonlinear Bishop-Phelps-Bollobás type theorem*, Q. J. Math. **70** (2019), no. 1, 7–16, DOI 10.1093/qmath/hay031. MR3927841
[43] S. Dantas, V. Kadets, S. K. Kim, H. J. Lee, and M. Martín, *On the pointwise Bishop-Phelps-Bollobás property for operators*, Canad. J. Math. **71** (2019), no. 6, 1421–1443, DOI 10.4153/s0008414x18000032. MR4028464
[44] S. Dantas, S. K. Kim, and H. J. Lee, *The Bishop-Phelps-Bollobás point property*, J. Math. Anal. Appl. **444** (2016), no. 2, 1739–1751, DOI 10.1016/j.jmaa.2016.07.009. MR3535786
[45] S. Dantas, S. K. Kim, H. J. Lee, and M. Mazzitelli, *Strong subdifferentiability and local Bishop-Phelps-Bollobás properties*, Rev. R. Acad. Cienc. Exactas Fís. Nat. Ser. A Mat. RACSAM **114** (2020), no. 2, Paper No. 47, 16, DOI 10.1007/s13398-019-00741-1. MR4049872
[46] I. K. Daugavet, *A property of completely continuous operators in the space C* (Russian), Uspehi Mat. Nauk **18** (1963), no. 5(113), 157–158. MR157225
[47] W. J. Davis, D. J. H. Garling, and N. Tomczak-Jaegermann, *The complex convexity of quasinormed linear spaces*, J. Funct. Anal. **55** (1984), no. 1, 110–150, DOI 10.1016/0022-1236(84)90021-1. MR733036
[48] R. Deville, G. Godefroy, and V. Zizler, *A smooth variational principle with applications to Hamilton-Jacobi equations in infinite dimensions*, J. Funct. Anal. **111** (1993), no. 1, 197–212, DOI 10.1006/jfan.1993.1009. MR1200641
[49] S. J. Dilworth, *Complex convexity and the geometry of Banach spaces*, Math. Proc. Cambridge Philos. Soc. **99** (1986), no. 3, 495–506, DOI 10.1017/S0305004100064446. MR830363
[50] P. N. Dowling, Z. Hu, and D. Mupasiri, *Complex convexity in Lebesgue-Bochner function spaces*, Trans. Amer. Math. Soc. **348** (1996), no. 1, 127–139, DOI 10.1090/S0002-9947-96-01508-5. MR1327255
[51] J. Ferrera, *Norm-attaining polynomials and differentiability*, Studia Math. **151** (2002), no. 1, 1–21, DOI 10.4064/sm151-1-1. MR1891537
[52] J. Globevnik, *On complex strict and uniform convexity*, Proc. Amer. Math. Soc. **47** (1975), 175–178, DOI 10.2307/2040227. MR355564

[53] J. Globevnik, *On interpolation by analytic maps in infinite dimensions*, Math. Proc. Cambridge Philos. Soc. **83** (1978), no. 2, 243–252, DOI 10.1017/S0305004100054505. MR475859

[54] L. A. Harris, *The numerical range of holomorphic functions in Banach spaces*, Amer. J. Math. **93** (1971), 1005–1019, DOI 10.2307/2373743. MR301505

[55] A. Iwanik, *Norm attaining operators on Lebesgue spaces*, Pacific J. Math. **83** (1979), no. 2, 381–386. MR557939

[56] M. Jiménez Sevilla and R. Payá, *Norm attaining multilinear forms and polynomials on preduals of Lorentz sequence spaces*, Studia Math. **127** (1998), no. 2, 99–112, DOI 10.4064/sm-127-2-99-112. MR1488146

[57] J. Johnson and J. Wolfe, *Norm attaining operators*, Studia Math. **65** (1979), no. 1, 7–19, DOI 10.4064/sm-65-1-7-19. MR554537

[58] V. Kadets, M. Martín, and R. Payá, *Recent progress and open questions on the numerical index of Banach spaces* (English, with English and Spanish summaries), RACSAM. Rev. R. Acad. Cienc. Exactas Fís. Nat. Ser. A Mat. **100** (2006), no. 1-2, 155–182. MR2267407

[59] V. M. Kadets, R. V. Shvidkoy, G. G. Sirotkin, and D. Werner, *Banach spaces with the Daugavet property*, Trans. Amer. Math. Soc. **352** (2000), no. 2, 855–873, DOI 10.1090/S0002-9947-99-02377-6. MR1621757

[60] J. Kim and H. J. Lee, *Strong peak points and strongly norm attaining points with applications to denseness and polynomial numerical indices*, J. Funct. Anal. **257** (2009), no. 4, 931–947, DOI 10.1016/j.jfa.2008.11.024. MR2535458

[61] S. G. Kim, *Numerical peak points and numerical Šilov boundary for holomorphic functions*, Proc. Amer. Math. Soc. **136** (2008), no. 12, 4339–4347, DOI 10.1090/S0002-9939-08-09402-1. MR2431048

[62] S. G. Kim and H. J. Lee, *Numerical peak holomorphic functions on Banach spaces*, J. Math. Anal. Appl. **364** (2010), no. 2, 437–452, DOI 10.1016/j.jmaa.2009.10.046. MR2576195

[63] S. K. Kim and H. J. Lee, *Uniform convexity and Bishop-Phelps-Bollobás property*, Canad. J. Math. **66** (2014), no. 2, 373–386, DOI 10.4153/CJM-2013-009-2. MR3176146

[64] S. K. Kim and H. J. Lee, *A Urysohn-type theorem and the Bishop-Phelps-Bollobás theorem for holomorphic functions*, J. Math. Anal. Appl. **480** (2019), no. 2, 123393, 8, DOI 10.1016/j.jmaa.2019.123393. MR4000081

[65] S. K. Kim, H. J. Lee, and M. Martín, *Bishop-Phelps-Bollobás property for bilinear forms on spaces of continuous functions*, Math. Z. **283** (2016), no. 1-2, 157–167, DOI 10.1007/s00209-015-1593-6. MR3489063

[66] V. L. Klee Jr., *Extremal structure of convex sets. II*, Math. Z. **69** (1958), 90–104, DOI 10.1007/BF01187394. MR92113

[67] H. J. Lee, *Generalized numerical index of function algebras*, J. Funct. Spaces, posted on 2019, Art. ID 9080867, 6, DOI 10.1155/2019/9080867. MR3939016

[68] H. J. Lee, *Monotonicity and complex convexity in Banach lattices*, J. Math. Anal. Appl. **307** (2005), no. 1, 86–101, DOI 10.1016/j.jmaa.2005.01.017. MR2138977

[69] H. J. Lee, *Complex convexity and monotonicity in quasi-Banach lattices*, Israel J. Math. **159** (2007), 57–91, DOI 10.1007/s11856-007-0038-2. MR2342473

[70] H. J. Lee, *Banach spaces with polynomial numerical index 1*, Bull. Lond. Math. Soc. **40** (2008), no. 2, 193–198, DOI 10.1112/blms/bdm113. MR2414778

[71] H. J. Lee and H.-J. Tag, *Diameter two properties in some vector-valued function spaces*, Rev. R. Acad. Cienc. Exactas Fís. Nat. Ser. A Mat. RACSAM **116** (2022), no. 1, Paper No. 17, 19, DOI 10.1007/s13398-021-01165-6. MR4326043

[72] H. J. Lee and H. Tag, *Remark on the Daugavet property for complex Banach spaces*, `arXiv:2302.11153`, 2023.

[73] J. Lindenstrauss, *On operators which attain their norm*, Israel J. Math. **1** (1963), 139–148, DOI 10.1007/BF02759700. MR160094

[74] V. Lomonosov, *The Bishop-Phelps theorem fails for uniform non-selfadjoint dual operator algebras*, J. Funct. Anal. **185** (2001), no. 1, 214–219, DOI 10.1006/jfan.2001.3754. MR1853756

[75] G. Lumer, *Semi-inner-product spaces*, Trans. Amer. Math. Soc. **100** (1961), 29–43, DOI 10.2307/1993352. MR133024

[76] C. M. McGregor, *Finite-dimensional normed linear spaces with numerical index* 1, J. London Math. Soc. (2) **3** (1971), 717–721, DOI 10.1112/jlms/s2-3.4.717. MR287288

[77] L. Narici and E. Beckenstein, *Topological vector spaces*, 2nd ed., Pure and Applied Mathematics (Boca Raton), vol. 296, CRC Press, Boca Raton, FL, 2011. MR2723563

[78] J. R. Partington, *Norm attaining operators*, Israel J. Math. **43** (1982), no. 3, 273–276, DOI 10.1007/BF02761947. MR689984

[79] R. Payá and Y. Saleh, *Norm attaining operators from $L_1(\mu)$ into $L_\infty(\nu)$*, Arch. Math. (Basel) **75** (2000), no. 5, 380–388, DOI 10.1007/s000130050519. MR1785447

[80] L. A. Moraes and L. Romero Grados, *Boundaries for algebras of holomorphic functions*, J. Math. Anal. Appl. **281** (2003), no. 2, 575–586, DOI 10.1016/S0022-247X(03)00150-1. MR1982675

[81] D. Sain, *Smooth points in operator spaces and some Bishop-Phelps-Bollobás type theorems in Banach spaces*, Oper. Matrices **13** (2019), no. 2, 433–445, DOI 10.7153/oam-2019-13-32. MR3980452

[82] D. Sain, A. Mal, K. Mandal, and K. Paul, *On uniform Bishop-Phelps-Bollobás type approximations of linear operators and preservation of geometric properties*, J. Math. Anal. Appl. **494** (2021), no. 1, Paper No. 124582, 22, DOI 10.1016/j.jmaa.2020.124582. MR4151570

[83] W. Schachermayer, *Norm attaining operators and renormings of Banach spaces*, Israel J. Math. **44** (1983), no. 3, 201–212, DOI 10.1007/BF02760971. MR693659

[84] W. Schachermayer, *Norm attaining operators on some classical Banach spaces*, Pacific J. Math. **105** (1983), no. 2, 427–438. MR691613

[85] G. Šilov, *Sur la théorie des idéaux dans les anneaux normés de fonctions* (French), C. R. (Doklady) Acad. Sci. URSS (N.S.) **27** (1940), 900–903. MR3464

[86] E. Thorp and R. Whitley, *The strong maximum modulus theorem for analytic functions into a Banach space*, Proc. Amer. Math. Soc. **18** (1967), 640–646, DOI 10.2307/2035432. MR214794

[87] O. Toeplitz, *Das algebraische Analogon zu einem Satze von Fejér* (German), Math. Z. **2** (1918), no. 1-2, 187–197, DOI 10.1007/BF01212904. MR1544315

[88] J. J. Uhl Jr., *Norm attaining operators on $L^1[0,1]$ and the Radon-Nikodým property*, Pacific J. Math. **63** (1976), no. 1, 293–300. MR405076

[89] P. Wojtaszczyk, *Some remarks on the Daugavet equation*, Proc. Amer. Math. Soc. **115** (1992), no. 4, 1047–1052, DOI 10.2307/2159353. MR1126202

Department of Mathematics, POSTECH, 77 Cheongam-ro, Nam-gu, Pohang, Gyeongbuk 37673, Republic of Korea
Email address: mathchoi@postech.ac.kr

Department of Mathematics, Chungbuk National University, 1 Chungdae-ro, Seowon-gu, Cheongju, Chungbuk 28644, Republic of Korea
Email address: skk@chungbuk.ac.kr

Department of Mathematics Education, Dongguk University - Seoul, 30 Pildong-ro Jung-gu, Seoul 04620, Republic of Korea
Email address: hanjulee@dongguk.edu

Proceedings of Symposia in Pure Mathematics
Volume **106**, 2024
https://doi.org/10.1090/pspum/106/01928

Translation Invariant Banach Function Spaces and Fourier Multiplier Operators

Ben de Pagter and Werner J. Ricker

Dedicated to Per Enflo on the occasion of his 80th birthday

ABSTRACT. In this survey article we consider various aspects of translation invariant Banach function spaces E over an infinite compact abelian group G and order bounded operators acting on E. The aim is to expose the interaction between certain features of the space E and properties of convolution operators acting on E via regular complex Borel measures on G.

1. Introduction

The aim of this survey article is to expose some connections on the interplay between a class of classical operators (convolution operators via measures on compact abelian groups) and the translation invariant function spaces in which these operators act. Since the setting here goes well beyond the usual family of translation invariant spaces considered (mainly L^p -spaces and $L^{p,q}$-spaces), it is necessary to review various aspects of such spaces. Moreover, the approach that we adopt involves a significant use of the theory of order bounded operators in complex Banach lattices, in particular, in Banach function spaces. Because certain aspects of this theory are not satisfactorily developed over the complex numbers for the purposes that we require, it is also necessary to review this topic. Due to the more general setting than usual, it is necessary and inevitable to include some proofs, especially because they do not occur elsewhere and the classical arguments may not apply.

Concerning general functional analysis and operator theory, we refer the reader to [**4**], [**5**], [**7**], for example, and for the aspects of integration and measure theory needed, see [**6**], [**10**], [**11**], [**27**], [**30**]. For the material needed involving classical, commutative harmonic analysis we refer to [**9**], [**12**],[**14**], [**15**], [**26**]. Last, but not least, the general theory of Banach lattices (and Banach function spaces) together with the theory of order bounded operators acting on such spaces can be found, for example, in [**1**], [**17**], [**18**], [**19**], [**20**], [**29**], [**31**]. It is time to be more precise.

A large and well studied subclass of the Banach spaces are the Banach lattices which, in turn, contain the Banach function spaces (briefly, B.f.s.) arising in many areas within classical analysis. A traditional topic here is commutative harmonic analysis, where a natural family of spaces consists of the L^p-spaces, $1 \leq p \leq \infty$, over

2020 *Mathematics Subject Classification.* Primary 46E30, 43A22; Secondary 47B38.
Key words and phrases. Compact abelian group, translation invariant Banach function space, Fourier multiplier operator, convolution operator, homogeneous measure.

a locally compact abelian group relative to Haar measure. Somewhat larger is the family of rearrangement invariant spaces over the group (e.g. Lorentz spaces, Orlicz spaces, Marcinkiewicz spaces, etc.). The common and essential feature is that all such spaces E are *translation invariant*, meaning if $f \in E$, then all translates of f by elements of the group also belong to E and have the same norm in E as f. The class of all such translation invariant B.f.s.' E (over a compact, abelian and infinite group G with respect to normalized Haar measure μ) was introduced in [**23**] and in the recent article [**24**], where a type of Fuglede theorem (well known for normal operators in Hilbert space) was established for Fourier multiplier operators. Various properties of such spaces were exposed which indicates that they are well suited for a further investigation of certain aspects of operator theory in them, concerning convolution operators, Fourier multiplier operators, and so on; see also [**25**]. The aim of this survey article is to present further aspects of the theory of translation invariant B.f.s.' over G, at least to the extent that we are able to exhibit (with proofs) some non-trivial extensions of classical operator theoretic results known to hold in the L^p-spaces.

Parallel to this class of spaces we also investigate a natural family of operators acting on such spaces E, namely, those which commute with all translation operators in E via elements of G. Such operators are traditionally called (Fourier) multiplier operators. The class of all such multiplier operators is in general rather unwieldy. A more tractable subclass consists of those operators which act in E via convolution with a regular complex Borel measure in G; these automatically commute with translations.

Our intention is to highlight the interaction between certain features of the underlying translation invariant B.f.s.' E over G and various properties of the convolution operators acting in E. We have made a selection of such interactions which we hope will be interesting and informative. For a more comprehensive treatment of this topic we refer to [**25**].

As a sample, we mention that two important closed subspaces of E are E_a (the order continuous part of E) and E_h (the homogeneous part of E). The space E_a is again a translation invariant B.f.s. over G (possibly $\{0\}$) whereas E_h is a translation invariant subspace of E and a sublattice but, it need not be a B.f.s. Indeed, for $E = L^\infty(G)$ it is known that $E_a = \{0\}$ and $E_h = C(G)$. It is always the case that $E_a \subseteq E_h$. Moreover, $E_h \neq \{0\}$ if and only if $L^\infty(G) \subseteq E$, in which case E_h is equal to the norm closure of $C(G)$ in E. If $E_a \neq \{0\}$, then $E_a = E_h$. In the event that $E = E_a$ the space E is said to have order continuous norm (briefly, o.c.-norm). We mention that always $E \subseteq L^1(G)$ but, there exist spaces E such that $L^\infty(G) \nsubseteq E$.

Consider a translation invariant B.f.s. E over G and $\lambda \in M(G)$, the space of all regular complex Borel measures on G. It is always the case that $T_\lambda^{(1)}$ maps $L^1(G)$ continuously into itself, where $T_\lambda^{(1)}$ denotes the convolution operator $f \mapsto f * \lambda$, for $f \in L^1(G)$. Since $E \subseteq L^1(G)$, it is meaningful to say that the convolution operator $T_\lambda^E : E \to E$ obtained by restricting $T_\lambda^{(1)}$ to E *exists for* E if $T_\lambda^{(1)}(E) \subseteq E$; it is then necessarily continuous. We address the question of which $\lambda \in M(G)$ satisfy $T_\lambda^{(1)}(E) \subseteq E$ and, in addition, which subclass of these operators T_λ^E is order bounded, that is, maps order bounded subsets of E to order bounded sets. It turns out that *every* $\lambda \in M(G)$ satisfies $T_\lambda^{(1)}(E_h) \subseteq E_h$. In the event that E has either o.c.-norm or the Fatou property, the operator T_λ^E exists for *every* $\lambda \in M(G)$ and T_λ^E is necessarily order bounded. There also exist spaces E for which the *only* $\lambda \in M(G)$ for which T_λ^E exists are the discrete measures.

On the other hand, the existence of certain convolution operators in E can affect the properties of E itself. It is well known that every $\lambda \in M(G)$ convolves $L^\infty(G)$ into itself. Henceforth, let E be a translation invariant B.f.s. over G satisfying $L^\infty(G) \subseteq E$. Then $\chi_A * \lambda \in L^\infty(G) \subseteq E$ for every $A \in \mathcal{B}(G)$, the Borel σ-algebra of G, that is, the set function $m_\lambda^E : A \mapsto \chi_A * \lambda$ is a finitely additive, E-valued vector measure. Note that $m_\lambda^E(A) = T_\lambda^E(\chi_A)$, for $A \in \mathcal{B}(G)$, whenever T_λ^E exists. The set function m_λ^E necessarily commutes with every translation operator τ_y, for all $y \in G$, that is, $m_\lambda^E(A+y) = \tau_y\left(m_\lambda^E(A)\right)$ for all $A \in \mathcal{B}(G)$ and $y \in G$. A measure $\lambda \in M(G)$ is called E-homogeneous if the range $m_\lambda^E(\mathcal{B}(G))$ of m_λ^E is contained in E_h. Denote the collection of all E-homogeneous measures on G by $M_h(E;G)$. The set of all $\lambda \in M(G)$ such that $m_\lambda^E : \mathcal{B}(G) \to E$ is actually (norm) countably additive is denoted by $M_{ca}(E;G)$. It is always the case that $L^1(G) \subseteq M_{ca}(E;G) \subseteq M_h(E;G)$. Somewhat remarkably, it turns out that actually $M_{ca}(E;G) = M_h(E;G)$. That is, $\lambda \in M(G)$ satisfies $m_\lambda^E(\mathcal{B}(G)) \subseteq E_h$ if and only if m_λ^E is countably additive. As a *consequence* of this fact, it can be established that:

(a) $E = E_h$ if and only if $E = E_a$;
(b) E_h is an ideal in E if and only if $E_h = E_a$;
(c) $C(G)$ is dense in E if and only if $E = E_a$.

The properties of E recorded in (a)-(c) have no (direct) relation to convolution operators yet, they follow from the identity $M_h(E;G) = M_{ca}(E;G)$, which is a statement about certain convolution operators in E via regular Borel measures on G.

2. Preliminaries

In this section we first collect together some definitions and facts concerning various aspects of the theory of Banach function spaces, such as the associate space, order continuity of the norm, the Fatou property and the order continuous part of the space. Let (Ω, Σ, μ) be a measure space. Since we are interested in the case where $\Omega = G$ is an infinite compact abelian group with normalized Haar measure μ, we will assume that $\mu(\Omega) = 1$ and that μ is atomless. We denote by $L^0(\mu)$ the space of all $\mathbb{C}$-valued, Σ-measurable functions (with the usual identification of functions which are equal μ-a.e. on Ω). Then $L^0(\mu)$ is a Dedekind complete complex Riesz space (or, vector lattice). Recall that a subset $A \subseteq L^0(\mu)$ is called an (order) *ideal* whenever A is a linear subspace of $L^0(\mu)$ with the property that $|f| \leq |g|$, with $f \in L^0(\mu)$ and $g \in A$, implies that $f \in A$.

In the latter part of the section we recall various properties and results concerning linear maps between B.f.s.' Of particular relevance are the order bounded and the order continuous operators. In addition to the usual operator norm we will also require the regular norm. An important notion is the modulus of an order bounded operator, which can be described by the well known Schaefer-De Schipper formula.

DEFINITION 2.1. A *Banach function space* (briefly B.f.s.) over (Ω, Σ, μ) is a Banach space $(E, \|\cdot\|_E)$, where $E \subseteq L^0(\mu)$ is an ideal and $\|\cdot\|_E$ is an absolutely monotone norm on E (that is, $\|f\|_E \le \|g\|_E$ whenever $|f| \le |g|$ in E).

Evidently, B.f.s.' are complex (Dedekind complete) Banach lattices. Therefore, we can freely apply the general theory of Banach lattices to B.f.s.' For the theory of Banach lattices we refer the reader to [**1**], [**20**], [**29**], [**31**], and for the theory of B.f.s.' to [**30**] (Chapter 15), [**3**]. The class of B.f.s.' includes the L^p-spaces $(1 \le p \le \infty)$, Orlicz spaces, Lorentz spaces, Marcinkiewicz spaces and many more. If E and F are two B.f.s.' over (Ω, Σ, μ) satisfying $E \subseteq F$, then the natural embedding of E into F is a positive linear operator between Banach lattices (that is, it maps E^+ into F^+) and hence, it is automatically continuous (see e.g. Theorem 12.3 in [**1**] or Proposition 1.3.5 in [**20**]). Here, $E^+ = \{f \in E : f \ge 0 \ \mu\text{-a.e.}\}$ is the positive cone of a B.f.s. E.

In Chapter 15 of [**30**] and in the series of papers [**18**], B.f.s.' are introduced via function norms. For a summary of this approach we refer to Remark 2.2 in [**24**]. In particular, the approach in [**30**] is equivalent to ours and so, all results from [**30**] may be used in our setting (see also [**31**], Section 112).

Given a B.f.s. $(E, \|\cdot\|_E)$ over (Ω, Σ, μ), there exists a set $A_0 \in \Sigma$ with the property that $f = 0$ μ-a.e. on $\Omega \backslash A_0$ for each $f \in E$ and, for every $A \in \Sigma$ satisfying $A \subseteq A_0$ and $\mu(A) > 0$, there exists $B \in \Sigma$ such that $B \subseteq A$, $\mu(B) > 0$ and $\chi_B \in E$ (see also [**30**], Section 72 or [**31**], Section 86). The set A_0 is uniquely determined (modulo μ-null sets) and is called the *carrier* of E. Moreover, there exists a sequence $(A_n)_{n=1}^{\infty}$ in Σ such that $A_n \subseteq A_0$ and $\chi_{A_n} \in E$, for all n, and $A_n \uparrow_n A_0$. Any B.f.s. that we encounter in this article has its carrier equal to Ω. In general, if $E \ne \{0\}$, then one can replace, without loss of generality, Ω by A_0.

Recall that a B.f.s. $(E, \|\cdot\|_E)$ over (Ω, Σ, μ) has *order continuous norm* (briefly o.c.-norm) if $\|f_\tau\|_E \downarrow_\tau 0$ whenever $f_\tau \downarrow_\tau 0$ in E (that is, (f_τ) is a downwards directed net in the positive cone E^+ with infimum 0). Since E is Dedekind complete, it follows from Theorem 103.6 in [**31**] that E has o.c.-norm if and only if E has σ-o.c.-norm (that is, $\|f_n\|_E \downarrow_n 0$ for any sequence $(f_n)_{n=1}^{\infty}$ in E^+ satisfying $f_n \downarrow_n 0$). For a sequence $(f_n)_{n=1}^{\infty}$ in E^+ the condition $f_n \downarrow_n 0$ is equivalent with $f_n(x) \downarrow 0$ for μ-a.e. $x \in \Omega$. By way of example, for $1 \le p < \infty$ the space $E = L^p(\mu)$ has o.c.-norm but, $L^\infty(\mu)$ does not have o.c.-norm.

Let $(E, \|\cdot\|_E)$ be a B.f.s. over (Ω, Σ, μ) with carrier Ω. The *Köthe dual* (or *associate space*) $E^\times$ of E is defined by

$$E^\times = \left\{g \in L^0(\mu) : \|g\|_{E^\times} < \infty\right\},$$

where

$$\|g\|_{E^\times} = \sup\left\{\int_\Omega |fg| \, d\mu : f \in E, \|f\|_E \le 1\right\}, \quad g \in L^0(\mu). \tag{2.1}$$

Then $(E^\times, \|\cdot\|_{E^\times})$ is a B.f.s. over (Ω, Σ, μ) with carrier equal to Ω; see [**30**], Sections 68, 69 and also [**31**], p. 418. From the definition it is also clear that *Hölder's inequality* holds, that is,

$$\left|\int_\Omega fg \, d\mu\right| \le \int_\Omega |fg| \, d\mu \le \|f\|_E \|g\|_{E^\times}, \quad f \in E, \ g \in E^\times. \tag{2.2}$$

According to [**30**], Section 69, Theorem 1, we also have

$$\|g\|_{E^\times} = \sup\left\{\left|\int_\Omega fg\,d\mu\right| : f \in E, \|f\|_E \le 1\right\}, \quad g \in E^\times. \tag{2.3}$$

For each $g \in E^\times$, define the linear functional $\varphi_g : E \to \mathbb{C}$ by setting

$$\varphi_g(f) = \int_\Omega fg\,d\mu, \quad f \in E.$$

It is easy to see that $\varphi_g \in E^*$ (the dual Banach space of E) and that $\|\varphi_g\|_{E^*} = \|g\|_{E^\times}$ for all $g \in E^\times$ (see (2.3)). Consequently, the map $g \longmapsto \varphi_g$, for $g \in E^\times$, is an isometric isomorphism from $E^\times$ into E^*. Via this map, $E^\times$ may be identified with a closed subspace of E^*. We frequently denote the *dual pairing* of E and $E^\times$ by $\langle\cdot,\cdot\rangle$, that is,

$$\langle f, g\rangle = \int_\Omega fg\,d\mu, \quad f \in E, \;\; g \in E^\times. \tag{2.4}$$

Since the carrier of $E^\times$ is equal to Ω (whenever the carrier of E is Ω), it follows that $E^\times$ separates the points of E (but, in general, $E^\times$ is not norming for E).

The functionals in E^* which are of the form φ_g for some $g \in E^\times$ may be identified with the band E^*_n in E^* consisting of all order continuous functionals on E; see [**30**], Section 69, Theorem 3, and also [**31**], Section 112. In particular, $E^* = E^\times$ if and only if E has o.c.-norm (cf. Theorem 2.4.2 in [**20**]).

A B.f.s. E over (Ω, Σ, μ) has the *Fatou property* if, for any net $0 \le u_\alpha \uparrow_\alpha$ in E satisfying $\sup_\alpha \|u_\alpha\|_E < \infty$, there exists $u \in E$ such that $0 \le u_\alpha \uparrow_\alpha u$ and $\|u_\alpha\|_E \uparrow_\alpha \|u\|_E$. In this case, sequences also suffice; cf. [**31**], Theorem 113.2. By a theorem of Halperin and Luxemburg (see [**30**], Section 71, Theorem 1), a B.f.s. E has the Fatou property if and only if $E = E^{\times\times}$ isometrically. We point out that in [**3**] the Fatou property is included as part of the definition of a B.f.s.

For B.f.s.' which do not have o.c.-norm, the following notion is useful; see [**30**], Section 72.

DEFINITION 2.2. Let $(E, \|\cdot\|_E)$ be a B.f.s. over (Ω, Σ, μ) with carrier Ω. A function $f \in E$ is said to have *absolutely continuous norm* if it follows, for every sequence $(f_n)_{n=1}^\infty$ in E^+ satisfying $|f| \ge f_n \downarrow_n 0$ in E, that $\|f_n\|_E \downarrow_n 0$. Note that f has absolutely continuous norm if and only if $|f|$ has absolutely continuous norm.

The collection of all $f \in E$ having absolutely continuous norm is denoted by E_a and is called the *absolutely continuous part* of E. Evidently, E has o.c.-norm if and only if $E = E_a$.

A useful characterization of functions with absolutely continuous norm is the following one (see Theorem 1 in Section 72 of [**30**]).

LEMMA 2.3. *Let E be a B.f.s. and $f \in E$. Then $f \in E_a$ if and only if $\|f\chi_{A_n}\|_E \downarrow_n 0$ whenever $A_n \downarrow_n \emptyset$ in Σ.*

The main features of the set E_a are exhibited in the next proposition; see [**30**], Section 72, Theorem 3. This result is actually valid in any Banach lattice (see, for example, Proposition 2.4.10 of [**20**]).

PROPOSITION 2.4. *Let $(E, \|\cdot\|_E)$ be a B.f.s. over (Ω, Σ, μ) with carrier Ω. The collection of functions E_a is a norm closed ideal in E. Consequently, $(E_a, \|\cdot\|_E)$ is also a B.f.s. over (Ω, Σ, μ) with o.c.-norm.*

We warn the reader that, in general, the carrier of E_a need not be equal to Ω, as is illustrated by the following example.

EXAMPLE 2.5. Recall that μ is assumed to be atomless. Let $E = L^\infty(\mu)$. It is routine to show that $E_a = \{0\}$. Hence, the carrier of E is Ω but, the carrier of E_a is $\emptyset$.

In the remainder of this section we discuss some properties of linear maps between B.f.s.' Let E and F be two B.f.s.' over (Ω, Σ, μ) or, more generally, two Dedekind complete, complex Banach lattices. As before, it is assumed that the carriers of both E and F are equal to Ω. The space of all $\mathbb{R}$-valued functions in E is denoted by $E_\mathbb{R}$, which is a Dedekind complete, real Banach lattice. Then $E = E_\mathbb{R} \oplus iE_\mathbb{R}$ is the *Banach lattice complexification* of $E_\mathbb{R}$. Similarly, $F = F_\mathbb{R} \oplus iF_\mathbb{R}$. For a detailed discussion of the complexification of vector lattices and Banach lattices we refer, for instance, to Section II.11 of [**29**] and Sections 91, 92 of [**31**].

The space of *all* linear maps from E into F is denoted by $\mathcal{L}^\sharp(E, F)$. Similarly, $\mathcal{L}^\sharp(E_\mathbb{R}, F_\mathbb{R})$ denotes the real vector space of all real linear maps from $E_\mathbb{R}$ into $F_\mathbb{R}$. Then $\mathcal{L}^\sharp(E_\mathbb{R}, F_\mathbb{R})$ can be identified with a real linear subspace of $\mathcal{L}^\sharp(E, F)$ via the identification

$$\mathcal{L}^\sharp(E_\mathbb{R}, F_\mathbb{R}) = \left\{T \in \mathcal{L}^\sharp(E, F) : T(E_\mathbb{R}) \subseteq F_\mathbb{R}\right\},$$

in which case $\mathcal{L}^\sharp(E, F) = \mathcal{L}^\sharp(E_\mathbb{R}, F_\mathbb{R}) \oplus i\mathcal{L}^\sharp(E_\mathbb{R}, F_\mathbb{R})$; see Section 92 of [**31**]. In particular, every $T \in \mathcal{L}^\sharp(E, F)$ has a unique representation

$$T = \operatorname{Re} T + i \operatorname{Im} T, \quad \operatorname{Re} T, \operatorname{Im} T \in \mathcal{L}^\sharp(E_\mathbb{R}, F_\mathbb{R}). \tag{2.5}$$

An operator $T \in \mathcal{L}^\sharp(E, F)$ is termed *order bounded* if, for every $u \in E^+$, there exists $v \in F^+$ such that $|Tf| \leq v$ whenever $f \in E$ satisfies $|f| \leq u$. The linear subspace of $\mathcal{L}^\sharp(E, F)$ consisting of all order bounded operators is denoted by $\mathcal{L}_b(E, F)$. Setting $\mathcal{L}_b(E_\mathbb{R}, F_\mathbb{R}) = \mathcal{L}_b(E, F) \cap \mathcal{L}^\sharp(E_\mathbb{R}, F_\mathbb{R})$, it is clear that $\mathcal{L}_b(E_\mathbb{R}, F_\mathbb{R})$ is a (real) linear subspace of $\mathcal{L}^\sharp(E_\mathbb{R}, F_\mathbb{R})$ and it is readily verified that

$$\mathcal{L}_b(E, F) = \mathcal{L}_b(E_\mathbb{R}, F_\mathbb{R}) \oplus i\mathcal{L}_b(E_\mathbb{R}, F_\mathbb{R}). \tag{2.6}$$

As usual, for $S, T \in \mathcal{L}_b(E_\mathbb{R}, F_\mathbb{R})$ we define $S \leq T$ whenever $Su \leq Tu$ for all $u \in E^+$. With respect to this partial ordering, $\mathcal{L}_b(E_\mathbb{R}, F_\mathbb{R})$ is a partially ordered vector space. The positive cone of $\mathcal{L}_b(E_\mathbb{R}, F_\mathbb{R})$ is denoted by $\mathcal{L}_b(E_\mathbb{R}, F_\mathbb{R})^+$. The elements of $\mathcal{L}_b(E_\mathbb{R}, F_\mathbb{R})^+$ will be referred to as *positive operators*. Note that if $T \in \mathcal{L}^\sharp(E, F)$, then $T \in \mathcal{L}_b(E_\mathbb{R}, F_\mathbb{R})^+$ if and only if $T(E^+) \subseteq F^+$. Since $F_\mathbb{R}$ is a Dedekind complete vector lattice, it follows that $\mathcal{L}_b(E_\mathbb{R}, F_\mathbb{R})$ is also a Dedekind complete vector lattice, with the supremum $S \vee T$ in $\mathcal{L}_b(E_\mathbb{R}, F_\mathbb{R})$ of $S, T \in \mathcal{L}_b(E_\mathbb{R}, F_\mathbb{R})$ given by

$$(S \vee T)(u) = \sup\left\{Sv + Tw : v, w \in E^+,\ u = v + w\right\}, \quad u \in E^+;$$

see Theorems 83.4 and 83.6 in [**31**]. In particular, for each $T \in \mathcal{L}_b(E_\mathbb{R}, F_\mathbb{R})$ its *modulus* $|T| = T \vee (-T)$ exists in $\mathcal{L}_b(E_\mathbb{R}, F_\mathbb{R})^+$ and is given by

$$|T|(u) = \sup\{|Tf| : f \in E_\mathbb{R},\ |f| \leq u\}, \quad u \in E^+. \tag{2.7}$$

It follows from (2.6) that $\mathcal{L}_b(E, F)$ has the structure of a complex vector lattice. In particular, for each $T \in \mathcal{L}_b(E, F)$ its modulus $|T| \in \mathcal{L}_b(E_\mathbb{R}, F_\mathbb{R})^+$ exists (see Section 92 of [**31**]) and is also given, analogous to (2.7), by the so-called *Schaefer-De Schipper formula*

$$|T|(u) = \sup\{|Tf| : f \in E,\ |f| \leq u\}, \quad u \in E^+; \tag{2.8}$$

see [**29**], Theorem IV.1.8 or [**31**], Theorem 92.6. We also write $\mathcal{L}_b(E,F)^+ = \mathcal{L}_b(E_{\mathbb{R}},F_{\mathbb{R}})^+$. It also follows that every $T \in \mathcal{L}_b(E,F)$ is a linear combination of at most four positive operators. In other words, $\mathcal{L}_b(E,F)$ coincides with the space of all *regular operators*, that is, $\mathcal{L}_b(E,F) = \mathcal{L}_r(E,F)$.

Recall that an operator $T \in \mathcal{L}_b(E_{\mathbb{R}},F_{\mathbb{F}})$ is called *order continuous* if $\inf_\tau |Tu_\tau| = 0$ holds in $F_{\mathbb{R}}$ for every downwards directed system $u_\tau \downarrow_\tau 0$ in $E_{\mathbb{R}}$ (see Section 86 of [**31**]). Since the space $E_{\mathbb{R}}$ is order separable, this is equivalent to the condition that $\inf_n |Tu_n| = 0$ holds in $F_{\mathbb{R}}$ for every decreasing sequence $u_n \downarrow_n 0$ in $E_{\mathbb{R}}$. The set $\mathcal{L}_n(E_{\mathbb{R}},F_{\mathbb{R}})$ of all order continuous operators from $E_{\mathbb{R}}$ into $F_{\mathbb{R}}$ is a band in $\mathcal{L}_b(E_{\mathbb{R}},F_{\mathbb{F}})$; see Theorem 84.2 in [**31**]. An operator $T \in \mathcal{L}_b(E,F)$ is termed order continuous whenever $\operatorname{Re} T, \operatorname{Im} T \in \mathcal{L}_n(E_{\mathbb{R}},F_{\mathbb{R}})$ and so $\mathcal{L}_n(E,F) = \mathcal{L}_n(E_{\mathbb{R}},F_{\mathbb{R}}) \oplus i\mathcal{L}_n(E_{\mathbb{R}},F_{\mathbb{R}})$ is a band in $\mathcal{L}_b(E,F)$; see Section 92 of [**31**]. In particular, if $T \in \mathcal{L}_b(E,F)$, then $T \in \mathcal{L}_n(E,F)$ if and only if $|T| \in \mathcal{L}_n(E_{\mathbb{R}},F_{\mathbb{R}})$. Consequently, it is readily verified that an operator $T \in \mathcal{L}_b(E,F)$ is order continuous if and only if $\inf_n |Tu_n| = 0$ holds in $F_{\mathbb{R}}$ for any decreasing sequence $u_n \downarrow_n 0$ in $E_{\mathbb{R}}$.

Since every positive linear operator between complex Banach lattices is automatically norm continuous (which follows, for instance, from [**1**], Theorem 12.3 or [**20**], Proposition 1.3.5), it follows that $\mathcal{L}_b(E,F) \subseteq \mathcal{L}(E,F)$, where $\mathcal{L}(E,F)$ denotes the space of all continuous linear maps from E into F equipped with the usual operator norm $\|\cdot\|$. In general, the inclusion $\mathcal{L}_b(E,F) \subseteq \mathcal{L}(E,F)$ is proper.

The *regular norm* $\|\cdot\|_r$ on $\mathcal{L}_b(E,F)$ is defined by

$$\|T\|_r = \| \, |T| \, \|, \quad T \in \mathcal{L}_b(E,F). \tag{2.9}$$

If $T \in \mathcal{L}_b(E,F)$, then it follows from (2.8) that $|Tf| \le |T|(|f|)$, for $f \in E$, and so $\|T\| \le \|T\|_r$. The norms $\|\cdot\|$ and $\|\cdot\|_r$ are, in general, not equivalent on $\mathcal{L}_b(E,F)$. The space $(\mathcal{L}_b(E,F), \|\cdot\|_r)$ is a complex Banach lattice (it is the Banach lattice complexification of $(\mathcal{L}_b(E_{\mathbb{R}},F_{\mathbb{R}}), \|\cdot\|_r)$); see Corollary IV.1.1 in [**29**]. The next proposition exhibits two important cases in which $\mathcal{L}_b(E,F) = \mathcal{L}(E,F)$; see, for instance, [**20**], Theorem 1.5.11 and [**29**], Corollary IV.1.2.

PROPOSITION 2.6. *Let E and F be B.f.s.' over (Ω,Σ,μ).*

(i) *Let $F = L^\infty(\mu)$. Then $\mathcal{L}_b(E,F) = \mathcal{L}(E,F)$ and $\|T\|_r = \|T\|$ for all $T \in \mathcal{L}(E,F)$.*
(ii) *Let $E = L^1(\mu)$ and F have the Fatou property. Then $\mathcal{L}_b(E,F) = \mathcal{L}(E,F)$ and $\|T\|_r = \|T\|$ for all $T \in \mathcal{L}(E,F)$.*

In particular, $\mathcal{L}_b(L^\infty(\mu)) = \mathcal{L}(L^\infty(\mu))$ and $\mathcal{L}_b(L^1(\mu)) = \mathcal{L}(L^1(\mu))$, with equality of the regular norm and the operator norm.

REMARK 2.7. Let E and F be two B.f.s.' over (Ω,Σ,μ). A linear map $\Phi : \mathcal{L}_b(E,F) \to \mathcal{L}_b(E,F)$ is called a *lattice homomorphism* if

$$|\Phi(T)| = \Phi(|T|), \quad T \in \mathcal{L}_b(E,F). \tag{2.10}$$

Any such lattice homomorphism Φ satisfies $\Phi(\mathcal{L}_b(E_{\mathbb{R}},F_{\mathbb{R}})) \subseteq \mathcal{L}_b(E_{\mathbb{R}},F_{\mathbb{R}})$ and its restriction $\Phi_{\mathbb{R}} : \mathcal{L}_b(E_{\mathbb{R}},F_{\mathbb{R}}) \to \mathcal{L}_b(E_{\mathbb{R}},F_{\mathbb{R}})$ is a real lattice homomorphism, that is, $\Phi_{\mathbb{R}}(S \vee T) = \Phi_{\mathbb{R}}(S) \vee \Phi_{\mathbb{R}}(T)$, for $S,T \in \mathcal{L}_b(E_{\mathbb{R}},F_{\mathbb{R}})$.

Conversely, let $\Phi : \mathcal{L}_b(E,F) \to \mathcal{L}_b(E,F)$ be a linear map that satisfies $\Phi(\mathcal{L}_b(E_{\mathbb{R}},F_{\mathbb{R}})) \subseteq \mathcal{L}_b(E_{\mathbb{R}},F_{\mathbb{R}})$ and suppose that its restriction $\Phi_{\mathbb{R}}$ to $\mathcal{L}_b(E_{\mathbb{R}},F_{\mathbb{R}})$ is a (real) lattice homomorphism. Then Φ is a lattice homomorphism on $\mathcal{L}_b(E,F)$. This can be readily verified, for instance, by using the construction of the complex modulus as is indicated in Exercise 91.12 of [**31**].

In particular, if $\Phi : \mathcal{L}_b(E,F) \to \mathcal{L}_b(E,F)$ is a bijective linear map which is bi-positive (that is, $\Phi\left(\mathcal{L}_b(E,F)^+\right) = \mathcal{L}_b(E,F)^+$), then Φ is a lattice homomorphism.

A linear subspace $K \subseteq \mathcal{L}_b(E,F)$ is called a *vector sublattice* of $\mathcal{L}_b(E,F)$ if both $\operatorname{Re} T$ and $|T|$ belong to K for all $T \in K$. Writing $K_{\mathbb{R}} = K \cap \mathcal{L}_b(E_{\mathbb{R}}, F_{\mathbb{R}})$, this implies that $K = K_{\mathbb{R}} \oplus iK_{\mathbb{R}}$ and that $K_{\mathbb{R}}$ is a (real) vector sublattice of $\mathcal{L}_b(E_{\mathbb{R}}, F_{\mathbb{R}})$.

Let $\Phi : \mathcal{L}_b(E,F) \to \mathcal{L}_b(E,F)$ be a lattice homomorphism and define

$$K = \{T \in \mathcal{L}_b(E,F) : \Phi(T) = T\}.$$

Then K is a vector sublattice of $\mathcal{L}_b(E,F)$. In fact, from (2.10) it is clear that $|T| \in K$ whenever $T \in K$. Furthermore, from (2.10) it is also follows that $\Phi\left(\mathcal{L}_b(E,F)^+\right) \subseteq \mathcal{L}_b(E,F)^+$, because $T = |T|$ for all $T \in \mathcal{L}_b(E,F)^+$. This implies that $\Phi(\mathcal{L}_b(E_{\mathbb{R}}, F_{\mathbb{R}})) \subseteq \mathcal{L}_b(E_{\mathbb{R}}, F_{\mathbb{R}})$ and so (cf. (2.5)) $\operatorname{Re}(\Phi(T)) = \Phi(\operatorname{Re} T)$ for all $T \in \mathcal{L}_b(E,F)$. Consequently, $\operatorname{Re} T \in K$ whenever $T \in K$.

3. Translation invariant Banach function spaces

In this short section, for the sake of self-containment, we recall the notion of a translation invariant B.f.s. and record some relevant properties (taken from [**24**]) which will be needed in the sequel.

Let G be an infinite compact abelian group and denote by μ normalized Haar measure on the Borel σ-algebra $\mathcal{B}(G)$. It is known that μ is atomless; see, for instance, [**22**], Lemma 7.97. The Banach space of all regular, complex Borel measures in G, equipped with the total variation norm $\|\cdot\|_{M(G)}$, is denoted by $M(G)$. The group operation is denoted by $+$ and the identity element of G is denoted by 0. Moreover, $\mu(-A) = \mu(A)$ for all $A \in \mathcal{B}(G)$. For the general theory of locally compact abelian groups we refer the reader, for instance, to [**12**], [**14**], [**15**], [**26**]. We denote $L^0(\mu)$ by the more traditional notation $L^0(G)$ and the corresponding L^p-spaces by $L^p(G)$, $1 \le p \le \infty$. A B.f.s. over $(G, \mathcal{B}(G), \mu)$ is simply called a B.f.s. over G. For each $y \in G$, define the *translation operator* $\tau_y : L^0(G) \to L^0(G)$ by setting

$$(\tau_y f)(x) = f(x-y), \quad x \in G,$$

for all $f \in L^0(G)$.

Definition 3.1. A *translation invariant B.f.s* over G is a B.f.s. $E \subseteq L^0(G)$ such that $\tau_y f \in E$ with $\|\tau_y f\|_E = \|f\|_E$ for all $y \in G$, whenever $f \in E$.

Such B.f.s.' include all spaces $L^p(G)$, $1 \le p \le \infty$, all Orlicz spaces, Lorentz spaces and Marcinkiewicz spaces over G. Actually, any rearrangement invariant B.f.s. over G is translation invariant. However, there exist plenty of translation invariant B.f.s.' which are *not* rearrangement invariant. We present an example (see also Example 3.11 in [**24**]).

Example 3.2. Let G_1 and G_2 be two infinite compact abelian groups (with normalized Haar measure μ_1 and μ_2, respectively) and consider the product group $G = G_1 \times G_2$ (with normalized Haar measure $\mu = \mu_1 \times \mu_2$). Let $1 < p, q < \infty$. Define the function norm $\|\cdot\|_{p\times q}$ on $L^0(G)$ by

$$\|f\|_{p\times q} = \left(\int_{G_1}\left(\int_{G_2} |f(x,y)|^q \, d\mu_2(y)\right)^{p/q} d\mu_1(x)\right)^{1/p}, \quad f \in L^0(G),$$

and let

$$E_{p\times q}=\left\{f\in L^{0}\left(G\right):\left\|f\right\|_{p\times q}<\infty\right\}.$$

Equipped with the norm $\|\cdot\|_{p\times q}$ the space $E_{p\times q}$ is a translation invariant B.f.s. over G with the Fatou property and o.c.-norm. It is readily verified that $E_{p\times q}$ is *not* rearrangement invariant whenever $p\neq q$.

We proceed to record various facts which occur in Section 3 of [**24**].

LEMMA 3.3. *Let $E\neq\{0\}$ be a translation invariant B.f.s. over G.*

(i) *The carrier of E is G.*
(ii) *The space $E\subseteq L^{1}\left(G\right)$ with a continuous embedding.*
(iii) *The Köthe dual $E^{\times}$ is a translation invariant B.f.s. and satisfies $L^{\infty}\left(G\right)\subseteq E^{\times}\subseteq L^{1}\left(G\right)$, with continuous embeddings.*
(iv) *If, in addition, E has the Fatou property, then $L^{\infty}\left(G\right)\subseteq E\subseteq L^{1}\left(G\right)$, with continuous embeddings.*

In connection to the above facts, we recall that if $E\neq\{0\}$ is any rearrangement invariant B.f.s. over G, then $L^{\infty}\left(G\right)\subseteq E\subseteq L^{1}\left(G\right)$ (as μ is atomless with $\mu\left(G\right)<\infty$). However, if E is a non-zero translation invariant B.f.s. over G, then the inclusion $L^{\infty}\left(G\right)\subseteq E$ may fail to hold in general (see Example 3.6 in [**24**], which is based on the results of W. Rudin in [**28**]). The following result shows, in particular, that $L^{\infty}\left(G\right)\subseteq E$ whenever $E\neq\{0\}$ is a translation invariant B.f.s. with o.c.-norm (see Proposition 3.8 in [**24**]).

PROPOSITION 3.4. *Let $E\neq\{0\}$ be a translation invariant B.f.s. over G with o.c.-norm. The following statements hold.*

(i) *For each $f\in E$ it is the case that* $\lim_{y\to 0}\left\|f-\tau_{y}f\right\|_{E}=0$.
(ii) *$L^{\infty}\left(G\right)\subseteq E\subseteq L^{1}\left(G\right)$ with continuous inclusions.*
(iii) *$C\left(G\right)$ is a dense subspace of E.*

In Theorem 7.11 below it will be shown, with quite some effort, that condition (i) of Proposition 3.4 actually implies that E has o.c.-norm.

REMARK 3.5. For each $f\in L^{0}\left(G\right)$, its *reflection* $\tilde{f}\in L^{0}\left(G\right)$ is defined by

$$\tilde{f}\left(x\right)=f\left(-x\right),\quad x\in G.$$

Evidently, if E is any rearrangement invariant B.f.s. over G, then $\tilde{f}\in E$ whenever $f\in E$. Also the translation invariant spaces $E_{p\times q}$ of Example 3.2 have this property. However, there do exist translation invariant B.f.s.' E for which there exist functions $f\in E$ with $\tilde{f}\notin E$; see Example 3.11 in [**24**].

4. The homogeneous part of a translation invariant B.f.s.

As indicated in the Introduction, there are two important closed subspaces of a translation invariant B.f.s. E over G. Namely, its order continuous part E_{a} (cf. Section 2) and its homogeneous part E_{h}, which is the topic of this section. It is shown that $E_{h}\neq\{0\}$ if and only if $L^{\infty}\left(G\right)\subseteq E$ (cf. Proposition 4.3) and that E_{a} is always a translation invariant B.f.s. over G satisfying $E_{a}\subseteq E_{h}$, with equality whenever $E_{a}\neq\{0\}$, in which case $C\left(G\right)$ is dense in E_{a}.

Let E be a translation invariant B.f.s. over an infinite compact abelian group G (equipped with normalized Haar measure μ). As follows from Proposition 3.4, if E has o.c.-norm, then $\lim_{y\to 0}\left\|f-\tau_{y}f\right\|_{E}=0$ for all $f\in E$. This motivates the following definition (the terminology is inspired by the notion of a homogeneous Banach space; see [**14**], Ch. I, Section 2).

DEFINITION 4.1. Let $E \neq \{0\}$ be a translation invariant B.f.s. over G. An element $f \in E$ is called *homogeneous* if $\lim_{x\to 0} \|f - \tau_x f\|_E = 0$. The set of all homogeneous elements in E is denoted by E_h and is called the *homogeneous part* of E.

If $f \in E_h$, then the function $x \longmapsto \tau_x f$, for $x \in G$, is continuous from G into E and $\tau_y f$ is homogeneous for all $y \in G$. It is routine to check that E_h is a closed, translation invariant, linear subspace of E (see also [**14**], p. 14, where the notation E_c is used). Since the translation operators $\tau_x : E \to E$, for $x \in G$, are linear lattice isomorphisms in E, it follows that E_h is also a complex Banach sublattice of E. For $E = L^\infty(G)$, we note that $E_a = \{0\}$ whereas $E_h = C(G)$; see [**8**]. This example also shows that E_h need not be an ideal in E. In Proposition 4.3 below some necessary and sufficient conditions for E_h to be non-trivial are presented. First we require the following lemma. Given $f, g \in L^1(G)$, their *convolution* $f * g \in L^1(G)$ is defined by

$$(f * g)(x) = \int_G f(x - y)\, g(y)\, d\mu(y), \quad \mu\text{-a.e. } x \in G;$$

for further details we refer to [**26**], for example.

LEMMA 4.2. *Let $E \neq \{0\}$ be a translation invariant B.f.s. over G and $f \in E_h$. For every $\varepsilon > 0$ there exists $w \in E \cap C(G)$ satisfying $\|w - f\|_E \leq \varepsilon$.*

PROOF. Let $\varepsilon > 0$ be given. Since f is homogeneous, there exists an open neighbourhood U of $0 \in G$ such that $\|f - \tau_y f\|_E \leq \varepsilon$ for all $y \in U$. Defining $g = \mu(U)^{-1} \chi_U \in L^\infty(G)$, the convolution $f * g \in C(G)$, [**26**], p. 4 (recall that $f \in L^1(G)$ as $E \subseteq L^1(G)$; see Lemma 3.3(ii)). On the other hand, since $f \in E_h$, the function $y \longmapsto \tau_y f$, for $y \in G$, is continuous from G into E and so its range is a compact (hence, separable) subset of E. Therefore, via Pettis' measurability theorem (see Theorem II.1.2 in [**6**]), this function is strongly μ-measurable and has bounded range. Hence, the Bochner μ-integral $\int_U^{(B)} \tau_y f d\mu(y) \in E$ exists; see Theorem II.2.2 in [**6**]. Fix $h \in E^\times$. Then

$$\begin{aligned}
\left\langle \int_U^{(B)} \tau_y f d\mu(y), h \right\rangle &= \int_U \langle \tau_y f, h \rangle\, d\mu(y) \\
&= \int_U \left(\int_G f(x - y)\, h(x)\, d\mu(x) \right) d\mu(y) \\
&= \int_G \left(\int_U f(x - y)\, d\mu(y) \right) h(x)\, d\mu(x) \\
&= \int_G (f * \chi_U)(x)\, h(x)\, d\mu(x) = \langle f * \chi_U, h \rangle .
\end{aligned}$$

Since $E^\times$ separates the points of E, this shows that

$$f * g = \mu(U)^{-1} \int_U^{(B)} \tau_y f d\mu(y).$$

Consequently,

$$(f * g) - f = \mu(U)^{-1} \int_U^{(B)} (\tau_y f - f)\, d\mu(y)$$

and so $f * g$ (which belongs to $C(G)$, as observed above) satisfies

$$\|(f * g) - f\|_E \leq \mu(U)^{-1} \int_U \|\tau_y f - f\|_E\, d\mu(y) \leq \varepsilon.$$

Therefore, the function $w = f * g \in E \cap C(G)$ has the desired properties. □

PROPOSITION 4.3. *Let E be a translation invariant B.f.s. over G. The following four statements are equivalent.*

(i) $E \cap C(G) \neq \{0\}$.
(ii) $L^\infty(G) \subseteq E$.
(iii) $C(G) \subseteq E$ *and* $E_h = \overline{C(G)}^E$ *(the norm closure of $C(G)$ in E).*
(iv) $E_h \neq \{0\}$.

PROOF. (i)⇒(ii). Choose any $0 \neq f \in C(G) \cap E$ and any $0 < \varepsilon \in \mathbb{R}$ such that the open set $O = \{x \in G : |f(x)| > \varepsilon\}$ is non-empty. Since $0 \leq \varepsilon\chi_O \leq |f|$, it follows that $\chi_O \in E$ and so, $\chi_{O+y} = \tau_y \chi_O \in E$ for all $y \in G$. The compactness of G implies that there exist elements $\{y_j\}_{j=1}^n$ in G such that $G = \bigcup_{j=1}^n (O + y_j)$. Hence, $\chi_G \leq \sum_{j=1}^n \chi_{O+y_j} = \sum_{j=1}^n \tau_{y_j}\chi_O \in E$ and so, $\chi_G \in E$. Consequently, $L^\infty(G) \subseteq E$.

(ii)⇒(iii). Evidently, $C(G) \subseteq E$. Let $f \in C(G) = L^\infty(G)_h$. Then

$$\lim_{x\to 0} \|\tau_x f - f\|_\infty = 0.$$

Since the inclusion of $L^\infty(G)$ into E is continuous (actually, $\|g\|_E \leq \|\chi_G\|_E \|g\|_\infty$ for all $g \in L^\infty(G)$), this implies that also $\lim_{x\to 0} \|\tau_x f - f\|_E = 0$, that is, $f \in E_h$. This shows that $C(G) \subseteq E_h$. Since E_h is closed in E, it follows that $\overline{C(G)}^E \subseteq E_h$. The reverse inclusion follows immediately from Lemma 4.2. This shows that (iii) holds.

That (iii) implies (iv) is evident. The implication (iv)⇒(i) is a consequence of Lemma 4.2. The proof is thereby complete. □

REMARK 4.4. Proposition 4.3 implies that $E_h = \{0\}$ if and only if $E \cap C(G) = \{0\}$, if and only if $L^\infty(G) \nsubseteq E$. The existence of non-trivial translation invariant B.f.s.' with this property is exhibited in Example 3.6 of [**24**].

What is the relation between the absolutely continuous part E_a (see Definition 2.2) and the homogeneous part E_h of a translation invariant B.f.s. $E \neq \{0\}$ over G? It is readily verified that $\tau_x f \in E_a$ for all $x \in G$ whenever $f \in E_a$. Consequently, E_a is a translation invariant B.f.s. with o.c.-norm over G (cf. Proposition 2.4). Proposition 3.4 (i) implies that $E_a \subseteq E_h$. Moreover, it follows from Lemma 3.3 (i) that the carrier of E_a is equal to G whenever $E_a \neq \{0\}$. For later reference, these observations are collected together in the next lemma.

LEMMA 4.5. *Let $E \neq \{0\}$ be a translation invariant B.f.s. over G. Then E_a is also a translation invariant B.f.s. over G and satisfies $E_a \subseteq E_h$. Furthermore, if $E_a \neq \{0\}$, then the carrier of E_a is equal to G.*

Actually, in the case that $E_a \neq \{0\}$, more can be said.

PROPOSITION 4.6. *Let E be translation invariant B.f.s. over G such that $E_a \neq \{0\}$. Then $L^\infty(G) \subseteq E_a$ and*

$$E_a = E_h = \overline{C(G)}^E = \overline{L^\infty(G)}^E. \tag{4.1}$$

PROOF. It follows from Proposition 3.4 (applied to the space E_a in place of E) that $L^\infty(G) \subseteq E_a$. Since E_a is closed in E, this implies that $\overline{L^\infty(G)}^E \subseteq E_a$. Furthermore, it follows from Lemma 4.5 that $E_a \subseteq E_h$ and so $E_h \neq \{0\}$. Then Proposition 4.3 yields that $E_h = \overline{C(G)}^E$. It is now clear that (4.1) holds. The proof is thereby complete. □

The following fact will be needed later.

LEMMA 4.7. *Let E be a translation invariant B.f.s. over G. For each $f \in E_h$ and each $g \in C(G)$ the product $gf \in E_h$.*

PROOF. First note that $gf \in E$ whenever $f \in E$ and $g \in C(G)$ and that

$$\|gf\|_E \leq \|g\|_\infty \|f\|_E. \tag{4.2}$$

Indeed, $|g| \leq \|g\|_\infty \chi_G$ and so $|gf| \leq \|g\|_\infty |f|$ from which it follows that both $gf \in E$ and (4.2) holds.

Let $f \in E_h$ and $g \in C(G)$. For each $x \in G$ observe that

$$\begin{aligned}\tau_x(gf) - gf &= (\tau_x g)(\tau_x f) - gf \\ &= [(\tau_x g) - g](\tau_x f) + g[(\tau_x f) - f].\end{aligned}$$

Using (4.2), the previous identity implies that

$$\begin{aligned}\|\tau_x(gf) - gf\|_E &\leq \|(\tau_x g) - g\|_\infty \|\tau_x f\|_E + \|g\|_\infty \|(\tau_x f) - f\|_E \\ &= \|(\tau_x g) - g\|_\infty \|f\|_E + \|g\|_\infty \|(\tau_x f) - f\|_E.\end{aligned}$$

Since $g \in C(G) = L^\infty(G)_h$ and $f \in E_h$, we have that $\lim_{x\to 0} \|(\tau_x g) - g\|_\infty = 0$ and $\lim_{x\to 0} \|(\tau_x f) - f\|_E = 0$. This suffices for the proof. □

5. Multiplier operators

Multiplier operators (that is, those commuting with all translation operators) on classical spaces such as, for instance, L^p-spaces, have been studied extensively; see, for example, [**15**] or Chapter 16 of [**9**]. The aim of this section is to expose a few of the basic properties of multiplier operators in more general translation invariant B.f.s.' which will be needed in the following sections.

As before, G is an infinite compact abelian group and μ denotes normalized Haar measure on G. Let E and F be non-zero translation invariant B.f.s.' over G. Suppose that $\lambda \in M(G)$ has the property that the convolution $f * \lambda \in F$ whenever $f \in E$ (note that always $f * \lambda \in L^1(G)$ exists for each $f \in E$ as $E \subseteq L^1(G)$; cf. Lemma 3.3 (ii)). The linear operator $T_\lambda^{(E,F)} : E \to F$ can then be defined by $T_\lambda^{(E,F)} f = f * \lambda$, for $f \in E$. It is readily verified that $T_\lambda^{(E,F)}$ is continuous (via the closed graph theorem) and commutes with translations, that is, $\tau_y T_\lambda^{(E,F)} = T_\lambda^{(E,F)} \tau_y$ for all $y \in G$. This motivates the following definition.

DEFINITION 5.1. Let E and F be non-zero translation invariant B.f.s.' over G. A continuous linear operator $T : E \to F$ is called an *(E,F)-multiplier* if $\tau_y T = T\tau_y$ for all $y \in G$ where, in the left-side $\tau_y : F \to F$ and in the right-side $\tau_y : E \to E$. The set of all (E,F) -multipliers is denoted by $\mathfrak{m}(E,F)$. The collection of all *order bounded (E,F)-multipliers* is denoted by $\mathfrak{m}_b(E,F)$, that is, $\mathfrak{m}_b(E,F) = \mathfrak{m}(E,F) \cap \mathcal{L}_b(E,F)$. In case $E = F$, we simply write $\mathfrak{m}(E)$ and $\mathfrak{m}_b(E)$ for $\mathfrak{m}(E,E)$ and $\mathfrak{m}_b(E,E)$, respectively.

The following facts are readily verified. Note that part (ii) in this lemma follows immediately from part (i), because $\|T\| \le \|T\|_r$, for $T \in \mathcal{L}_b(E,F)$; see Section 2.

LEMMA 5.2. *Let E and F be non-zero translation invariant B.f.s.' over G.*

(i) *The set $\mathfrak{m}(E,F)$ is a closed linear subspace of $\mathcal{L}(E,F)$.*
(ii) *The set $\mathfrak{m}_b(E,F)$ is a closed linear subspace of $\mathcal{L}_b(E,F)$ with respect to the regular norm $\|\cdot\|_r$.*

The space $\mathfrak{m}_b(E,F)$ has even more structure, as indicated by the next result.

PROPOSITION 5.3. *Let E and F be non-zero translation invariant B.f.s.' over G. Then $\mathfrak{m}_b(E,F)$ is a vector sublattice of $\mathcal{L}_b(E,F)$. Consequently, $\mathfrak{m}_b(E,F)$ is a Banach sublattice of $\mathcal{L}_b(E,F)$, equipped with the regular norm.*

PROOF. For each $y \in G$, the maps $\tau_y : E \to E$ and $\tau_y : F \to F$ are lattice isomorphisms. Given any $y \in G$, define the linear map S_y from $\mathcal{L}_b(E,F)$ into itself by

$$S_y(T) = \tau_{-y} \circ T \circ \tau_y, \qquad T \in \mathcal{L}_b(E,F).$$

Since $S_y^{-1} = S_{-y}$, it is clear that S_y is bi-positive and hence (see Remark 2.7), is a lattice isomorphism on $\mathcal{L}_b(E,F)$ for all $y \in G$. Therefore, for each $y \in G$, the fixed point space

$$\{T \in \mathcal{L}_b(E,F) : T = S_y(T)\} = \{T \in \mathcal{L}_b(E,F) : \tau_y T = T\tau_y\}$$

of S_y is a vector sublattice of $\mathcal{L}_b(E,F)$; see Remark 2.7. Consequently,

$$\mathfrak{m}_b(E,F) = \bigcap\nolimits_{y\in G} \{T \in \mathcal{L}_b(E,F) : T = S_y(T)\}$$

is a vector sublattice of $\mathcal{L}_b(E,F)$. The final statement follows immediately from Lemma 5.2 (ii). □

As is well known, given $\lambda \in M(G)$ and $f \in L^1(G)$, the convolution $f * \lambda \in L^1(G)$ exists and satisfies $\|f * \lambda\|_1 \le \|\lambda\|_{M(G)} \|f\|_1$; see Chapter 1 of [**26**]. Consequently, for each $\lambda \in M(G)$, the continuous linear operator $T_\lambda^{(1)} : L^1(G) \to L^1(G)$, defined by $T_\lambda^{(1)} f = f * \lambda$, for $f \in L^1(G)$, satisfies $\left\|T_\lambda^{(1)}\right\| \le \|\lambda\|_{M(G)}$. It is a classical result that all operators $T \in \mathfrak{m}\left(L^1(G)\right)$ are of the form $T = T_\lambda^{(1)}$ for some unique $\lambda \in M(G)$; see, for example, Theorem 3.1.1 in [**15**]. As a consequence, we obtain the following result, which goes back to [**2**] (in the setting of locally compact groups). It should be kept in mind that $\mathcal{L}\left(L^1(G)\right) = \mathcal{L}_b\left(L^1(G)\right)$ (see Proposition 2.6) and so $\mathfrak{m}\left(L^1(G)\right) = \mathfrak{m}_b\left(L^1(G)\right)$.

PROPOSITION 5.4. *The map $\Phi : M(G) \to \mathfrak{m}(L^1(G))$, given by $\Phi(\lambda) = T_\lambda^{(1)}$, for $\lambda \in M(G)$, is a bijective, linear, lattice isomorphism and norm isometry. In particular, the modulus*

$$\left|T_\lambda^{(1)}\right| = T_{|\lambda|}^{(1)}, \quad \lambda \in M(G).$$

Moreover, $\left\|T_\lambda^{(1)}\right\| = \left\|T_\lambda^{(1)}\right\|_r = \|\lambda\|_{M(G)}$, *for* $\lambda \in M(G)$.

PROOF. Recall from Proposition 5.3 that $\mathfrak{m}(L^1(G)) = \mathfrak{m}_b(L^1(G))$ is a Banach sublattice of $\mathcal{L}_b(L^1(G))$. Evidently, Φ is linear. The map Φ is bijective and both Φ and Φ^{-1} are positive. Therefore, Φ is a linear lattice isomorphism from $M(G)$ onto $\mathfrak{m}(L^1(G))$, that is, $\left|T_\lambda^{(1)}\right| = T_{|\lambda|}^{(1)}$ for $\lambda \in M(G)$.

Let $\lambda \in M(G)$. It has already been observed that $\left\|T_\lambda^{(1)}\right\| \le \|\lambda\|_{M(G)}$. Since $\left\|T_\lambda^{(1)}\right\| = \left\|T_\lambda^{(1)}\right\|_r$ (see Proposition 2.6) and $\|\chi_G\|_1 = 1$, it follows that

$$\begin{aligned}\left\|T_\lambda^{(1)}\right\| &= \left\|T_\lambda^{(1)}\right\|_r = \left\| \left|T_\lambda^{(1)}\right| \right\| \\ &= \left\|T_{|\lambda|}^{(1)}\right\| \ge \|\chi_G * |\lambda|\|_1 = |\lambda|(G) = \|\lambda\|_{M(G)}.\end{aligned}$$

Hence, $\left\|T_\lambda^{(1)}\right\| = \left\|T_\lambda^{(1)}\right\|_r = \|\lambda\|_{M(G)}$ and so Φ is an isometry. □

6. Convolution operators via measures

According to Section 5, every measure $\lambda \in M(G)$ convolves $L^1(G)$ continuously into itself, that is, $\lambda * f \in L^1(G)$ for every $f \in L^1(G)$ and the corresponding convolution operator $T_\lambda^{(1)} : f \mapsto \lambda * f$ so defined is continuous from $L^1(G)$ into itself. So, given any pair of translation invariant B.f.s.' over G, necessarily satisfying $E, F \subseteq L^1(G)$, the operator $T_\lambda^{(1)}$ always has the property that $T_\lambda^{(1)}(E) \subseteq L^1(G)$. The aim of this section is to investigate the space of those measures λ satisfying $T_\lambda^{(1)}(E) \subseteq F$, in which case the corresponding convolution operator from E into F is denoted by $T_\lambda^{(E,F)}$, and to expose various properties of the subspace of $\mathcal{L}(E,F)$ consisting of all such operators $T_\lambda^{(E,F)}$. It is time to be more precise.

Throughout this section, G is an infinite compact abelian group with normalized Haar measure μ. Let E and F be non-zero translation invariant B.f.s.' over G. Suppose that $\lambda \in M(G)$ satisfies $f * \lambda \in F$ for all $f \in E$, briefly $E * \lambda \subseteq F$ or $T_\lambda^{(1)}(E) \subseteq F$. Then the linear map $T_\lambda = T_\lambda^{(E,F)} : E \to F$ is defined by $T_\lambda f = f * \lambda$, for $f \in E$, and we say that *the operator of convolution $T_\lambda : E \to F$ exists.*

Let E and F be non-zero translation invariant B.f.s.' over G. Denote by $M^{(E,F)}(G)$ the set of all $\lambda \in M(G)$ for which the convolution operator $T_\lambda = T_\lambda^{(E,F)} : E \to F$ exists. In other words,

$$M^{(E,F)}(G) = \left\{\lambda \in M(G) : T_\lambda^{(1)}(E) \subseteq F\right\}$$

and $T_\lambda^{(E,F)} = T_\lambda^{(1)}|_E : E \to F$, for $\lambda \in M^{(E,F)}(G)$, is the restriction to E of $T_\lambda^{(1)}$ considered to be taking its values in $F \subseteq L^1(G)$. Furthermore, we set

$$M_b^{(E,F)}(G) = \left\{\lambda \in M^{(E,F)}(G) : T_\lambda^{(E,F)} \text{ is order bounded}\right\}.$$

If $E = F$, then we denote the spaces $M^{(E,E)}(G)$ and $M_b^{(E,E)}(G)$ simply by $M^E(G)$ and $M_b^E(G)$, respectively, and the operator $T_\lambda^{(E,E)}$ briefly by T_λ^E, for each $\lambda \in M^E(G)$. For the special case when $E = L^p(G)$ and $F = L^r(G)$, with $1 \le p, r \le \infty$, the sets $M^{(L^p(G),L^r(G))}(G)$ and $M_b^{(L^p(G),L^r(G))}(G)$ are denoted by $M^{p,r}(G)$ and $M_b^{p,r}(G)$, respectively (and simply by $M^p(G)$ and $M_b^p(G)$ in the event that $p = r$). Of course, for $\lambda \in M^{p,r}(G)$, the corresponding convolution operator from $L^p(G)$ into $L^r(G)$ is denoted by $T_\lambda^{p,r}$ (briefly, $T_\lambda^{(p)}$ in the event that $p = r$).

It is clear that $M^{(E,F)}(G)$ and $M_b^{(E,F)}(G)$ are linear subspaces of $M(G)$. If $E = F$, then $M^E(G)$ and $M_b^E(G)$ are also unital subalgebras of $M(G)$ with respect to the convolution product. With the notation just introduced, it follows from the discussion prior to Proposition 5.4 that

$$M^1(G) = M_b^1(G) = M(G).$$

We begin by establishing some basic properties of convolutions.

LEMMA 6.1. *Let E and F be non-zero translation invariant B.f.s.'s over G and $\lambda \in M^{(E,F)}(G)$. Then $T_\lambda^{(E,F)}$ is norm continuous. Moreover, if $\lambda \in M_b^{(E,F)}(G)$, then $T_\lambda^{(E,F)}$ is also order continuous, that is, $T_\lambda^{(E,F)} \in \mathcal{L}_n(E,F)$.*

PROOF. For simplicity, write $T_\lambda = T_\lambda^{(E,F)}$. Since both E and F are continuously embedded in $L^1(G)$ (see Lemma 3.3 (i)) and the map $f \longmapsto f*\lambda$ is continuous from $L^1(G)$ into itself, the continuity of T_λ follows via the closed graph theorem.

Let $\lambda \in M_b^{(E,F)}(G)$ and $f_n \downarrow_n 0$ in E. Since $E \subseteq L^1(G)$, also $f_n \downarrow_n 0$ in $L^1(G)$, which implies that $\|f_n\|_1 \to 0$ as $n \to \infty$. Hence, $\|T_\lambda(f_n)\|_1 = \left\|T_\lambda^{(1)}(f_n)\right\|_1 \to 0$ as $n \to \infty$ and so $\inf_n |T_\lambda f_n| = 0$. We can conclude that $T_\lambda^{(E,F)} \in \mathcal{L}_n(E,F)$. □

The previous result, that is, $T_\lambda^{(E,F)} \in \mathcal{L}(E,F)$, together with $\tau_y T_\lambda^{(E,F)} = T_\lambda^{(E,F)}\tau_y$, for all $y \in G$ (which follows from the properties of convolution), imply that $T_\lambda^{(E,F)} \in \mathfrak{m}(E,F)$ whenever $\lambda \in M^{(E,F)}(G)$.

Another simple observation is the following result.

LEMMA 6.2. *Let E and F be non-zero translation invariant B.f.s.'s over G and $\lambda \in M^{(E,F)}(G)$. Then $T_\lambda^{(E,F)} \ge 0$ if and only if $\lambda \ge 0$. In particular, $T_\lambda^{(E,F)} = 0$ if and only if $\lambda = 0$.*

PROOF. Evidently, $T_\lambda^{(E,F)} \ge 0$ whenever $\lambda \ge 0$.

Suppose that $T_\lambda^{(E,F)} \ge 0$, that is, $f * \lambda \ge 0$ for all $f \in E^+$. Lemma 3.3(i) implies that the carrier of E is G. Hence, for a fixed $f \in L^1(G)^+$, there exists a sequence $(f_n)_{n=1}^\infty$ in E^+ such that $0 \le f_n \uparrow_n f$ and so $\lim_{n\to\infty} \|f - f_n\|_1 = 0$. The continuity of $T_\lambda^{(1)}$ implies that $f_n * \lambda \to f * \lambda$ in $L^1(G)$ as $n \to \infty$. Since $\{f_n * \lambda\}_{n=1}^\infty \subseteq L^1(G)^+$, it follows that $f * \lambda \in L^1(G)^+$. So, we have shown that $f * \lambda \in L^1(G)^+$ for every $f \in L^1(G)^+$. In particular, let $h \in C(G)^+ \subseteq L^1(G)^+$. Then $h * \lambda \in C(G)^+$ and hence,

$$\int_G h(-y)\, d\lambda(y) = (h * \lambda)(0) \ge 0.$$

Since $h \in C(G)^+$ is arbitrary, the Riesz representation theorem implies that $\lambda \ge 0$ in $M(G)$. □

Next we consider the space $M_b(E,F)$ in more detail. The following observation will be used. The spaces E and F need not be translation invariant. Recall that $\mathcal{L}(L^1(G)) = \mathcal{L}_b(L^1(G))$; see Proposition 2.6(ii).

LEMMA 6.3. *Let E and F be B.f.s.' over G with $E,F \subseteq L^1(G)$ and let $T : L^1(G) \to L^1(G)$ be a continuous linear operator. The following two statements are equivalent.*

(i) *$|T|(E) \subseteq F$, where $|T|$ is the modulus of T in $\mathcal{L}(L^1(G))$.*
(ii) *$T(E) \subseteq F$ and the restriction $T_E : E \to F$ of T to E is order bounded.*

Furthermore, in this case $|T_E| = |T|_E$ (the restriction of $|T|$ to E).

PROOF. Since $|Tf| \leq |T|(|f|)$ for all $f \in E$ and F is an ideal in $L^1(G)$, it is clear that (i) implies (ii).

Suppose now that (ii) holds. Since F is Dedekind complete, the modulus $|T_E|$ of T_E in $\mathcal{L}_b(E,F)$ is given by

$$|T_E|(u) = \sup\{|Tf| : f \in E, |f| \leq u\}, \quad u \in E^+,$$

where the supremum is taken in F (see (2.8)). Since F is an ideal in $L^1(G)$, this supremum coincides with the supremum taken in $L^1(G)$. Moreover, since E is an ideal in $L^1(G)$, we have $\{f \in E : |f| \leq u\} = \{f \in L^1(G) : |f| \leq u\}$ for all $u \in E^+$. Consequently,

$$|T_E|(u) = \sup\{|Tf| : f \in L^1(G), |f| \leq u\} = |T|(u), \quad u \in E^+,$$

from which it follows that $|T|(E) \subseteq F$ and $|T_E| = |T|_E$. The proof is complete. $\square$

PROPOSITION 6.4. *Let E and F be non-zero translation invariant B.f.s.' over G. The set $M_b^{(E,F)}(G)$ is given by*

$$M_b^{(E,F)}(G) = \left\{\lambda \in M(G) : |\lambda| \in M^{(E,F)}(G)\right\} \tag{6.1}$$

and it is an ideal in $M(G)$. The map $\lambda \longmapsto T_\lambda^{(E,F)}$, for $\lambda \in M_b^{(E,F)}(G)$, is a lattice isomorphism from $M_b^{(E,F)}(G)$ into $\mathfrak{m}_b(E,F)$ and hence, also into $\mathcal{L}_b(E,F)$.

PROOF. Recall from Proposition 5.3 that $\mathfrak{m}_b(E,F)$ is a vector sublattice of $\mathcal{L}_b(E,F)$. As observed in Proposition 5.4, $\left|T_\lambda^{(1)}\right| = T_{|\lambda|}^{(1)}$ for all $\lambda \in M(G)$. Let $\lambda \in M(G)$ satisfy $|\lambda| \in M^{(E,F)}(G)$, that is, $T_{|\lambda|}^{(E,F)} : E \to F$ exists. This means that $T_{|\lambda|}^{(1)}(E) \subseteq F$, that is, $\left|T_\lambda^{(1)}\right|(E) \subseteq F$. It follows from Lemma 6.3 (applied with $T = T_\lambda^{(1)}$) that also $T_\lambda^{(E,F)} : E \to F$ exists and is order bounded, that is, $\lambda \in M_b^{(E,F)}(G)$, with $\left|T_\lambda^{(E,F)}\right| = T_{|\lambda|}^{(E,F)}$. So, the right-side of (6.1) is contained in $M_b^{(E,F)}(G)$.

Let $\lambda \in M_b^{(E,F)}(G)$, that is, $T_\lambda^{(E,F)} : E \to F$ exists and is order bounded. It follows from Lemma 6.3 that $\left|T_\lambda^{(1)}\right| = T_{|\lambda|}^{(1)}$ maps E into F. Hence, $T_{|\lambda|}^{(E,F)} : E \to F$ exists, that is, $|\lambda| \in M^{(E,F)}(G)$. This proves that (6.1) holds.

Given $\lambda \in M(G)$, it follows from (6.1) that $\lambda \in M_b^{(E,F)}(G)$ if and only if $|\lambda| \in M_b^{(E,F)}(G)$. Fix $\lambda \in M_b^{(E,F)}(G)^+$ and suppose that $0 \leq \nu \leq \lambda$ in $M(G)$. From the definition of convolution it is clear that $\left|T_\nu^{(1)} f\right| \leq T_\lambda^{(1)}(|f|)$, for $f \in L^1(G)$. Consequently, if $f \in E$, then

$$\left|T_\nu^{(1)} f\right| \leq T_\lambda^{(1)}(|f|) = T_\lambda^{(E,F)}(|f|) \in F,$$

and so $T_\nu^{(1)} f \in F$. This shows that $T_\nu^{(1)}(E) \subseteq F$, that is, $\nu \in M^{(E,F)}(G)^+ = M_b^{(E,F)}(G)^+$. We can conclude that $M_b^{(E,F)}(G)$ is an ideal in $M(G)$. By the first part of the current proof $\left|T_\lambda^{(E,F)}\right| = T_{|\lambda|}^{(E,F)}$, for each $\lambda \in M_b^{(E,F)}(G)$, and so the map $\lambda \longmapsto T_\lambda^{(E,F)}$ is a lattice homomorphism from $M_b^{(E,F)}(G)$ into $\mathfrak{m}_b(E,F)$. Finally, by Lemma 6.2, this map is also injective. □

Next we exhibit some large classes of translation invariant B.f.s.' E for which the convolution operator $T_\lambda^E : E \to E$ exists for *every* $\lambda \in M(G)$. Recall from Section 4 that E_h is the homogeneous part of E, which is a translation invariant Banach sublattice of E but, not necessarily an ideal in E.

LEMMA 6.5. *Let $E \neq \{0\}$ be a translation invariant B.f.s. over G. For each $\lambda \in M(G)$, the operator of convolution $T_\lambda : E_h \to E_h \subseteq E$ exists and satisfies $\|T_\lambda\| \leq \|\lambda\|_{M(G)}$, where $\|T_\lambda\|$ is the operator norm of T_λ in $\mathcal{L}(E_h)$.*

PROOF. Fix $\lambda \in M(G)$ and let $f \in E_h$. Then the function $F(y) = \tau_y f$, for $y \in G$, is continuous from G into the Banach space E_h and so it is strongly $|\lambda|$ -measurable and has bounded range. By the Radon-Nikodym theorem there exists a $\mathcal{B}(G)$-measurable function $\varphi : G \to \mathbb{C}$ satisfying $|\varphi| = \chi_G$ and $\lambda = \varphi \cdot |\lambda|$. Then the function $\varphi F : G \to E_h$ is also strongly $|\lambda|$-measurable and has bounded range. So, φF is Bochner $|\lambda|$-integrable over G (see Theorem II.2.2 in [**6**]) and we define the Bochner λ-integral by $\int_G^{(B)} \tau_y f d\lambda(y) = \int_G^{(B)} \varphi F d|\lambda| \in E_h$. Since $E_h \subseteq E \subseteq L^1(G)$, it follows that

$$\int_G \left(\int_G |f(x-y) g(y)| \, d\mu(x) \right) d|\lambda|(y) \leq |\lambda|(G) \|f\|_1 \|g\|_\infty < \infty,$$

for all $g \in L^\infty(G)$. Fubini's theorem implies that

$$\left\langle \int_G^{(B)} \tau_y f d\lambda(y), g \right\rangle = \int_G \langle \tau_y f, g \rangle \, d\lambda(y) = \langle f * \lambda, g \rangle, \quad g \in L^\infty(G),$$

and so $f * \lambda = \int_G^{(B)} \tau_y f d\lambda$. This shows that $T_\lambda : E_h \to E_h$ exists. Moreover, it follows via Theorem II.2.4 (ii) in [**6**] that

$$\|T_\lambda f\|_{E_h} = \left\| \int_G^{(B)} \tau_y f d\lambda(y) \right\|_{E_h} \leq \int_G \|\tau_y f\|_{E_h} d|\lambda|(y) = |\lambda|(G) \|f\|_{E_h}.$$

Since $f \in E_h$ is arbitrary, we can conclude that $\|T_\lambda\| \leq |\lambda|(G)$. □

If E is a translation invariant B.f.s. over G with o.c.-norm, then it follows from Proposition 3.4 (i) that $E = E_h$. Accordingly, the next result is an immediate consequence of Proposition 6.4 and Lemma 6.5.

THEOREM 6.6. *Let $E \neq \{0\}$ be a translation invariant B.f.s. over G with o.c.-norm. Then $M^E(G) = M_b^E(G) = M(G)$ and*

$$\left\|T_\lambda^E\right\| \leq \|\lambda\|_{M(G)}, \quad \lambda \in M(G).$$

The map $\lambda \longmapsto T_\lambda^E$, for $\lambda \in M(G)$, is a lattice isomorphism from $M(G)$ into $\mathfrak{m}_b(E)$ and hence, also into $\mathcal{L}_b(E)$.

For spaces with the Fatou property there is a similar result.

THEOREM 6.7. *Let $E \neq \{0\}$ be a translation invariant B.f.s. over G with the Fatou property. Then $M^E(G) = M_b^E(G) = M(G)$. Furthermore,*

$$\left\|T_\lambda^E\right\| \leq \|\lambda\|_{M(G)}, \quad \lambda \in M(G),$$

and the map $\lambda \longmapsto T_\lambda^E$, for $\lambda \in M(G)$, is a lattice isomorphism from $M(G)$ into $\mathfrak{m}_b(E)$.

PROOF. Assume first that $\lambda \in M(G)^+$. Fix $f \in E$ and define $h = |f| * \lambda \in L^1(G)^+$. Let $0 \leq g \in E^\times$ satisfy $\|g\|_{E^\times} \leq 1$. Then

$$\begin{aligned}\int_G h(x)\,g(x)\,d\mu(x) &= \int_G\left(\int_G |f(x-y)|\,g(x)\,d\lambda(y)\right)d\mu(x)\\ &= \int_G\left(\int_G |f(x-y)|\,g(x)\,d\mu(x)\right)d\lambda(y)\\ &= \int_G\left(\int_G \tau_y|f|(x)\,g(x)\,d\mu(x)\right)d\lambda(y).\end{aligned}$$

Since $\tau_y|f| \in E$ and $\|\tau_y|f|\|_E = \|f\|_E$ for all $y \in G$, it follows that

$$\int_G (\tau_y|f|)(x)\,g(x)\,d\mu(x) \leq \|\tau_y|f|\|_E\,\|g\|_{E^\times} \leq \|f\|_E, \quad y \in G.$$

Accordingly, $\int_G h(x)\,g(x)\,d\mu(x) \leq \|f\|_E\,\lambda(G)$. This shows that $h \in E^{\times\times}$ with $\|h\|_{E^{\times\times}} \leq \|f\|_E\,\lambda(G)$. By the Fatou property of E we have $E^{\times\times} = E$ isometrically; see Section 2. So, $h \in E$ with $\|h\|_E \leq \|f\|_E\,\lambda(G)$. But, $|f * \lambda| \leq h$ and so also $f * \lambda \in E$ with $\|f * \lambda\|_E \leq \|f\|_E\,\lambda(G)$. Hence, $T_\lambda^E : E \to E$ exists and satisfies $\left\|T_\lambda^E\right\| \leq \|\lambda\|_{M(G)}$.

It now follows from Proposition 6.4 that $M^E(G) = M_b^E(G) = M(G)$ and that the map $\lambda \longmapsto T_\lambda^E$, for $\lambda \in M(G)$, is a lattice isomorphism from $M(G)$ into $\mathcal{L}_b(E)$. Finally, if $\lambda \in M(G)$, then

$$\left\|T_\lambda^E\right\| \leq \left\|\,\left|T_\lambda^E\right|\,\right\| = \left\|T_{|\lambda|}^E\right\| \leq \|\,|\lambda|\,\|_{M(G)} = \|\lambda\|_{M(G)}.$$

The proof is thereby complete. □

REMARK 6.8. Denote by $M_d(G)$ the space of all discrete (that is, purely atomic) measures in $M(G)$. Given *any* translation invariant B.f.s. E over G, it can be shown that always $M_d(G) \subseteq M_b^E(G)$. There exist translation invariant B.f.s.' E with the property that $M^E(G) = M_b^E(G) = M_d(G)$. Of course, such B.f.s.' E necessarily satisfy $L^\infty(G) \nsubseteq E$. For details we refer to [**25**].

We end this section with a number of observations related to convolution operators that will be needed in the sequel.

The dual group of G is denoted by Γ (see, for example, Section 1.2 of [**26**]). For $x \in G$ and $\gamma \in \Gamma$ we write $\gamma(x) = (x, \gamma)$. Since Γ is frequently considered as a subset of $C(G)$, we shall write the group operation in Γ as *multiplication*, that is, if $\gamma_1, \gamma_2 \in \Gamma$, then $\gamma_1\gamma_2 \in \Gamma$ is given by

$$(x, \gamma_1\gamma_2) = (x, \gamma_1)(x, \gamma_2), \quad x \in G.$$

The identity element of Γ is, of course, χ_G. In particular,

$$(x, \gamma^{-1}) = \overline{(x, \gamma)} = (-x, \gamma), \quad x \in G, \quad \gamma \in \Gamma. \tag{6.2}$$

The linear span of Γ in $C(G)$ is denoted by $\tau(G)$ and is called the space of *trigonometric polynomials* on G.

Let E and F be two non-zero translation invariant B.f.s.' over G. Given $\gamma \in \Gamma$ and $T \in \mathcal{L}(E, F)$, the linear map $T^\gamma : E \to F$ is defined by

$$T^\gamma f = \gamma T(\bar{\gamma} f), \quad f \in E. \tag{6.3}$$

It is readily verified that $T^\gamma \in \mathcal{L}(E, F)$ with $\|T^\gamma\| = \|T\|$.

For every $\lambda \in M(G)$ and $f \in L^1(|\lambda|)$, the measure $f \cdot \lambda \in M(G)$ is defined by

$$(f \cdot \lambda)(A) = \int_A f d\lambda, \quad A \in \mathcal{B}(G).$$

Its variation measure satisfies $|f \cdot \lambda| = |f| \cdot |\lambda|$ and hence,

$$\|f \cdot \lambda\|_{M(G)} = \int_G |f|\, d|\lambda| = \|f\|_{L^1(|\lambda|)}, \quad \lambda \in M(G), \ f \in L^1(|\lambda|); \tag{6.4}$$

see e.g. Theorems 6.12 and 6.13 in [**27**].

The following fact follows via a straightforward computation. See, for example, the discussion after Lemma 1 in [**32**], where T_γ is the operator $f \mapsto \overline{T^\gamma f}$ with T^γ given by (6.3).

LEMMA 6.9. *Let* $\lambda \in M^{(E,F)}(G)$ *and* $\gamma \in \Gamma$. *Then* $\left(T_\lambda^{(E,F)}\right)^\gamma = T_{\gamma\cdot\lambda}^{(E,F)}$. *In particular,* $\gamma \cdot \lambda \in M^{(E,F)}(G)$. *Consequently,* $g \cdot \lambda \in M^{(E,F)}(G)$ *whenever* $\lambda \in M^{(E,F)}(G)$ *and* $g \in \tau(G)$.

The following general fact will also be needed.

LEMMA 6.10. *Let* B *be a closed linear subspace of* $M(G)$ *with the property that* $\gamma \cdot \lambda \in B$ *whenever* $\lambda \in B$ *and* $\gamma \in \Gamma$. *Then* B *is a band in* $M(G)$.

PROOF. Since B is a linear subspace of $M(G)$, it is clear that $f \cdot \lambda \in B$ whenever $\lambda \in B$ and $f \in \tau(G)$. Also note that the space $\tau(G)$ is dense in $C(G)$ with respect to the sup-norm by the Stone-Weierstrass theorem. For the fact that $\tau(G)$ separates the points of G see, for instance, Section 1.5.2 in [**26**] or Corollary 4.7.3 in [**16**].

Suppose that $0 \neq \lambda \in M(G)$. Fix $h \in L^1(|\lambda|)$ and let $\varepsilon > 0$. Since $C(G)$ is dense in $L^1(|\lambda|)$, there exists $g \in C(G)$ such that $\|h - g\|_{L^1(|\lambda|)} < \varepsilon/2$; see e.g. Theorem 3.14 in [**27**]. By the density of $\tau(G)$ in $C(G)$ there exists $f \in \tau(G)$ satisfying $\|g - f\|_\infty < \varepsilon/2\,|\lambda|(G)$ and hence, $\|g - f\|_{L^1(|\lambda|)} < \varepsilon/2$. Accordingly, $\|h - f\|_{L^1(|\lambda|)} < \varepsilon$. It follows that $\tau(G)$ is dense in $L^1(|\lambda|)$ for every non-zero $\lambda \in M(G)$.

To show that B is an ideal in $M(G)$, let $\lambda \in B$ and $\nu \in M(G)$ satisfy $|\nu| \leq |\lambda|$. There exists a function $h \in L^1(|\lambda|)$ such that $\nu = h \cdot \lambda$ (consider Theorem 6.12 in [**27**] in combination with the Radon-Nikodym theorem; actually, we can choose h such that $|h(x)| \leq 1$ for all $x \in G$). Let $(h_n)_{n=1}^{\infty}$ be a sequence in $\tau(G)$ such that $\|h - h_n\|_{L^1(|\lambda|)} \to 0$ as $n \to \infty$. As observed in the beginning of the proof, we have that $h_n \cdot \lambda \in B$ for all $n \in \mathbb{N}$. It follows from (6.4) that

$$\|\nu - (h_n \cdot \lambda)\|_{M(G)} = \|(h \cdot \lambda) - (h_n \cdot \lambda)\|_{M(G)} = \|h - h_n\|_{L^1(|\lambda|)}, \quad n \in \mathbb{N},$$

and hence, that $\|\nu - (h_n \cdot \lambda)\|_{M(G)} \to 0$ as $n \to \infty$. By hypothesis B is closed in $M(G)$ and so we can conclude that $\nu \in B$. Hence, B is an ideal in $M(G)$.

Since $M(G)$ has o.c.-norm, this implies that B is a band in $M(G)$ (cf. Theorem 105.2 in [**31**]). The proof is thereby complete. □

7. Vector measures associated with convolution

In this section we discuss some properties of, in general, finitely additive vector measures taking values in a translation invariant B.f.s. E over G and which arise via convolution with scalar measures. Hence, they necessarily commute with translations. In particular, a characterization is given of those scalar measures for which the corresponding finitely additive vector measure is actually countably additive; see Theorem 7.9 below. It turns out that the homogeneous part E_h of E plays a crucial role for those vector measures which are absolutely continuous with respect to μ. Due to the nature of the arguments involved it is necessary to first treat the case when G is metrizable and then, with some effort, to remove this restriction. For the general theory of vector measures we refer, for instance, to Chapter I in [**6**]; see also [**22**].

As before, G is an infinite compact abelian group with normalized Haar measure μ. Let E be a translation invariant B.f.s. over G with $L^\infty(G) \subseteq E$. Recall that the inclusion of $L^\infty(G)$ into E is continuous. Actually, if $f \in L^\infty(G)$, then $|f| \leq \|f\|_\infty \chi_G$ and so $\|f\|_E \leq \|\chi_G\|_E \|f\|_\infty$. *It will be convenient to assume throughout this section that* $\|\chi_G\|_E = 1$. This can always be achieved, by replacing the norm $\|\cdot\|_E$ with the translation invariant, B.f.s.-norm $\|\chi_G\|_E^{-1} \|\cdot\|_E$. For this norm it follows that

$$\|f\|_E \leq \|f\|_\infty, \quad f \in L^\infty(G). \tag{7.1}$$

As is well known (cf. also Theorem 6.7), if $\lambda \in M(G)$ and $f \in L^\infty(G)$, then $f * \lambda \in L^\infty(G)$ with $\|f * \lambda\|_\infty \leq \|f\|_\infty \|\lambda\|_{M(G)}$. In view of (7.1) also

$$\|f * \lambda\|_E \leq \|f\|_\infty \|\lambda\|_{M(G)}, \quad f \in L^\infty(G), \ \lambda \in M(G). \tag{7.2}$$

This shows that $\lambda \in M^{(L^\infty(G),E)}(G)$ with $\left\|T_\lambda^{(L^\infty(G),E)}\right\| \leq \|\lambda\|_{M(G)}$ for all $\lambda \in M(G)$. Therefore (cf. Proposition 6.4),

$$M^{(L^\infty(G),E)}(G) = M_b^{(L^\infty(G),E)}(G) = M(G). \tag{7.3}$$

Let $\lambda \in M(G)$. Define the *finitely additive vector measure* $m_\lambda^E : \mathcal{B}(G) \to E$ via convolution, that is,

$$m_\lambda^E(A) = \chi_A * \lambda, \quad A \in \mathcal{B}(G). \tag{7.4}$$

Note that

$$m_\lambda^E(A)(x) = \lambda(-A + x), \quad x \in G, \ A \in \mathcal{B}(G), \tag{7.5}$$

and that $m_\lambda^E(A) = 0$ μ-a.e. on G whenever $A \in \mathcal{B}(G)$ satisfies $\mu(A) = 0$. It is readily verified (for instance, by using (7.5)) that m_λ^E *commutes with translations*, that is, $m_\lambda^E(A+y) = \tau_y(m_\lambda^E(A))$ for all $A \in \mathcal{B}(G)$ and all $y \in G$. In general, m_λ^E is *not* countably additive (for example, take $E = L^\infty(G)$ and the Dirac point measure $\lambda = \delta_0$, in which case $m_\lambda^E(A) = \chi_A$, for $A \in \mathcal{B}(G)$). However, if E has o.c.-norm, then m_λ^E is necessarily countably additive for every $\lambda \in M(G)$. Indeed, fix $\lambda \in M(G)$ and let $(A_n)_{n=1}^\infty$ be any sequence in $\mathcal{B}(G)$ satisfying $A_n \downarrow_n \emptyset$. Then (7.5) yields that $m_{|\lambda|}^E(A_n)(x) = |\lambda|(-A_n + x)$, for $x \in G$, which implies that $m_{|\lambda|}^E(A_n) \downarrow_n 0$ in E and so $\left\| m_{|\lambda|}^E(A_n) \right\|_E \to 0$ as $n \to \infty$. From (7.5) it is also clear that $\left| m_\lambda^E(A_n) \right| \le m_{|\lambda|}^E(A_n)$ and hence, that $\left\| m_\lambda^E(A_n) \right\|_E \le \left\| m_{|\lambda|}^E(A_n) \right\|_E$ for all $n \in \mathbb{N}$. Consequently, $\left\| m_\lambda^E(A_n) \right\|_E \to 0$ as $n \to \infty$. We can conclude that m_λ^E is countably additive. Note, whenever T_λ^E exists (e.g. if E has o.c.-norm or the Fatou property), that $m_\lambda^E(A) = T_\lambda^E(\chi_A)$, for $A \in \mathcal{B}(G)$.

The set of all $\lambda \in M(G)$ for which the finitely additive vector measure $m_\lambda^E : \mathcal{B}(G) \to E$ is countably additive is denoted by $M_{ca}(E;G)$, that is,

$$M_{ca}(E;G) = \left\{ \lambda \in M(G) : m_\lambda^E \text{ is countably additive} \right\}.$$

In this connection, the following result is of some interest. Let $m : \mathcal{B}(G) \to E$ be a countably additive measure. We say that m is *absolutely continuous with respect to* μ if $\lim_{\mu(A)\to 0} m(A) = 0$, which is denoted by $m \ll \mu$. A set $A \in \mathcal{B}(G)$ is called m-null if $m(B) = 0$ for every set $B \in \mathcal{B}(G)$ with $B \subseteq A$. By a theorem of Pettis (see Theorem I.2.1 in [**6**]), $m \ll \mu$ if and only if every μ-null set in $\mathcal{B}(G)$ is also m-null.

LEMMA 7.1. *Let $E \neq \{0\}$ be a translation invariant B.f.s. over G. Let $m : \mathcal{B}(G) \to E$ be any countably additive vector measure which commutes with translations and satisfies $m \ll \mu$. Then its range $\{m(A) : A \in \mathcal{B}(G)\}$ is contained in the homogeneous part E_h of E.*

PROOF. Let $A \in \mathcal{B}(G)$ and fix $\varepsilon > 0$. Since $m \ll \mu$, there exists a $\delta > 0$ such that

$$B \in \mathcal{B}(G) \text{ with } \mu(B) \le \delta \quad \Rightarrow \quad \|m(B)\|_E \le \varepsilon. \tag{7.6}$$

Since $\mu(A\Delta(A+x)) = \|\chi_A - \tau_x\chi_A\|_1 \to 0$ for $x \to 0$ (as $L^1(G)_h = L^1(G)$), there exists a neighbourhood U of 0 in G such that

$$\mu(A\Delta(A+x)) \le \delta, \quad \forall\, x \in U.$$

Furthermore, for each $x \in G$, the identities

$$\begin{aligned} m(A) - \tau_x m(A) &= m(A) - m(A+x) \\ &= m(A\backslash(A+x)) - m((A+x)\backslash A) \end{aligned}$$

imply that

$$\|m(A) - \tau_x m(A)\|_E \le \|m(A\backslash(A+x))\|_E + \|m((A+x)\backslash A)\|_E. \tag{7.7}$$

Since both $A\backslash(A+x)$ and $(A+x)\backslash A$ are contained in $A\Delta(A+x)$, it follows from (7.6) and (7.7) that $\|m(A) - \tau_x m(A)\|_E \le 2\varepsilon$, for all $x \in U$. This shows that $m(A) \in E_h$. □

COROLLARY 7.2. *Let E be a translation invariant B.f.s. over G with $L^{\infty}(G) \subseteq E$ and let $\lambda \in M_{ca}(E;G)$. Then $m_{\lambda}^{E} \ll \mu$ and $m_{\lambda}^{E}(A) \in E_h$ for all $A \in \mathcal{B}(G)$.*

PROOF. As noted after (7.5), $m_{\lambda}^{E}(A) = 0$ μ -a.e. on G whenever $A \in \mathcal{B}(G)$ satisfies $\mu(A) = 0$, that is, each μ-null set is an m_{λ}^{E} -null set. By Pettis' theorem, this is equivalent to $m_{\lambda}^{E} \ll \mu$ as m_{λ}^{E} is countably additive. The final statement follows from Lemma 7.1. □

The above observations motivate the following definition.

DEFINITION 7.3. Let E be a translation invariant B.f.s. over G with $L^{\infty}(G) \subseteq E$. A measure $\lambda \in M(G)$ is called *E-homogeneous* if the range

$$m_{\lambda}^{E}(\mathcal{B}(G)) = \{\chi_A * \lambda : A \in \mathcal{B}(G)\} \subseteq E_h.$$

In this case, $m_{\lambda}^{E} : \mathcal{B}(G) \to E_h$ is a finitely additive vector measure. We denote the set of all E -homogeneous measures by $M_h(E;G)$.

Recall that the condition $L^{\infty}(G) \subseteq E$ implies that $E_h = \overline{C(G)}^{E}$ and so surely $E_h \neq \{0\}$ in this case; cf. Proposition 4.3.

It is clear that $M_{ca}(E;G)$ and $M_h(E;G)$ are both linear subspaces of $M(G)$. According to Corollary 7.2, it is always the case that $M_{ca}(E;G) \subseteq M_h(E;G)$. One of the main results in this section is that actually $M_{ca}(E;G) = M_h(E;G)$ and that this set is a band in $M(G)$; see Theorem 7.9 below.

In Theorem 1 of [**21**] the particular case when $E = L^{\infty}(G)$ is discussed in some detail. Recalling that $L^{\infty}(G)_h = C(G)$ (see the discussion after Definition 4.1), it was shown in [**21**] that $M_{ca}(L^{\infty}(G);G) = M_h(L^{\infty}(G);G)$ and that a measure $\lambda \in M(G)$ belongs to $M_{ca}(L^{\infty}(G);G)$ if and only if $\lambda \ll \mu$, that is, $M_{ca}(L^{\infty}(G);G) = L^1(G) \subseteq M(G)$ (where $L^1(G)$ is identified isometrically with a closed subspace of $M(G)$ via the map $f \mapsto f \cdot \mu$, for $f \in L^1(G)$). As observed above, $M_{ca}(E;G) = M(G)$ whenever E has o.c.-norm.

Let E be a translation invariant B.f.s. over G with $L^{\infty}(G) \subseteq E$. Since the inclusion of $L^{\infty}(G)$ into E is continuous, it is clear that $M_{ca}(L^{\infty}(G);G) \subseteq M_{ca}(E;G)$. Therefore, it is always the case that

$$L^1(G) \subseteq M_{ca}(E;G).$$

Our next objective is to establish that $M_h(E;G)$ is a band in $M(G)$; see Lemma 7.7 below. Some preparation is needed.

LEMMA 7.4. *Let E be a translation invariant B.f.s. over G with $L^{\infty}(G) \subseteq E$ and $\lambda \in M_h(E;G)$. Then*

$$T_{\lambda}^{(L^{\infty}(G),E)}(L^{\infty}(G)) \subseteq E_h.$$

PROOF. For convenience we write $T_{\lambda} = T_{\lambda}^{(L^{\infty}(G),E)}$. By hypothesis $T_{\lambda}(\chi_A) \in E_h$ for all $A \in \mathcal{B}(G)$. Since E_h is a linear subspace of E, this implies that $T_{\lambda}(s) \in E_h$ for all $\mathcal{B}(G)$-simple functions $s : G \to \mathbb{C}$. Given $f \in L^{\infty}(G)$, there exists a sequence $(s_n)_{n=1}^{\infty}$ of $\mathcal{B}(G)$-simple functions satisfying $\|f - s_n\|_{\infty} \to 0$ as $n \to \infty$. The operator $T_{\lambda} : L^{\infty}(G) \to E$ is continuous (cf. (7.2)) and hence, $T_{\lambda}(s_n) \to T_{\lambda}(f)$ as $n \to \infty$ in the norm of E. But, E_h is closed in E (cf. Section 4) and so we can conclude that $T_{\lambda}(f) \in E_h$. The proof is complete. □

LEMMA 7.5. *Let E be a translation invariant B.f.s. over G with $L^\infty(G) \subseteq E$. The linear subspace $M_h(E;G)$ is norm closed in $M(G)$.*

PROOF. Suppose that $\lambda \in M(G)$ and $(\lambda_n)_{n=1}^\infty$ is a sequence in $M_h(E;G)$ satisfying $\|\lambda - \lambda_n\|_{M(G)} \to 0$ as $n \to \infty$. Let $A \in \mathcal{B}(G)$. Via (7.2), it follows that

$$\left\|m_\lambda^E(A) - m_{\lambda_n}^E(A)\right\|_E = \|\chi_A * (\lambda - \lambda_n)\|_E \leq \|\lambda - \lambda_n\|_{M(G)}, \quad n \in \mathbb{N},$$

and so $m_{\lambda_n}^E(A) \to m_\lambda^E(A)$ as $n \to \infty$ in the norm of E. But, $m_{\lambda_n}^E(A) \in E_h$ for all n and E_h is closed in E. So, we can conclude that $m_\lambda^E(A) \in E_h$. Since $A \in \mathcal{B}(G)$ is arbitrary, this completes the proof. □

Recall from (6.3) that, for each $T \in \mathcal{L}(L^\infty(G), E)$ and $\gamma \in \Gamma$, the operator $T^\gamma \in \mathcal{L}(L^\infty(G), E)$ is defined by $T^\gamma f = \gamma T(\bar{\gamma} f)$, for $f \in L^\infty(G)$. Furthermore, whenever $L^\infty(G) \subseteq E$, it follows from (7.3) and Lemma 6.9 (applied in the present setting) that

$$\left(T_\lambda^{(L^\infty(G),E)}\right)^\gamma = T_{\gamma\cdot\lambda}^{(L^\infty(G),E)}, \quad \lambda \in M(G) = M^{(L^\infty(G),E)}(G), \ \gamma \in \Gamma. \tag{7.8}$$

LEMMA 7.6. *Let E be a translation invariant B.f.s. over G with $L^\infty(G) \subseteq E$ and $\lambda \in M_h(E;G)$. Then $\gamma \cdot \lambda \in M_h(E;G)$ for all $\gamma \in \Gamma$.*

PROOF. We again write $T_\lambda = T_\lambda^{(L^\infty(G),E)}$. Let $\gamma \in \Gamma$ and $A \in \mathcal{B}(G)$ be fixed. Then (7.8) yields that

$$m_{\gamma\cdot\lambda}^E(A) = T_{\gamma\cdot\lambda}(\chi_A) = (T_\lambda)^\gamma(\chi_A) = \gamma T_\lambda(\bar{\gamma}\chi_A).$$

Since $\bar{\gamma}\chi_A \in L^\infty(G)$, it follows from Lemma 7.4 that $T_\lambda(\bar{\gamma}\chi_A) \in E_h$. Now Lemma 4.7 implies that $\gamma T_\lambda(\bar{\gamma}\chi_A) \in E_h$, as $\gamma \in C(G)$. Hence, $m_{\gamma\cdot\lambda}(A) \in E_h$. The proof is thereby complete. □

The following result is an immediate consequence of Lemma 6.10 in combination with Lemmas 7.5 and 7.6.

LEMMA 7.7. *Let E be a translation invariant B.f.s. over G with $L^\infty(G) \subseteq E$. Then $M_h(E;G)$ is a band in $M(G)$.*

For the proof of the main result (cf. Theorem 7.9 below) some further preparation is needed. First we prove a special case of this result.

LEMMA 7.8. *Let G be metrizable and E be a translation invariant B.f.s. over G with $L^\infty(G) \subseteq E$. Then $M_h(E;G) = M_{ca}(E;G)$.*

PROOF. Since G is metrizable, the Banach space $C(G)$ is separable with respect to $\|\cdot\|_\infty$; see e.g. **[13]** , Theorem 26.15. Since $E_h = \overline{C(G)}^E$ (cf. Proposition 4.3) and the embedding of $C(G)$ into E is continuous (as the embedding of $L^\infty(G)$ into E is continuous), this implies that E_h is a norm separable Banach space. Therefore, E_h cannot contain an isomorphic Banach space copy of ℓ^∞. Fix $\lambda \in M_h(E;G)$. By **[6]**, Corollary I.4.7, to prove that m_λ^E is countably additive, it suffices to show that $\langle m_\lambda^E, x^*\rangle$ is countably additive for all $x^* \in \Lambda$, where $\Lambda \subseteq E^*$ is any total set.

Choose $\Lambda = E^{\times} \subseteq E^{*}$ and select $g \in E^{\times}$. For each $A \in \mathcal{B}(G)$ we have, via Fubini's theorem, that

$$\begin{aligned}\langle m_{\lambda}^{E}, g\rangle (A) = \langle m_{\lambda}^{E}(A), g\rangle &= \int_{G} m_{\lambda}^{E}(A)\, g\, d\mu \\ &= \int_{G}\left(\int_{G}\chi_{A}(x-y)\, d\lambda(y)\right) g(x)\, d\mu(x) \\ &= \int_{G}\left(\int_{G}\chi_{A}(x-y)\, g(x)\, d\mu(x)\right) d\lambda(y) \\ &= \int_{G}\left(\int_{G}\chi_{A}(x)\, g(x+y)\, d\mu(x)\right) d\lambda(y) \\ &= \int_{A}\left(\int_{G} g(x+y)\, d\lambda(y)\right) d\mu(x) = \int_{A}\left(g * \tilde{\lambda}\right) d\mu,\end{aligned}$$

where $\tilde{\lambda}$ is the reflection of λ, that is, $\tilde{\lambda}(B) = \lambda(-B)$ for $B \in \mathcal{B}(G)$. This shows that $\langle m_{\lambda}^{E}, g\rangle = \left(g * \tilde{\lambda}\right)\cdot \mu$. Noting that $g * \tilde{\lambda} \in L^{1}(G)$ (actually, $g * \tilde{\lambda} \in E^{\times}$) it follows that $\langle m_{\lambda}^{E}, g\rangle$ is countably additive. Since $E^{\times}$ is a total set, that is, it separates the points of E, we can conclude that m_{λ}^{E} is countably additive. This completes the proof. $\square$

To relax the condition of G being metrizable will require some effort. First some preliminary remarks. Let $H \subseteq G$ be a closed subgroup, G/H be its quotient group and $\pi : G \to G/H$ be the quotient map. Then G/H is a compact abelian group and π is a continuous, open and surjective group homomorphism. Normalized Haar measure $\mu_{G/H}$ on G/H is the image measure under π of normalized Haar measure μ on G, that is,

$$\mu_{G/H}(A) = \mu\left(\pi^{-1}(A)\right), \quad A \in \mathcal{B}(G/H).$$

Given $x \in G$, its coset $\pi x \in G/H$ is also denoted by $[x]$. The induced map $\pi_{*} : L^{0}(G/H) \to L^{0}(G)$, defined by $\pi_{*}g = g \circ \pi$, for $g \in L^{0}(G/H)$, is an order continuous, lattice isomorphism of $L^{0}(G/H)$ into $L^{0}(G)$ and $\pi_{*}(\chi_{G/H}) = \chi_{G}$.

Given any translation invariant B.f.s. $E \subseteq L^{0}(G)$, define

$$E_{H} = \pi_{*}^{-1}(E) = \left\{g \in L^{0}(G/H) : \pi_{*}g \in E\right\}. \tag{7.9}$$

Note that E_{H} is an ideal in $L^{0}(G/H)$; see [**19**], Theorem 59.2 (iv). Indeed, let $g_{1} \in L^{0}(G/H)$ and $g_{2} \in E_{H}$ satisfy $|g_{1}| \leq |g_{2}|$. Then the inequality

$$|\pi_{*}g_{1}| = \pi_{*}|g_{1}| \leq \pi_{*}|g_{2}| = |\pi_{*}g_{2}|, \tag{7.10}$$

together with $\pi_{*}g_{2} \in E$, yields $\pi_{*}g_{1} \in E$, that is, $g_{1} \in E_{H}$.

Next define

$$\|g\|_{E_{H}} = \|\pi_{*}g\|_{E}, \quad g \in E_{H}. \tag{7.11}$$

It is clear that $\|\cdot\|_{E_{H}}$ is a norm on E_{H} satisfying $\|g_{1}\|_{E_{H}} \leq \|g_{2}\|_{E_{H}}$ whenever $|g_{1}| \leq |g_{2}|$ in E_{H} (cf. (7.10)). Evidently, the map $\pi_{*} : E_{H} \to E$ is an isometric, lattice isomorphism of E_{H} into E. To show that E_{H} is complete, it suffices to verify the Riesz-Fischer property in E_{H} (see e.g. [**30**] , Section 64). So, let $(g_{n})_{n=1}^{\infty}$ be any sequence in E_{H}^{+} satisfying $\sum_{n=1}^{\infty}\|g_{n}\|_{E_{H}} < \infty$. This implies that $\sum_{n=1}^{\infty}\|\pi_{*}g_{n}\|_{E} < \infty$ (cf. (7.11)) and so, since E is complete, the pointwise sum $f = \sum_{n=1}^{\infty}\pi_{*}g_{n} \in E$. Defining $g([x]) = \sum_{n=1}^{\infty} g_{n}([x])$, for $[x] \in G/H$, and noting that

$g_n([x]) = (\pi_* g_n)(x)$ for $x \in G$, this implies that $g([x]) < \infty$, $\mu_{G/H}$-a.e. on G/H, that is, $g \in L^0(G/H)^+$. Since π_* is order continuous, it follows that $\pi_* g = f \in E$. Hence, $g \in E_H$, which shows that E_H has the Riesz-Fischer property. Consequently, E_H is a B.f.s. over G/H.

A direct computation yields, for each $y \in G$, that

$$(\tau_y \circ \pi_*)(g) = \left(\pi_* \circ \tau_{[y]}\right)(g), \quad g \in L^0(G/H), \tag{7.12}$$

from which it follows that E_H is a translation invariant B.f.s. over G/H. Moreover, if $L^\infty(G) \subseteq E$, then $L^\infty(G/H) \subseteq E_H$. In fact, $\pi_*(\chi_{G/H}) = \chi_G \in E$ and so $\chi_{G/H} \in E_H$. Also note that $\|\chi_{G/H}\|_{E_H} = 1$ whenever it is assumed that $\|\chi_G\|_E = 1$.

An important identity (via (7.11) and (7.12)) is that

$$\begin{aligned} \left\|g - \tau_{[y]} g\right\|_{E_H} &= \left\|\pi_* g - \pi_* \tau_{[y]} g\right\|_E \\ &= \left\|\pi_* g - \tau_y \pi_* g\right\|_E, \quad g \in E_H, \ y \in G. \end{aligned} \tag{7.13}$$

This identity implies that the homogeneous part $(E_H)_h$ of E_H is given by

$$(E_H)_h = \{g \in E_H : \pi_* g \in E_h\}. \tag{7.14}$$

Indeed, suppose that $g \in E_H$ satisfies $\pi_* g \in E_h$. Given $\varepsilon > 0$, there exists an open neighbourhood U of 0 in G such that $\|\pi_* g - \tau_y \pi_* g\|_E \leq \varepsilon$ for all $y \in U$. It follows from (7.13) that $\left\|g - \tau_{[y]} g\right\|_{E_H} \leq \varepsilon$ for all $[y] \in \pi(U)$, with $\pi(U)$ being an open neighbourhood of $[0]$ as π is an open map. Hence, $g \in E_H$. Conversely, suppose that $g \in (E_H)_h$. Given $\varepsilon > 0$, there exists an open neighbourhood V of $[0]$ in G/H such that $\left\|g - \tau_{[y]} g\right\|_{E_H} \leq \varepsilon$ for all $[y] \in V$. It follows from (7.13) that $\|\pi_* g - \tau_y \pi_* g\|_E \leq \varepsilon$ for all $y \in \pi^{-1}(V)$, with $\pi^{-1}(V)$ being an open neighbourhood of 0 in G as π is continuous. Hence, $\pi_* g \in E_h$. This proves (7.14).

Suppose that $E \subseteq L^0(G)$ is a translation invariant B.f.s. over G and $L^\infty(G) \subseteq E$. Let $\lambda \in M(G)^+$ be given. As before, define the finitely additive vector measure $m_\lambda^E : \mathcal{B}(G) \to L^\infty(G) \subseteq E$ by $m_\lambda^E(A) = \chi_A * \lambda$, for $A \in \mathcal{B}(G)$. Let ν be the image measure on $\mathcal{B}(G/H)$ of λ under π, that is,

$$\nu(A) = \lambda\left(\pi^{-1}(A)\right), \quad A \in \mathcal{B}(G/H).$$

Then ν is regular (see, for example, Theorem 418I in [**10**]), that is, $\nu \in M(G/H)^+$. The corresponding vector measure $m_\nu^{E_H} : \mathcal{B}(G/H) \to E_H$ is then defined by $m_\nu^{E_H}(A) = \chi_A * \nu$, for $A \in \mathcal{B}(G/H)$.

The claim is that

$$\pi_* m_\nu^{E_H}(A) = m_\lambda^E\left(\pi^{-1}(A)\right), \quad A \in \mathcal{B}(G/H). \tag{7.15}$$

Indeed, let $A \in \mathcal{B}(G/H)$. It is easy to verify that $\pi^{-1}(-A + [x]) = -\pi^{-1}(A) + x$ for every $x \in G$. It follows from (7.5), applied to ν in G/H and to λ in G that, for each $x \in G$, we have

$$\begin{aligned} \left(\pi_* m_\nu^{E_H}(A)\right)(x) &= \left(m_\nu^{E_H}(A)\right)([x]) = \nu(-A + [x]) \\ &= \lambda\left(\pi^{-1}(-A + [x])\right) = \lambda\left(-\pi^{-1}(A) + x\right) \\ &= m_\lambda^E\left(\pi^{-1}(A)\right)(x). \end{aligned}$$

This proves the claim, that is, the identity (7.15).

Suppose now that $\lambda \in M_h(E;G)^+$, that is, $\lambda \in M(G)^+$ is E-homogeneous. It follows from (7.15) that

$$\pi_* m_\nu^{E_H}(A) = m_\lambda^E\left(\pi^{-1}(A)\right) \in E_h, \quad A \in \mathcal{B}(G/H),$$

and so (7.14) implies that $m_\nu^{E_H}(A) \in (E_H)_h$ for every $A \in \mathcal{B}(G/H)$. Hence, ν is E_H-homogeneous.

We are now in a position to prove the main result.

THEOREM 7.9. *Let E be a translation invariant B.f.s. over G such that $L^\infty(G) \subseteq E$. Then $M_h(E;G) = M_{ca}(E;G)$ and this space is a band in $M(G)$.*

PROOF. The inclusion $M_{ca}(E;G) \subseteq M_h(E;G)$ follows from Corollary 7.2. So, we need to show that $M_h(E;G) \subseteq M_c(E;G)$. Since $M_h(E;G)$ is a band in $M(G)$ (see Lemma 7.7), it follows, in particular, that each $\lambda \in M_h(E;G)$ is a linear combination of at most four elements from $M_h(E;G)^+$. Consequently, we only have to show that m_λ^E is countably additive for every $\lambda \in M_h(E;G)^+$.

Fix $\lambda \in M_h(E;G)^+$. Clearly m_λ^E then takes its values in E^+. The proof is divided into five steps.

Step 1. Let $(K_n)_{n=1}^\infty$ be any sequence of compact subsets of G satisfying $\mu(K_n) \to 0$ as $n \to \infty$. *The claim is that* $\left\|m_\lambda^E(K_n)\right\|_E \to 0$ *as* $n \to \infty$.

The regularity of μ implies that there exist open sets O_n in G such that $K_n \subseteq O_n$ for each $n \in \mathbb{N}$ and $\mu(O_n) \to 0$ as $n \to \infty$. For each $n \in \mathbb{N}$, there exists an open neighbourhood U_n of 0 in G satisfying $K_n + U_n \subseteq O_n$. By Theorem 8.7 of [**12**], there exists a closed subgroup $H \subseteq G$ such that $H \subseteq \bigcap_{n=1}^\infty U_n$ and G/H is metrizable. Let E_H be the translation invariant B.f.s. over G/H as specified in (7.9). Let $\nu \in M(G/H)^+$ be the image measure of λ under π. As follows from the observations preceding the present theorem, ν is E_H-homogeneous, that is, $\nu \in M_h(E_H;G/H)$. Since G/H is metrizable, it follows from Lemma 7.8 that the corresponding vector measure $m_\nu^{E_H} : \mathcal{B}(G/H) \to E_H$ is countably additive. This implies, in particular, that $m_\nu^{E_H} \ll \mu_{G/H}$; see Corollary 7.2.

For each $n \in \mathbb{N}$, define the set $W_n = \pi(K_n)$, which is a compact subset of G/H satisfying $\pi^{-1}(W_n) = H + K_n$. Since $H + K_n \subseteq U_n + K_n \subseteq O_n$, for $n \in \mathbb{N}$, it follows that

$$\mu_{G/H}(W_n) = \mu\left(\pi^{-1}(W_n)\right) = \mu(H + K_n) \leq \mu(U_n + K_n) \leq \mu(O_n)$$

and hence, $\mu_{G/H}(W_n) \to 0$ as $n \to \infty$. But, $m_\nu^{E_H} \ll \mu_{G/H}$ and so $\left\|m_\nu^{E_H}(W_n)\right\|_{E_H} \to 0$ as $n \to \infty$. Using that m_λ^E is positive in E and that $K_n \subseteq \pi^{-1}(W_n)$, the identity (7.15) implies that

$$0 \leq m_\lambda^E(K_n) \leq m_\lambda^E\left(\pi^{-1}(W_n)\right) = \pi_* m_\nu^{E_H}(W_n), \quad n \in \mathbb{N},$$

and hence,

$$\left\|m_\lambda^E(K_n)\right\|_E \leq \left\|\pi_* m_\nu^{E_H}(W_n)\right\|_E = \left\|m_\nu^{E_H}(W_n)\right\|_{E_H}, \quad n \in \mathbb{N}.$$

Therefore, we can conclude that $\left\|m_\lambda^E(K_n)\right\|_E \to 0$ as $n \to \infty$. This proves the claim.

Step 2. Denote by

$$T_\lambda = T_\lambda^{(L^\infty(G),E)} : L^\infty(G) \to E$$

the convolution operator corresponding to λ. Let $(g_n)_{n=1}^{\infty}$ be any sequence in $C(G)$ such that $\|g_n\|_1 \to 0$ as $n \to \infty$ and for which there exists a constant $c > 0$ satisfying $|g_n| \le c\chi_G$ for all $n \in \mathbb{N}$. *The claim is that* $\|T_\lambda g_n\|_E \to 0$ *as* $n \to \infty$.

Indeed, given $\varepsilon > 0$, define the compact subsets $K_n \subseteq G$ by

$$K_n = \{x \in G : |g_n(x)| \ge \varepsilon\}, \quad n \in \mathbb{N}.$$

Since $\|g_n\|_1 \ge \int_{K_n} |g_n| \, d\mu \ge \varepsilon\mu(K_n)$, it is clear that $\mu(K_n) \to 0$ as $n \to \infty$. It follows from Step 1 that $\|T_\lambda(\chi_{K_n})\|_E = \left\|m_\lambda^E(K_n)\right\|_E \to 0$ as $n \to \infty$. Furthermore,

$$|g_n| \le c\chi_{K_n} + \varepsilon\chi_{G\setminus K_n} \le c\chi_{K_n} + \varepsilon\chi_G, \quad n \in \mathbb{N}.$$

Since T_λ is a positive linear operator, this implies that

$$|T_\lambda g_n| \le T_\lambda |g_n| \le cT_\lambda(\chi_{K_n}) + \varepsilon T_\lambda(\chi_G)$$

and so

$$\|T_\lambda g_n\|_E \le c\|T_\lambda(\chi_{K_n})\|_E + \varepsilon\|T_\lambda(\chi_G)\|_E, \quad n \in \mathbb{N}.$$

Since $\|T_\lambda(\chi_{K_n})\|_E \to 0$ as $n \to \infty$, it follows that $\limsup_{n\to\infty} \|T_\lambda g_n\|_E \le \varepsilon\|T_\lambda(\chi_G)\|_E$. But, $\varepsilon > 0$ is arbitrary, and so we can conclude that $\|T_\lambda g_n\|_E \to 0$ as $n \to \infty$.

Step 3. Let $(g_n)_{n=1}^{\infty}$ be a $\|\cdot\|_1$-Cauchy sequence in $C(G)$ satisfying $|g_n| \le c\chi_G$ for all $n \in \mathbb{N}$ and some constant $c > 0$. *The claim is that* $(T_\lambda g_n)_{n=1}^{\infty}$ *is a Cauchy sequence in* E.

Indeed, suppose that $(T_\lambda g_n)_{n=1}^{\infty}$ is *not* a Cauchy sequence in E. Then there exist $\varepsilon > 0$ and a sequence $n_1 < n_2 < \cdots$ in $\mathbb{N}$ such that $\left\|T_\lambda g_{n_k} - T_\lambda g_{n_{k+1}}\right\|_E \ge \varepsilon$ for all $k \in \mathbb{N}$. Define $h_k = g_{n_k} - g_{n_{k+1}}$, for $k \in \mathbb{N}$. Then $(h_k)_{k=1}^{\infty}$ is a sequence in $C(G)$ satisfying $|h_k| \le 2c\chi_G$ and $\|T_\lambda h_k\|_E \ge \varepsilon$ for all k. Furthermore, since $(g_n)_{n=1}^{\infty}$ is a $\|\cdot\|_1$-Cauchy sequence, it follows that $\|h_k\|_1 \to 0$ as $k \to \infty$. This contradicts the result of Step 2 applied to the sequence $(h_k)_{k=1}^{\infty}$.

Step 4. Let $f \in L^\infty(G)$ satisfy $0 \le f \le \chi_G$ and let $\varepsilon > 0$. *The claim is that there exists* $g \in C(G)$ *with* $0 \le g \le \chi_G$ *such that* $\|f - g\|_1 \le \varepsilon$ *and* $\|T_\lambda f - T_\lambda g\|_E \le \varepsilon$.

Since $C(G)$ is dense in $L^1(G)$ with respect to $\|\cdot\|_1$, there exists a sequence $(g_n)_{n=1}^{\infty}$ in $C(G)$ satisfying $\|f - g_n\|_1 \to 0$ as $n \to \infty$. Replacing, if necessary, g_n by $|g_n| \wedge \chi_G$, we may assume that $0 \le g_n \le \chi_G$ for all $n \in \mathbb{N}$. Indeed, clearly $|g_n| \wedge \chi_G \in C(G)$ and, moreover,

$$\begin{aligned}\|(|g_n| \wedge \chi_G) - f\|_1 &= \|(|g_n| \wedge \chi_G) - (f \wedge \chi_G)\|_1 \le \||g_n| - f\|_1 \\ &= \||g_n| - |f|\|_1 \le \|g_n - f\|_1,\end{aligned}$$

for $n \in \mathbb{N}$. Now, the sequence $(g_n)_{n=1}^{\infty}$ is $\|\cdot\|_1$-Cauchy (as it is convergent) and so it follows from Step 3 that $(T_\lambda g_n)_{n=1}^{\infty}$ is a Cauchy sequence in E. Therefore, there exists $h \in E$ such that $\|h - T_\lambda g_n\|_E \to 0$ as $n \to \infty$.

Since $E \subseteq L^1(G)$, the operator T_λ is the restriction to $L^\infty(G)$ of the continuous convolution operator $T_\lambda^{(1)} : L^1(G) \to L^1(G)$. Therefore, it follows from $\|f - g_n\|_1 \to 0$ that $\|T_\lambda f - T_\lambda g_n\|_1 \to 0$ as $n \to \infty$. But, the embedding of E into $L^1(G)$ is continuous and so we also have that $\|h - T_\lambda g_n\|_1 \to 0$ as $n \to \infty$. Hence, $h = T_\lambda f$ and so $\|T_\lambda f - T_\lambda g_n\|_E \to 0$ as $n \to \infty$. Consequently, $\|f - g_n\|_1 \le \varepsilon$ and $\|T_\lambda f - T_\lambda g_n\|_E \le \varepsilon$ whenever n is large enough. This proves the claim.

Step 5. Let $(f_n)_{n=1}^{\infty}$ be a sequence in $L^\infty(G)$ satisfying $f_n \downarrow_n 0$ and $0 \le f_n \le \chi_G$ for all $n \in \mathbb{N}$. *The claim is that* $\|T_\lambda f_n\|_E \to 0$ *as* $n \to \infty$.

Indeed, it follows from Step 4 that, for each $n \in \mathbb{N}$, there exists $0 \leq g_n \leq \chi_G$ in $C(G)$ satisfying $\|f_n - g_n\|_1 \leq 1/n$ and $\|T_\lambda f_n - T_\lambda g_n\|_E \leq 1/n$ for all $n \in \mathbb{N}$. Since $L^1(G)$ has o.c.-norm and $f_n \downarrow_n 0$, it follows that $\|f_n\|_1 \to 0$ and hence, also $\|g_n\|_1 \to 0$ as $n \to \infty$. By Step 2, this implies that $\|T_\lambda g_n\|_E \to 0$ as $n \to \infty$. Since also $\|T_\lambda f_n - T_\lambda g_n\|_E \to 0$ as $n \to \infty$, we can conclude that $\|T_\lambda f_n\|_E \to 0$ as $n \to \infty$. This proves the claim.

Finally, it is now clear that the finitely additive vector measure m_λ^E is actually countably additive. Indeed, suppose that $A_n \downarrow_n \emptyset$ in $\mathcal{B}(G)$. Then $\chi_{A_n} \downarrow_n 0$ in $L^\infty(G)$ and so it follows from Step 5 that $\|m_\lambda^E(A_n)\|_E = \|T_\lambda \chi_{A_n}\|_E \to 0$ as $n \to \infty$. The proof is thereby complete. □

REMARK 7.10. Let E be a translation invariant B.f.s. over G such that $L^\infty(\mu) \subseteq E$. Every $\lambda \in M_h(E;G)$ satisfies $m_\lambda^E \ll \mu$; see Corollary 7.2 and Theorem 7.9. If $\lambda \in M(G)$ and m_λ^E is not μ-continuous, then m_λ^E cannot be countably additive.

Using Theorem 7.9 we can establish the following result.

THEOREM 7.11. *Let $E \neq \{0\}$ be a translation invariant B.f.s. over G. Then $E = E_h$ if and only if E has o.c.-norm.*

PROOF. That $E = E_h$ whenever E has o.c.-norm follows from Proposition 3.4 and Definition 4.1.

Suppose now that $E = E_h$. Then $E_h \neq \{0\}$ and so Proposition 4.3 implies that $L^\infty(G) \subseteq E_h = E$. Accordingly, the Dirac measure δ_0 is E-homogeneous and so, by Theorem 7.9, the corresponding vector measure $m_{\delta_0}^E$ is countably additive. But, $m_{\delta_0}^E(A) = \chi_A * \delta_0 = \chi_A$, for $A \in \mathcal{B}(G)$. Consequently, whenever $A_n \downarrow_n \emptyset$ in $\mathcal{B}(G)$ we have that $\|\chi_{A_n}\|_E \to 0$ as $n \to \infty$. By Lemma 2.3 this implies that $\chi_G \in E_a$. Hence, $E_a \neq \{0\}$ and so $E_a = E_h$; see Proposition 4.6. Consequently, $E = E_a$, that is, E has o.c.-norm. □

The following consequences of Theorem 7.11 are worth mentioning.

COROLLARY 7.12. *Let $E \neq \{0\}$ be a translation invariant B.f.s. over G.*

(i) *E_h is an ideal in E if and only if $E_h = E_a$.*
(ii) *$C(G)$ is dense in E if and only if E has o.c.-norm.*

PROOF. (i). Evidently, if $E_h = E_a$, then E_h is an ideal (cf. Proposition 2.4). Conversely, suppose that E_h is an ideal in E. Then E_h is a homogeneous B.f.s. on G and so, by Theorem 7.11, E_h has o.c.-norm. This implies that $E_h \subseteq E_a$. By Lemma 4.5 , we always have $E_a \subseteq E_h$. This proves part (i).

(ii). If E has o.c.-norm, then it follows from Proposition 3.4 (iii) that $C(G)$ is a norm dense subspace of E.

Conversely, suppose that $C(G)$ is a dense subspace of E. Then $L^\infty(G) \subseteq E$ and so Proposition 4.3 implies that $\overline{C(G)}^E = E_h$. Hence, $E = E_h$, that is, E is homogeneous. Consequently, Theorem 7.11 implies that E has o.c.-norm. □

References

[1] C. D. Aliprantis and O. Burkinshaw, *Positive operators*, Pure and Applied Mathematics, vol. 119, Academic Press, Inc., Orlando, FL, 1985. MR809372

[2] W. Arendt, *On the* o*-spectrum of regular operators and the spectrum of measures*, Math. Z. **178** (1981), no. 2, 271–287, DOI 10.1007/BF01262044. MR631633

[3] C. Bennett and R. Sharpley, *Interpolation of operators*, Pure and Applied Mathematics, vol. 129, Academic Press, Inc., Boston, MA, 1988. MR928802

[4] S. K. Berberian, *Lectures in functional analysis and operator theory*, Graduate Texts in Mathematics, No. 15, Springer-Verlag, New York-Heidelberg, 1974. MR417727

[5] J. B. Conway, *A course in functional analysis*, Graduate Texts in Mathematics, vol. 96, Springer-Verlag, New York, 1985, DOI 10.1007/978-1-4757-3828-5. MR768926

[6] J. Diestel and J. J. Uhl Jr., *Vector measures*, Mathematical Surveys, No. 15, American Mathematical Society, Providence, RI, 1977. With a foreword by B. J. Pettis. MR453964

[7] N. Dunford and J. T. Schwartz, *Linear Operators. I. General Theory*, Pure and Applied Mathematics, Vol. 7, Interscience Publishers, Inc., New York; Interscience Publishers Ltd., London, 1958. With the assistance of W. G. Bade and R. G. Bartle. MR117523

[8] D. A. Edwards, *On translates of L^∞-functions*, J. London Math. Soc. **36** (1961), 431–432, DOI 10.1112/jlms/s1-36.1.431. MR137790

[9] R. E. Edwards, *Fourier series. Vol. 2: A modern introduction*, 2nd ed., Graduate Texts in Mathematics, vol. 85, Springer-Verlag, New York-Berlin, 1982. MR667519

[10] D. H. Fremlin, *Measure theory. Vol. 4: Topological measure spaces. Part I, II*, Torres Fremlin, Colchester, 2006. Corrected second printing of the 2003 original. MR2462372

[11] P. R. Halmos, *Measure Theory*, Graduate Texts in Mathematics 18, Springer-Verlag, New York-Heidelberg-Berlin, 1974.

[12] E. Hewitt and K. A. Ross, *Abstract harmonic analysis. Vol. I: Structure of topological groups, integration theory, group representations*, 2nd ed., Grundlehren der Mathematischen Wissenschaften, vol. 115, Springer-Verlag, Berlin-New York, 1979. MR551496

[13] G. J. O. Jameson, *Topology and normed spaces*, Chapman and Hall, London; Halsted Press [John Wiley & Sons, Inc.], New York, 1974. MR463890

[14] Y. Katznelson, *An introduction to harmonic analysis*, Second corrected edition, Dover Publications, Inc., New York, 1976. MR422992

[15] R. Larsen, *An introduction to the theory of multipliers*, Die Grundlehren der mathematischen Wissenschaften, Band 175, Springer-Verlag, New York-Heidelberg, 1971. MR435738

[16] R. Larsen, *Banach algebras: An introduction*, Pure and Applied Mathematics, No. 24, Marcel Dekker, Inc., New York, 1973. MR487369

[17] J. Lindenstrauss and L. Tzafriri, *Classical Banach spaces. II: Function spaces*, Ergebnisse der Mathematik und ihrer Grenzgebiete [Results in Mathematics and Related Areas], vol. 97, Springer-Verlag, Berlin-New York, 1979. MR540367

[18] W. A. J. Luxemburg and A. C. Zaanen, *Notes on Banach function spaces. I–XIII*, Indag. Math. Nederl. Akad. Wetensch. Proc. Ser. A **66–67**, (1963–1966). MR173168

[19] W. A. J. Luxemburg and A. C. Zaanen, *Riesz spaces. Vol. I*, North-Holland Mathematical Library, North-Holland Publishing Co., Amsterdam-London; American Elsevier Publishing Co., Inc., New York, 1971. MR511676

[20] P. Meyer-Nieberg, *Banach lattices*, Universitext, Springer-Verlag, Berlin, 1991, DOI 10.1007/978-3-642-76724-1. MR1128093

[21] S. Okada and W. J. Ricker, *Ideal properties and integral extension of convolution operators on $L^\infty(G)$*, Note Mat. **31** (2011), no. 1, 149–172. MR2963965

[22] S. Okada, W. J. Ricker, and E. A. Sánchez Pérez, *Optimal domain and integral extension of operators: Acting in function spaces*, Operator Theory: Advances and Applications, vol. 180, Birkhäuser Verlag, Basel, 2008. MR2418751

[23] B. de Pagter and W. J. Ricker, *Fourier multipliers and spectral measures in Banach function spaces*, Positivity **13** (2009), no. 1, 225–241, DOI 10.1007/s11117-008-2210-2. MR2466239

[24] B. de Pagter and W. J. Ricker, *A Fuglede type theorem for Fourier multiplier operators*, Indag. Math. (N.S.) **34** (2023), no. 2, 259–273, DOI 10.1016/j.indag.2022.10.005. MR4547862

[25] B. de Pagter and W.J. Ricker, *Translation invariant Banach function spaces, vector measures, and Fourier multipliers*, work in progress.

[26] W. Rudin, *Fourier analysis on groups*, Wiley Classics Library, John Wiley & Sons, Inc., New York, 1990. Reprint of the 1962 original; A Wiley-Interscience Publication, DOI 10.1002/9781118165621. MR1038803

[27] W. Rudin, *Real and complex analysis*, McGraw-Hill Book Co., New York-Toronto-London, 1966. MR210528

[28] W. Rudin, *Invariant means on L^∞*, Studia Math. **44** (1972), 219–227, DOI 10.4064/sm-44-3-219-227. MR304975

[29] H. H. Schaefer, *Banach lattices and positive operators*, Die Grundlehren der mathematischen Wissenschaften, Band 215, Springer-Verlag, New York-Heidelberg, 1974. MR423039

[30] A. C. Zaanen, *Integration*, North-Holland Publishing Co., Amsterdam; Interscience Publishers John Wiley & Sons, Inc., New York, 1967. Completely revised edition of An introduction to the theory of integration. MR222234

[31] A. C. Zaanen, *Riesz spaces. II*, North-Holland Mathematical Library, vol. 30, North-Holland Publishing Co., Amsterdam, 1983. MR704021

[32] A. Zappa, *The center of an algebra of operators*, Colloq. Math. **39** (1978), no. 2, 343–349, DOI 10.4064/cm-39-2-343-349. MR522377

Delft Institute of Applied Mathematics, Faculty EEMCS, Delft University of Technology, P.O. Box 5031, 2600 GA Delft, The Netherlands

Email address: b.depagter@tudelft.nl

Math.-Geogr. Fakultät, Katholische Universität Eichstätt-Ingolstadt, Ostenstr. 26-28, D-85072 Eichstätt, Germany

Email address: werner.ricker@ku.de

Proceedings of Symposia in Pure Mathematics
Volume **106**, 2024
https://doi.org/10.1090/pspum/106/01929

Cycle Spaces: Invariant Projections and Applications to Transportation Cost

Stephen J. Dilworth, Denka Kutzarova, and Mikhail I. Ostrovskii

This paper is dedicated to Professor Per H. Enflo

ABSTRACT. The paper starts with discussion of applications of cycle spaces to transportation cost. After a short survey of the known results on cycle spaces, we turn to the study of minimal projections onto cycle spaces in the corresponding ℓ_1-spaces. This study is naturally related to the study of invariant projections on the cycle space, which, in turn, are determined by the properties of representations of the automorphism group of the corresponding graph. The main focus is on discrete tori and Hamming graphs.

1. Introduction

1.1. Cycle and cut spaces. Consider a connected graph G with vertex set V and edge set E (disconnected graphs can be analyzed componentwise). We consider only *simple* graphs (that is, graphs without loops and parallel edges). We follow the graph theory terminology of [**6**, **9**].

In some cases, we consider *weighted graphs.* This means that there are positive numbers assigned to edges, and that these numbers satisfy the following condition: the weight of the edge uv does not exceed the total weight of any path joining u and v. We endow the vertex set of such graphs with the shortest path distance: the distance $d(u,v) = d_G(u,v)$ =(the minimal total weight of paths joining u and v). For graphs with no weights we assume that the weight (length) of an edge is equal to 1.

2020 *Mathematics Subject Classification.* Primary 46B04; Secondary 20B25, 20C15, 52A21, 90B06.

Key words and phrases. Cycle space, invariant projection, irreducible representation, transportation cost, Wasserstein distance.

The first author was supported by Simons Foundation Collaboration Grant No. 849142 and would like to thank the Isaac Newton Institute, Cambridge, for support and hospitality during the programme Approximation, Sampling and Compression in Data Science, where work on this paper was undertaken. This work was supported by ESPRC Grant no EP/K032208/1.

The second author was supported by Simons Foundation Collaboration Grant No. 636954 and would like to thank the Isaac Newton Institute, Cambridge, for support and hospitality during the programme Approximation, Sampling and Compression in Data Science, where work on this paper was undertaken. This work was supported by ESPRC Grant no EP/K032208/1.

The third-named author gratefully acknowledges the support by the National Science Foundation grant NSF DMS-1953773.

In the other direction, each finite metric space (X, d) can be identified with an apposite weighted graph. This weighted graph will be specified uniquely according to the procedure described below and - following [**1**] - will be called the *canonical graph associated with the metric space.* The graph is defined as follows. First, we consider a complete weighted graph with vertices in X, and the weight of an edge uv $(u, v \in X)$ defined as $d(u, v)$. After that we delete all edges uv for which there exists a vertex $w \notin \{u, v\}$ satisfying $d(u, w) + d(w, v) = d(u, v)$. It is easy to see that, for a finite metric space, the described procedure is well-defined and leads to a uniquely determined by (X, d) weighted graph, which we denote $G(X, E)$, where $E = E(X)$ is the edge set of the canonical graph, and X serves as a vertex set of G.

As a result, the constructed in this way graph $G(X, E)$ possesses the following feature, which is crucial for its application in the theory of the transportation cost spaces: the weighted graph distance of $G(X, E)$ on X coincides with the original metric on X. Note that if X is defined initially as a weighted graph with its weighted graph distance, then the corresponding canonical graph can be different because some edges can be dropped. However, if we start with a simple unweighted graph endowed with the graph distance, we will recover it as a canonical graph.

To proceed, we specify an orientation on the edge set E. This means that we fix a *direction* on the edges by selecting the *head* and *tail* of for each edge. The orientation can be chosen arbitrarily as the notions important for further reasoning do not depend on the choice of this orientation. We call this orientation a *reference orientation.*

Notation: if u is a tail and v is a head of a directed edge $e = uv$, we denote the directed edge either $\overrightarrow{uv}$ or $\overleftarrow{vu}$. We also write $u = \text{tail}(e) = e^-$ and $v = \text{head}(e) = e^+$.

NOTE 1. For some 'well-organized' graphs there exist convenient choices of reference orientations.

The *incidence matrix* D of an oriented (= directed) graph G is defined as a matrix whose rows are labelled using vertices of G, columns and labelled using edges of G and the ve-entry is given by

$$d_{ve} = \begin{cases} 1, & \text{if } v = e^+, \\ -1, & \text{if } v = e^-, \\ 0, & \text{if } v \text{ is not incident to } e. \end{cases}$$

Denote by $\mathbb{R}^E$ and $\mathbb{R}^V$ the space of real-valued functions on the edge set and the vertex set, respectively. Interpreting elements of $\mathbb{R}^E$ and $\mathbb{R}^V$ as column vectors, we may regard D as a matrix of a linear operator $D : \mathbb{R}^E \to \mathbb{R}^V$. We also consider the transpose matrix D^T and the corresponding operator $D^T : \mathbb{R}^V \to \mathbb{R}^E$.

It is easy to describe $\ker D^T$. In fact, for $f \in \mathbb{R}^V$ the value of $D^T(f) \in \mathbb{R}^E$ at an edge e is $f(e^+) - f(e^-)$, therefore $f \in \ker D^T$ if and only if it has the same value at the ends of each edge. Since we assumed that G is connected, this happens if and only if f is constant on $V(G)$. Therefore the ranks of the operators D^T and D are equal to $|V| - 1$.

Observe that

$$(DF)(v) = \sum_{e,\ e^+=v} F(e) - \sum_{e,\ e^-=v} F(e) \tag{1}$$

for $F \in \mathbb{R}^E$.

DEFINITION 2. We introduce the *cycle space* (also called the *flow space* or *cycle subspace*) of the graph G as $Z = Z(G) = \ker D \subseteq \mathbb{R}^E$. The name is chosen because, as we explain below, $Z(G)$ is spanned by the set of signed indicator functions (see (2)) of all cycles in G. The orthogonal complement of Z in $\mathbb{R}^E$ (with respect to the standard inner product) is called the *cut space* or *cut subspace* of G; we denote it $B = B(G)$. This name is chosen because $B(G)$ is spanned by *cuts*, that is, by the signed indicator functions of sets of the edges joining a subset U in $V(G)$ with its complement $\bar{U}$. These edge sets are oriented in such a way that heads of all edges are in U.

Cycle, cut, and flow spaces are standard notions of the algebraic graph theory; see [**4, 5, 16**].

Consider a cycle C in G (in a graph-theoretical sense) with one of the two possible orientations satisfying the following condition: each vertex of C is a head of exactly one edge and a tail of exactly one edge. We introduce the *signed indicator function* $\chi_C \in \mathbb{R}^E$ of the oriented cycle C (in an oriented graph G) by

$$\chi_C(e) = \begin{cases} 1 & \text{if } e \in C \text{ and its orientations in } C \text{ and } G \text{ are the same} \\ -1 & \text{if } e \in C \text{ but its orientations in } C \text{ and } G \text{ are different} \\ 0 & \text{if } e \notin C. \end{cases} \tag{2}$$

It is immediate from (1) that $\chi_C \in \ker D$. To show that signed indicator functions span $\ker D$ we observe that $\dim(\ker D) = |E(G)| - \operatorname{rank}(D) = |E(G)| - |V(G)| + 1$. To see that there are $|E(G)| - |V(G)| + 1$ linearly independent cycles we pick a spanning tree in G and consider all cycles formed by the spanning tree and edges of G which are not in the tree (there are $|E(G)| - |V(G)| + 1$ of them). It is easy to see that these cycles are linearly independent.

This computation also implies that the dimension of the cut space in G is $|E(G)| - (|E(G)| - |V(G)| + 1) = |V(G)| - 1$ (for a connected graph G).

There is a very useful standard basis in $B(G)$: consider all vertices of G except one, and consider cuts joining them with the rest of the graph (that is we use the definition of a cut joining U and $\bar{U}$ for $U = \{v\}$). Also, as for cycle spaces, there exist bases for $B(G)$ corresponding to spanning trees. Namely, for each spanning tree and each edge in it consider the cut $(U, \bar{U})$ consisting of vertices of two connected components of the spanning tree with a removed edge.

1.2. Transportation cost spaces. Let (X, d) be a finite metric space (as is noted above X can be identified with the canonical weighted graph having X as its vertex set). Consider a real-valued finitely supported function f on X with a zero sum, that is, $\sum_{v \in \operatorname{supp} f} f(v) = 0$. A natural and important interpretation of such a function, is considering it as a *transportation problem*: one needs to transport certain product from locations where $f(v) > 0$ to locations where $f(v) < 0$.

One can easily see that f can be represented as

$$f = a_1(\mathbf{1}_{x_1} - \mathbf{1}_{y_1}) + a_2(\mathbf{1}_{x_2} - \mathbf{1}_{y_2}) + \cdots + a_n(\mathbf{1}_{x_n} - \mathbf{1}_{y_n}), \tag{3}$$

where $a_i \geq 0$, $x_i, y_i \in X$, and $\mathbf{1}_u(x)$ for $u \in X$ is the *indicator function* of u, defined by

$$\mathbf{1}_u(x) = \begin{cases} 1 & \text{if } x = u, \\ 0 & \text{if } x \neq u. \end{cases}$$

We call each such representation a *transportation plan* for f because it can be interpreted as a plan of moving a_i units of the product from x_i to y_i. The *cost* of the transportation plan (3) is defined as $\sum_{i=1}^n a_i d(x_i, y_i)$.

We denote the real vector space of all transportation problems by $\mathrm{TP}(X)$. We introduce the *transportation cost norm* (or just *transportation cost*) $\|f\|_{\mathrm{TC}}$ of a transportation problem f as the infimum of costs of transportation plans satisfying (3). Using the triangle inequality and the fact that f is finitely supported it is easy to see that the infimum of costs of transportation plans for f is attained. A transportation plan for f whose cost is equal to $\|f\|_{\mathrm{TC}}$ is called an *optimal transportation plan.*

DEFINITION 3. The finite-dimensional normed space $(\mathrm{TP}(X), \|\cdot\|_{\mathrm{TC}})$ is called a *transportation cost space* and is denoted by $\mathrm{TC}(X)$.

Transportation cost spaces are of interest in many areas and are studied under many different names. We list some of them in the alphabetical order: *Arens-Eells space, earth mover distance, Kantorovich-Rubinstein distance, Lipschitz-free space, Wasserstein distance.* We prefer to use the term *transportation cost space* since it makes the subject of this work instantly clear to a wide circle of readers and it also reflects the historical approach leading to these notions (see [**21**, **22**]). Interested readers can find a review of the main definitions, notions, facts, terminology and historical notes pertinent to the subject in [**26**, Section 1.6].

Some more definitions which we use. Let μ and ν be probability measures on X. Then $\mu = \sum_{x\in X} p_x\delta_x$ and $\nu = \sum_{x\in X} q_x\delta_x$, where $p_x, q_x \geq 0$, $\sum_{x\in X} p_x = \sum_{x\in X} q_x = 1$, and δ_x denotes a unit point mass located at x. The associated transportation problem $f(\mu,\nu)$ is defined by $f(\mu,\nu)(x) = p_x - q_x$. Let $\mathcal{P}_1(X)$ be the set of all probability measures on X. The 1-*Wasserstein distance* d_{W_1} on $\mathcal{P}(X)$ is defined by

$$d_{W_1}(\mu,\nu) = \|f(\mu,\nu)\|_{\mathrm{TC}}.$$

By a *pointed metric space* we mean a metric space (X,d) with a *base point*, denoted by O. For a pointed metric space X by $\mathrm{Lip}_0(X)$ we denote the space of all Lipschitz functions $f : X \to \mathbb{R}$ satisfying $f(O) = 0$. It is not difficult to check that $\mathrm{Lip}_0(X)$ is a Banach space with respect to the norm $\|f\| = \mathrm{Lip}(f)$ ($\mathrm{Lip}(f)$ is the Lipschitz constant of f). As is well known $\mathrm{TC}(X)^* = \mathrm{Lip}_0(X)$ (see e.g. [**30**, Section 10.2]).

We use the standard terminology of Banach space theory [**3**] and the theory of metric embeddings [**30**].

1.3. Applications of cycle spaces to the theory of transportation cost. We introduce several norms on $\mathbb{R}^E$. Let $G = (V(G), E(G)) = (V, E)$ be a finite graph. Then $\ell_1(E)$, $\ell_2(E)$, and $\ell_\infty(E)$ are defined as $\mathbb{R}^E$ with the norms $\|f\|_1 = \sum_{e\in E}|f(e)|$, $\|f\|_2 = \left(\sum_{e\in E}|f(e)|^2\right)^{\frac{1}{2}}$, and $\|f\|_\infty = \max_{e\in E}|f(e)|$, respectively. We also consider the inner product $\langle f, g\rangle$ associated with $\|f\|_2$.

If $G = (V(G), E(G)) = (V, E)$ is the canonical (weighted) graph for a metric space (X, d), with the weight of each edge $uv \in E(X)$ given as $d(u, v)$, we introduce also the norm $\|f\|_{1,d} = \sum_{uv\in E}|f(uv)|d(u,v)$, and denote the corresponding normed space $\ell_{1,d}(E)$.

The following important observations establish relations between the cycle spaces and transportation cost spaces.

(1) The space $\mathrm{TC}(X)$ is isometric to the quotient space $\ell_{1,d}(E)/Z(G)$. In the case of an unweighted graph G, the space $\mathrm{TC}(G)$ is isometric to

$\ell_1(E)/Z(G)$. The result was proved in the latter case in [**30**, Proposition 10.10]. It was generalized to the weighted case in [**27**]. Different proofs were obtained in [**1, 11**]. We recommend the readers to check the simple transparent proof in [**28**, Proposition 1.5].

(2) Each isometry of $\mathrm{TC}(X)$ and $\mathrm{TC}(Y)$ is induced by a cycle-preserving bijection of $E(X)$ and $E(Y)$ [**1**]. This result allowed the authors of [**1**] to use Whitney's result [**37**] on cycle-preserving bijections and classify isometries of transportation cost spaces on finite metric spaces. In particular, this result implies that if the canonical graphs associated with X and Y are 3-connected, then each isometry between $\mathrm{TC}(X)$ and $\mathrm{TC}(Y)$ is induced by an isomorphism of the corresponding graphs. (In the other direction the result holds if the graphs are unweighted, or the weights are distributed according to one of the isomorphisms. See a similar result for Sobolev spaces on graphs in [**29**].)

The result **(1)** shows that for a better understanding of the structure of $\mathrm{TC}(X)$ it is crucial to understand the structure of $Z(G)$ (as a subspace of $\ell_{1,d}(G)$) and its position in $\ell_{1,d}(G)$.

1.4. Old and new results and observations on cycle spaces. In this section and in the rest of the paper we restrict our attention to unweighted graphs.

(A) It seems that the oldest result which can be regarded as a result about cycle spaces is the following result of Erdős and Pósa [**13**] (reprinted in [**12**]): The cycle space of dimension d contains at least roughly $d/\log d$ edge-disjointly supported vectors, and thus a 1-complemented subspace in $\ell_1(E)$ of the corresponding dimension.

(B) The well-known observation about quotient spaces (see [**10**, Fact 4.4]) implies that in order to show that $\mathrm{TC}(G)$ is "far" from ℓ_1^n with respect to the Banach-Mazur distance, it suffices to show that the relative projection constant $\lambda(Z(G),\ell_1(E))$ is large. This inspired the work of [**10, 11**] on estimates for $\lambda(Z(G),\ell_1(E))$ from below.

(C) Another tool which was systematically used in this context, was the following observation which goes back at least to Grünbaum [**17**] (see also later papers [**2, 33**]). The observation states that the projection of minimal norm onto a given subspace (in cases when existence is obvious) can be found among so-called invariant projections onto the subspace. We mean the following situation. Let Y_0 be a subspace of a finite-dimensional Banach space Y_1 (there exist infinite-dimensional generalizations, but we do not need them in this paper). Let Ω be a finite group of isometries of Y_1 leaving invariant the subspace Y_0. A projection $P: Y_1 \to Y_0$ is called *invariant with respect to* Ω if $P\omega = \omega P$ for every $\omega \in \Omega$. It is easy to construct an invariant projection starting with any projection $P_0: Y_1 \to Y_0$. Namely, one can easily check that

$$P_{\mathrm{inv}} := \frac{1}{|\Omega|}\sum_{\omega\in\Omega}\omega^{-1}P_0\omega \tag{4}$$

is an invariant projection onto Y_0 and that $\|P_{\mathrm{inv}}\| \le \|P_0\|$.

If we start by choosing a projection P with the minimal possible norm (which may not be unique), then $\|P_{\rm inv}\| = \|P\|$. This leads us to a conclusion: To find/estimate the relative projection constant $\lambda(Y_0, Y_1) := \inf\{\|P\| : P : Y_1 \to Y_0 \text{ is a projection}\}$, it suffices to compute/estimate norms of invariant projections. In this connection it becomes important to understand:

PROBLEM 4. In what cases is an invariant projection $P : \ell_1(E) \to Z(G)$ unique?

In cases where the invariant projection $P : \ell_1(E) \to Z(G)$ is not unique, it is important to understand the structure of the set of invariant projections.

(D) We present a condition for uniqueness of invariant projections in terms of group representations (Theorem 5 and Corollary 6 below). Our sources for basic facts of representation theory are [**35, 36**].

We introduce the group $\mathrm{Cyc}(G)$ of isometries of $\ell_1(E)$ for which $Z(G)$ is an invariant subspace. If the graph G is 3-connected, then, by the mentioned in **(2)** result of Whitney [**37**], the group $\mathrm{Cyc}(G)$ is canonically isomorphic to the group $\mathrm{Aut}(G)$ of automorphisms of the graph G.

Since isometries of $\ell_1(E)$ are signed permutations of the unit vector basis, they can also be regarded as orthogonal operators on $\ell_2(E)$. This observation implies that not only $Z(G)$, but also $B(G)$ is an invariant subspace of $\mathrm{Cyc}(G)$ and $\mathrm{Aut}(G)$.

Thus, we have two natural representations of the groups $\mathrm{Cyc}(G)$ and $\mathrm{Aut}(G)$: one on $Z(G)$ and the other on $B(G)$. Both $Z(G)$ and $B(G)$ are regarded as subspaces of $\ell_2(E)$.

Note that representation theory is mostly devoted to unitary representations in Hilbert spaces over complex field. However, some of the results of this theory continue to hold for orthogonal representations in Hilbert spaces over $\mathbb{R}$. This aspect of the representation theory is presented in [**36**].

To state our next theorem, recall that two representations of a group Γ on vector spaces V and W are *equivalent* if there exists a linear isomorphism $\phi\colon V \to W$ which commutes with the actions of G on V and W.

THEOREM 5. *The invariant projection $P : \ell_2(E) \to Z(G)$ is unique if and only if the natural representations of* $\mathrm{Cyc}(G)$ *on $Z(G)$ and $B(G)$ do not have equivalent restrictions to non-zero subspaces.*

Combining this theorem with the mentioned in **(2)** result of Whitney [**37**], we get the following corollary.

COROLLARY 6. *For a 3-connected graph G, the invariant projection $P : \ell_2(E) \to Z(G)$ is unique if and only if the natural representations of* $\mathrm{Aut}(G)$ *on $Z(G)$ and $B(G)$ do not have equivalent restrictions to non-zero subspaces.*

PROOF OF THEOREM 5. Observe that the orthogonal projection $P_{\rm orth} : \ell_2(E) \to Z(G)$ is invariant. The best way to see it is to write decompositions of $P_{\rm orth}$ and an arbitrary $A \in \mathrm{Cyc}(G)$ with respect to the orthogonal decomposition $\ell_2(E) = Z(G) \oplus B(G)$. In fact, invariance of $Z(G)$ and $B(G)$ for any $A \in \mathrm{Cyc}(G)$, shows that

$$A = \begin{pmatrix} A_Z & 0 \\ 0 & A_B \end{pmatrix},$$

where A_Z and A_B are the corresponding restrictions. Also, we have

$$P_{\text{orth}} = \begin{pmatrix} I_{Z(G)} & 0 \\ 0 & 0 \end{pmatrix},$$

where $I_{Z(G)}$ is the identity operator on $Z(G)$. The validity of $P_{\text{orth}}A = AP_{\text{orth}}$ is immediate.

Observe that any projection of $\ell_2(E)$ onto $Z(G)$ is of the form

$$P = \begin{pmatrix} I_{Z(G)} & Q \\ 0 & 0 \end{pmatrix},$$

where Q is some linear operator, $Q : B(G) \to Z(G)$.

It is easy to see that $PA = AP$ is equivalent to

$$QA_B = A_Z Q. \tag{5}$$

Next we prove the 'if' part of the theorem. It suffices to show that if $Q \neq 0$, then (5) implies that A_Z and A_B have restrictions to non-zero subspaces which are equivalent representations.

Note, that the observation at the beginning of [**36**, Appendix to III.5, p 50] shows that in the real case one can also decompose an orthogonal representation into an orthogonal sum of irreducible over the reals representations. Therefore, the representations A_Z and A_B of $\text{Cyc}(G)$ can be decomposed into irreducible over the reals representations. Let

$$B(G) = V_1 \oplus V_2 \oplus \cdots \oplus V_k$$

be the corresponding decomposition of $B(G)$. If $Q \neq 0$, then there exists $i \in \{1, \ldots, k\}$ such that $Q|_{V_i} \neq 0$.

Since V_i is an invariant subspace for all restrictions A_B of $A \in \text{Cyc}(G)$, the formula (5) implies that $Q(V_i)$ is an invariant subspace for A_Z, $A \in \text{Cyc}(G)$.

Then the usual proof of the first part of Schur's lemma (as in [**35**]) shows that the restriction of A_B to V_i and the restriction of A_Z to $Q(V_i)$ are equivalent representations.

To prove the 'only if' part of the theorem, it suffices to show that if A_Z and A_B have equivalent restrictions, then there exists $Q \neq 0$ satisfying (5).

In fact, let the equivalent restrictions be to the subspaces $V \subset B(G)$ and $U \subset Z(G)$. Let $R : V \to U$ be the isomorphism establishing the equivalence, and let $P_V : B(G) \to V$ be the orthogonal projection. It is easy to see that $Q : B(G) \to Z(G)$ given by $Q = RP_V$ is the desired operator. □

1.5. A general problem on representations of automorphism groups. Corollary 6 makes the following problem very important. This problem can be regarded as a part of the general problem of descriptions of representations of the groups $\text{Aut}(G)$ which are among the most important groups in algebraic combinatorics; see [**4, 16, 24**].

PROBLEM 7. Given a graph G, when do the restrictions of the natural representation of $\text{Aut}(G)$ on $\ell_2(E)$ to $Z(G)$ and $B(G)$ have equivalent restrictions to non-zero subspaces?

Of course, the same problem is of importance also for groups $\text{Cyc}(G)$, but these groups are used much less frequently.

We have to mention that we did not find any publications explicitly stating any results on Problem 7.

1.6. Relations between the norm of minimal projection and other geometric parameters. Henceforth $\|T\|_1$ (resp. $\|T\|_\infty$) denotes the operator norm of a linear operator $T\colon \ell_1(E) \to \ell_1(E)$ (resp. $T\colon \ell_\infty(E) \to \ell_\infty(E)$).

The comments in Section 1.4 (C) show that the collection of invariant projections onto $Z(G)$ is non-empty as the orthogonal projection P_{orth} is invariant. Also they imply that there exist invariant projections of minimal possible norm.

Minimal projections onto $Z(G)$ yield information about the projection constants of $\text{Lip}_0(G)$ and the L_1-distortion of $(\mathcal{P}(G), d_{W_1})$. Let us recall the relevant definitions.

Let Y be a finite-dimensional normed space and let Y_1 be any subspace of ℓ_∞ that is isometrically isomorphic to Y. Recall that the *projection constant* of Y, denoted $\lambda(Y)$, is defined by

$$\lambda(Y) = \inf\{\|P\|\colon P\colon \ell_\infty \to \ell_\infty \text{ is a projection with range } Y_1\}.$$

It follows from the Hahn-Banach extension property of ℓ_∞ that the definition of $\lambda(Y)$ is independent of the choice of Y_1 (see e.g. [**19**, Prop. 0.19]). Let (X, d) and (Y, ρ) be metric spaces. An injective map $F : X \to Y$ is called a *bilipschitz embedding* if there exist constants $C_1, C_2 > 0$ so that for all $u, v \in X$

$$C_1 d(u, v) \le \rho(F(u), F(v)) \le C_2 d(u, v).$$

The *distortion* of F is defined as $\text{Lip}(F) \cdot \text{Lip}(F^{-1}|_{F(X)})$, where $\text{Lip}(\cdot)$ denotes the Lipschitz constant.

The L_1-*distortion* $c_1(X)$ is defined as the infimum of distortions of all bilipschitz embeddings of (X, d) into any space $L_1(\Omega, \Sigma, \mu)$.

Differentiability of Lipschitz maps (see [**3**, Theorem 7.9]) implies that if X is a finite-dimensional normed space, then $c_1(X)$ is equal to the infimum of distortions of all linear embeddings of X into $L_1[0, 1]$.

The *Banach-Mazur distance* between normed spaces Y_1 and Y_2 of the same finite dimension is defined by

$$d_{\text{BM}}(Y_1, Y_2) = \inf\{\|T\| \cdot \|T^{-1}\|\colon T\colon Y_1 \to Y_2 \text{ is a linear isomorphism}\}.$$

The connection between L_1-distortion and projection constants on the one hand and minimal projections onto $Z(G)$ on the other is summarized in the following essentially known result.

PROPOSITION 8. *Let $P_{\min}$ be a minimal projection of $\ell_1(E)$ onto $Z(G)$.*

(i)
$$c_1(\text{TC}(G)) \le \|I - P_{\min}\|_1.$$

(ii)
$$c_1((\mathcal{P}(G), d_{W_1})) \le \|I - P_{\min}\|_1.$$

(iii)
$$\lambda(\text{Lip}_0(G)) \ge \|P_{\min}\|_1 - 1.$$

(iv) *For $N = \dim(\text{TC}(G)) = |V(G)| - 1$,*
$$d_{\text{BM}}(\text{TC}(G), \ell_1^N) \ge \|P_{\min}\|_1 - 1.$$

PROOF. (i) Let Q be any projection from $\ell_1(E)$ onto $Z(G)$. For all $x \in \ker(Q)$ and $y \in Z(G)$, we have

$$\|x\|_1 = \|(I-Q)(x+y)\|_1 \leq \|I-Q\|_1\|x+y\|_1.$$

Hence, identifying $\mathrm{TC}(G)$ with $\ell_1(E)/Z(G)$, we get

$$\|x+Z(G)\|_{\mathrm{TC}} \leq \|x\|_1 \leq \|I-Q\|_1\|x+Z(G)\|_{\ell_1(E)/Z(G)} = \|I-Q\|_1\|x+Z(G)\|_{\mathrm{TC}}.$$

Setting $Q = P_{\min}$, the linear mapping $T\colon \ell_1(E)/Z(G) \to \ker(Q)$ given by $x + Z(G) \mapsto x$ yields (i).

It was proved in [**25**, Lemma 3.1] that $c_1((\mathcal{P}(G), d_{W_1})) = c_1(\mathrm{TC}(G))$, so (i) and (ii) are equivalent.

(iii) Note that $\mathrm{Lip}_0(G) = \mathrm{TC}(G)^*$ is isometrically isomorphic to $(B(G), \|\cdot\|_\infty) \subset \ell_\infty(E)$ by $\mathrm{TC}(G) = \ell_1(E)/Z(G)$, since $B(G) = Z(G)^\perp$. Now suppose Q is any projection from $\ell_\infty(E)$ onto $B(G)$. Then $I - Q^*$ is a projection from $\ell_1(E)$ onto $Z(G)$. So

$$\|Q\|_\infty \geq \|I-Q\|_\infty - 1 = \|I-Q^*\|_1 - 1 \geq \|P_{\min}\|_1 - 1,$$

and hence

$$\lambda(\mathrm{Lip}_0(G)) = \inf \|Q\|_\infty \geq \|P_{\min}\|_1 - 1.$$

(iv) By duality,

$$d_{\mathrm{BM}}(\mathrm{TC}(G), \ell_1^N) = d_{\mathrm{BM}}(\mathrm{Lip}_0(G), \ell_\infty^N) \geq \lambda(\mathrm{Lip}_0(G)) \geq \|P_{\min}\|_1 - 1. \qquad \square$$

2. Invariant and minimal projections onto cycle spaces for different families of graphs

2.1. Summary of results. This article is a continuation of [**10**] and [**11**] in which invariant projections for the diamond, Laakso, and somewhat more general recursive families of graphs were investigated, with applications to the corresponding transportation cost spaces.

At this point, the established relation between the uniqueness of invariant projections and properties of representations of automorphism groups (Theorem 5 and Corollary 6) does not play a significant role in this study. Increase of this role requires the development of the representation theory of $\mathrm{Aut}(G)$ in the corresponding direction (we have not found suitable results in the literature).

We have only one example of such application (see Section 2.3). On the other hand, in view of Theorem 5, our results on invariant projections immediately imply the corresponding consequences about representations of $\mathrm{Aut}(G)$ for G in question.

In Section 2.2 we review definitions of Cayley graphs, generators and relators of groups, and introduce a useful system of cycles in Cayley graphs.

Section 2.3 establishes non-uniqueness of invariant projections for Cayley graphs of a group Γ for which there are no automorphisms except multiplications by elements of Γ.

In Section 2.4 we characterize invariant projections on the 2-dimensional torus $\mathbb{T}_n := \mathbb{Z}_n \times \mathbb{Z}_n$ and compute $\|P_{\min}\|_1$ for $n \leq 6$. In Section 2.5 and Section 2.6 we extend the uniqueness results of the previous section to tori in higher dimensions and to Hamming graphs. In the final section we show that the orthogonal projection P_{orth} is minimal and that $\|I - P_{\mathrm{orth}}\|_1 = \dfrac{n+1}{2}$ for the Hamming cube $\mathbb{Z}_2^n$.

2.2. Remarks on Cayley graphs. The graphs which we consider are the Cayley graphs of finite groups. Let us recall the definition. Relevant material of Group Theory can be found in the introductory chapters of [**32**], on Cayley graphs - in [**16**, Chapter 3].

Let Γ be a finite group with the identity element 1 or 1_Γ if the binary operation is multiplication, and 0 or 0_Γ if the binary operation is addition.

NOTE 9. We shall use multiplicative notation for general groups. On the other hand, for $\mathbb{Z}_n$, considered as a group with respect to addition, we use additive notation (to avoid the confusion with the multiplication in the ring $\mathbb{Z}_n$).

We say that a subset $\mathcal{X} = \{x_i\}_{i=1}^k$ *generates* Γ or is a *generating set* for Γ if every element of Γ can be written as a (finite) product of $\{x_i\}_{i=1}^k \bigcup \{x_i^{-1}\}_{i=1}^k$. Such product is also called a *word* formed from $\mathcal{X}$ and $\mathcal{X}^{-1} := \{x^{-1} : x \in \mathcal{X}\}$.

A finite group Γ with generating set $\mathcal{X}$ can be described in terms of *relations* between elements of $\mathcal{X} \cup \mathcal{X}^{-1}$ or *relators*. The set of all relators is denoted $\mathcal{R}$.

EXAMPLE 10. Group $\mathbb{Z}_n^2$, in additive notation, has generating set $\{x, y\}$ and

(1) relations $x + y = y + x$, $\underbrace{x + \cdots + x}_{n \text{ times}} = 0$, $\underbrace{y + \cdots + y}_{n \text{ times}} = 0$

(2) or relators $-x - y + x + y$, $\underbrace{x + \cdots + x}_{n \text{ times}}$, $\underbrace{y + \cdots + y}_{n \text{ times}}$.

So *relations* are equalities between words and *relators* are words which have to be equal to the group identity element.

Let $\mathcal{X}$ be a generating subset of a group Γ satisfying $\mathcal{X}^{-1} = \mathcal{X}$. The *right-invariant Cayley graph* $\mathrm{Cay}(\Gamma, \mathcal{X})$ of Γ corresponding to the generating set $\mathcal{X}$ is the graph whose vertices are elements of Γ and elements $g, h \in \Gamma$ are connected by an edge if and only if $hg^{-1} \in \mathcal{X}$. The graph distance of this graph is called the *word metric* because the distance between group elements g and h is the shortest representation of hg^{-1} as a word in terms of elements of $\mathcal{X}$.

We use the term *right-invariant* because multiplication of all elements of Γ on the right with a fixed element of Γ is an isomorphism of the graph.

Similarly, one can define a *left-invariant Cayley graph.* It is different, but has similar properties, see [**30**, Remark 5.10].

The following proposition will play an important role in our results.

PROPOSITION 11. *The set of all signed indicator functions of directed cycles of the form*

$$y, g_n y, g_{n-1} g_n y, \ldots, g_2 \cdots g_n y, g_1 \cdots g_n y = y,$$

where $y \in \Gamma$ *and* $r = g_1 \ldots g_n \in \mathcal{R}$, $g_1, \ldots, g_n \in \mathcal{X}$, *spans the cycle space of* $\mathrm{Cay}(\Gamma, \mathcal{X})$.

PROOF. It is easy to see that if we describe a group Γ in terms of generators $\mathcal{X}$ and relators $\mathcal{R}$, then the set of all words which are equal to 1 is Γ can be described as so-called *normal closure* of $\mathcal{R}$, that is, the set of all finite products of the form $w^{-1} r^{\pm 1} w$, where w is any word and $r \in \mathcal{R}$. In fact, the images of all other words in the quotient of the free group generated by $\mathcal{X}$ over the normal closure are not equal to the identity.

Therefore, any word which represents a closed walk is obtained by reducing the word of the form:

$$w_1^{-1}r_1^{\pm 1}w_1w_2^{-1}r_2^{\pm 1}w_2\dots w_p^{-1}r_p^{\pm 1}w_p; \tag{6}$$

reducing means deleting products of the forms $x_ix_i^{-1}$ (see [**32**, Sections 2.1-2.2] if you never studied this).

This implies that each cycle in $\mathrm{Cay}(\Gamma,\mathfrak{X})$ is represented by the same vector on the edge set $E(\mathrm{Cay}(\Gamma,\mathfrak{X}))$ as the closed walk (6), and this walk, because of cancellation of terms corresponding to $w_i^{\pm 1}$ is represented by a linear combination of images in $\ell_1(E(\mathrm{Cay}(\Gamma,\mathfrak{X})))$ of cycles corresponding to r_i (note that the cycle corresponding to r_i^{-1} is just the negation of the cycle corresponding to r_i). □

2.3. Groups with graphical regular representation. Godsil [**15**] proved the following result: Each non-solvable finite group Γ (see [**32**, Chapter 5] for the definition and basic facts on solvable groups) has a subset $\mathfrak{X}=\mathfrak{X}^{-1}$ such that the only automorphisms of the Cayley graph $G:=\mathrm{Cay}(\Gamma,\mathfrak{X})$ are shifts by elements of Γ, so $\mathrm{Aut}(G)=\Gamma$.

THEOREM 12. *For this G there is no uniqueness of invariant projection onto $Z(G)$.*

The proof of this theorem is based on two lemmas.

LEMMA 13. *The representation of $\Gamma=\mathrm{Aut}(G)$ on the cut space $B(G)$ contains all irreducible representations of Γ except the unit representation.*

PROOF. In fact, consider the natural basis $\{x_g\}_{g\neq 1_\Gamma}$ in $B(G)$ introduced at the end of Section 1.1. If we write the matrices of the representation of Γ in this basis, we get that the character has value $n-1$ (where $n=|\Gamma|$) on 1_Γ and value -1 on all other elements. So it is the difference of the characters of the regular representation and the unit representation. Therefore (by general theory, see [**35,36**]) the representation of Γ on $B(G)$ contains all irreducible representations except the unit representation. □

LEMMA 14. *The representation of $\Gamma=\mathrm{Aut}(G)$ on the cycle space $Z(G)$ contains irreducible representations of Γ which are not equivalent to the unit representation.*

PROOF. It suffices to prove that there is a vector $x\in Z(G)$ which is mapped by one of $g\in\Gamma$ onto a vector which is linearly independent with x.

So we need to exhibit a cycle which is not mapped onto itself by some element of the group. To see the existence of such cycle, observe that only the cycle joining elements of a cyclic group in a cyclic order is mapped onto itself by all elements of the group. There are many reasons for which our situation is different. □

2.4. Two-dimensional torus.

2.4.1. *Notation.* The torus $\mathbb{T}_n=\mathbb{Z}_n\times\mathbb{Z}_n$ viewed as a graph consists of vertices (i,j) with $0\le i,j<n$. We assume that $n\ge 3$ in the following definitions. The edge set E_n consists of edges oriented as follows : $(i,j)\to(i+1,j)$ (horizontal left to right) and $(i,j)\to(i,j+1)$ (vertical bottom to top) with addition mod n. The corresponding "row" and "column" edge vectors which form a basis of the edge space are denoted $e_{i,j}^R$ and $e_{i,j}^C$. Let $\ell_1(E_n)$, $\ell_2(E_n)$, and $\ell_\infty(E_n)$ denote the edge space equipped with the usual ℓ_1, ℓ_2, and ℓ_∞ norms with the edge vectors as the

standard unit vector basis. Let $Z_n = Z(\mathbb{T}_n)$ and $B_n = B(\mathbb{T}_n)$ denote the cycle and cut spaces.

Let $\Gamma_n = \Gamma(\mathbb{T}_n)$ be the group of linear isometries of $\ell_1(E_n)$ for which Z_n is invariant.

Let $H_n = H(\mathbb{T}_n)$ be the automorphism group of $\mathbb{T}_n$. As mentioned above, each $g \in H_n$ induces a $\hat{g} \in \Gamma_n$. Let U_n be the subgroup of H_n generated by α and β, defined as follows:

$$\alpha((i,j)) = (-i,j); \qquad \beta((i,j)) = (j,i).$$

Let V_n be the subgroup of H_n consisting of translations $T_{(a,b)}$ $((a,b) \in \mathbb{T}_n)$ defined by

$$T_{(a,b)}(i,j) = (a+i, b+j).$$

A projection P from $\ell_1(E_n)$ onto Z_n or B_n is Γ_n*-invariant* if $Pg = gP$ for all $g \in \Gamma_n$.

Corresponding to each $(i,j) \in \mathbb{T}^n$ there are a "small square" cycle vector

$$S((i,j)) := e^R_{i,j} + e^C_{i+1,j} - e^R_{i,j+1} - e^C_{i,j}$$

and a "cross" cut vector corresponding to the cut $(\{(i,j)\}, \mathbb{T}^n \setminus \{(i,j)\})$

$$X((i,j)) := e^R_{i,j} + e^C_{i,j} - e^C_{i-1,j} - e^R_{i,j-1}.$$

For $n \geq 3$, we define "row" and "column" cycle vectors

$$R := \sum_{j=0}^{n-1} e^R_{j,0} \quad \text{and} \quad K := \sum_{j=0}^{n-1} e^C_{0,j}.$$

2.4.2. *Invariant projections onto Z_n.*

LEMMA 15.

$$\mathcal{B}^1_n = \{S((i,j)) \colon (i,j) \neq (0,0)\} \cup \{R, K\}$$

and

$$\mathcal{B}^2_n = \{X((i,j)) \colon (i,j) \neq (0,0)\}$$

are algebraic bases for Z_n and B_n respectively.

PROOF. Clearly, $\mathcal{B}^1_n$ is a linearly independent subset of Z_n and $|\mathcal{B}^1_n| = n^2+1 = \dim(Z_n)$. Similarly, $\mathcal{B}^2_n$ is a linearly independent subset of B_n and $|\mathcal{B}^2_n| = n^2-1 = \dim(B_n)$. □

LEMMA 16. *Let $n \geq 3$ and $\theta \in \Gamma_n$. Then either θ or $-\theta$ is induced by a graph automorphism ϕ.*

PROOF. It is easy to see that each $\theta \in \Gamma_n$ induces an orthogonal operator on $\ell_2(E)$. Therefore θ leaves the cut space B_n invariant. Therefore each $\theta(X((i,j)))$ is a sum of four signed edge vectors and is orthogonal to each $S((a,b))$. To ensure orthogonality, the 4 edges in the edge support of $\theta(X((i,j)))$ must intersect the edge support of each $S((a,b))$ in either 0 or 2 edges, which readily implies that $\theta(X((i,j))) = \varepsilon((i,j))X(\phi((i,j)))$, where $\varepsilon((i,j)) = \pm 1$ and ϕ is a bijection of the vertex set of $\mathbb{T}_n$. If (i_1,j_1) is a neighbor of (i,j) then $\theta(X((i_1,j_1)))$ and $\theta(X((i,j)))$ share a signed edge vector. Hence $\phi((i_1,j_1))$ is a neighbor of $\phi((i,j))$, i.e., ϕ is a graph automorphism, and $\varepsilon((i_1,j_1))) = \varepsilon((i,j))$. By a connectedness argument, $\varepsilon((i,j))$ is constant. If $\epsilon((i,j)) \equiv 1$ then $\theta = \hat{\phi}$ and if $\varepsilon((i,j)) \equiv -1$ then $-\theta = \hat{\phi}$. □

REMARK 17. By a classical result of Whitney [**37**] (see also [**31**, Section 5.3]), Lemma 16 is valid for any 3-connected graph. For graphs which are not 3-connected there can be a rich set of cycle-preserving bijections which do not correspond to an automorphism of the graph. The cases of some of such graphs (Diamond graphs, Laakso graphs, etc.) were studied in [**10, 11**].

LEMMA 18. *Let* $n \geq 5$. *Then* $H_n = U_n \cdot V_n = V_n \cdot U_n$.

PROOF. Let $g \in H_n$. Since $n \geq 5$, g maps the vertices $\{(i,j), (i+1,j), (i+1,j+1), (i+1,j)\}$ of a 'small square' cycle onto the vertices of another small square. Hence g maps collinear triples $\{(i,j), (i+1,j), (i+2,j)\}$ and $\{(i,j), (i,j+1), (i,j+2)\}$ onto other collinear triples with 'midpoints' mapped to midpoints. There exists $v \in V_n$ such that $vg((0,0)) = (0,0)$. Since vg maps collinear triples onto collinear triples, it follows that there exists $u \in U_n$ such that uvg fixes $(0,0)$, $(1,0)$ and $(0,1)$. Thus, uvg also fixes $(1,1)$. Using the observation about collinear triples again, it follows easily that $uvg = I$, i.e., $g \in U_n \cdot V_n$.
Similarly, we prove $g \in V_n \cdot U_n$. □

REMARK 19. Lemma 18 is false for $n = 4$. This is because there are automorphisms which do not map collinear triples onto collinear triples: consider $\phi : \mathbb{T}_4 \to \mathbb{T}_4$ given by

$$(2,0) \mapsto (1,3), (3,0) \mapsto (0,3), (2,1) \mapsto (1,2), (3,1) \mapsto (0,2)$$
$$(0,2) \mapsto (3,1), (1,2) \mapsto (2,1), (0,3) \mapsto (3,0), (1,3) \mapsto (2,0),$$

with all other vertices fixed by ϕ. Then ϕ is easily seen to be a graph automorphism which maps the "small square" cycle $((0,1),(1,1),(1,2),(0,2),(0,1))$ onto the "row" cycle $((0,1),(1,1),(2,1),(3,1),(0,1))$.

We require the following simple observation.

LEMMA 20. *For* $g \in H_n$, $\hat{g}(X((i,j))) = X(g((i,j)))$.

LEMMA 21. *Let* $n \geq 5$. *Suppose* $Y \in Z_n$ *is* U_n*-invariant (i.e.,* $\hat{u}(Y) = Y$ *for all* $u \in U_n$*). Then* Y *induces a unique* Γ_n*-invariant mapping* $A\colon B_n \to Z_n$ *satisfying* $A(X((0,0))) = Y$. *Conversely, if* A *is* Γ_n*-invariant then* $Y = A(X((0,0)))$ *satisfies* $\hat{u}(Y) = Y$ *for all* $u \in U_n$.

PROOF. Let

$$Y = \sum_{(i,j)\neq(0,0)} a(i,j)S((i,j)) + bR + cK$$

be the basis expansion of Y. Note that $\hat{\alpha}(K) = -K$, $\hat{\alpha}(R) = R$, and $\hat{\alpha}(S((i,j))) = -S((-i-1,j))$. Hence

$$\hat{\alpha}(Y) = -\sum_{(i,j)\neq(0,0)} a(i,j)S((-i-1,j)) + bR - cK.$$

Since $Y = \hat{\alpha}(Y)$, it follows that $c = 0$. Similarly, $b = 0$. Hence

$$\begin{aligned} \sum_{(i,j)\neq(0,0)} \hat{T}_{(i,j)}(Y) &= \sum_{(k,l)\neq(0,0)} a(k,l)\Big(\sum_{(i,j)\neq(0,0)} \hat{T}_{(i,j)}(S((k,l)))\Big) \\ &= -\sum_{(k,l)\neq(0,0)} a(k,l)S((k,l)) = -Y. \end{aligned} \tag{7}$$

For $(i,j) \neq (0,0)$, let $A(X((i,j))) = \hat{T}_{(i,j)}(Y)$ and extend linearly to define A on B_n. Note that by (7)

$$A(X((0,0))) = A(-\sum_{(i,j)\neq(0,0)} X((i,j))) = -\sum_{(i,j)\neq(0,0)} \hat{T}_{(i,j)}(Y) = Y$$

as desired. Now let us check that A is Γ_n-invariant. Let $\gamma \in \Gamma_n$. Then $\gamma = \pm\hat{g}$ for some $g \in H_n$. So it suffices to check that $\hat{g}A = A\hat{g}$. Fix $(i,j) \in \mathbb{T}_n$. Since $H_n = V_n \cdot U_n$, it follows that $gT_{(i,j)} = T_{g((i,j))}u$ for some $u \in U_n$. Hence

$$\begin{aligned}\hat{g}A(X((i,j))) &= \hat{g}\hat{T}_{(i,j)}(Y)\\ &= \hat{T}_{g((i,j))}\hat{u}(Y)\\ &= \hat{T}_{g((i,j))}(Y)\\ &= A(X(g((i,j))))\\ &= A\hat{g}(X((i,j))).\end{aligned}$$

Now suppose that $A\colon B_n \to Z_n$ is Γ_n-invariant. By Lemma 20, for all $u \in U_n$,

$$\hat{u}(A(X(0,0))) = A(\hat{u}(X((0,0)))) = A(X((0,0)))$$

and

$$A(X((i,j))) = A(\hat{T}_{(i,j)}(X((0,0)))) = \hat{T}_{(i,j)}(A(X(0,0))).$$

So $Y = A(X((0,0)))$ is U_n-invariant and A is uniquely determined by $A(X((0,0)))$. □

LEMMA 22. *Suppose $n \geq 3$. Let $Y = \sum_{(i,j)\neq(0,0)} a(i,j)S((i,j)) + bR + cK \in Z_n$. Then Y is U_n-invariant if and only if*

1. $b = c = 0$,
2. $a(i,i) = 0$ *and* $a(i,-i-1) = 0$ *for* $0 \leq i \leq n-1$,
3. *if* $n = 2k+1$ *is odd then* $a(i,k) = a(k,j) = 0$ *for all* $0 \leq i,j \leq n-1$,
4. *if* $j \neq i$ *and* $j \neq -i-1$ *(and, if* $n = 2k+1$*, then* $i \neq k$ *and* $j \neq k$*) then*

$$\begin{aligned}a(i,j) &= -a(-i-1,j) = -a(j,i) = a(-j-1,i) = -a(i,-j-1)\\ &= a(-i-1,-j-1) = -a(-j-1,-i-1) = a(j,-i-1).\end{aligned}$$

PROOF. It was proved above that if Y is U_n-invariant then $b = c = 0$. For the remaining conditions, consider the orbit of $S((i,j))$ under the action of U_n as given by the following table:

$u \in U_n$	$\hat{u}(S((i,j)))$
I	$S((i,j))$
α	$-S((-i-1,j))$
β	$-S((j,i))$
$\alpha\beta$	$S((-j-1,i))$
$\beta\alpha\beta$	$-S((i,-j-1))$
$(\alpha\beta)^2$	$S((-i-1,-j-1))$
$\beta(\alpha\beta)^2$	$-S((-j-1,-i-1))$
$(\alpha\beta)^3$	$S((j,-i-1))$

We determine necessary conditions for $\hat{u}(Y) = Y$ to hold. Referring to the table, setting $u = \beta$ implies $a(i,i) = -a(i,i)$, i.e., $a(i,i) = 0$ which is the first part of (2). Setting $u = \beta(\alpha\beta)^2$, and taking into account that

$$\hat{u}(S((n-1,n-1)) = -S((0,0)) = \sum_{(i,j)\neq(0,0)} S((i,j)),$$

yields

$$a(i,-i-1) = -a(i,-i-1) + a(n-1,n-1) = -a(i,-i-1) + 0 \Rightarrow a(i,-i-1) = 0,$$

which is the second part of (2). Now suppose that $n = 2k+1$ is odd. Setting $u = \alpha$, and taking into account that $\hat{\alpha}(S((n-1,0))) = -S((0,0)) = \sum_{(i,j)\neq(0,0)} S((i,j))$ and that (from (2)) $a(n-1,0) = 0$, yields

$$a(k,j) = -a(k,j) + a(n-1,0) = -a(k,j) + 0 = -a(k,j) \Rightarrow a(k,j) = 0.$$

Setting $u = \beta$ gives $a(i,k) = -a(k,j) = 0$. So (3) is proved. Finally, if (i,j) satisfies the assumptions of (4) then the orbit of $S((i,j))$ under U_n has cardinality 8 and (4) follows from the table by considering each $u \in U_n$ in turn. (Note that if $\hat{u}(S((i,j))) = \pm S((0,0))$ then $(i,j) \in \{(0,n-1),(n-1,0),(n-1,n-1)\}$ and so $a(i,j) = 0$ from (1) and (2).)

This completes the proof of necessity of (1)-(4). Sufficiency is easily checked. □

COROLLARY 23. *Let $n \geq 3$. Then the collection of U_n-invariant vectors $Y \in Z_n$ is a linear subspace of dimension $[n/2]([n/2]-1)/2$ with basis consisting of the vectors of the form*

$$\begin{aligned} y = S((i,j)) &- S((-i-1,j)) - S((j,i)) \\ &+ S((-j-1,i)) - S((i,-j-1)) + S((-i-1,-j-1)) \\ &- S((-j-1,-i-1)) + S((j,-i-1)), \end{aligned}$$

where (i,j) satisfies the assumptions of (4) in Lemma 22.

THEOREM 24. *Let $n \geq 5$. The collection of all Γ_n-invariant projections P on $\ell_1(E_n)$ with range Z_n is parameterized by the linear space of Γ_n-invariant mappings $A\colon B_n \to Z_n$ which has dimension $[n/2]([n/2]-1)/2$.*

PROOF. Note that $\ell_2(E_n) = Z_n \oplus B_n$. With respect to this decomposition the Γ_n-invariant projections are of the form

$$P = \begin{bmatrix} I & A \\ 0 & 0 \end{bmatrix},$$

where $A\colon B_n \to Z_n$ is Γ_n-invariant. The result follows from the previous lemma. □

COROLLARY 25. *The orthogonal projection P_{orth} is the unique Γ_n-invariant projection if and only if $n \leq 4$.*

PROOF. We omit the easy proof of uniqueness in the case $n = 2$. Note that $[n/2]([n/2]-1)/2 > 0$ if and only if $n \geq 4$, so uniqueness for $n = 3$ follows from Corollary 23 and non-uniqueness for $n \geq 5$ follows from Theorem 24. So suppose

$n = 4$. By Lemma 22 the linear space of U_4-invariant vectors is one-dimensional with basis vector

$$\begin{aligned} Y = &S((1,0)) - S((2,0)) + S((3,1)) - S((3,2)) \\ &+ S((2,3)) - S((1,3)) + S((0,2)) - S((0,1)). \end{aligned}$$

Consider the automorphism $\phi\colon \mathbb{T}_4 \to \mathbb{T}_4$ described in Remark 19. Suppose, to derive a contradiction, that there exists a Γ_4-invariant projection P with $P((X(0,0))) = Y$. Note that $\hat{\phi}(X((0,0))) = X((0,0))$, and so

$$\hat{\phi}(Y) = \hat{\phi}(P(X((0,0)))) = P(\hat{\phi}(X((0,0)))) = P(X((0,0))) = Y.$$

However, a direct calculation shows that

$$\begin{aligned} \hat{\phi}(Y) = &S((1,0)) + 2S((2,0)) - S((0,1)) - 2S((3,1)) \\ &- 2S((0,2)) - S((3,2)) + 2S((1,3)) + S((2,3)). \end{aligned}$$

So $\hat{\phi}(Y) \neq Y$, which is the desired contradiction. This proves uniqueness in the case $n = 4$. □

2.4.3. *Minimal projections.* If P is a Γ_n-invariant projection onto Z_n, then $\|P\|_1 = \|P(e)\|_1$ for any edge vector e since the action of Γ_n on E_n is transitive. So Theorem 24 can be used to compute the norm of an invariant minimal projection $P_{\min}$. The problem reduces to minimizing a convex piecewise-linear function of $[n/2]([n/2]-1)/2$ variables. In the case $n = 5$, the family of Γ_n-invariant projections is one-dimensional but P_{orth} is the unique minimal Γ_n-invariant projection.

For $n = 6$, however, P_{orth} is not a minimal projection. The corresponding piecewise-linear function of 3 variables was minimized with the aid of MAPLE. The results are as follows.

PROPOSITION 26. *There is a unique Γ_6-invariant minimal projection $P_{\min}$ from $\ell_1(E_6)$ onto Z_6 with*

$$\|P_{\min}\|_1 = \frac{109}{36} \quad \textit{and} \quad \|I - P_{\min}\|_1 = 3.$$

On the other hand,

$$\|P_{\text{orth}}\|_1 = \frac{3839}{1260} \quad \textit{and} \quad \|I - P_{\text{orth}}\|_1 = \frac{317}{105}.$$

For $n \leq 5$, P_{orth} is the unique Γ_n-invariant minimal projection and the relevant norms are as summarized in the following table:

n	$\|P_{\text{orth}}\|_1$	$\|I - P_{\text{orth}}\|_1$
2	1	3/2
3	19/9	2
4	41/16	5/2
5	69/25	68/25

PROPOSITION 27. *Let $n \geq 3$ and let $P_{\min}$ be a minimal Γ_n-invariant projection from $\ell_1(E_n)$ onto Z_n. Then*

$$\lambda(\text{Lip}_0(\mathbb{T}_n)) = \|I - P_{\min}\|_1 = \|P_{\min}\|_1 - \frac{1}{n^2}.$$

PROOF. Let P be any Γ_n-invariant projection onto Z_n. Note that $\langle P(e), e\rangle$ is independent of $e \in E_n$. So

$$\langle P(e), e\rangle = \frac{\mathrm{trace}(P)}{|E_n|} = \frac{\dim(Z_n)}{2n^2} = \frac{n^2+1}{2n^2}.$$

From (8), we obtain

$$\|I-P\|_1 = \|e - P(e)\|_1 = \|P(e)\|_1 + 1 - 2\langle P(e), e\rangle = \|P\|_1 - \frac{1}{n^2}. \tag{8}$$

Now suppose that Q is a minimal projection from $\ell_\infty(E_n)$ onto B_n. We may assume that Q is Γ_n-invariant. Then $P = I - Q^*$ is a Γ_n-invariant projection onto Z_n (see Corollary 62 below for a proof of this in greater generality). Thus,

$$\|Q\|_\infty = \|Q^*\|_1 = \|I-P\|_1 = \|P\|_1 - \frac{1}{n^2} \geq \|P_{\min}\|_1 - \frac{1}{n^2} = \|I - P_{\min}\|.$$

The reverse inequality $\|Q\|_\infty \leq \|I - P_{\min}\|_1$ follows from the minimality of Q and the fact that $I - P^*_{\min}$ is a projection onto B_n. Hence

$$\lambda(\mathrm{Lip}_0(\mathbb{T}_n)) = \|Q\|_\infty = \|I - P_{\min}\|_1 = \|P_{\min}\|_1 - \frac{1}{n^2}.$$

□

COROLLARY 28. *(a) For $n \leq 5$, $\lambda(\mathrm{Lip}_0(\mathbb{T}_n)) = \|I - P_{\mathrm{orth}}\|_1$ and $c_1((\mathcal{P}(\mathbb{T}_n), d_{W_1})) \leq \|I - P_{\mathrm{orth}}\|_1$ as given in the table above.*

(b) $\lambda(\mathrm{Lip}_0(\mathbb{T}_6)) = 3$ and $c_1((\mathcal{P}(\mathbb{T}_6), d_{W_1})) \leq 3$.

PROOF. The estimates for $c_1(\mathcal{P}(\mathbb{T}_n), d_{W_1}))$ in (a) and (b) follow from Proposition 8 and those for $\mathrm{Lip}_0(\mathbb{T}_n)$ follow from Propositions 26 and 27 and the table above. □

REMARK 29. The estimates for $c_1((\mathcal{P}(\mathbb{T}_n), d_{W_1}))$ above improve somewhat the estimates for small n given by John's theorem [**20**] asserting that if Y is an n-dimensional normed space then $d_{\mathrm{BM}}(Y, \ell_2^n) \leq \sqrt{n}$:

$$c_1((\mathcal{P}(\mathbb{T}_n), d_{W_1})) = c_1(\mathrm{TC}(\mathbb{T}_n)) \leq d_{\mathrm{BM}}(\mathrm{TC}(\mathbb{T}_n), \ell_2^{n^2-1}) \leq \sqrt{n^2-1},$$

where the first equality follows from [**25**, Lemma 3.1].

COROLLARY 30. *For all $n \geq 3$, $d_{\mathrm{BM}}(\mathrm{TC}(\mathbb{T}_n), \ell_1^{n^2-1}) \geq \|P_{\min}\|_1 - 1/n^2$.*

PROOF. By the proof of Proposition 8(iv),

$$d_{\mathrm{BM}}(\mathrm{TC}(\mathbb{T}_n), \ell_1^{n^2-1}) \geq \lambda(\mathrm{Lip}_0(\mathbb{T}_n)) = \|P_{\min}\|_1 - \frac{1}{n^2}.$$ □

Unfortunately, we do not have sharp asymptotic estimates for $\|P_{\min}\|_1$ or even $\|P_{\mathrm{orth}}\|_1$. The following summarizes what is known.

PROPOSITION 31. *There exist positive constants a_1 and a_2 such that for $n \geq 2$,*

$$a_1\sqrt{\log(n)} \leq \|P_{\min}\|_1 \leq a_2 \log(n).$$

PROOF. The lower estimate follows from the estimate $c_1((\mathcal{P}(\mathbb{T}_n), d_{W_1})) \geq a_1\sqrt{\log(n)}$ due to Naor and Schechtman [**25**] combined with Proposition 8(ii). The upper estimate is valid for all finite graphs $G = (V, E)$, with $|V| = n$, equipped with the graph distance. It is proved using stochastic tree embeddings [**34**]. □

We close this section with a central open question related to our results.

QUESTION 32. Find good asymptotic estimates (improving on Proposition 31) for $\|P_{\min}\|_1$ and/or $\|P_{\mathrm{orth}}\|_1$.

2.5. m-dimensional tori. In this section we consider invariant projections on the cycle spaces of m-dimensional tori $\mathbb{Z}_n^m$ for $m, n \geq 2$. These graphs are 3-connected, so by Remark 17 the cycle-preserving isometries of the edge space are induced by graph automorphisms.

The cases $n = 2, 3$ can be placed in a more general setting of Hamming graphs considered in Section 2.6. Here we start with the case $n = 4$.

2.5.1. *The case $n = 4$.* Next we prove uniqueness for $\mathbb{Z}_4^m$. In the next lemma we use the notation introduced in Section 2.4.

LEMMA 33. *For each $(i, j) \in \mathbb{T}_4$ there exists $\psi \in H_4$ such that*

$$\psi((0,0)) = (0,0) \quad \textit{and} \quad \hat{\psi}(S((i,j))) = -S((i,j)).$$

PROOF. This follows from the table in the proof of Lemma 22 and from the definition of ϕ in Remark 19. Since $\hat{\beta}(S((i,j))) = -S((j,i)))$, it suffices to check the claim in the range $0 \leq i < j \leq 3$:

$$\hat{\psi}(S((0,1))) = -S((0,1)) \quad \text{for} \quad \psi = \phi^{-1}\alpha\phi;$$
$$\hat{\psi}(S((0,2))) = -S((0,2)) \quad \text{for} \quad \psi = \beta\phi.$$
$$\hat{\psi}(S((0,3))) = -S((0,3)) \quad \text{for} \quad \psi = \phi^{-1}\alpha\phi;$$
$$\hat{\psi}(S((1,2))) = -S((1,2)) \quad \text{for} \quad \psi = \phi^{-1}\beta\alpha\beta\phi;$$
$$\hat{\psi}(S((1,3))) = -S((1,3)) \quad \text{for} \quad \psi = \phi;$$
$$\hat{\psi}(S((2,3))) = -S((2,3)) \quad \text{for} \quad \psi = \phi^{-1}\alpha\phi.$$

□

LEMMA 34. *Suppose $\psi \in H_4$ is given by*

$$\psi((i_1, i_2)) = (j_1, j_2) \qquad ((i_1, i_2) \in \mathbb{T}_4).$$

Then, for each $m \geq 2$, ψ generates a graph automorphism Ψ of $\mathbb{Z}_4^m$ given by

$$\Psi((i_1, i_2, i_3, \ldots, i_m)) = (j_1, j_2, i_3, \ldots, i_m) \qquad ((i_1, \ldots, i_m) \in \mathbb{Z}_4^m).$$

Moreover, if $\psi((0,0)) = (0,0)$, then $\Psi((0,\ldots,0)) = (0,\ldots,0)$.

PROOF. Immediate from the definitions. □

LEMMA 35. *Let $m \geq 2$. The cycle space $Z(\mathbb{Z}_n^m)$ is spanned by the collection of all vectors v_S corresponding to (i,j)-parallel small square cycles and (for $n \geq 3$) all i-parallel cycle vectors v_R corresponding to an i-parallel row cycle R with vertices:*

$$(x_1, \ldots, x_{i-1}, \{0, 1, 2, \ldots, n-1\}, x_{i+1}, \ldots, x_m),$$

where $x_1, \ldots, x_m$ are arbitrary.

PROOF. As in Example 10 we view the graph associated to $\mathbb{Z}_n^m$ as the Cayley graph of a group with generators g_i $(1 \leq i \leq m)$ satisfying the following relations:

$$g_i + g_j = g_j + g_i \qquad (i \neq j), \tag{9}$$

and

$$\underbrace{g_i + \cdots + g_i}_{n \text{ times}} = 0 \qquad (i = 1, \ldots, m). \tag{10}$$

The cycles corresponding to (9) are (i, j)-parallel small squares and those corresponding to (10) are i-parallel row cycles. Hence the result follows from Proposition 11. □

Let $\Gamma_0(\mathbb{Z}_n^m)$ denote the group of automorphisms of $\mathbb{Z}_n^m$ which fix $(0, \ldots, 0)$.

THEOREM 36. *Let $m \geq 2$. Then P_{orth} is the unique invariant projection on $Z(\mathbb{Z}_4^m)$.*

PROOF. Let S be a small square cycle. Without loss of generality, we may assume that S is $(1, 2)$-parallel. By Lemmas 33 and 34 there exists $\psi_S \in \Gamma_0(\mathbb{Z}_4^m)$ such that $\hat{\psi}_S(v_S) = -v_S$. Similarly, if R is a 1-parallel row cycle (without loss of generality) then $\hat{\psi}_R(v_R) = -v_R$, where $\psi_R \in \Gamma_0(\mathbb{Z}_4^m)$ is given by $\psi_R(x_1, \ldots, x_n) = (-x_1, \ldots, x_n)$. Hence, as in the proof of Lemma 42,

$$\sum_{\psi \in \Gamma_0(\mathbb{Z}_4^m)} \hat{\psi}(v_S) = \sum_{\psi \in \Gamma_0(\mathbb{Z}_4^m)} \hat{\psi}(v_R) = 0.$$

By Lemma 35 it follows that if $v \in Z(\mathbb{Z}_4^m)$ is $\Gamma_0(\mathbb{Z}_4^m)$-invariant, then $v = 0$.

Let X be the sum of all edge vectors joining the vertex $(0, 0, \ldots, 0)$ to its neighbors oriented with $(0, 0, \ldots, 0)$ as initial vertex. Note that X is a $\Gamma_0(\mathbb{Z}_4^m)$-invariant vector belonging to the cut space $B(\mathbb{Z}_4^m)$. Let P be an invariant projection onto $Z(\mathbb{Z}_4^m)$. Then, for all $\phi \in \Gamma_0(\mathbb{Z}_4^m)$,

$$\hat{\phi}(P(X)) = P(\hat{\phi}(X)) = P(X).$$

Hence $P(X) \in Z(\mathbb{Z}_4^m)$ is $\Gamma_0(\mathbb{Z}_4^m)$-invariant, so $P(X) = 0$ from the above. The argument given in Lemma 21 for $m = 2$ readily extends to $m \geq 2$ to show that P annihilates $B(\mathbb{Z}_4^m)$, i.e., $P = P_{\text{orth}}$. □

2.5.2. *The case $n \geq 5$.* Next we show that for $n \geq 5$ and $m \geq 2$ there exist nonorthogonal invariant projections onto $Z(\mathbb{Z}_n^m)$.

PROPOSITION 37. *Let $n \geq 5$ and $m \geq 2$. For each $\psi \in \Gamma_0(\mathbb{Z}_n^m)$ there exist a permutation σ of $\{1, \ldots, m\}$ and signs $\alpha_j = \pm 1$ $(1 \leq j \leq m)$ such that*

$$\psi((i_1, \ldots, i_m)) = (\alpha_1 i_{\sigma(1)}, \ldots, \alpha_m i_{\sigma(m)}). \tag{11}$$

PROOF. Since ψ fixes $(0, \ldots, 0)$ it is clear that there is a unique choice of σ and $\alpha_j = \pm 1$ such that (11) holds for all points of the form $(0, \ldots, 0, 1, 0, \ldots, 0)$.

Without loss of generality, we may assume that σ is the identity permutation and $\alpha_j = 1$ $(1 \leq j \leq m)$. Hence ψ fixes each point of the form $(0, \ldots, 0, 1, 0, \ldots, 0)$. By induction on $\operatorname{card}\{1 \leq j \leq m \colon i_j = 1\}$, it follows that

$$\psi(x) = x \text{ for } x = (i_j)_{j=1}^m, \text{ where } i_j \in \{0, 1\}. \tag{12}$$

We shall prove by induction that $\psi(x) = x$ for all $x \in \mathbb{Z}_n^m$. To that end, let $0 \leq r \leq m$ and suppose that

$$\psi(x) = x \text{ if } i_j \in \{0, 1\} \text{ for all } j > r \text{ and } i_j \in \mathbb{Z}_n \text{ for all } j \leq r. \tag{13}$$

Note that, by (12), $r = 0$ satisfies (13). Suppose that (13) is satisfied for some r, where $0 \leq r < m$. Fix values $x_j \in \mathbb{Z}_n$ for $1 \leq j \leq r$ and $x_j \in \{0, 1\}$ for $j \geq r + 2$. Since (13) is satisfied, it follows that

$$\psi((x_1, \ldots, x_r, i_{r+1}, x_{r+2}, \ldots x_m)) = (x_1, \ldots, x_r, i_{r+1}, x_{r+2}, \ldots, x_m)$$

provided $i_{r+1} \in \{0, 1\}$. Since $n \geq 5$, it follows as in the proof of Lemma 18 that ψ maps 'collinear triples' onto collinear triples. Hence (13) is satisfied by $r+1$, which completes the proof of the inductive hypothesis. So (13) is satisfied by $r = m$, i.e., ψ is the identity mapping as claimed. □

THEOREM 38. *Let $n \geq 5$ and $m \geq 2$. There exist nontrivial invariant projections onto $Z(\mathbb{Z}_n^m)$.*

PROOF. Let S be the small square cycle oriented as

$$(0, 1, 0, \ldots, 0) \to (1, 1, 0, \ldots, 0) \to (1, 2, 0, \ldots, 0) \to$$
$$(0, 2, 0, \ldots, 0) \to (0, 1, 0, \ldots, 0)$$

and let $v_S \in Z(\mathbb{Z}_n^m)$ be the corresponding cycle vector. Let

$$w = \sum_{\psi \in \Gamma_0(\mathbb{Z}_n^m)} \hat{\psi}(v_S).$$

Using Proposition 37, the coefficient of w on the edge vector $(1, 1, 0, \ldots, 0) \to (1, 2, 0, \ldots, 0)$ equals

$$\operatorname{card}\{\psi \in \Gamma_0(\mathbb{Z}_n^m) \colon \sigma(1) = 1, \sigma(2) = 2, \alpha_1 = \alpha_2 = 1\} = 2^{m-2}(m-2)! \neq 0.$$

Hence $w \in Z(\mathbb{Z}_n^m)$ is a nontrivial $\Gamma_0(\mathbb{Z}_n^m)$-invariant vector.

Let X be the sum of all edge vectors joining the vertex $(0, 0, \ldots, 0)$ to its neighbors oriented with $(0, 0, \ldots, 0)$ as initial vertex. Note that X is a $\Gamma_0(\mathbb{Z}_n^m)$-invariant vector belonging to the cut space of $\mathbb{Z}_n^m$. The argument given in Lemma 21 for $m = 2$ readily extends to $m \geq 2$ to show that there is a unique invariant projection P onto $Z(\mathbb{Z}_n^m)$ satisfying $P(X) = w$. □

REMARK 39. For $m > 2$, we do not have a simple description of the invariant projections similar to Theorem 24. This is related to the fact that we do not have a simple algebraic basis for $Z(\mathbb{Z}_n^m)$ analogous to the basis of small square cycle vectors of Lemma 15 for $m = 2$.

2.6. Hamming graphs. For $n \geq 2$, let $A_n = \{0, \ldots, n-1\}$ be an 'alphabet' of size n. Equip A_n^m with the Hamming metric:

$$d((a_i)_{i=1}^m, (b_i)_{i=1}^m) = \operatorname{card}\{i \colon a_i \neq b_i\}.$$

We call A_n^m the *Hamming graph* because we regard it as the vertex set of a graph with edges $\{(a_i)_{i=1}^m, (b_i)_{i=1}^m\}$ such that $d((a_i)_{i=1}^m, (b_i)_{i=1}^m) = 1$. Note that A_n^m and $\mathbb{Z}_n^m$ are isometric if and only if $n \leq 3$. Hamming graphs are well-known objects in algebraic combinatorics (see [**7**, Section 9.2]).

Let $\Gamma(n, m)$ denote the group of graph automorphisms of A_n^m and let $\Gamma_0(n, m)$ denote its subgroup consisting of the automorphisms which fix $(0, 0, \ldots, 0)$.

LEMMA 40. *Each $\phi \in \Gamma_0(n, m)$ is uniquely determined by a permutation σ of $\{1, 2, \ldots, m\}$ and permutations ϕ_i $(1 \leq i \leq m)$ of A_n such that $\phi_i(0) = 0$ and such that*

$$\phi((a_i)_{i=1}^m) = (b_i)_{i=1}^m, \quad \text{where} \quad b_{\sigma(i)} = \phi_i(a_i). \tag{14}$$

PROOF. Since ϕ fixes $(0, 0, \ldots, 0)$, it follows that σ and ϕ_i exist such that (14) holds provided $\operatorname{card}\{i \colon a_i \neq 0\} \leq 1$. Hence (14) holds for all $(a_i)_{i=1}^m$ by a straightforward induction on $\operatorname{card}\{i \colon a_i \neq 0\}$. □

A 'small square' cycle is said to be (i, j)-parallel, where $1 \le i < j \le m$, if the vertices of the corresponding square S are of the form:

$$(x_1, \ldots, x_{i-1}, \{a_1, a_2\}, x_{i+1} \ldots, x_{j-1}, \{b_1, b_2\}, x_{j+1}, \ldots, x_m).$$

Let v_S denote the corresponding cycle vector belonging to the cycle space $Z(A_n^m)$. A 'small triangle' cycle is said to be i-parallel if the vertices of the corresponding triangle T are of the form

$$(x_1, \ldots, x_{i-1}, \{a_1, a_2, a_3\}, x_{i+1}, \ldots, x_m).$$

Let $v_T \in Z(A_n^m)$ denote the corresponding cycle vector.

LEMMA 41. *$Z(A_n^m)$ is spanned by the collection of signed indicator functions of all small square and all small triangle cycles.*

PROOF. Clearly, it suffices to prove the following statement: Let $f \in Z(A_n^m)$ be the signed indicator function of a directed cycle C of length k. Then, adding to f some signed indicator functions of triangular and square cycles described above, we get a sum of signed indicator functions of directed cycles of length $\le k-1$ each.

We need to consider the following cases:

(1) The cycle C contains two consecutive edges for which the ends differ in the same coordinate. Let the edges be $\overrightarrow{uv}$ and $\overrightarrow{vw}$. In this case $\overrightarrow{vu}, \overrightarrow{uw}$, and $\overrightarrow{wv}$ form a triangle of the described type, and its "addition" decreases the "length" of C.

(2) Any two consecutive edges change different coordinates. Note that each coordinate which ever changes in the cycle C, has to be changed at least twice. Let l be the smallest number of edges between any two edges changing the same coordinate. Let $\overrightarrow{uv}$ and $\overrightarrow{xy}$ be these two edges changing the same coordinate, and $\overrightarrow{vw}$ be an edge in C on the shortest path from v to x.

We prove the following statement (as is easy to see, it achieves the goal described above). There is a square cycle of the described type such that adding its signed indicator function to f we get one of the following four outcomes:

(a) A signed indicator function of a cycle of length k in which the smallest number of edges between two edges changing the same coordinate is equal to $l-1$,

(b) Empty cycle,

(c) A signed indicator function of a cycle of length $k-2$,

(d) A union of two nontrivial cycles of total length k.

In fact, without loss of generality, assume that

$$\begin{aligned} u &= (x_1, x_2, x_3, \ldots, x_m), \\ v &= (\tilde{x}_1, x_2, x_3, \ldots, x_m), \\ w &= (\tilde{x}_1, \tilde{x}_2, x_3, \ldots, x_m). \end{aligned}$$

We introduce $\bar{v} = (x_1, \tilde{x}_2, x_3, \ldots, x_m)$. There are four possibilities:

(i) $\bar{v}$ is not a vertex of the cycle C.

In this case, we add to f the signed indicator function of the cycle $w, v, u, \bar{v}, w$, and get a cycle of length k in which the edge $\overrightarrow{\bar{v}w}$ and $\overrightarrow{xy}$ change the same coordinate, and the number of edges between them is $l-1$.

(ii) $\bar{v}$ is a vertex of the cycle C, and the cycle has four vertices $w, v, u, \bar{v}, w$. In this case, adding to f the signed indicator function of the cycle $w, v, u, \bar{v}, w$, we get 0.

(iii) $\bar{v}$ is a vertex of the cycle C, and exactly one of the edges $u\bar{v}$ and $\bar{v}w$ is an edge of C.
In this case, adding to f the signed indicator function of the cycle $w, v, u, \bar{v}, w$, we get the signed indicator function of a cycle with $k-2$ edges.

(iv) $\bar{v}$ is a vertex of the cycle C, and none of the edges $u\bar{v}$ and $\bar{v}w$ is an edge of C.
In this case, adding to f the signed indicator function of the cycle $w, v, u, \bar{v}, w$, we get the sum of signed indicator function of two nontrivial cycles with total length of k edges.

□

LEMMA 42. *Suppose $v \in Z(A_n^m)$ is $\Gamma_0(n,m)$-invariant (i.e., $\hat{\phi}(v) = v$ for all $\phi \in \Gamma_0(n,m)$). Then $v = 0$.*

PROOF. Suppose v is $\Gamma_0(n,m)$-invariant. Then

$$\frac{1}{\operatorname{card}(\Gamma_0(n,m))} \sum_{\phi \in \Gamma_0(n,m)} \hat{\phi}(v) = v.$$

So by Lemma 41 it suffices to check that

$$\sum_{\phi \in \Gamma_0(n,m)} \hat{\phi}(v_S) = \sum_{\phi \in \Gamma_0(n,m)} \hat{\phi}(v_T) = 0$$

for all small squares S and small triangles T.

Let S be a small square. Without loss of generality, we may assume that S is (1,2)-parallel.

CASE I. S has edges oriented as follows:

$$\begin{aligned}&(0,0,x_3,\dots,x_n) \to (a,0,x_3,\dots,x_n) \to (a,b,x_3,\dots,x_n) \to \\ &(0,b,x_3,\dots,x_n) \to (0,0,x_3,\dots,x_n).\end{aligned}$$

where $a \neq 0, b \neq 0$ and $x_3,\dots,x_n$ are arbitrary. Consider $\psi_S \in \Gamma_0(n,m)$ corresponding to permutations σ and ψ_i ($1 \le i \le m$), where σ is the transposition which interchanges 1 and 2, and $\psi_1 = \psi_2$ is the transposition which interchanges a and b if $a \neq b$ or the identity if $a = b$, and ψ_i is the identity for $3 \le i \le m$. Then ψ_S maps S onto the same square but with the opposite orientation:

$$\begin{aligned}&(0,0,x_3,\dots,x_n) \to (0,b,x_3,\dots,x_n) \to (a,b,x_3,\dots,x_n) \to \\ &(a,0,x_3,\dots,x_n) \to (0,0,x_3,\dots,x_n),\end{aligned}$$

So $\hat{\psi_S}(v_S) = -v_S$. Hence, using the group property of $\Gamma_0(n,m)$,

$$\begin{aligned}\sum_{\phi \in \Gamma_0(n,m)} \hat{\phi}(v_S) &= \frac{1}{2} \sum_{\phi \in \Gamma_0(n,m)} (\hat{\phi}(v_S) + \hat{\phi}\hat{\psi_S}(v_S)) \\ &= \frac{1}{2} \sum_{\phi \in \Gamma_0(N,m)} \hat{\phi}(v_S + \hat{\psi_S}(v_S)) = 0.\end{aligned}$$

CASE II. S has edges oriented as follows:

$$(a_1, b_1, x_3, \ldots, x_n) \to (a_2, b_1, x_3, \ldots, x_n) \to (a_2, b_2, x_3, \ldots, x_n) \to$$
$$(a_1, b_2, x_3, \ldots, x_n) \to (a_1, b_1, x_3, \ldots, x_n),$$

where $a_1 \neq 0$, $a_2 \neq 0$, and $x_3, \ldots, x_n$ are arbitrary. Consider $\psi_S \in \Gamma(n,m)$, where ψ_1 is the transposition (a_1, a_2) and ψ_i $(2 \leq i \leq n)$ and σ are the identity. Then $\hat{\psi_S}(v_S) = -v_S$ and hence $\sum_{\phi \in \Gamma_0(n,m)} \hat{\phi}(v_S) = 0$.

Let T be a small triangle. Without loss of generality, T is 1-parallel with oriented edges as follows:

$$(a_1, x_2, \ldots, x_n) \to (a_2, x_2, \ldots, x_n) \to (a_3, x_2, \ldots, x_n) \to (a_1, x_2, \ldots, x_n),$$

where $a_1 \neq 0$, $a_2 \neq 0$, and $x_2, \ldots, x_n$ are arbitrary. Consider $\psi_T \in \Gamma(n,m)$, where ψ_1 is the transposition (a_1, a_2) and ψ_i $(2 \leq i \leq n)$ and σ are the identity. Then $\hat{\psi_T}(v_S) = -v_S$ and hence $\sum_{\phi \in \Gamma(n,m)} \hat{\phi}(v_T) = 0$.

□

THEOREM 43. *The orthogonal projection P_{orth} is the unique $\Gamma(n,m)$-invariant projection onto $Z(A_n^m)$.*

PROOF. Let X be the sum of all edge vectors joining the vertex $(0, 0, \ldots, 0)$ to its neighbours oriented with $(0, 0, \ldots, 0)$ as initial vertex. Note that X is a $\Gamma_0(n,m)$-invariant vector belonging to the cut space of A_n^m.

Let P be a $\Gamma(n,m)$-invariant projection onto $Z(A_n^m)$. Then, for all $\phi \in \Gamma_0(n,m)$,

$$\hat{\phi}(P(X)) = P(\hat{\phi}(X)) = P(X).$$

So $P(X) \in Z(A_n^m)$ is $\Gamma_0(n,m)$-invariant. By Lemma 42, $P(X) = 0$. Let $a = (a_i)_{i=1}^m \in A_n^m$. Clearly, $\Gamma(n,m)$ is transitive, so there exists $\psi_a \in \Gamma(n,m)$ such that $\psi_a((0, 0, \ldots, 0)) = a$. Hence, by $\Gamma(n,m)$-invariance,

$$P(\hat{\psi_a}(X)) = \hat{\psi_a}(P(X)) = \hat{\psi_a}(0) = 0.$$

Since $\{\hat{\psi_a}(X) \colon a \in A_n^m\}$ spans the cut space, it follows that P annihilates the cut space, i.e., $P = P_{\text{orth}}$. □

COROLLARY 44. *Let $m \geq 2$. Then P_{orth} is the unique invariant projection onto $Z(\mathbb{Z}_n^m)$ for $n = 2$ and $n = 3$.*

2.7. The Hamming Cube. The main goal of this section is the following result.

THEOREM 45. *Let Q_n be the orthogonal projection from the edge space of the Hamming cube $\mathbb{Z}_2^n$ onto the cut space. Then $\|Q_n\|_1 = \dfrac{n+1}{2}$.*

2.7.1. *Notation.* First, let us settle on some notation for $\mathbb{Z}_2^n$ to simplify the discussion:

(1) Let $V_n = \{(\varepsilon_i)_{i=1}^n \colon \varepsilon_i \in \{0, 1\}\}$ denote the vertex set and let E_n denote the edge set. The group of graph automorphisms is denoted Γ_n. (Note that the notation introduced here differs from that of Section 2.4.)
(2) A directed edge e with initial vertex $v = e_-$ and terminal vertex $w = e_+$ will be denoted as $\overrightarrow{vw}$ or $\overrightarrow{e_-e_+}$. We will also use the same notation $\overrightarrow{vw}$ to denote the corresponding edge vector belonging to $E(\mathbb{Z}_2^n)$.

(3) The edge, cycle, and cut spaces are denoted $E(\mathbb{Z}_2^n)$, $Z(\mathbb{Z}_2^n)$, and $B(\mathbb{Z}_2^n)$.
(4) We say that an oriented edge e is j-parallel if $\varepsilon_j = 0$ at e_- and $\varepsilon_j = 1$ at e_+.
(5) Let e_0 be the edge with initial vertex $(0, 0, \ldots, 0)$ and terminal vertex $(1, 0, \ldots, 0)$.
(6) For $0 \le k \le n-1$, let v_k denote any vertex $(\varepsilon_i)_{i=1}^n$ such that

$$\varepsilon_1 = 0 \quad \text{and} \quad \operatorname{card}\{i \colon \varepsilon_i = 1\} = k.$$

For $1 \le k \le n$, let w_k be any vertex $(\varepsilon_i)_{i=1}^n$ such that

$$\varepsilon_1 = 1 \quad \text{and} \quad \operatorname{card}\{i \colon \varepsilon_i = 1\} = k.$$

(7) There are three types of edges:

$$\begin{aligned} &\overrightarrow{w_k w_{k+1}} \quad (1 \le k \le n-1); \\ &\overrightarrow{v_k w_{k+1}} \quad (0 \le k \le n-1); \\ &\overrightarrow{v_k v_{k+1}} \quad (0 \le k \le n-2). \end{aligned}$$

Note that e_0 is the unique edge of the form $\overrightarrow{v_0 w_1}$. We use arrows to indicate our choice of the reference orientation of $\mathbb{Z}_2^n$.

We will investigate the basis expansion of $Q_n(e_0)$ with respect to the standard edge basis E_n of $E(\mathbb{Z}_2^n)$. To that end, let us define the basis coefficients corresponding to the three edge types:

$$\begin{aligned} a_k &= \langle Q_n(e_0), \overrightarrow{w_k w_{k+1}} \rangle \quad (1 \le k \le n-1), \\ b_k &= \langle Q_n(e_0), \overrightarrow{v_k w_{k+1}} \rangle \quad (0 \le k \le n-1), \\ c_k &= \langle Q_n(e_0), \overrightarrow{v_k v_{k+1}} \rangle \quad (0 \le k \le n-2), \end{aligned}$$

Note that, by symmetry, a_k does not depend on the choice of edge of type $\overrightarrow{w_k w_{k+1}}$, and similarly for b_k and c_k. More precisely, we use the fact that for any two vectors of the same type there is an automorphism of $\mathbb{Z}_2^n$ mapping one of them onto the other and fixing e_0.
(8) For each $v \in V_n$, let $X(v) \in B(\mathbb{Z}_2^n)$ denote the cut vector which is the sum of the n edge vectors emanating from v oriented so that v is the initial vertex.

2.7.2. *Technical lemmas about the basis coefficients of* $Q_n(e_0)$. Throughout this section we assume that $n \ge 3$.

LEMMA 46. $a_k = -c_{k-1}$ *for* $1 \le k \le n-1$.

PROOF. Consider the translation of $\mathbb{Z}_2^n$ given by $\phi(v) = v + (1, 0, \ldots, 0)$ $(v \in \mathbb{Z}_2^n)$. Then $Q_n \hat{\phi} = \hat{\phi} Q_n$ by Γ_n-invariance of Q_n. Hence

$$\begin{aligned} a_k &= \langle Q_n(e_0), \overrightarrow{w_k w_{k+1}} \rangle \\ &= \langle \hat{\phi}(Q_n(e_0)), \hat{\phi}(\overrightarrow{w_k w_{k+1}}) \rangle \\ &= \langle Q_n(\hat{\phi}(e_0)), \overrightarrow{v_{k-1} v_k} \rangle \\ &= \langle Q_n(-e_0), \overrightarrow{v_{k-1} v_k} \rangle \\ &= -c_{k-1}. \end{aligned}$$

□

LEMMA 47. $b_0 = \langle Q_n(e_0), e_0 \rangle = \dfrac{2}{n} - \dfrac{1}{n 2^{n-1}}$.

PROOF. By symmetry, $\langle Q_n(f), f\rangle$ is independent of the choice of edge vector f. Hence

$$\begin{aligned}\langle Q_n(e_0), e_0\rangle &= \frac{1}{\operatorname{card}(E_n)} \sum_{f\in E_n} \langle Q_n(f), f\rangle \\ &= \frac{\operatorname{trace}(Q_n)}{\operatorname{card}(E_n)} \\ &= \frac{\dim(B_n)}{\operatorname{card}(E_n)} \\ &= \frac{2^n - 1}{n2^{n-1}} \\ &= \frac{2}{n} - \frac{1}{n2^{n-1}}.\end{aligned}$$

□

LEMMA 48. $c_0 = \dfrac{1}{n-1} - \dfrac{2}{n(n-1)} + \dfrac{1}{2^{n-1}n(n-1)}$.

PROOF.

$$\begin{aligned}1 &= \langle X((0,0,\ldots,0)), e_0)\rangle \\ &= \langle X((0,0,\ldots,0)), Q_n(e_0)\rangle \\ &= \langle e_0, Q_n(e_0)\rangle + (n-1)\langle \overrightarrow{v_0v_1}, Q_n(e_0)\rangle \\ &= b_0 + (n-1)c_0.\end{aligned}$$

Hence, by Lemma 47,

$$c_0 = \frac{1-b_0}{n-1} = \frac{1}{n-1} - \frac{2}{n(n-1)} + \frac{1}{2^{n-1}n(n-1)}.$$

□

LEMMA 49. $a_k = \dfrac{b_k - b_{k-1}}{2}$ *for* $1 \le k \le n-1$.

PROOF. Consider a small square cycle with directed edges of the form

$$v_{k-1} \to w_k \to w_{k+1} \to v_k \to v_{k-1}.$$

The corresponding cycle vector is given by

$$\overrightarrow{v_{k-1}w_k} + \overrightarrow{w_kw_{k+1}} - \overrightarrow{v_kw_{k+1}} - \overrightarrow{v_{k-1}v_k} \in Z(\mathbb{Z}_2^n).$$

Since $Q_n(e_0) \in B(\mathbb{Z}_2^n) = Z(\mathbb{Z}_2^n)^\perp$, we have

$$\begin{aligned}0 &= \langle Q_n(e_0), \overrightarrow{v_{k-1}w_k} + \overrightarrow{w_kw_{k+1}} - \overrightarrow{v_kw_{k+1}} - \overrightarrow{v_{k-1}v_k}\rangle \\ &= b_{k-1} + a_k - b_k - c_{k-1} \\ &= b_{k-1} + a_k - b_k + a_k\end{aligned}$$

by Lemma 46. Hence the result follows. □

LEMMA 50. $b_1 = \dfrac{2}{n(n-1)} - \dfrac{1}{2^{n-2}n(n-1)} - \dfrac{1}{n2^{n-1}}$.

PROOF.

$$b_1 = 2a_1 + b_0$$

(by Lemma 49)

$$= -2c_0 + b_0$$

(by Lemma 46)

$$= -2[\frac{1}{n-1} - \frac{2}{n(n-1)} + \frac{1}{2^{n-1}n(n-1)}] + \frac{2}{n} - \frac{1}{n2^{n-1}}$$

(by Lemmas 48 and 47)

$$= \frac{2}{n(n-1)} - \frac{1}{2^{n-2}n(n-1)} - \frac{1}{n2^{n-1}}.$$

□

LEMMA 51. *For* $2 \leq k \leq n-1$,

$$(n-k)a_k - (k-1)a_{k-1} - b_{k-1} = 0.$$

PROOF. For $2 \leq k \leq n-1$, consider a cut vector of the form $X(w_k)$ for a particular choice of w_k. There are $n-k$ edges of form $\overrightarrow{w_k w_{k+1}}$ with initial vertex w_k, and $k-1$ edges of form $\overrightarrow{w_{k-1} w_k}$, and 1 of form $\overrightarrow{v_{k-1} w_k}$, with terminal vertex w_k. Moreover, the support of $X(w_k)$ is disjoint form e_0. Hence

$$\begin{aligned}
0 &= \langle e_0, X(w_k)\rangle \\
&= \langle Q_n(e_0), X(w_k)\rangle \\
&= (n-k)\langle Q_n(e_0), \overrightarrow{w_k w_{k+1}}\rangle - (k-1)\langle Q_n(e_0), \overrightarrow{w_{k-1} w_k}\rangle - \langle Q_n(e_0), \overrightarrow{v_{k-1} w_k}\rangle \\
&= (n-k)a_k - (k-1)a_{k-1} - b_{k-1}.
\end{aligned}$$

□

LEMMA 52. *For* $2 \leq k \leq n-1$,

$$(n-k)b_k - (n+1)b_{k-1} + (k-1)b_{k-2} = 0.$$

PROOF.

$$0 = (n-k)a_k - (k-1)a_{k-1} - b_{k-1}$$

(by Lemma 51)

$$= \frac{(n-k)}{2}(b_k - b_{k-1}) - \frac{(k-1)}{2}(b_{k-1} - b_{k-2}) - b_{k-1}$$

(by Lemma 49)

$$= \frac{1}{2}[(n-k)b_k - (n+1)b_{k-1} + (k-1)b_{k-2}].$$ □

LEMMA 53. $b_{n-2} = \dfrac{(n+1)}{n-1} b_{n-1}$.

PROOF. Note that $(0,1,1,\dots,1)$ is the unique vertex of form v_{n-1}. Since e_0 is not contained in the support of $X(v_{n-1})$,

$$\begin{aligned}0 &= \langle e_0, X(v_{n-1})\rangle \\ &= \langle Q_n(e_0), X(v_{n-1})\rangle \\ &= \langle Q_n(e_0), \overrightarrow{v_{n-1}w_n}\rangle - (n-1)\langle Q_n(e_0), \overrightarrow{v_{n-2}v_{n-1}}\rangle \\ &= b_{n-1} - (n-1)c_{n-2} \\ &= b_{n-1} + (n-1)a_{n-1}\end{aligned}$$

(by Lemma 46)

$$= b_{n-1} + \frac{(n-1)}{2}(b_{n-1} - b_{n-2})$$

(by Lemma 49)

$$= \frac{(n+1)}{2}b_{n-1} - \frac{(n-1)}{2}b_{n-2}.$$

Hence the result follows. □

LEMMA 54. *The sequence $(b_k)_{k=0}^{n-1}$ is positive and strictly decreasing.*

PROOF. First we prove positivity. By Lemmas 47 and 50,

$$b_0 > b_1 > 0.$$

Suppose, to derive a contradiction, that $b_k \le 0$ and $b_k < b_{k-1}$ for some k satisfying $2 \le k \le n-1$. Note that, in fact, $k \le n-2$ since by Lemma 53

$$b_{n-2} = \frac{(n+1)}{n-1}b_{n-1},$$

which cannot hold if $b_{n-1} \le 0$ and $b_{n-1} < b_{n-2}$.

By Lemma 52, with k replaced by $k+1$ (as we may since $k \le n-2$),

$$\frac{(n+1)}{n-1}b_k = \frac{(n-k-1)}{n-1}b_{k+1} + \frac{k}{n-1}b_{k-1}. \tag{15}$$

Note that the right-hand side is a proper convex combination of b_{k-1} and b_{k+1}. However, by assumption, $b_k \le 0$ and $b_k < b_{k-1}$, so

$$\frac{(n+1)}{n-1}b_k \le b_k < b_{k-1}.$$

Hence, from the strict convexity of (15) and our assumption that $b_k \le 0$,

$$0 \ge b_k \ge \frac{(n+1)}{n-1}b_k > b_{k+1}.$$

It follows that b_{k+1} satisfies the same assumptions as b_k. So, by iteration,

$$0 \ge b_k > b_{k+1} > \cdots > b_{n-2} > b_{n-1},$$

which contradicts Lemma 53. It follows that $b_k > 0$ for all $0 \le k \le n-1$.

Next, let us prove that $(b_k)_{k=0}^{n-1}$ is strictly decreasing. By Lemma 53 and positivity,

$$b_{n-2} = \frac{(n+1)}{n-1}b_{n-1} > b_{n-1} > 0.$$

Hence, if $(b_k)_{k=0}^{n-1}$ is not strictly decreasing, then there exists k satisfying $1 \le k \le n-2$ such that $b_k > b_{k+1}$ and $b_k \ge b_{k-1}$. But then by (15)

$$0 < b_k < \frac{(n+1)}{n-1} b_k \le \max(b_{k-1}, b_{k+1}),$$

which is a contradiction. □

COROLLARY 55. *$a_k < 0$ for $1 \le k \le n-1$ and $c_k > 0$ for $0 \le k \le n-2$.*

PROOF. By Lemmas 49 and 54,

$$a_k = \frac{1}{2}(b_k - b_{k-1}) < 0,$$

and, by Lemma 46, $c_k = -a_{k+1} > 0$. □

2.7.3. *Calculation of $\|Q_n\|_1$.*

LEMMA 56. *$\|Q_n\|_1 = F(n) + 2G(n)$, where*

$$F(n) = \sum_{k=0}^{n-1} \binom{n-1}{k} b_k, \quad G(n) = \sum_{k=0}^{n-2} (n-1-k) \binom{n-1}{k} c_k.$$

PROOF. Since the graph automorphism group of $\mathbb{Z}_2^n$ is edge-transitive, the distribution of $Q_n(e)$ is independent of $e \in E_n$. Hence

$$\|Q_n\|_1 = \max_{e \in E_n} \|Q_n(e)\|_1 = \|Q_n(e_0)\|_1.$$

There are $\binom{n-1}{k}$ edges of the form $\overrightarrow{v_k w_{k+1}}$, $(n-1-k)\binom{n-1}{k}$ of form $\overrightarrow{v_k v_{k+1}}$, and $(n-k)\binom{n-1}{k-1}$ of form $\overrightarrow{w_k w_{k+1}}$. Hence

$$\begin{aligned}
\|Q_n(e_0)\|_1 &= \sum_{k=0}^{n-1} \binom{n-1}{k} |b_k| + \sum_{k=0}^{n-2} (n-1-k) \binom{n-1}{k} |c_k| \\
&\quad + \sum_{k=0}^{n-2} (n-k) \binom{n-1}{k-1} |a_k| \\
&= \sum_{k=0}^{n-1} \binom{n-1}{k} |b_k| + 2 \sum_{k=0}^{n-2} (n-1-k) \binom{n-1}{k} |c_k|
\end{aligned}$$

(by Lemma 46)

$$= \sum_{k=0}^{n-1} \binom{n-1}{k} b_k + 2 \sum_{k=0}^{n-2} (n-1-k) \binom{n-1}{k} c_k$$

by positivity of b_k (Lemma 54) and c_k (Lemma 55). □

LEMMA 57. *$F(n) = 1$.*

PROOF. Let x be the sum over all edge vectors of the form $\overrightarrow{v_k w_{k+1}}$ as k ranges from 0 to $n-1$. Note that x is the sum over all 1-parallel edge vectors. Hence x is orthogonal to every small square cycle vector. So $x \in B(\mathbb{Z}_2^n)$, i.e., $Q_n(x) = x$.

Note that $F(n) = \sum_u \langle Q_n(e_0), u\rangle$, where u ranges over all 1-parallel edge vectors. Hence

$$\begin{aligned} F(n) &= \sum_u \langle Q_n(e_0), u\rangle \\ &= \sum_u \langle e_0, Q_n(u)\rangle \\ &= \langle e_0, Q_n(\sum_u u)\rangle \\ &= \langle e_0, Q_n(x)\rangle \\ &= \langle e_0, x\rangle \\ &= 1. \end{aligned}$$

□

To state the next lemma, let us introduce some notation. For $2 \leq j \leq n$:

(1) let x_j be the sum over all j-parallel edge vectors of form $\overrightarrow{v_k v_{k+1}}$ (i.e., $\varepsilon_1 = 0$ at the initial and terminal vertices);
(2) let u_j be the sum over all j-parallel edge vectors of form $\overrightarrow{w_k w_{k+1}}$ (i.e., $\varepsilon_1 = 1$ at the initial and terminal vertices);
(3) let y_j be the sum of all 1-parallel edge vectors with $\varepsilon_j = 0$ at the initial and terminal vertices;
(4) let z_j be the sum of all 1-parallel edge vectors with $\varepsilon_j = 1$ at the initial and terminal vertices.

LEMMA 58. *For* $2 \leq j \leq n$, $Q_n(x_j) = \frac{3}{4}x_j + \frac{1}{4}(u_j + y_j - z_j)$.

PROOF. It suffices to verify that

$$\frac{3}{4}x_j + \frac{1}{4}(u_j + y_j - z_j) \in B(\mathbb{Z}_2^n) \tag{16}$$

and

$$x_j - [\frac{3}{4}x_j + \frac{1}{4}(u_j + y_j - z_j)] = \frac{1}{4}(x_j - u_j - y_j + z_j) \in Z(\mathbb{Z}_2^n). \tag{17}$$

To verify (16), we check that $\frac{3}{4}x_j + \frac{1}{4}(u_j + y_j - z_j)$ is orthogonal to every small square cycle vector. It suffices to consider small square cycles with 1-parallel or j-parallel (or both) edges as the support of $\frac{3}{4}x_j + \frac{1}{4}(u_j + y_j - z_j)$ is the union of the 1-parallel and j-parallel edges:

Case 1: A small square cycle S with 1-parallel and i-parallel edges, where $i \neq j$. Let v_S be the corresponding cycle vector.

Case 1(α): If $\varepsilon_j = 0$ on the edges of S, then $x_j = u_j = z_j \equiv 0$ on the edges of S and $y_j = 1$ on the 1-parallel pair of edges. Hence

$$\langle \frac{3}{4}x_j + \frac{1}{4}(u_j + y_j - z_j), v_S\rangle = \frac{1}{4}\langle y_j, v_S\rangle = \frac{1}{4} - \frac{1}{4} = 0.$$

Case 1(β): If If $\varepsilon_j = 1$ on the edges of S, then $x_j = u_j = y_j \equiv 0$ on the edges of S and $z_j = 1$ on the 1-parallel pair of edges. Hence

$$\langle \frac{3}{4}x_j + \frac{1}{4}(u_j + y_j - z_j), v_S\rangle = -\frac{1}{4}\langle z_j, v_S\rangle = -\frac{1}{4} + \frac{1}{4} = 0.$$

Case 2: A small square cycle S with j-parallel and i-parallel edges, where $i \neq 1$. Let v_S be the corresponding cycle vector.

Case 2(α): If $\varepsilon_1 = 0$ on the edges of S, then $u_j = y_j = z_j \equiv 0$ on the edges of S and $x_j = 1$ on the j-parallel pair of edges. Hence

$$\langle \frac{3}{4}x_j + \frac{1}{4}(u_j + y_j - z_j), v_S\rangle = \frac{3}{4}\langle x_j, v_S\rangle = \frac{3}{4} - \frac{3}{4} = 0.$$

Case 2(β): If If $\varepsilon_1 = 1$ on the edges of S, then $x_j = y_j = z_j \equiv 0$ on the edges of S and $u_j = 1$ on the j-parallel pair of edges. Hence

$$\langle \frac{3}{4}x_j + \frac{1}{4}(u_j + y_j - z_j), v_S\rangle = \frac{1}{4}\langle u_j, v_S\rangle = \frac{1}{4} - \frac{1}{4} = 0.$$

Cases 3: A small square cycle S with 1-parallel and j-parallel edges. First consider the 1-parallel edges. On the 1-parallel edge with $\varepsilon_j = 0$, we have

$$\frac{3}{4}x_j + \frac{1}{4}(u_j + y_j - z_j) = 0 + 0 + \frac{1}{4} - 0 = \frac{1}{4}.$$

On the 1-parallel edge with $\varepsilon_j = 1$, we have

$$\frac{3}{4}x_j + \frac{1}{4}(u_j + y_j - z_j) = 0 + 0 + 0 - \frac{1}{4} = -\frac{1}{4}.$$

Now consider the j-parallel edges. On the j-parallel edge with $\varepsilon_1 = 0$, we have

$$\frac{3}{4}x_j + \frac{1}{4}(u_j + y_j - z_j) = \frac{3}{4} + 0 + 0 - 0 = \frac{3}{4}.$$

On the j-parallel edge with $\varepsilon_1 = 1$, we have

$$\frac{3}{4}x_j + \frac{1}{4}(u_j + y_j - z_j) = 0 + \frac{1}{4} + 0 - 0 = \frac{1}{4}.$$

Hence

$$\langle \frac{3}{4}x_j + \frac{1}{4}(u_j + y_j - z_j), v_S\rangle = \frac{1}{4} + \frac{1}{4} - (-\frac{1}{4}) - \frac{3}{4} = 0.$$

To verify (17), we check that $x_j - u_j - y_j + z_j$ is orthogonal to each cut vector $X(v)$ for $v = (\varepsilon_i)_{i=1}^n \in \mathbb{Z}_2^n$.

CASE 1 ($\varepsilon_1 = \varepsilon_j = 0$).

$$\langle x_j - u_j - y_j + z_j, X(v)\rangle = 1 \cdot 1 - 0 - 1 \cdot 1 + 0 = 0.$$

CASE 2 ($\varepsilon_1 = 0, \varepsilon_j = 1$).

$$\langle x_j - u_j - y_j + z_j, X(v)\rangle = 1 \cdot (-1) - 0 - 0 + 1 \cdot 1 = 0 = 0.$$

CASE 3 ($\varepsilon_1 = 1, \varepsilon_j = 0$).

$$\langle x_j - u_j - y_j + z_j, X(v)\rangle = 0 - 1 \cdot 1 - 1 \cdot (-1) + 0 = 0.$$

CASE 4 ($\varepsilon_1 = \varepsilon_j = 1$).

$$\langle x_j - u_j - y_j + z_j, X(v)\rangle = 0 - 1 \cdot (-1) - 0 + 1 \cdot (-1) = 0.$$

□

LEMMA 59. $G(n) = \dfrac{n-1}{4}$.

PROOF. Note that

$$\sum_{j=2}^{n} x_j = \sum_{\overrightarrow{v_k v_{k+1}}} e,$$

where the sum is taken over all edges e of the form $\overrightarrow{v_k v_{k+1}}$. Hence

$$\begin{aligned} G(n) &= \sum_{\overrightarrow{v_k v_{k+1}}} \langle Q_n(e_0), e\rangle \\ &= \langle Q_n(e_0), \sum_{\overrightarrow{v_k v_{k+1}}} e\rangle \\ &= \langle Q_n(e_0), \sum_{j=2}^{n} x_j\rangle \\ &= \sum_{j=2}^{n} \langle e_0, Q_n(x_j)\rangle \\ &= \sum_{j=2}^{n} \langle e_0, \frac{3}{4}x_j + \frac{1}{4}(u_j + y_j - z_j)\rangle \end{aligned}$$

(by Lemma 58)

$$\begin{aligned} &= \sum_{j=2}^{n} (\frac{3}{4}\langle e_0, x_j\rangle + \frac{1}{4}[\langle e_0, u_j\rangle + \langle e_0, y_j\rangle - \langle e_0, z_j\rangle]) \\ &= \sum_{j=2}^{n} [0 + 0 + \frac{1}{4} - 0] \\ &= \frac{n-1}{4}. \end{aligned}$$

□

PROOF OF THEOREM 45. By Lemmas 56, 57, and 59,

$$\|Q_n\|_1 = F(n) + 2G(n) = 1 + 2\frac{(n-1)}{4} = \frac{n+1}{2}.$$

□

COROLLARY 60. *Let P_n be the orthogonal projection onto $Z(\mathbb{Z}_2^n)$. Then*

$$\|P_n\|_1 = \frac{n+3}{2} - \frac{4}{n} + \frac{1}{n2^{n-2}}.$$

PROOF.

$$\begin{aligned} \|P_n\|_1 &= \|I - Q_n\|_1 \\ &= \sum_{e\in E_n} |\langle e_0 - Q_n(e_0), e\rangle| \\ &= \sum_{e\in E_n} |\langle Q_n(e_0), e\rangle| + 1 - 2\langle Q_n(e_0), e_0\rangle \\ &= \|Q_n\|_1 + 1 - 2b_0 \\ &= \frac{n+1}{2} + 1 - 2(\frac{2}{n} - \frac{1}{n2^{n-1}}) \end{aligned}$$

(by Lemma 47 and Theorem 45)

$$= \frac{n+3}{2} - \frac{4}{n} + \frac{1}{n2^{n-2}}.$$

□

2.7.4. *The projection constant of* $\mathrm{Lip}_0(\mathbb{Z}_2^n)$.

LEMMA 61. *Let $G = (V, E)$ be a finite connected graph with edge space $\ell_1(E)$ and automorphism group Γ. Then $\hat{\phi}^* = \hat{\phi}^{-1}$ for all $\phi \in \Gamma$.*

PROOF. For all $x \in \ell_1(E)$,

$$\hat{\phi}(x) = \sum_{e\in E} \langle x, \overrightarrow{e_-e_+}\rangle \overrightarrow{\phi(e_-)\phi(e_+)} = \sum_{e\in E} \langle x, \overrightarrow{\phi^{-1}(e_-)\phi^{-1}(e_+)}\rangle \overrightarrow{e_-e_+}.$$

Hence, for all $x, y \in \ell_1(E)$,

$$\begin{aligned}\langle \hat{\phi}(x), y\rangle &= \sum_{e\in E} \langle x, \overrightarrow{\phi^{-1}(e_-)\phi^{-1}(e_+)}\rangle \langle y, \overrightarrow{e_-e_+}\rangle \\ &= \sum_{e\in E} \langle x, \overrightarrow{e_-e_+}\rangle \langle y, \overrightarrow{\phi(e_-)\phi(e_+)}\rangle.\end{aligned}$$

Thus,

$$\begin{aligned}\hat{\phi}^*(y) &= \sum_{e\in E} \langle y, \overrightarrow{\phi(e_-)\phi(e_+)}\rangle \overrightarrow{e_-e_+} \\ &= \sum_{e\in E} \langle y, \overrightarrow{e_-e_+}\rangle \overrightarrow{\phi^{-1}(e_-)\phi^{-1}(e_+)} \\ &= \hat{\phi}^{-1}(y).\end{aligned}$$

□

COROLLARY 62. *If $T\colon \ell_1(E) \to \ell_1(E)$ is Γ-invariant then T^* is also Γ-invariant.*

THEOREM 63. $\lambda(\mathrm{Lip}_0(\mathbb{Z}_2^n)) = \dfrac{n+1}{2}$.

PROOF. Let Q be a Γ_n-invariant projection from $\ell_1(E(\mathbb{Z}_2^n))$ onto $B(\mathbb{Z}_2^n)$. By Corollary 62, $I - Q^*$ is a Γ_n-invariant projection onto $B(\mathbb{Z}_2^n)^\perp = Z(\mathbb{Z}_2^n)$. Hence, by Corollary 44, $I - Q^* = P_n$. So $Q = I - P_n = Q_n$. So Q_n is a minimal projection from $\ell_\infty(E(\mathbb{Z}_2^n))$ onto $(B(\mathbb{Z}_2^n), \|\cdot\|_\infty)$. Since $(B(\mathbb{Z}_2^n), \|\cdot\|_\infty)$ is isometrically isomorphic to $\mathrm{Lip}_0(\mathbb{Z}_2^n)$, Theorem 45 gives

$$\lambda(\mathrm{Lip}_0(\mathbb{Z}_2^n)) = \|Q_n\|_\infty = \|Q_n\|_1 = \frac{n+1}{2}.$$

□

COROLLARY 64.

$$\frac{n+1}{2} \le d_{\mathrm{BM}}(\mathrm{TC}(\mathbb{Z}_2^n), \ell_1^{2^n-1}) \le 2n.$$

PROOF. Suppose that $f\colon \mathbb{Z}_2^n \to \mathbb{R}$ satisfies $f((0,0,\dots,0)) = 0$. Then

$$\begin{aligned}\frac{\|f\|_\infty}{n} &\le \sup_{x\neq y\in\mathbb{Z}_2^n} \frac{|f(x)-f(y)|}{d(x,y)} \\ &\le 2\|f\|_\infty.\end{aligned}$$

Hence $d_{\mathrm{BM}}(\mathrm{Lip}_0(\mathbb{Z}_2^n), \ell_\infty^{2^n-1}) \leq 2n$. On the other hand,

$$d_{\mathrm{BM}}(\mathrm{Lip}_0(\mathbb{Z}_2^n), \ell_\infty^{2^n-1}) \geq \lambda(\mathrm{Lip}_0(\mathbb{Z}_2^n)) = \frac{n+1}{2}.$$

By duality, $d_{\mathrm{BM}}(\mathrm{Lip}_0(\mathbb{Z}_2^n), \ell_\infty^{2^n-1}) = d_{\mathrm{BM}}(\mathrm{TC}(\mathbb{Z}_2^n), \ell_1^{2^n-1})$. □

REMARK 65. It is known that $cn \leq c_1(\mathrm{TC}(\mathbb{Z}_2^n)) \leq Cn$ for some absolute constants $0 < c \leq C < \infty$. The left-hand side inequality is due to [**23**, Corollary 2], the right-hand side inequality is due to [**8, 14, 18**]. The left-hand side of the inequality implies a weaker version of the left-hand side inequality in Corollary 64.

Acknowledgment

We thank Kevin Ford for numerical results related to the recurrence relation in Lemma 52, and the referees for their helpful comments.

References

[1] M. Alexander, M. Fradelizi, L. C. García-Lirola, and A. Zvavitch, *Geometry and volume product of finite dimensional Lipschitz-free spaces*, J. Funct. Anal. **280** (2021), no. 4, Paper No. 108849, 38, DOI 10.1016/j.jfa.2020.108849. MR4181163

[2] A. D. Andrew, *On subsequences of the Haar system in $C(\Delta)$*, Israel J. Math. **31** (1978), no. 1, 85–90, DOI 10.1007/BF02761382. MR506384

[3] Y. Benyamini and J. Lindenstrauss, *Geometric nonlinear functional analysis. Vol. 1*, American Mathematical Society Colloquium Publications, vol. 48, American Mathematical Society, Providence, RI, 2000, DOI 10.1090/coll/048. MR1727673

[4] N. Biggs, *Algebraic graph theory*, 2nd ed., Cambridge Mathematical Library, Cambridge University Press, Cambridge, 1993. MR1271140

[5] N. Biggs, *Algebraic potential theory on graphs*, Bull. London Math. Soc. **29** (1997), no. 6, 641–682, DOI 10.1112/S0024609397003305. MR1468054

[6] J. A. Bondy and U. S. R. Murty, *Graph theory*, Graduate Texts in Mathematics, vol. 244, Springer, New York, 2008, DOI 10.1007/978-1-84628-970-5. MR2368647

[7] A. E. Brouwer, A. M. Cohen, and A. Neumaier, *Distance-regular graphs*, Ergebnisse der Mathematik und ihrer Grenzgebiete (3) [Results in Mathematics and Related Areas (3)], vol. 18, Springer-Verlag, Berlin, 1989, DOI 10.1007/978-3-642-74341-2. MR1002568

[8] M. S. Charikar, *Similarity estimation techniques from rounding algorithms*, Proceedings of the Thirty-Fourth Annual ACM Symposium on Theory of Computing, ACM, New York, 2002, pp. 380–388, DOI 10.1145/509907.509965. MR2121163

[9] R. Diestel, *Graph theory*, 5th ed., Graduate Texts in Mathematics, vol. 173, Springer, Berlin, 2017, DOI 10.1007/978-3-662-53622-3. MR3644391

[10] S. J. Dilworth, D. Kutzarova, and M. I. Ostrovskii, *Lipschitz-free spaces on finite metric spaces*, Canad. J. Math. **72** (2020), no. 3, 774–804, DOI 10.4153/s0008414x19000087. MR4098600

[11] S. J. Dilworth, D. Kutzarova, and M. I. Ostrovskii, *Analysis on Laakso graphs with application to the structure of transportation cost spaces*, Positivity **25** (2021), no. 4, 1403–1435, DOI 10.1007/s11117-021-00821-w. MR4301142

[12] P. Erdős, *The art of counting: Selected writings*, Mathematicians of Our Time, Vol. 5, MIT Press, Cambridge, Mass.-London, 1973. Edited by Joel Spencer and with a dedication by Richard Rado. MR532670

[13] P. Erdős and L. Pósa, *On the maximal number of disjoint circuits of a graph*, Publ. Math. Debrecen **9** (1962), 3–12, DOI 10.5486/pmd.1962.9.1-2.02. MR150756

[14] J. Fakcharoenphol, S. Rao, and K. Talwar, *A tight bound on approximating arbitrary metrics by tree metrics*, J. Comput. System Sci. **69** (2004), no. 3, 485–497, DOI 10.1016/j.jcss.2004.04.011. MR2087946

[15] C. D. Godsil, *GRRs for nonsolvable groups*, Algebraic methods in graph theory, Vol. I, II (Szeged, 1978), Colloq. Math. Soc. János Bolyai, vol. 25, North-Holland, Amsterdam-New York, 1981, pp. 221–239. MR642043

[16] C. Godsil and G. Royle, *Algebraic graph theory*, Graduate Texts in Mathematics, vol. 207, Springer-Verlag, New York, 2001, DOI 10.1007/978-1-4613-0163-9. MR1829620

[17] B. Grünbaum, *Projection constants*, Trans. Amer. Math. Soc. **95** (1960), 451–465, DOI 10.2307/1993567. MR114110

[18] P. Indyk, N. Thaper, *Fast image retrieval via embeddings*, ICCV 03: Proceedings of the 3rd International Workshop on Statistical and Computational Theories of Vision, 2003.

[19] G. J. O. Jameson, *Summing and nuclear norms in Banach space theory*, London Mathematical Society Student Texts, vol. 8, Cambridge University Press, Cambridge, 1987, DOI 10.1017/CBO9780511569166. MR902804

[20] F. John, *Extremum problems with inequalities as subsidiary conditions*, Studies and Essays Presented to R. Courant on his 60th Birthday, January 8, 1948, Interscience Publishers, New York, 1948, pp. 187–204. MR30135

[21] L. V. Kantorovich, *On mass transportation* (Russian), Doklady Acad. Naus SSSR, (N.S.) **37**, (1942), 199–201; English transl.: *J. Math. Sci.* (N. Y.), **133** (1942), no. 4, 1381–1382.

[22] L. V. Kantorovich and M. K. Gavurin, *Application of mathematical methods in the analysis of cargo flows* (Russian), Problems of improving of transport efficiency, USSR Academy of Sciences Publishers, Moscow, 1949, pp. 110–138.

[23] S. Khot and A. Naor, *Nonembeddability theorems via Fourier analysis*, Math. Ann. **334** (2006), no. 4, 821–852, DOI 10.1007/s00208-005-0745-0. MR2209259

[24] J. Lauri and R. Scapellato, *Topics in graph automorphisms and reconstruction*, 2nd ed., London Mathematical Society Lecture Note Series, vol. 432, Cambridge University Press, Cambridge, 2016, DOI 10.1017/CBO9781316669846. MR3496604

[25] A. Naor and G. Schechtman, *Planar earthmover is not in L_1*, SIAM J. Comput. **37** (2007), no. 3, 804–826, DOI 10.1137/05064206X. MR2341917

[26] S. Ostrovska and M. I. Ostrovskii, *Generalized transportation cost spaces*, Mediterr. J. Math. **16** (2019), no. 6, Paper No. 157, 26, DOI 10.1007/s00009-019-1433-8. MR4029537

[27] S. Ostrovska and M. I. Ostrovskii, *On relations between transportation cost spaces and ℓ_1*, J. Math. Anal. Appl. **491** (2020), no. 2, 124338, 12, DOI 10.1016/j.jmaa.2020.124338. MR4121095

[28] S. Ostrovska and M. I. Ostrovskii, *Isometric structure of transportation cost spaces on finite metric spaces*, Rev. R. Acad. Cienc. Exactas Fís. Nat. Ser. A Mat. RACSAM **116** (2022), no. 4, Paper No. 153, 19, DOI 10.1007/s13398-022-01301-w. MR4455126

[29] M. I. Ostrovskii, *Isometric classification of Sobolev spaces on graphs*, Colloq. Math. **109** (2007), no. 2, 287–295, DOI 10.4064/cm109-2-10. MR2318524

[30] M. I. Ostrovskii, *Metric embeddings: Bilipschitz and coarse embeddings into Banach spaces*, De Gruyter Studies in Mathematics, vol. 49, De Gruyter, Berlin, 2013, DOI 10.1515/9783110264012. MR3114782

[31] J. Oxley, *Matroid theory*, 2nd ed., Oxford Graduate Texts in Mathematics, vol. 21, Oxford University Press, Oxford, 2011, DOI 10.1093/acprof:oso/9780198566946.001.0001. MR2849819

[32] D. J. S. Robinson, *A course in the theory of groups*, 2nd ed., Graduate Texts in Mathematics, vol. 80, Springer-Verlag, New York, 1996, DOI 10.1007/978-1-4419-8594-1. MR1357169

[33] W. Rudin, *Projections on invariant subspaces*, Proc. Amer. Math. Soc. **13** (1962), 429–432, DOI 10.2307/2034952. MR138012

[34] Th. Schlumprecht, personal communication, 2022.

[35] J.-P. Serre, *Linear representations of finite groups*, Graduate Texts in Mathematics, Vol. 42, Springer-Verlag, New York-Heidelberg, 1977. Translated from the second French edition by Leonard L. Scott. MR450380

[36] B. Simon, *Representations of finite and compact groups*, Graduate Studies in Mathematics, vol. 10, American Mathematical Society, Providence, RI, 1996, DOI 10.1090/gsm/010. MR1363490

[37] H. Whitney, *2-isomorphic graphs*, Amer. J. Math. **55** (1933), no. 1-4, 245–254, DOI 10.2307/2371127. MR1506961

Department of Mathematics, University of South Carolina, Columbia, South Carolina 29208
Email address: dilworth@math.sc.edu

Department of Mathematics, University of Illinois Urbana-Champaign, Urbana, Illinois 61807; and Institute of Mathematics and Informatics, Bulgarian Academy of Sciences, Sofia, Bulgaria
Email address: denka@illinois.edu

Department of Mathematics and Computer Science, St. John's University, 8000 Utopia Parkway, Queens, New York 11439
Email address: ostrovsm@stjohns.edu

Proceedings of Symposia in Pure Mathematics
Volume **106**, 2024
https://doi.org/10.1090/pspum/106/01930

The Finest Locally Convex Module Topology

Francisco Javier García-Pacheco

This paper is dedicated to Professor Per H. Enflo.

ABSTRACT. Recently, a new concept in Associative Ring Theory has been introduced: unit segments. By means of unit segments, the classical notion of convexity can be transported to modules in a natural way. This allows to define locally convex module topologies, which extend the classical locally convex vector topologies in the real or complex case. We then survey recent results on this topic by constructing the finest locally convex module topology. Geometric properties of the convex sets are also studied within the manuscript.

1. Introduction and preliminaries

1.1. Historical context. The study of topological rings and modules is increasingly present and relevant in Functional Analysis [**23**]. For instance, classical Operator Theory on complex normed spaces [**16–19, 44**] has already been transported to the scope of normed modules over absolutely valued rings [**26, 33**], leading to a Representation Theory of groups in the group of topological isomorphisms on a topological module over a topological ring [**26**]. Later in [**29**], the basic notions of Operator Theory were further taken to the scope of seminormed modules over seminormed rings, and in some cases, to topological modules over topological rings by making use of convergence linear topologies in order to define dual module topologies. Basic notions of classical Geometry of Banach spaces, such as inner structure and extremal properties, have also been transposed to the general ambience of topological rings and modules [**27**].

On the other hand, classical Measure Theory [**14, 25, 38, 47, 49**] deals with positive real-valued measures defined on Boolean algebras of sets. Due to the famous Stone Representation Theorem for Boolean algebras [**48**], every Boolean algebra is isomorphic (in the category of Boolean algebras) to a Boolean algebra of sets. This way, classical Measure Theory retained full generality when it came to the definition of classical measures until the birth of effect algebras [**37**]. However, these classical measures were always real or complex valued. In the remarkable book

2020 *Mathematics Subject Classification.* Primary 46H25, 16W80; Secondary 54H13.

Key words and phrases. Topological module, topological ring, absorbing set, closed unit neighborhood of zero, unit segment.

The author was supported in part by Consejería de Universidad, Investigación e Innovación de la Junta de Andalucía: ProyExcel00780 (Operator Theory: An interdisciplinary approach), and ProyExcel01036 (Multifísica y optimización multiobjetivo de estimulación magnética transcraneal).

[**21**], vector measures are deeply studied. These measures are again defined on a Boolean algebra of sets; however, they are valued on a real or complex Banach space. After the irruption of effect algebras [**37, 39, 45**], modern measures were no longer studied in Boolean algebras. Classical properties of Measure Theory were then developed for effect algebras [**7–11**]. In [**37**], the authors consider measures on effect algebras with values on a commutative group, introducing the interesting concept of "universal group" for an effect algebra. In [**13, 41, 43, 57**], measures on effect algebras with values on a commutative topological group are considered. Later, in [**40, 56**], the authors study vector-valued measures on effect algebras with values on real or complex Hausdorff locally convex topological vector spaces. A few years later, measures on effect algebras with values on a real or complex normed space were considered in [**3, 4**]. In the recent manuscript [**12**], measures on effect algebras with values on a commutative normed group are studied. Later, in [**31**], measures on effect algebras with values on (topological) modules are analysed by taking advantage of the algebraic structure of modules over rings. Finally, measures with values on partially ordered modules are studied in [**32**], for which a deep study of the order topology in an ordered ring/module is first entailed (it is worth mentioning that the order topology is typically considered only on totally ordered sets). In our work, the order topology is considered in partially ordered rings/modules. This allows the definition of classical notions such as Radon and regularity for measures valued on ordered modules. Classically, Radon and regularity for measures have been studied for positive real-valued measures [**38**]. In [**21**], the notion of Radon measure was taken to the scope of vector measures (Banach space valued). In [**32**], Radon measures are defined in a much more general context for measures valued in a topological module. When the order topology in an ordered module is a module topology, it is proved then a relationship between inner and outer regular measures and between inner regular measures and Radon measures.

Finally, another approach where topological rings and modules are surely present is in Spectral Theory. The First Postulate of Quantum Mechanics [**20**] establishes that a Quantum Mechanical System is represented by an infinite-dimensional separable complex Hilbert space. The selfadjoint (bounded or unbounded) operators represent the observable magnitudes. The positive selfadjoint bounded operators lying below the identity represent the unsharp measurements, also called effects (this is first example of effect algebra [**37**]). Every unit vector of the Hilbert space is by definition a q-bit. The states of the system are presented by the "rays", that is, the one dimensional complex subspaces generated by each q-bit. The study of the algebraic properties of Quantum Systems led to the irruption of C^*-algebras and effect algebras. Other algebraic structures following the previous ones are the $*$-rings and the Hilbert C^*-modules [**1, 2, 24**].

This survey gathers several main, as well as recent, aspects of Geometry of Topological Rings and Modules, such as closed unit neighborhoods of zero, unit segments, balanced sets, absorbing sets, and extremal sets. It also gathers several main aspects of Operator Theory on Topological Rings and Modules. Finally, this manuscript contains the following novel result: The construction of the Finest Locally Convex Module Topology (Theorem 2.67).

1.2. Affine manifolds. By default, all rings and modules considered throughout this manuscript will be nonzero. In general, all objects of any category in this manuscript will not be the null object by default. Also, all actions will be considered as left actions unless otherwise is explicitly stated. In relation with the last statement, it is worth mentioning that the dual a left module is a right module. Also, all rings considered in this paper will be associative and unitary by default.

DEFINITION 1.1 (Affine manifold). An affine manifold in a module $\mathcal{X}$ over a ring R is a translation of a submodule of $\mathcal{X}$, that is, $x+\mathcal{Y}$ where $x \in \mathcal{X}$ and $\mathcal{Y}$ is a submodule of $\mathcal{X}$.

Let $\mathcal{X}$ be a module over a ring R. Observe that an affine manifold $x+\mathcal{Y}$ contains 0 if and only if $x+\mathcal{Y}=\mathcal{Y}$. Also, if $(x_i+\mathcal{Y}_i)_{i\in I}$ is a family of affine manifolds of $\mathcal{X}$ such that $\bigcap_{i\in I}(x_i+\mathcal{Y}_i)\neq\varnothing$, then $\bigcap_{i\in I}(x_i+\mathcal{Y}_i)=x+\operatorname{span}\left(\bigcap_{i\in I}(x_i+\mathcal{Y}_i)-x\right)$ for every $x\in\bigcap_{i\in I}(x_i+\mathcal{Y}_i)$. In other words, $\bigcap_{i\in I}(x_i+\mathcal{Y}_i)$ is an affine manifold of $\mathcal{X}$. This fact motivates the following definition.

DEFINITION 1.2 (Affine hull). The affine hull of a subset A of a module $\mathcal{X}$ over a ring R is defined as the intersection of all affine manifolds of $\mathcal{X}$ containing A. The affine hull of A is denoted by $\operatorname{aff}(A)$.

Observe that aff is in a fact a hull operator. Since $\operatorname{span}(A)$ is trivially an affine manifold containing A, it holds that $\operatorname{aff}(A)\subseteq\operatorname{span}(A)$. Basic properties of the affine hull follow. Let $\mathcal{X},\mathcal{Y}$ be modules over a ring R. Let $T:\mathcal{X}\to\mathcal{Y}$ be a linear operator. Let $A\subseteq\mathcal{X}$ and $B\subseteq\mathcal{Y}$ be nonempty. Let $u\in\mathcal{U}(R)$ and $x\in\mathcal{X}$. Then:

- $\operatorname{aff}(A)=x+\operatorname{span}(A-x)$ for every $x\in A$.
- $\operatorname{aff}(A)=\operatorname{span}(A)$ if and only if $0\in\operatorname{aff}(A)$.
- $\operatorname{aff}(x+A)=x+\operatorname{aff}(A)$.
- $\operatorname{aff}(uA)=u\operatorname{aff}(A)$.
- $\operatorname{aff}(T(A))=T\left(\operatorname{aff}(A)\right)$.
- $\operatorname{aff}\left(T^{-1}(B)\right)\subseteq T^{-1}\left(\operatorname{aff}(B)\right)$.

There are several simple examples of affine manifolds. For instance, singletons are trivially affine manifolds since they are translations of the null submodule. The whole module is itself an affine manifold. Other examples follow.

EXAMPLE 1.3 (Straight line). In a module $\mathcal{X}$ over a ring R, the line passing through $x,y\in\mathcal{X}$ is defined as the affine manifold $\operatorname{stl}(x,y):=\operatorname{aff}(\{x,y\})=x+R(y-x)$. Special attention will be paid to a notable subset of $\operatorname{stl}(x,y)$ given by $\operatorname{stl}_{\mathcal{U}}(x,y):=x+(\mathcal{U}(R)\cup\{0\})(y-x)$.

Under the settings of the previous example and definition, $\mathcal{U}(R)$ is standard notation for the multiplicative group of invertibles of R. Other standard notation for $\mathcal{U}(R)$ is $R^{\times}$, which will not be employed here.

EXAMPLE 1.4 (Hyperplane). In a module $\mathcal{X}$ over a ring R, the affine manifold given by $(x')^{-1}(\{s\})$, for $x'\in\mathcal{X}'$ and $s\in R$ in the range of x', represents a hyperplane. Note that $(x')^{-1}(\{s\})=x+\ker(x')$ for each $x\in(x')^{-1}(\{s\})$.

The following technical lemma, whose proof we include for the sake of completion, will be used later on.

LEMMA 1.5. *Let $\mathcal{X}$ be a module over a division ring R. Let $f,g\in\mathcal{L}(\mathcal{X},R)\setminus\{0\}$. Then:*

(1) $\ker(f)$ *is a maximal submodule of* $\mathcal{X}$.
(2) *If* $\ker(f)=\ker(g)$, *then there exists* $r\in R$ *such that* $g=fr$.

PROOF.

(1) Let $\mathcal{Y}$ be a submodule of $\mathcal{X}$ strictly containing $\ker(f)$ and fix an arbitrary $y \in \mathcal{Y} \setminus \ker(f)$. We will show that $\mathcal{X} = \mathcal{Y}$. Indeed, if $x \in \mathcal{X}$, then $x = f(x)f(y)^{-1}y + \left(-f(x)f(y)^{-1}y + x\right) \in R\mathcal{Y} + \ker(f) \subseteq \mathcal{Y}$.
(2) Fix an arbitrary $x \in \mathcal{X} \setminus \ker(f)$. Take $r := f(x)^{-1}g(x)$. Take any $y \in \mathcal{X}$. Note that $y - f(y)f(x)^{-1}x \in \ker(f) = \ker(g)$, therefore $g(y) = f(y)r$.

□

1.3. Internal points and absorbance. If X is a topological space and A is a subset of X, then $\operatorname{int}(A), \operatorname{cl}(A), \operatorname{bd}(A)$ stand for the interior, the closure and the boundary of A, respectively. If $x \in X$, then by $\mathcal{N}_x(X)$ we intend to denote the filter of neighborhoods of x. If there is no confusion with X, then we will simply denote it by $\mathcal{N}_x$.

The notions of topological ring and topological module will be very much employed throughout this manuscript. The following well-known theorems [**6**, **52**–**54**] characterize ring and module topologies in terms of filter bases.

THEOREM 1.6. *Let R be a ring. If τ is a ring topology on R and $\mathcal{B}$ is a base of neighborhoods of* 0*, then the following is verified:*

(1) *For every $V \in \mathcal{B}$, there exists $U \in \mathcal{B}$ with $U + U \subseteq V$.*
(2) *For every $V \in \mathcal{B}$, there exists $U \in \mathcal{B}$ with $-U \subseteq V$.*
(3) *For every $V \in \mathcal{B}$, there exists $U \in \mathcal{B}$ with $UU \subseteq V$.*
(4) *For every $V \in \mathcal{B}$ and every $r \in R$, there exists $U \in \mathcal{B}$ with $rU \cup Ur \subseteq V$.*

Conversely, if $\mathcal{B}$ is a filter base of $\mathcal{P}(R)$ verifying all four properties above, then there exists a unique ring topology on R such that $\mathcal{B}$ is a base of 0*-neighborhoods. This topology is given by $\tau := \{A \subseteq R : \forall r \in A \ \exists U \in \mathcal{B} \ \ r + U \subseteq A\} \cup \{\varnothing\}$.*

Deeper properties of topological rings for a more algebraic viewpoint can be found in [**5**, **50**, **51**, **55**].

THEOREM 1.7. *Let R be a topological ring and $\mathcal{X}$ an R-module. If τ is a module topology on R and $\mathcal{B}$ is a base of neighborhoods of* 0 *in $\mathcal{X}$, then the following is verified:*

(1) *For every $V \in \mathcal{B}$, there exists $U \in \mathcal{B}$ with $U + U \subseteq V$.*
(2) *For every $V \in \mathcal{B}$, there exists $U \in \mathcal{B}$ with $-U \subseteq V$.*
(3) *For every $V \in \mathcal{B}$, there exist $U \in \mathcal{B}$ and $W \in \mathcal{N}_0(R)$ with $WU \subseteq V$.*
(4) *For every $V \in \mathcal{B}$ and every $r \in R$, there exists $U \in \mathcal{B}$ with $rU \subseteq V$.*
(5) *For every $V \in \mathcal{B}$ and every $x \in \mathcal{X}$, there exists $W \in \mathcal{N}_0(R)$ with $Wx \subseteq V$.*

Conversely, if $\mathcal{B}$ is a filter base of $\mathcal{P}(\mathcal{X})$ verifying all five properties above, then there exists a unique module topology on $\mathcal{X}$ for which $\mathcal{B}$ is a base of zero-neighborhoods. This topology is given by $\tau := \{A \subseteq \mathcal{X} : \forall x \in A \ \exists U \in \mathcal{B} \ \ x + U \subseteq A\} \cup \{\varnothing\}$.

Internal point is a geometrical concept originally introduced in [**15**, TVS II.26 p.66] for real or complex topological vector spaces to study the geometry of convex bodies, however this notion probably finds its origins in [**22**]. For generalizations of internal points in the setting of topological vector spaces, we refer to [**28**], where a solution to the Faceless Problem is provided. In [**36**], this notion was extended to topological rings and modules.

DEFINITION 1.8 (Internal point). Let $\mathcal{X}$ be a module over a topological ring R. Let $A \subseteq \mathcal{X}$. A point $a \in A$ is said to be an (affine) internal point of A provided for every $x \in \mathcal{X}$ ($x \subset \operatorname{aff}(A)$) there exists a neighborhood V of 0 in R such that $a + V(x-a) \subseteq A$. The set of (affine) internal points of A is denoted by $\operatorname{ainter}(A)$ ($\operatorname{inter}(A)$).

Another concept on which we will strongly rely is the one of practical ring [**33**, Definition 6]. A topological ring R is said to be practical provided that $0 \in \operatorname{cl}(\mathcal{U}(R))$, that is, 0 belongs to the closure of the invertibles $\mathcal{U}(R)$ of R. Practical rings serve to extend the classical Operator Theory to the topological module setting. An extensive study on practical topological rings can be found in [**33**].

THEOREM 1.9. *Let $\mathcal{X}$ be a topological module over a topological ring R. Let $A \subseteq \mathcal{X}$. Then* $\operatorname{int}(A) \subseteq \operatorname{inter}(A) \subseteq \operatorname{ainter}(A) \subseteq A$. *If, in addition, R is practical and* $\operatorname{inter}(A) \neq \varnothing$, *then* $\operatorname{aff}(A) = \mathcal{X}$ *and* $\operatorname{inter}(A) = \operatorname{ainter}(A)$.

PROOF. If $a \in \operatorname{int}(A)$, then we can find a neighborhood U of 0 in $\mathcal{X}$ such that $a + U \subseteq A$. Now let $x \in \mathcal{X}$. There exists V a neighborhood of 0 in R such that $V(x-a) \subseteq U$. Then $a + V(x-a) \subseteq a + U \subseteq A$. This shows that $\operatorname{int}(A) \subseteq \operatorname{inter}(A)$. The rest of the inclusions are trivial. Suppose next that R is practical. Fix $a \in \operatorname{inter}(A)$. If $x \in \mathcal{X}$, then there exists a neighborhood V of 0 in R such that $a + V(x-a) \subseteq A$. By hypothesis, R is practical, therefore we can find $u \in \mathcal{U}(R)$ with $u \in V$. Then $a + u(x-a) \in A$ so take $b \in A$ with $a + u(x-a) = b$. Note that $x = a + u^{-1}(b-a) \in a + \operatorname{span}(A-a) = \operatorname{aff}(A)$. $\square$

A direct consequence of Theorem 1.9 is the fact that if A is an open subset of a topological module, then $\operatorname{int}(A) = \operatorname{inter}(A) = \operatorname{ainter}(A) = A$. Theorem 1.9 motivates the following definition.

DEFINITION 1.10 (Linearly open set). Let $\mathcal{X}$ be a module over a topological ring R. A subset A of $\mathcal{X}$ is said to be linearly open provided that $A = \operatorname{inter}(A)$.

According to Theorem 1.9(2), every open subset of a topological module is linearly open. The converse is not true as shown by Theorem 1.11.

THEOREM 1.11. *Let R be a Hausdorff topological ring. Let $\mathcal{X}$ be a topological R-module such that there exists a noncontinuous R-linear map $f : \mathcal{X} \to R$. Then $\mathcal{X} \setminus \ker(f)$ is linearly open.*

PROOF. Take any $x \in \mathcal{X} \setminus \ker(f)$. Fix an arbitrary $y \in \mathcal{X}$. Since R is Hausdorff and $f(x) \neq 0$, there exists an additively symmetric 0-neighborhood $U \subseteq R$ such that $f(x) \notin U$. There exists an additively symmetric 0-neighborhood $V \subseteq R$ such that $Vf(y) \subseteq U$. Let us show that $x + Vy \subseteq \mathcal{X} \setminus \ker(f)$. Suppose on the contrary that there exists $v \in V$ such that $f(x+vy) = 0$. Then $f(x) = -vf(y) \in Vf(y) \subseteq U$, which is a contradiction. As a consequence, $x + Vy \subseteq \mathcal{X} \setminus \ker(f)$, hence $x \in \operatorname{inter}(\mathcal{X} \setminus \ker(f))$. Thus, $\mathcal{X} \setminus \ker(f) = \operatorname{inter}(\mathcal{X} \setminus \ker(f))$, meaning that $\mathcal{X} \setminus \ker(f)$ is linearly open. $\square$

The notion of internal points allows to extend the one of absorbing set to the scope of topological rings and modules [**34–36**].

DEFINITION 1.12 (Absorbing set). Let $\mathcal{X}$ be a module over a topological ring R. A nonempty subset A of $\mathcal{X}$ is said to be absorbing provided that $0 \in \operatorname{inter}(A)$.

In what follows, if R is a ring, then the set of left divisors of an element $s \in R$ is defined and denoted by $\ell d(s) := \{r \in R : \exists t \in R \setminus \{0\} \quad rt = s\}$.

Similarly, $rd(s)$ stands for the set of right divisors of s. Recall that R is called an integral domain if $\ell d(0) = rd(0) = \{0\}$. Notice that $\mathcal{U}(R) = \ell d(1) \cap rd(1)$. Also, $\ell d(0) \cap rd(1) = \varnothing = rd(0) \cap \ell d(1)$ and thus $(\ell d(0) \cup rd(0)) \cap \mathcal{U}(R) = \varnothing$.

An R-module $\mathcal{X}$ is called torsionfree provided that the equality $rx = 0$ implies that either $x = 0$ or $r \in \ell d(0) \cup rd(0)$. Note that a torsionfree module $\mathcal{X}$ over an integral domain R satisfies that $rx = 0$ implies $r = 0$ or $x = 0$. We refer the reader to [**33**, Section 2.4] for a wider perspective on separation properties on modules such as torsionfree and torsionless properties.

1.4. Ordered rings. Ordered rings are just rings endowed with a partial order compatible with the ring operations.

LEMMA 1.13. *Let R be a ring endowed with a partial order $\leq$ compatible with the addition, that is, translation-invariant in the sense that for all $r, s \in R$ with $r \leq s$, then $r + t \leq s + t$ for all $t \in R$. Let $R^+ := \uparrow 0 = \{t \in R : t \geq 0\}$. The following left- and right-homotecy invariances are equivalent:*

(1) *For all $r, s \in R$ with $r \leq s$, we have that $tr \leq ts$ for all $t \in R^+$.*
(2) *For all $r, s \in R$ with $r \leq s$, we have that $rt \leq st$ for all $t \in R^+$.*
(3) *For all $r, s \in R^+$, $rs \in R^+$.*

PROOF.

1⇒2 Note that $s - r \in R^+$, so $0 = (s - r)0 \leq (s - r)t = st - rt$, meaning that $rt \leq st$.

2⇒3 Note that $0 \leq r$ and $s \in R^+$, so $0 = 0s \leq rs$, meaning that $rs \in R^+$.

3⇒1 Note that $s - r, t \in R^+$, so $ts - tr = t(s - r) \in R^+$, meaning that $tr \leq ts$.

□

Lemma 1.13 has strong consequences in the definition of ordered ring since left- and right-homotecy invariances are equivalent.

DEFINITION 1.14 (Ordered ring). An ordered ring is a ring R endowed with a partial order $\leq$ compatible with the addition and multiplication operations, that is, for all $r, s \in R$ such that $r \leq s$, then

- Translation invariance: $r + t \leq s + t$ for all $t \in R$.
- Homotecy invariance: $rs \in R^+$ for all $r, s \in R^+$.

Some basic, and easily verifiable, properties of ordered rings are displayed in the next proposition.

PROPOSITION 1.15. *Let R be a nonzero ordered ring. For all $r, s, t, u \in R$:*

(1) *If $r < s$ and $t \leq u$, then $r + t < s + u$.*
(2) *If $r < s$ and $t > 0$ is not a left zero-divisor, then $tr < ts$.*
(3) *If $r < s$ and $t > 0$ is not a right zero-divisor, then $rt < st$.*
(4) *If $r < s$, then $s - r > 0$ and $-s < -r$. In particular, $R^- := \downarrow 0 = \{a \in R : a \leq 0\} = -R^+$. Also, $-(\downarrow r) = \uparrow(-r)$, $-(\uparrow r) = \downarrow(-r)$, $\downarrow r = r + R^-$, and $\uparrow r = r + R^+$.*
(5) *If $0, 1$ are comparable, then $0 < 1$.*
(6) *If $r, 0$ are comparable, then $r^2 \geq 0$.*
(7) *If $0, 1$ are comparable, $r \in \mathcal{U}(R)$, $r > 0$, and $r^{-1}, 0$ are comparable, then $r^{-1} > 0$.*

(8) *If* $0, 1$ *are comparable,* $r, s \in \mathcal{U}(R)$, $0 < r < s$, *and* r^{-1},0 *and* s^{-1},0 *are comparable, then* $0 < s^{-1} < r^{-1}$.
(9) *If* $0, 1$ *are comparable, then* $\mathrm{char}(R) = 0$.
(10) R *is totally ordered if and only if* $R^+ \cup R^- = R$.

PROOF.

(1) By the translation invariance, we know that $r + t \leq s + t \leq s + u$. If $r + t = s + u$, then $r + t = s + t$, thus we obtain the contradiction that $r = s$. Hence $r + t < s + u$.
(2) By the left-homotecy invariance, $tr \leq ts$. Suppose on the contrary that $tr = ts$. Then $t(r - s) = 0$, meaning the contradiction that $t \in \ell d(0)$ in view of the fact that $r - s \neq 0$.
(3) By the right-homotecy invariance, $rt \leq st$. Suppose on the contrary that $rt = st$. Then $(r - s)t = 0$, meaning the contradiction that $t \in rd(0)$ in view of the fact that $r - s \neq 0$.
(4) By the translation invariance, $0 = r - r < s - r$. If we apply again the translation invariance, $-s = -s + 0 < -s + (s - r) = -r$.
(5) Suppose on the contrary that $1 < 0$. By Proposition 1.15(4), $0 < -1$. By the left-homotecy invariance, $-1 = (-1)1 < (-1)0 = 0$, obtaining a contradiction.
(6) Suppose first that $r \geq 0$. By the left-homotecy invariance, $0 = r0 \leq rr = r^2$. Suppose next that $r \leq 0$. By Proposition 1.15(4), $0 \leq -r$. Then $0 \leq (-r)^2 = r^2$.
(7) Suppose on the contrary that $r^{-1} < 0$. By the left-homotecy invariance, $1 = rr^{-1} < r0 = 0$, contradicting Proposition 1.15(5).
(8) By Proposition 1.15(7), $r^{-1} > 0$ and $s^{-1} > 0$. By the left-homotecy invariance, $0 = r^{-1}0 < r^{-1}r < r^{-1}s$. The right-homotecy invariance assures that $0 = 0s^{-1} < 1s^{-1} < r^{-1}ss^{-1} = r^{-1}$.
(9) Suppose on the contrary that there exists $n \in \mathbb{N}$ for which $1 + \overset{n}{\cdots} + 1 = 0$. By Proposition 1.15(5), $0 < 1$. By Proposition 1.15(1), $0 = 0 + \overset{n}{\cdots} + 0 < 1 + \overset{n}{\cdots} + 1 = 0$, which is a contradiction.
(10) If R is totally ordered, then every $r \in R$ is comparable to 0, so either $r \geq 0$ or $r \leq 0$, hence $r \in R^+ \cup R^-$. Conversely, suppose that $R^+ \cup R^- = R$ and consider any $r, s \in R$. Then $r - s \in R^+ \cup R^-$, that is, either $r - s \geq 0$ or $r - s \leq 0$, resulting in either $r \geq s$ or $s \geq r$.

□

Ring orderings for which $0, 1$ are comparable are called unital ring orderings. There are other examples of ordered rings for which $0, 1$ are not comparable than the rings of characteristic different from 0. Indeed, the trivial partial order $r \leq s \Leftrightarrow r = s$ is clearly a ring order for which $0, 1$ are not comparable (unless $1 = 0$).

PROPOSITION 1.16. *Let* R *be a ring. If* $\leq$ *is a ring order on* R, *then the positive cone* R^+ *satisfies the following properties:*

(1) $0 \in R^+$.
(2) $R^+ \cap R^- = \{0\}$.
(3) $R^+ + R^+ \subseteq R^+$.
(4) $R^+R^+ \subseteq R^+$.

Conversely, if $S \subseteq R$ is a subset of R satisfying the four conditions above, then there exists a unique ring order on R, given by $r \le s \Leftrightarrow s - r \in S$ for all $r, s \in R$, for which $S = R^+$.

PROOF. If $\le$ is a ring order on R, then the positive cone R^+ trivially satisfies all four conditions above in view of Proposition 1.15. Conversely, let $S \subseteq R$ satisfying all four conditions above. We will check first that $r \le s \Leftrightarrow s - r \in S$ is a partial order in R. Indeed, fix arbitrary elements $r, s, t \in R$:

- Reflexivity: $r - r = 0 \in S \Rightarrow r \le r$.
- Antisymmetricity: If $r \le s$ and $s \le r$, then $s - r \in S$ and $r - s \in S$, so $s - r \in S \cap (-S) = \{0\}$, meaning that $s = r$.
- Transitivity: If $r \le s$ and $s \le t$, then $s - r \in S$ and $t - s \in S$, so $t - r = (t - s) + (s - r) \in S + S \subseteq S$, meaning that $r \le t$.

Next, let us check that this is a ring order. Indeed, fix arbitrary elements $r, s, t \in R$:

- Translation invariance: If $r \le s$, then $s - r \in S$, so $(s + t) - (r + t) = s - r \in S$, meaning that $r + t \le s + t$.
- Homotecy invariance: If $r, s \ge 0$, then $r, s \in S$, so $rs \in SS \subseteq S$, meaning that $0 \le rs$.

It is immediate to check that $S = R^+$. Finally, if $\preceq$ is another ring order whose positive cone is S, then $r \le s \Leftrightarrow s - r \in S \Leftrightarrow r \preceq s$ for all $r, s \in R$, reaching the conclusion that $\le$ and $\preceq$ are in fact the same ordering. □

Proposition 1.16 motivates the following definition.

DEFINITION 1.17 (Ordering set). Let R be a ring. A subset $S \subseteq R$ satisfying all four conditions of Proposition 1.16 will be called an ordering set. If $\text{char}(R) = 0$ and $1 \in S$, then S is called a unital ordering set.

Note that the intersection of any family of (unital) ordering subsets is again a (unital) ordering subset, and the union of any increasing family of (unital) ordering subsets is again a (unital) ordering subset. By means of Zorn's Lemma, we will demonstrate the existence of ring orderings.

LEMMA 1.18. *Let R be a ring. There exists a maximal ordering subset $S \subseteq R$. If, in addition, $\text{char}(R) = 0$, then there exists a maximal unital ordering subset.*

PROOF. Let $\mathcal{L} := \{T \subseteq R : T \text{ is an ordering subset}\}$. Notice that $\mathcal{L} \neq \varnothing$ since $\{0\} \in \mathcal{L}$. If $\mathcal{C} \subseteq \mathcal{L}$ is a chain of $\mathcal{L}$, then $\bigcup \mathcal{C} \in \mathcal{L}$. Finally, Zorn's Lemma guarantees the existence of maximal elements in $\mathcal{L}$. If $\text{char}(R) = 0$, then we can consider $\mathcal{L}_1 := \{T \in \mathcal{L} : 1 \in T\}$. Notice that $\mathcal{L}_1 \neq \varnothing$ since $\{0\} \cup \{1 + \overset{n}{\cdots} + 1 : n \in \mathbb{N}\} \in \mathcal{L}_1$. Also, any chain $\mathcal{C} \subseteq \mathcal{L}_1$ satisfies that $\bigcup \mathcal{C} \in \mathcal{L}_1$. Zorn's Lemma guarantees again the existence of maximal elements in $\mathcal{L}_1$. □

Notice that $\{0\}$ is the smallest ordering subset and $\{0, 1\}$ is the smallest unital ordering subset (in characteristic 0). A special kind of ordered rings will serve in further chapters, among other things, to construct regular measures.

REMARK 1.19. Let R be an ordered ring. The following conditions are equivalent:

(1) R^+ is totally ordered.
(2) $R^+ \cup R^-$ is a subring of R.
(3) $R^+ + R^- \subseteq R^+ \cup R^-$.
(4) $R^+ \cup R^-$ is totally ordered.

The previous remark motivates the definition of faithful ordered ring.

DEFINITION 1.20 (Faithful ordered ring). Let R be a ring. A ring order on R is called faithful provided that any of the equivalent conditions of Remark 1.19 is satisfied.

By means of Zorn's Lemma, maximal faithful ring orderings can be constructed.

LEMMA 1.21. *Let R be a ring. If* $\operatorname{char}(R) = 0$, *then there exists a maximal unital and faithful ordering subset.*

PROOF. Let $\mathcal{L} := \{T \subseteq R : T \text{ is an ordering subset}\}$ and $\mathcal{L}_{\mathsf{F}} := \{T \in \mathcal{L} : 1 \in T \text{ and } T - T \subseteq T \cup -T\}$. Notice that $\mathcal{L}_{\mathsf{F}} \neq \varnothing$ since $\{0\} \cup \{1 + \overset{n}{\cdots} + 1 : n \in \mathbb{N}\} \in \mathcal{L}_{\mathsf{F}}$. Also, any chain $\mathcal{C} \subseteq \mathcal{L}_{\mathsf{F}}$ satisfies that $\bigcup \mathcal{C} \in \mathcal{L}_{\mathsf{F}}$. Zorn's Lemma guarantees again the existence of maximal elements in $\mathcal{L}_{\mathsf{F}}$. □

Observe that $\{0\} \cup \{1 + \overset{n}{\cdots} + 1 : n \in \mathbb{N}\}$ is the smallest faithful unital ordering subset (in characteristic 0).

1.5. Ordered modules. Ordered modules are just modules, over ordered rings, endowed with a partial order compatible with the module operations. Let $\mathcal{X}$ be a module, over an ordered ring R, endowed with a partial order $\leq$ compatible with the addition, that is, translation-invariant in the sense that for all $x, y \in \mathcal{X}$ with $x \leq y$, then $x + z \leq y + z$ for all $x \in \mathcal{X}$. Let $\mathcal{X}^+ := \uparrow 0 = \{x \in \mathcal{X} : x \geq 0\}$. The following left- and right-homotecy invariances are equivalent:

- For all $x, y \in \mathcal{X}$ with $x \leq y$, we have that $rx \leq ry$ for all $r \in R^+$.
- For all $r, s \in R$ with $r \leq s$, we have that $rx \leq sx$ for all $x \in \mathcal{X}^+$.

DEFINITION 1.22 (Ordered module). An ordered module is a module $\mathcal{X}$, over an ordered ring R, endowed with a partial order $\leq$ compatible with the addition and left-multiplication operations, that is, for all $x, y \in \mathcal{X}$ with $x \leq y$, then

- Translation invariance: $x + z \leq y + z$ for all $z \in \mathcal{X}$.
- Left-homotecy invariance: $rx \leq ry$ for all $r \in R^+$.

Notice that ordered rings are ordered modules over themselves. The following are basic properties satisfied by module orderings. Let R be a ordered ring and $\mathcal{X}$ a ordered R-module. For all $x, y, z, u \in \mathcal{X}$ and all $r, s \in R$:

- If $x < y$ and $z \leq u$, then $x + z < y + u$.
- If $\mathcal{X}$ is torsionfree, $x < y$, and $r > 0$ is not a left zero-divisor, then $rx < ry$.
- If $\mathcal{X}$ is torsionfree, $x > 0$, $r < s$, and $s - r$ is not a left zero-divisor, then $rx < sx$.
- If $x < y$, then $x - y > 0$ and $-y < -x$. In particular, $\mathcal{X}^- := \downarrow 0 = \{x \in \mathcal{X} : x \leq 0\} = -\mathcal{X}^+$. Also, $-(\downarrow x) = \uparrow(-x)$, $-(\uparrow x) = \downarrow(-x)$, $\downarrow x = x + \mathcal{X}^-$, and $\uparrow x = x + \mathcal{X}^+$.
- $\mathcal{X}$ is totally ordered if and only if $\mathcal{X}^+ \cup \mathcal{X}^- = \mathcal{X}$.

The trivial ordering $x \leq y \iff x = y$ is clearly a module order on any module over an ordered ring.

Let R be an ordered ring. Let $\mathcal{X}$ be an R-module. If $\leq$ is a module order on $\mathcal{X}$, then the positive cone $\mathcal{X}^+$ satisfies that $0 \in \mathcal{X}^+$, $\mathcal{X}^+ \cap \mathcal{X}^- = \{0\}$, $\mathcal{X}^+ + \mathcal{X}^+ \subseteq \mathcal{X}^+$, and $R^+\mathcal{X}^+ \subseteq \mathcal{X}^+$. Conversely, if $S \subseteq \mathcal{X}$ is a subset of $\mathcal{X}$ satisfying the previous four conditions, then there exists a unique module order on $\mathcal{X}$, given by $x \leq y \Leftrightarrow x - y \in S$ for all $x, y \in \mathcal{X}$, for which $S = \mathcal{X}^+$. Such a set S will be called an ordering set. The intersection of any family of ordering subsets is again an ordering subset, and the union of any increasing family of ordering subsets is again an ordering subset. Thus, the collection of all ordering subsets of a module over an ordered ring is inductive when endowed with the inclusion, so by Zorn's Lemma, every module over an ordered ring has a maximal ordering subset.

1.6. Order topology. Recall that if X is a partially ordered set (in what follows a poset), then $\uparrow x := \{y \in X : y \geq x\}$, $\downarrow x := \{y \in X : y \leq x\}$, $\uparrow_\times x := (\uparrow x) \setminus \{x\} = \{y \in X : y > x\}$, $\downarrow^\times x := (\downarrow x) \setminus \{x\} = \{y \in X : y < x\}$ for every $x \in X$. The interval notation will be used in the manuscript, that is, $(-\infty, y] := \downarrow y$, $[x, \infty) := \uparrow x$, and $[x, y] = \uparrow x \cap \downarrow y$ (similarly for open and semiopen intervals).

If X is a set and $\mathcal{S}$ is a nonemtpy subset of $\mathcal{P}(X)$, then the set of finite intersections of $\mathcal{S}$, $\mathcal{B}(\mathcal{S}) := \left\{\bigcap_{T \in \mathcal{T}} T : \mathcal{T} \subseteq \mathcal{S} \text{ finite}\right\}$, is trivially closed under finite intersections, thus $\mathcal{B}(\mathcal{S})$ is a base for a topology on X if and only if $\bigcup_{S \in \mathcal{S}} S = X$.

THEOREM 1.23. *Let X be a poset. Then $\{\uparrow_\times x, \downarrow^\times x : x \in X\}$ is a subbase for a topology on X if and only if for every $x \in X$ either $\uparrow_\times x \neq \varnothing$ or $\downarrow^\times x \neq \varnothing$.*

PROOF. Observe that $\{\uparrow_\times x, \downarrow^\times x : x \in X\}$ is a subbase for a topology on X if and only if $\bigcup_{x \in X} \uparrow_\times x \cup \bigcup_{x \in X} \downarrow^\times x = X$, if and only if for every $x \in X$ there exists $y \in X$ such that either $x \in \uparrow_\times y$ or $x \in \downarrow^\times y$, if and only if for every $x \in X$ there exists $y \in X$ such that either $y \in \downarrow^\times x$ or $y \in \uparrow_\times x$. □

Theorem 1.23 makes possible the definition of order topology.

DEFINITION 1.24 (Order topology). Let X be a poset such that for every $x \in X$ either $\uparrow_\times x \neq \varnothing$ or $\downarrow^\times x \neq \varnothing$. The order topology on X is the topology generated by the subbase $\{\uparrow_\times x, \downarrow^\times x : x \in X\}$.

There is a way to construct the order topology in those posets for which there exists an element x satisfying that $\uparrow_\times x = \downarrow^\times x = \varnothing$, which consists in considering the following subbase: $\{\uparrow_\times x, \downarrow^\times x : x \in X\} \cup \{X\}$. It is immediate to observe that open intervals are open in the order topology, however closed intervals are not necessarily closed in the order topology

LEMMA 1.25. *Let X be a poset such that $\{\uparrow_\times x, \downarrow^\times x : x \in X\}$ is a subbase for a topology on X. If $x_0 \in X$ satisfies that $\uparrow_\times x_0 \neq \varnothing$ is downward directed and $\downarrow^\times x_0 \neq \varnothing$ is upward directed, then $\{(a, b) : a < x_0 < b\}$ is a base of neighborhoods of x_0 for the order topology.*

PROOF. Let $W \subseteq X$ be an x_0-neighborhood for the order topology. There can be found $a_1, \ldots, a_n, b_1, \ldots, b_m \in X$ satisfying that $x_0 \in \uparrow_\times a_1 \cap \cdots \cap \uparrow_\times a_n \cap \downarrow^\times b_1 \cap \cdots \cap \downarrow^\times b_m \subseteq W$. Observe that $a_i < x_0$ and $b_j > x_0$ for all $i \in \{1, \ldots, n\}$ and all $j \in \{1, \ldots, m\}$. Since $\downarrow^\times x_0$ is upward directed, there exists $a_0 \in \downarrow^\times x_0$ such that $a_0 \geq a_i$ for all $i \in \{1, \ldots, n\}$. Similarly, since $\uparrow_\times x_0$ is downward directed, there exists $b_0 \in \uparrow_\times x_0$ such that $b_0 \leq b_j$ for all $j \in \{1, \ldots, m\}$. Finally, notice that $x_0 \in (a_0, b_0) \subseteq \uparrow_\times a_1 \cap \cdots \cap \uparrow_\times a_n \cap \downarrow^\times b_1 \cap \cdots \cap \downarrow^\times b_m \subseteq W$. □

2. Main results

2.1. Unit neighborhoods of zero. A subset U of a topological space X is called regular open provided that $U = \operatorname{int}(\operatorname{cl}(U))$. A subset B of X is called regular closed provided that $B = \operatorname{cl}(\operatorname{int}(B))$. If $U \subseteq X$ is open and $B \subseteq X$ is closed, then $\operatorname{cl}(U)$ and $\operatorname{int}(B)$ are regular closed and regular open, respectively.

Unit neighborhoods of zero, also known as unit zero-neighborhoods, constitute a new concept in Associative Ring Theory with plenty of applications in topological modules. We refer the reader to [**30**, **33**–**36**] for a wider perspective on this notion.

DEFINITION 2.1 (Unit zero-neighborhood). Let R be a topological ring. Let U, B be subsets of R. Then:

- U is called an open unit neighborhood of 0 provided that U is an additively symmetric, $U = -U$, and multiplicatively idempotent, $UU = U$, regular open neighborhood of 0 such that $1 \in \mathrm{cl}(U)$.
- B is called a closed unit neighborhood of 0 provided that B is regular closed and its interior is an open unit neighborhood of 0.

Observe that, if U is an open unit neighborhood of 0, then $\mathrm{cl}(U)$ is a closed unit neighborhood of 0, and, if B is a closed unit neighborhood of 0, then $\mathrm{int}(B)$ is an open unit neighborhood of 0. In [**34**, Theorem 3.7], it is proved that the only proper closed unit neighborhood of 0 in $\mathbb{R}$ is $[-1, 1]$ and the only proper closed unit neighborhood of 0 in $\mathbb{C}$ containing the unit complex numbers $\mathbb{T}$ is the closed unit disk $\mathbb{D}$. In [**46**], other proper closed unit neighborhoods of 0 in $\mathbb{C}$ are provided.

PROPOSITION 2.2. *Let R be a topological ring. The following conditions are equivalent for a subset B of R:*

(1) *B is a closed unit neighborhood of* 0.
(2) *B is an additively symmetric and multiplicatively idempotent regular closed neighborhood of* 0 *containing* 1 *such that* $\mathrm{int}(B)$ *is multiplicatively idempotent as well.*
(3) *B is an additively symmetric regular closed neighborhood of* 0 *containing* 1 *such that* $\mathrm{int}(B)$ *is multiplicatively idempotent.*

PROOF. Suppose first that B is a closed unit neighborhood of 0. By definition, B is regular closed and $\mathrm{int}(B)$ is an open unit neighborhood of 0, therefore B is a neighborhood of 0. Also, $1 \in \mathrm{cl}(\mathrm{int}(B)) = B$ because $\mathrm{int}(B)$ is an open unit neighborhood of 0. Since $\mathrm{int}(B)$ is additively symmetric, we have that $B = \mathrm{cl}(\mathrm{int}(B)) = \mathrm{cl}(-\mathrm{int}(B)) = -\mathrm{cl}(\mathrm{int}(B)) = -B$, which implies that B is additively symmetric as well. Next, $1 \in \mathrm{cl}(\mathrm{int}(B)) = B$, so $B \subseteq BB$. On the other hand, $\mathrm{int}(B)$ is multiplicatively idempotent because it is an open unit neighborhood of 0, therefore the continuity of the multiplication allows that $BB = \mathrm{cl}(\mathrm{int}(B))\mathrm{cl}(\mathrm{int}(B)) \subseteq \mathrm{cl}(\mathrm{int}(B)\mathrm{int}(B)) = \mathrm{cl}(\mathrm{int}(B)) = B$, so B is multiplicatively idempotent. Conversely, assume that B is an additively symmetric regular closed neighborhood of 0 containing 1 such that $\mathrm{int}(B)$ is multiplicatively idempotent. Since B is regular closed, all that remains to prove is that $\mathrm{int}(B)$ is an open unit neighborhood of 0. Since B is closed, we have that $\mathrm{int}(B)$ is regular open. Also, B is a neighborhood of 0, so $\mathrm{int}(B)$ is a neighborhood of 0 as well. Next, since B is additively symmetric, we have that $\mathrm{int}(B) = \mathrm{int}(-B) = -\mathrm{int}(B)$, therefore $\mathrm{int}(B)$ is additively symmetric as well. Also, $\mathrm{int}(B)$ is multiplicatively idempotent by hypothesis. Finally, $1 \in B = \mathrm{cl}(\mathrm{int}(B))$. As a consequence, $\mathrm{int}(B)$ is an open unit neighborhood of 0.

□

Topological rings in which multiplicative inversion is continuous will be called topological inversion rings. Notice that topological division rings are topological inversion rings by definition. However, division topological rings need not be topological inversion rings.

LEMMA 2.3. *Let R be a topological inversion ring. Let U be an open subset of R. If $1 \in \mathrm{cl}(U \cap \mathcal{U}(R))$, then $U \subseteq UU$.*

PROOF. Fix an arbitrary $u \in U$. Let $(u_i)_{i\in I} \subseteq U \cap \mathcal{U}(R)$ be a net converging to 1. By the continuity of the inversion, we deduce that $\left(u_i^{-1}\right)_{i\in I}$ also converges to 1, therefore $\left(u_i^{-1}u\right)_{i\in I}$ converges to u, which means that we can find $j \in I$ with $u_j^{-1}u \in U$. Finally, $u = u_j\left(u_j^{-1}u\right) \in UU$. □

LEMMA 2.4. *Let R be a topological ring. Let B be an additively symmetric regular closed neighborhood of 0 containing 1. Then $1 \in \mathrm{cl}\left(\mathrm{int}(B) \setminus \{0\}\right)$.*

PROOF. Observe that $1 \in B = \mathrm{cl}(\mathrm{int}(B))$ because B is regular closed. Notice that $1 \in \mathrm{cl}(\mathrm{int}(B)) = \mathrm{cl}(\mathrm{int}(B)\setminus\{0\}) \cup \mathrm{cl}(\{0\})$, thus we conclude that $1 \in \mathrm{cl}(\mathrm{int}(B)\setminus\{0\})$ whenever $1 \notin \mathrm{cl}(\{0\})$. Finally, if $1 \in \mathrm{cl}(\{0\})$, then $0 \in 1+\mathrm{int}(B)$, so $-1 \in \mathrm{int}(B)$ and hence $1 \in \mathrm{int}(B)$ because $\mathrm{int}(B)$ is additively symmetric, hence $1 \in \mathrm{int}(B) \setminus \{0\} \subseteq \mathrm{cl}(\mathrm{int}(B) \setminus \{0\})$. □

The two previous technical lemmas together with Proposition 2.2 allow a useful characterization of closed unit zero-neighborhoods in topological division rings.

COROLLARY 2.5. *Let R be a topological division ring. A subset B of R is a closed unit neighborhood of 0 if and only if B is an additively symmetric and multiplicatively idempotent regular closed neighborhood of 0 containing 1.*

PROOF. By taking into consideration Proposition 2.2, if B is a closed unit neighborhood of 0, then B is an additively symmetric and multiplicatively idempotent regular closed neighborhood of 0 containing 1. Conversely, suppose that B is an additively symmetric and multiplicatively idempotent regular closed neighborhood of 0 containing 1. By bearing in mind again Proposition 2.2, all we need to show is that $\mathrm{int}(B)$ is multiplicatively idempotent. In first place, the multiplication of two open sets is open in topological division rings, therefore $\mathrm{int}(B)\mathrm{int}(B)$ is open. Since $\mathrm{int}(B)\mathrm{int}(B) \subseteq BB = B$, we deduce that $\mathrm{int}(B)\mathrm{int}(B) \subseteq \mathrm{int}(B)$. Next, according to Lemma 2.4, $1 \in \mathrm{cl}\left(\mathrm{int}(B) \setminus \{0\}\right) = \mathrm{cl}(\mathrm{int}(B) \cap \mathcal{U}(R))$. In view of Lemma 2.3, $\mathrm{int}(B) \subseteq \mathrm{int}(B)\mathrm{int}(B)$. Now, by Proposition 2.2, we conclude that B is a closed unit neighborhood of 0. □

By means of the notion of total closed unit zero-neighborhood, we will be able to construct nonempty linearly open sets with empty interior.

DEFINITION 2.6 (Total closed unit zero-neighborhood). Let R be a topological ring. A closed unit neighborhood B of 0 in R is called total provided that, for every $u \in \mathcal{U}(R)$, either $u \in B$ or $u^{-1} \in B$.

Keep in mind that the existence of a left- or right-bounded closed unit neighborhood of zero implies that the corresponding ring is practical.

THEOREM 2.7. *Let $\mathcal{X}$ be a topological module over a topological division ring R. Let $B \subseteq R$ a total right-bounded closed unit neighborhood of 0. If $f \in \mathcal{L}(\mathcal{X}, R)$ is so that $\ker(f)$ is closed, then f is continuous.*

PROOF. Assume to contrary that f is not continuous. There exists a net $(x_i)_{i\in I} \subseteq \mathcal{X}$ converging to 0 and an open neighborhood V of 0 in R such that $f(x_i) \notin V$ for all $i \in I$. By hypothesis, $\{Bu : u \in \mathcal{U}(R)\}$ is a base of neighborhoods of 0. Fix $u \in R \setminus \{0\}$ such that $Bu \subseteq V$. Observe that $f(x_i)u^{-1} \notin B$ for all $i \in I$. Since B is total, $uf(x_i)^{-1} \in B$ for all $i \in I$. By bearing in mind Theorem 2.62(1), we conclude that $\left(uf(x_i)^{-1}x_i\right)_{i\in I}$ converges to 0. Now, fix $x \in \mathcal{X}$ such that $f(x) \neq 0$ (notice that such x must exist since constant functions are continuous). Finally, notice that the net $\left(uf(x)^{-1}x - uf(x_i)^{-1}x_i\right)_{i\in I}$ is contained in $\ker(f)$ and converges to $uf(x)^{-1}x \notin \ker(f)$. □

The following corollary, when combined with Theorem 1.11, constitutes an example of a linearly open set with void topological interior.

COROLLARY 2.8. *Let R be a topological division ring endowed with a total right-bounded closed unit neighborhood B of 0. Let $\mathcal{X}$ be a topological R-module such that there exists a noncontinuous R-linear map $f : \mathcal{X} \to R$. Then $\mathcal{X} \setminus \ker(f)$ has empty interior in $\mathcal{X}$.*

PROOF. According to Lemma 1.5(1), $\ker(f)$ is a maximal submodule of $\mathcal{X}$. Therefore, $\mathrm{cl}(\ker(f))$ is either $\ker(f)$ or $\mathcal{X}$. If $\mathrm{cl}(\ker(f)) = \ker(f)$, then $\ker(f)$ is closed in $\mathcal{X}$, hence f is continuous in view of Theorem 2.7, which is a contradiction. As a consequence, $\mathcal{X} = \mathrm{cl}(\ker(f))$. This implies that $\ker(f)$ is dense in $\mathcal{X}$, meaning that $\mathcal{X} \setminus \ker(f)$ has empty interior in $\mathcal{X}$. □

2.2. Unit segments. We will rely on this new concept of unit segment [**30**, Definition 3.2] to define convexity on modules.

DEFINITION 2.9 (Unit segment). Let R be a ring. A subset $B^+ \subseteq R$ is called a unit segment if it verifies the following:

- $0, 1 \in B^+$.
- $B^+ \cap B^- = \{0\}$.
- $1 + B^- \subseteq B^+$.
- $B^+B^+ \subseteq B^+$.

Here, $B^- := -B^+$ is called negative unit segment.

Under the settings of the previous definition, it actually holds that $B^+B^+ = B^+$ and $1 + B^- = B^+$. Trivial examples of unit segments are $[0,1]$ and $\mathbb{Q} \cap [0,1]$ in $\mathbb{R}$. Observe also that the intersection of any family of unit segments is again a unit segment, and the union of any increasing family of unit segments is also a unit segment. The existence of unit segments is directly linked to the characteristic of the ring via Zorn's Lemma.

LEMMA 2.10. *If R is a ring, then there exists a unit segment of R if and only if $\mathrm{char}(R) \neq 2$. In this situation, it can be found a maximal unit segment of R.*

PROOF. Suppose that there exists a unit segment $B^+ \subseteq R$. If $\mathrm{char}(R) = 2$, then $1 + 1 = 0$, so $1 = -1 \in -B^+ = B^-$ reaching the contradiction that $1 \in B^+ \cap B^- = \{0\}$. Conversely, suppose that $\mathrm{char}(R) \neq 2$. Consider the set $\mathcal{L} := \{A \subseteq R : A \text{ is a unit segment}\}$ partially ordered by the inclusion. Notice that $\mathcal{L}$ is not empty because $\{0,1\} \in \mathcal{L}$ due to the fact that $\mathrm{char}(R) \neq 2$. Note also that, if $(A_i)_{i\in I}$ is a chain of $\mathcal{L}$, then it is easy to verify that $\bigcup_{i\in I} A_i \in \mathcal{L}$. By the Zorn's Lemma, there exists a maximal element. □

Note that $\{0,1\}$ is the smallest unit segment (in characteristic different from 2).

PROPOSITION 2.11. *Let A be a torsionfree algebra over an integral domain ring R. If B_R^+ is a unit segment of R, then $B_A^+ := B_R^+1$ is a unit segment of A.*

PROOF. It is straight forward to check that $0, 1 \in B_R^+1$ and $\left(B_R^+1\right)\left(B_R^+1\right) = B_R^+1$. Also, $1 + B_R^-1 = \left(1 + B_R^-\right)1 = B_R^+1$. Finally, $B_R^+1 \cap B_R^-1 = \{0\}$ because A is torsionfree and R is an integral domain. □

A direct consequence of the previous proposition is the following corollary.

COROLLARY 2.12. *If A is a real algebra, then $[0,1]1$ is a unit segment of A.*

Basic properties of unit segments are presented in the next lemma [**31**, Lemma 2].

LEMMA 2.13. *Let R be a ring. Let $B^+ \subseteq R$ be a unit segment. Then:*

(1) $(1 + B^+) \cap B^+ = \{1\}$.
(2) *If $r_1, \ldots, r_k \in B^+$ with $r_1 + \cdots + r_k = 1$, then $r_1 + \cdots + r_{k-1} \in B^+$.*
(3) *If $a, b \in B^+$ and $ab = 0$, then $a + b \in B^+$.*
(4) *If R is endowed with a ring topology, then $0 \in \mathrm{cl}\,(B^+ \setminus \{0,1\})$ if and only if $1 \in \mathrm{cl}\,(B^+ \setminus \{0,1\})$.*

PROOF.

(1) If $a, b \in B^+$ satisfy that $1 + a = b$, then $1 - b = -a \in B^+ \cap B^- = \{0\}$.
(2) Simply notice that $r_1 + \cdots + r_{k-1} = 1 - r_k \in 1 + B^- = B^+$.
(3) Note that $1 - a, 1 - b \in B^+$, therefore $1 - (a + b) = 1 - (a + b) + ab = (1 - a)(1 - b) \in B^+B^+ = B^+$, hence $a + b = 1 - [1 - (a + b)] \in B^+$.
(4) Indeed, $1 - B^+ = B^+$ and $1 - \{0,1\} = \{0,1\}$, therefore $0 \in \mathrm{cl}\,(B^+ \setminus \{0,1\}) \Leftrightarrow 1 \in 1 - \mathrm{cl}\,(B^+ \setminus \{0,1\}) \Leftrightarrow 1 \in \mathrm{cl}\,(1 - B^+ \setminus \{0,1\}) \Leftrightarrow 1 \in \mathrm{cl}\,(B^+ \setminus \{0,1\})$.

□

Lemma 2.13(2) motivates the following definition [**31**, Definition 1].

DEFINITION 2.14 (Full unit segment). Let R be a ring. A unit segment $B^+ \subseteq R$ is called full if for every $r_1, \ldots, r_k \in B^+$ with $r_1 + \cdots + r_k \in B^+$, then $r_1 + \cdots + r_{k-1} \in B^+$.

Basic properties of full unit segments are displayed in the next technical lemma.

LEMMA 2.15. *Let R be a ring. Let $B^+ \subseteq R$ be a full unit segment. Then:*

(1) *If $r_1, \ldots, r_k \in B^+$ with $r_1 + \cdots + r_k \in B^+$, then $\sum_{j \in J} r_j \in B^+$ for every $J \subseteq \{1, \ldots, k\}$.*
(2) *If $r_1, \ldots, r_k \in B^+ \setminus \{0\}$ with $r_1 + \cdots + r_k \in B^+$, then $r_1 + \cdots + r_k \in B^+ \setminus \{0\}$.*
(3) $\mathrm{char}(R) = 0$.

PROOF. By recursively applying the definition of full unit segment, one can see that if $r_1, \ldots, r_k \in B^+$ with $r_1 + \cdots + r_k \in B^+$, then $\sum_{j \in J} r_j \in B^+$ for every $J \subseteq \{1, \ldots, k\}$. Suppose next that $r_i \neq 0$ for all $i \in \{1, \ldots, k\}$. If $r_1 + \cdots + r_k = 0$, then $r_1 + \cdots + r_{k-1} = -r_k \in B^+ \cap B^- = \{0\}$, which contradicts that $r_k \neq 0$. Finally, let us prove that $\mathrm{char}(R) = 0$. Suppose that there exists $n \in \mathbb{N}$ such that $1 + \overset{n}{\cdots} + 1 = 0$. Notice that $n > 1$ since by default we are assuming that $R \neq \{0\}$.

On the one hand, $1+\overset{n-1}{\cdots}+1=-1$. On the other hand, $1+\overset{n}{\cdots}+1=0\in B^+$ so $-1=1+\overset{n-1}{\cdots}+1\in B^+$, that is, $-1\in B^-\cap B^+=\{0\}$, reaching the contradiction that $-1=0$. $\square$

The intersection of any family of full unit segments is again a full unit segment, and the union of any increasing family of full unit segments is also a full unit segment. Zorn's Lemma assures the existence of maximal full unit segments in rings of characteristic 0.

LEMMA 2.16. *If R is a ring with $\operatorname{char}(R)=0$, then there exists a maximal full unit segment of R.*

PROOF. Consider the set $\mathcal{L}:=\{A\subseteq R : A \text{ is a full unit segment}\}$ partially ordered by the inclusion. Notice that $\mathcal{L}$ is not empty because $\{0,1\}\in\mathcal{L}$ due to the fact that $\operatorname{char}(R)=0$. Note also that, if $(A_i)_{i\in I}$ is a chain of $\mathcal{L}$, then it is easy to verify that $\bigcup_{i\in I}A_i\in\mathcal{L}$. By the Zorn's Lemma, there exists a maximal element. $\square$

Note that $\{0,1\}$ is the smallest full unit segment (in characteristic 0). Full unit segments endow rings with a unital ring ordering.

THEOREM 2.17. *Let R be a ring. Let $B^+\subseteq R$ be a full unit segment. The set $S:=\bigcup_{n\in\mathbb{N}}B^++\overset{n}{\cdots}+B^+$ is a unital ordering subset, hence it endows R with a unique ring ordering whose positive cone is S and for which $0,1$ are comparable. Even more, this ring ordering satisfies that $B^+=[0,1]$.*

PROOF. Let us verify all four conditions of Proposition 1.16 one by one:

(1) $0\in S$. Indeed, $0\in B^+\subseteq S$.
(2) $S\cap(-S)=\{0\}$. Indeed, take $b_1,\dots,b_n,c_1,\dots,c_m\in B^+\setminus\{0\}$ such that $b_1+\cdots+b_m=-c_1-\cdots-c_m$. Then $b_1+\cdots+b_m+c_1+\cdots+c_m=0$, contradicting Lemma 2.15(2).
(3) $S+S\subseteq S$. This is automatically satisfied by construction of S.
(4) $SS\subseteq S$. This is automatically satisfied since $B^+B^+=B^+$.

Finally, let us prove that $B^+=[0,1]$. Indeed, if $r\in B^+$, then $r\in S$, so $r\geq 0$. Also, $1-r\in 1+B^-=B^+\subseteq S$, hence $1-r\geq 0$, meaning that $r\leq 1$. As a consequence, $B^+\subseteq[0,1]$. Conversely, take $r\in[0,1]$. Then $r\geq 0$ so $r\in S$. Also, $r\leq 1$, hence $1-r\in S$. We can then find $a_1,\dots,a_n\in B^+$ and $b_1,\dots,b_m\in B^+$ such that $r=a_1+\cdots+a_n$ and $1-r=b_1+\cdots+b_m$. Therefore, $a_1+\cdots+a_n+b_1+\cdots+b_m=1\in B^+$. By Lemma 2.15(1), $r=a_1+\cdots+a_n\in B^+$. $\square$

Notice that, in Theorem 2.17, the fullness of the unit segment has only been used to show the second condition of Proposition 1.16. From now on, whenever we have a full unit segment, we will implicitly consider its induced unital ring ordering.

LEMMA 2.18. *Let R be a ring. Let $B^+\subseteq R$ be a full unit segment. If $a,b\in B^+$ and $a\leq b$, then $b-a\in B^+$.*

PROOF. According to Theorem 2.17, $b-a\in\bigcup_{n\in\mathbb{N}}B^++\overset{n}{\cdots}+B^+$, so take $b_1,\dots,b_n\in B^+$ such that $a+b_1+\cdots+b_n=b\in B^+$. By Lemma 2.15(1), $b_1+\cdots+b_n\in B^+$. $\square$

According to [**30**, Theorem 3.3], if R is a totally ordered ring, then $B^+:=[0,1]$ is a unit segment of R. The next result is an improvement of [**30**, Theorem 3.3] by showing that if R is a partially ordered ring with $0, 1$ comparable, then $B^+ := [0,1]$ is a full unit segment of R.

THEOREM 2.19. *If R is an ordered ring such that $0, 1$ are comparable, then $B^+ := [0,1]$ is a full unit segment of R.*

PROOF. Let us first check that all four properties of Definition 2.9 are satisfied:

(1) By choice of B^+, we clearly have that $0, 1 \in B^+$.
(2) It suffices to show that $B^+B^+ \subseteq B^+$. Indeed, if $r, s \in B^+$, then $0 \leq r \leq 1$ and $0 \leq s \leq 1$, therefore $0 \leq rs \leq 1$. Thus, $rs \in B^+$.
(3) $B^+ \cap B^- = \{0\}$. Notice that if $r, s \in [0,1]$ and $r = -s$, then $-1 \leq -s \leq 0$, so we obtain that $r = -s \leq 0 \leq r$. In other words, $r = 0$.
(4) $1 + B^- \subseteq B^+$. Indeed, if $r \in [0,1]$, then $-1 \leq -r \leq 0$, hence $0 \leq 1 - r \leq 1$, that is, $1 - r \in [0,1] = B^+$.

Finally, let us prove that $B^+ := [0,1]$ is full. Indeed, let $r_1, \ldots, r_k \in B^+$ with $r_1 + \cdots + r_k \in B^+$. Since $0 \leq r_i \leq 1$ for all $i \in \{1, \ldots, k\}$, we obtain that $0 \leq r_1 + \cdots + r_{k-1}$. Next, $r_1 + \cdots + r_k \leq 1$, so $r_1 + \cdots + r_{k-1} \leq 1 - r_k \leq 1$ since $1 - r_k \in 1 - B^+ = B^+$. As a consequence, $r_1 + \cdots + r_{k-1} \in B^+$. □

In order to define convexity on modules, it only suffices to rely on unit segments. However, in order to achieve certain geometrical properties, we will have to consider special cases of unit segments.

DEFINITION 2.20. Let R be a ring. Let $B^+ \subseteq R$ be a unit segment. Then B^+ is called:

- Completely full if B^+ is full, R^+ is multiplicatively symmetric and $B^+ \setminus \{0\} \subseteq \mathcal{U}(R)$.
- Totally full if B^+ is full and totally ordered.
- Strong if for every $r \in B^+$, either $2r \in B^+$ or $2(1-r) \in B^+$.
- Commuting if $rB^+ = B^+r$ for all $r \in R$.
- Centered if $2 \in \mathcal{U}(R)$ and $2^{-1} \in B^+$.

Recall that a subset A of a (multiplicative) monoid G is said to be (multiplicatively) symmetric provided that $(A \cap G^\times)^{-1} \subseteq A$.

THEOREM 2.21. *Let R be a ring. Let $B^+ \subseteq R$ be a completely full unit segment. If $a, b \in B^+ \setminus \{0\}$ satisfy that $a \leq b$, then $b^{-1}a, ab^{-1} \in B^+$.*

PROOF. Take $a, b \in B^+ \setminus \{0\}$ with $a \leq b$. By hypothesis, $(R^+ \cap \mathcal{U}(R))^{-1} \subseteq R^+$, thus $b^{-1} > 0$, hence $0 = b^{-1}0 \leq b^{-1}a \leq b^{-1}b = 1$. As a consequence, $b^{-1}a \in [0,1]$. By relying on Theorem 2.17, we conclude that $b^{-1}a \in [0,1] = B^+$. On the other hand, $b^{-1}a \leq 1$ and $a > 0$ so $ab^{-1}a \leq a$, hence $ab^{-1}aa^{-1} \leq 1$ because $a^{-1} > 0$, meaning that $ab^{-1} \leq 1$. Finally, it is clear that $ab^{-1} \geq 0$, so it only suffices to apply again Theorem 2.17 to conclude that $ab^{-1} \in [0,1] = B^+$. □

COROLLARY 2.22. *Let R be a ring. Let $B^+ \subseteq R$ be a full unit segment. If B^+ is centered and completely and totally full, then B^+ is strong.*

PROOF. Fix an arbitrary $r \in B^+$. Notice that $2^{-1}, r, 1 - r$ are all comparable because B^+ is totally ordered. If $2^{-1} < r$ and $2^{-1} < 1 - r$, then $1 = 2^{-1} + 2^{-1} < r + (1 - r) = 1$, which is a contradiction. As a consequence, either $r \leq 2^{-1}$ or $1 - r \leq 2^{-1}$. In accordance with Theorem 2.21, either $2r \in B^+$ or $2(1-r) \in B^+$. □

The notion of totally full unit segment is directly linked to the concept of faithfullness for unital ordering sets.

THEOREM 2.23. *Let R be a ring. If $B^+ \subseteq R$ is a totally full unit segment, then the unital ring ordering induced by B^+ on R is faithful. Conversely, if R is endowed with a faithful unital ring ordering, then $B^+ := [0,1]$ is a totally full unit segment.*

PROOF. If R is endowed with a faithful unital ring ordering, then R^+ is totally ordered, so in particular, $B^+ := [0,1] \subseteq R^+$ is totally ordered, hence a totally full unit segment. Conversely, suppose that $B^+ \subseteq R$ is a totally full unit segment. The unital ring ordering induced by B^+ on R is given by the unital ordering subset $S := \bigcup_{n\in\mathbb{N}} B^+ + \overset{n}{\cdots} + B^+$ in view of Theorem 2.17. We will prove that S is faithful by showing that $S - S \subseteq S \cup -S$ (see Remark 1.19). Indeed, let $s_1, s_2 \in S$. We can find $a_1, \ldots, a_n, b_1, \ldots, b_m \in B^+$ such that $s_1 = a_1 + \cdots + a_n$ and $s_2 = b_1 + \cdots + b_m$. Since $0 \in B^+$, by adding zeros if necessary, we may assume that $m = n$. Then $s_1 - s_2 = (a_1 - b_1) + \cdots + (a_n - b_n)$. Next, for all $i \in \{1, \ldots, n\}$, either $a_i \leq b_i$ or $b_i \leq a_i$, because by hypothesis B^+ is totally full. As a consequence and by relying on Lemma 2.18, for all $i \in \{1, \ldots, n\}$, either $b_i - a_i \in B^+$ or $a_i - b_i \in B^+$. If $b_i - a_i \in B^+$ for all $i \in \{1, \ldots, n\}$, then $s_1 - s_2 \in B^- + \overset{n}{\cdots} + B^- \subseteq -S$. If $a_i - b_i \in B^+$ for all $i \in \{1, \ldots, n\}$, then $s_1 - s_2 \in B^+ \overset{n}{\cdots} + B^+ \subseteq S$. Otherwise, again we have $s_1 - s_2 = c_1 + \cdots + c_k - d_1 - \cdots - d_l$, where $c_1, \ldots, c_k, d_1, \ldots, d_l \in B^+$ and $k + l < n$. By repeating this process as finitely many times as necessary, we end up having that either $s_1 - s_2 \in S$ or $s_1 - s_2 \in -S$, obtaining the desired result. □

The following lemma in this subsection shows the existence of maximal unit segments via Zorn's Lemma.

LEMMA 2.24. *Let R be a ring with* $\operatorname{char}(R) = 0$. *There exists a maximal completely full unit segment, a maximal totally full unit segment, a maximal strong unit segment, and a maximal commuting unit segment. There also exists a maximal completely, totally, strong and commuting full unit segment.*

PROOF. Consider the sets $\mathcal{L}_c := \{A \subseteq R : A \text{ is a completely full unit segment of} R\}$, $\mathcal{L}_t := \{A \subseteq R : A \text{ is a totally full unit segment of } R\}$, $\mathcal{L}_s := \{A \subseteq R : A \text{ is a strong unit segment of } R\}$, and $\mathcal{L}_{co} := \{A \subseteq R : A \text{ is a commuting unit segment of } R\}$. Also, consider $\mathcal{L} := \mathcal{L}_c \cap \mathcal{L}_t \cap \mathcal{L}_s \cap \mathcal{L}_{co}$, partially ordered by the inclusion. Notice that $\mathcal{L}$ is not empty because $\{0,1\} \in \mathcal{L}$. If $\mathcal{C}$ is a chain in any of the previous sets, then it is easy to verify that $\bigcup \mathcal{C}$ also lies in that set. By Zorn's Lemma, there exists a maximal element. □

The next technical lemma will be crucial towards accomplishing our goals.

LEMMA 2.25. *Let $\mathcal{X}$ be a module over a topological inversion ring R. Let B^+ be a unit segment of R such that $B^+ \setminus \{0\} \subseteq \mathcal{U}(R)$ and $0 \in \operatorname{cl}(B^+ \setminus \{0,1\})$. For every $x \neq y$ in $\mathcal{X}$ and every neighborhood V of 0 in R, there exists $r_0 \in B^+ \setminus \{0,1\}$ such that $v_0 := -(1-r_0)^{-1} r_0 \in V$ and $x = r_0 y + (1-r_0)(x + v_0(y-x))$.*

PROOF. The following function

$$\begin{array}{rcl} B^+ \setminus \{1\} & \to & R \\ r & \mapsto & -(1-r)^{-1} r \end{array}$$

is well defined and continuous at 0. Therefore, there exists a neighborhood U of 0 in R such that, if $r \in U \cap (B^+ \setminus \{1\})$, then $-(1-r)^{-1}r \in V$. Since $0 \in \mathrm{cl}\,(B^+ \setminus \{0,1\})$, we can find $r_0 \in U \cap (B^+ \setminus \{0,1\})$, hence $v_0 := -(1-r_0)^{-1}r_0 \in V$. Then

$$\begin{aligned} x &= x + (r_0 + (1-r_0)v_0)\,(y-x) \\ &= x + r_0(y-x) + (1-r_0)v_0(y-x) \\ &= r_0 y + (1-r_0)\,(x + v_0(y-x))\,. \end{aligned}$$

□

2.3. Convexity. In [**42**], the concept of C^*-convexity was introduced. This notion is based upon positive elements of a C^*-algebra (keep in mind that positive elements endow the real vector subspace of selfadjoint elements of a C^*-algebra with an ordering of real vector spaces). By relying upon unit segments, a more natural notion of convexity for modules can be defined [**30**, Definition 3.4].

DEFINITION 2.26 (Geometric simplex). Let $\mathcal{X}$ be a module over a ring R. Let B^+ be a unit segment of R. A geometric simplex is defined as the set

$$\Delta(x_1,\ldots,x_k) := \left\{\sum_{i=1}^k r_i x_i : r_1,\ldots,r_k \in B^+, \sum_{i=1}^k r_1 = 1\right\}$$

for $x_1,\ldots,x_k \in \mathcal{X}$. The interior of $\Delta(x_1,\ldots,x_k)$ is defined as

$$\Theta(x_1,\ldots,x_k) := \left\{\sum_{i=1}^k r_i x_i : r_1,\ldots,r_k \in B^+ \setminus \{0,1\}, \sum_{i=1}^k r_1 = 1\right\}.$$

It is clear that $\Theta(x_1,\ldots,x_k) \subseteq \Delta(x_1,\ldots,x_k)$. If $B^+ = \{0,1\}$ is the trivial unit segment, then $\Theta(x_1,\ldots,x_k) = \varnothing$ for all $x_1,\ldots,x_k \in \mathcal{X}$. On the other hand, notice that $\Theta(x_1,\ldots,x_k)$ is not necessarily the topological interior of $\Delta(x_1,\ldots,x_k)$. If $x \neq y$, then the geometric simplex $\Delta(x,y)$ is called a segment. Observe that, since $1 + B^- = B^+$, $\Delta(x,y) = x + B^+(y-x) \subseteq \mathrm{stl}(x,y)$.

REMARK 2.27. Let $\mathcal{X}$ be a module over a ring R. Let B^+ be a unit segment of R. For every $x,y \in \mathcal{X}$, $\Delta(x,y) \setminus \{x,y\} \subseteq \Theta(x,y)$. If, in addition, $\mathcal{X}$ is a torsionfree module, R is an integral domain, B^+ is nontrivial and $x \neq y$, then $\Delta(x,y) \setminus \{x,y\} = \Theta(x,y)$. If $x = y$, then $\Delta(x,x) = \{x\} = \Theta(x,x)$, hence $\Delta(x,x) \setminus \{x\} = \varnothing \neq \Theta(x,x)$.

DEFINITION 2.28 (Convex set). Let $\mathcal{X}$ be a module over a ring R. Let B^+ be a unit segment of R. A subset A of $\mathcal{X}$ is said to be convex if, for every $x_1,\ldots,x_k \in A$, $\Delta(x_1,\ldots,x_k) \subseteq A$.

Whenever we talk about convexity in the ring R, we will be considering R as a left R-module.

LEMMA 2.29. *Let R be a ring. Let $r \in R \setminus (\ell d(0) \cup rd(0))$ and $A \subseteq R$ such that $rA = Ar$. Let $a_1,\ldots,a_k \in A$ such that $a_1 + \cdots + a_k = 1$. For every $i \in \{1,\ldots,k\}$, let $b_i, c_i \in A$ with $rb_i = a_i r$ and $c_i r = ra_i$. Then $b_1 + \cdots + b_k = 1 = c_1 + \cdots + c_k$.*

PROOF. Notice that

$$r\left(\sum_{i=1}^k b_i\right) = \sum_{i=1}^k rb_i = \sum_{i=1}^k a_i r = \left(\sum_{i=1}^k a_i\right) r = r,$$

thus, $r\left(\sum_{i=1}^k b_i - 1\right) = 0$. Since r is not a zero divisor, we conclude that $\sum_{i=1}^k b_i = 1$. A similar reasoning applies for the c_i's. □

LEMMA 2.30. *Let $\mathcal{X}, \mathcal{Y}$ be modules over a ring R. Let B^+ be a unit segment of R. Let $a_1, \ldots, a_k \in \mathcal{X}$. Then:*

(1) *If $x \in X$, then $x + \Delta(x_1, \ldots, x_k) = \Delta(x + x_1, \ldots, x + x_k)$.*
(2) *If $T : \mathcal{X} \to \mathcal{Y}$ is linear, then $T(\Delta(x_1, \ldots, x_k)) = \Delta(T(x_1), \ldots, T(x_k))$.*
(3) *If $r \in R \setminus (\ell d(0) \cup rd(0))$ and $B^+ r = rB^+$, then $r\Delta(x_1, \ldots, x_k) = \Delta(rx_1, \ldots, rx_k)$.*

PROOF. We will only prove the last item. The others are trivial. Let $r_1, \ldots, r_k \in B^+$ such that $\sum_{i=1}^k r_i = 1$. For every $i \in \{1, \ldots, k\}$, there exists $s_i \in B^+$ with $r_i r = r s_i$. According to Lemma 2.29, we have that $\sum_{i=1}^k s_i = 1$. Finally

$$\sum_{i=1}^k r_i(rx_i) = \sum_{i=1}^k (r_i r)x_i = \sum_{i=1}^k (rs_i)x_i = r\sum_{i=1}^k s_i x_i.$$

This shows that $\Delta(rx_1, \ldots, rx_k) \subseteq r\Delta(x_1, \ldots, x_k)$. The reverse inclusion can be proved similarly. □

Next corollary, the details of whose we spare to the reader, is an immediate consequence of Lemma 2.30.

COROLLARY 2.31. *Let $\mathcal{X}, \mathcal{Y}$ be modules over a ring R. Let B^+ be a unit segment of R. Let $A \subseteq \mathcal{X}$ be convex. Then:*

(1) *If $x \in \mathcal{X}$, then $x + A$ is convex.*
(2) *If $T : \mathcal{X} \to \mathcal{Y}$ is linear, then $T(A)$ is also convex.*
(3) *If $r \in R \setminus (\ell d(0) \cup rd(0))$ and $B^+ r = rB^+$, then rA is convex as well.*

The following technical lemma relates geometric simplices with affine intervals.

LEMMA 2.32. *Let $\mathcal{X}$ be a module over a ring R. Let B^+ be a unit segment of R. If $x, y \in \mathcal{X}$, then $2\Delta(x, y) = (x + y) + (1 + 2B^-)(y - x)$.*

PROOF. Fix an arbitrary $s = 1 - 2r \in 1 + 2B^-$ with $r \in B^+$. Then $(x + y) + s(y - x) = (1 - s)x + (1 + s)y = (2r)x + (2 - 2r)y = 2(rx + (1 - r)y) \in 2\Delta(x, y)$. Conversely, let $r \in B^+$. Then $2(rx + (1 - r)y) = (2r)x + (2 - 2r)y = (1 - s)x + (1 + s)y = (x + y) + s(y - x) \in (x + y) + (1 + 2B^-)(y - x)$, where $s = 1 - 2r \in 1 + 2B^-$. □

The following proposition characterizes convexity in the classical way.

PROPOSITION 2.33. *Let $\mathcal{X}$ be a module over a ring R. Let B^+ be a completely full unit segment of R. Let $A \subseteq \mathcal{X}$. The following conditions are equivalent:*

(1) *A is convex.*
(2) *For every $x, y \in A$, $\Delta(x, y) \subseteq A$.*

PROOF. Assume that for every $x, y \in A$, $\Delta(x, y) \subseteq A$. Take $x_1, \ldots, x_k \in A$ and $r_1, \ldots, r_k \in B^+$ with $\sum_{i=1}^k r_1 = 1$. We have to prove that $\sum_{i=1}^k r_i x_i \in A$. We will proceed by induction on $k \in \mathbb{N}$. For $k = 1$, it is trivial and, for $k = 2$, it is our hypothesis. So, assume that the result holds for $k \geq 2$ and let us prove it for $k + 1$. We may assume without any loss of generality that $r_1 \neq 1$. Otherwise, if $r_1 = \cdots = r_{k+1} = 1$, then $\operatorname{char}(R) \neq 0$, contradicting Lemma 2.15. Since B^+ is completely full, $1-r_1=r_2+\cdots+r_{k+1}\in B^+\setminus\{0\}\subseteq\mathcal{U}(R)$ and $(1-r_1)^{-1}r_i\in B^+$ for all $i\in\{2,\ldots,k+1\}$ in virtue of Theorem 2.21. On the other hand,

$$\sum_{i=2}^{k+1}(1-r_1)^{-1}r_i=(1-r_1)^{-1}\sum_{i=2}^{k+1}r_i=(1-r_1)^{-1}(1-r_1)=1.$$

This means that $\sum_{i=2}^{k+1}(1-r_1)^{-1}r_ix_i\in A$ by the induction hypothesis. Finally,

$$\sum_{i=1}^{k+1}r_ix_i=r_1x_1+(1-r_1)\sum_{i=2}^{k+1}(1-r_1)^{-1}r_ix_i\in\Delta\left(x_1,\sum_{i=2}^{k+1}(1-r_1)^{-1}r_ix_i\right)\subseteq A.$$

□

LEMMA 2.34. *Let $\mathcal{X}$ be a module over a ring R. Let B^+ be a full unit segment of R. Let $x,y\in\mathcal{X}$. Then*

(1) *$\Delta(z_1,z_2)\subseteq\Delta(x,y)$ for all $z_1,z_2\in\Delta(x,y)$.*
(2) *If B^+ is completely and totally full, then $\Delta(x,y)=\Delta(x,z)\cup\Delta(z,y)$ for all $z\in\Delta(x,y)$.*

PROOF.

(1) Let $r_i,s_i\in B^+$ such that $1=r_i+s_i$ and $z_i=r_ix+s_iy$ for $i=1,2$. Let $r,s\in B^+$ with $r+s=1$. Then $r(r_1x+s_1y)+s(r_2x+s_2y)=(rr_1+sr_2)x+(rs_1+ss_2)y$. Note that $rr_1+sr_2+rs_1+ss_2=1$ and $rr_1,sr_2,rs_1,ss_2\in B^+B^+=B^+$. According to Lemma 2.15(1), $rr_1+sr_2,rs_1+ss_2\in B^+$. As a consequence, $rz_1+sz_2\in\Delta(x,y)$.
(2) By Lemma 2.34(1), we have that $\Delta(x,y)\supseteq\Delta(x,z)\cup\Delta(z,y)$ for all $z\in\Delta(x,y)$. Let us prove the reverse inclusion. Fix an arbitrary $z\in\Delta(x,y)$ and write $z=r_0x+s_0y$ with $r_0,s_0\in B^+$ and $r_0+s_0=1$. Take any $p=rx+sy\in\Delta(x,y)$ with $r,s\in B^+$ and $r+s=1$. Since B^+ is completely and totally full, and by bearing in mind Theorem 2.21, we have that either $ss_0^{-1}\in B^+$ or $rs_0^{-1}\in B^+$ or $sr_0^{-1}\in B^+$ or $rr_0^{-1}\in B^+$. We may assume without any loss of generality that $ss_0^{-1}\in B^+$. Then

$$\begin{aligned}p=rx+sy&=\left(1-ss_0^{-1}s_0\right)x+\left(ss_0^{-1}s_0\right)y=\left(1-ss_0^{-1}(1-r_0)\right)x+\left(ss_0^{-1}s_0\right)y\\&=\left(1-ss_0^{-1}+ss_0^{-1}r_0\right)x+\left(ss_0^{-1}s_0\right)y=\left(1-ss_0^{-1}\right)x+\left(ss_0^{-1}r_0\right)x+\left(ss_0^{-1}s_0\right)y\\&=\left(1-ss_0^{-1}\right)x+ss_0^{-1}\left(r_0x+s_0y\right)=\left(1-ss_0^{-1}\right)x+\left(ss_0^{-1}\right)z\in\Delta(x,z).\end{aligned}$$

□

By combining together Proposition 2.33 and Lemma 2.34(1), we immediately deduce the following corollary.

COROLLARY 2.35. *Let $\mathcal{X}$ be a module over a ring R. Let B^+ be a completely full unit segment of R. Then $\Delta(x,y)$ is convex for all $x,y\in\mathcal{X}$.*

Let $\mathcal{X}$ be a module over a ring R. Let B^+ be a unit segment of R. Observe that the nonempty intersection of convex sets is convex. This allows to define the convex hull of a subset $A\subseteq\mathcal{X}$ as the intersection of all convex subsets containing A. The convex hull of A will be denoted by $\operatorname{conv}(A)$ is clearly convex since it is an intersection of convex sets. Also, A is convex if and only if $A=\operatorname{conv}(A)$. For all these reasons the convex hull is a hull operator.

THEOREM 2.36. *Let $\mathcal{X}$ be a module over a ring R. Let B^+ be a unit segment of R. If $A \subseteq \mathcal{X}$, then*

$$\begin{aligned}\operatorname{conv}(A) &= \bigcup \{\Delta(x_1, \ldots, x_k) : x_1, \ldots, x_k \in A, k \in \mathbb{N}\} \\ &= \left\{\sum_{i=1}^{k} r_i x_i : x_1, \ldots, x_k \in A, r_1, \ldots, r_k \in B^+, \sum_{i=1}^{k} r_i = 1, k \in \mathbb{N}\right\}.\end{aligned}$$

PROOF. Denote $H := \bigcup \{\Delta(x_1, \ldots, x_k) : x_1, \ldots, x_k \in A, k \in \mathbb{N}\}$. If D is a convex subset of $\mathcal{X}$ containing A, then $H \subseteq D$ by definition. In particular, $H \subseteq \operatorname{conv}(A)$. Notice that $A \subseteq H$, so if we prove that H is convex, then we can conclude that $H = \operatorname{conv}(A)$. Consider $x_{ij} \in A$ and $s_j, r_{ij} \in B^+$ for $i = 1, \ldots, k_j$ and $j = 1, \ldots, l$ such that $\sum_{j=1}^{l} s_j = 1 = \sum_{i=1}^{k_j} r_{ij}$ for all $j = 1, \ldots, l$. Notice that

$$\sum_{j=1}^{l} s_j \sum_{i=1}^{k_j} r_{ij} x_{ij} = \sum_{j=1}^{l} \sum_{i=1}^{k_j} s_j r_{ij} x_{ij} \in \Delta\left(x_{11}, \ldots, x_{k_1 1}, \ldots, x_{1l}, \ldots, x_{k_l l}\right) \subseteq H,$$

because $s_j r_{ij} \in B^+ B^+ = B^+$ for all $i = 1, \ldots, k_j$ and all $j = 1, \ldots, l$ and

$$\sum_{j=1}^{l} \sum_{i=1}^{k_j} s_j r_{ij} = \sum_{j=1}^{l} s_j \sum_{i=1}^{k_j} r_{ij} = \sum_{j=1}^{l} s_j = 1.$$

□

An immediate consequence of Theorem 2.36 together with Lemma 2.30(3) is the following corollary.

COROLLARY 2.37. *Let $\mathcal{X}, \mathcal{Y}$ be modules over a ring R. Let B^+ be a unit segment of R. Let $A \subseteq \mathcal{X}$. Then:*

(1) *If $x \in \mathcal{X}$, then* $\operatorname{conv}(x + A) = x + \operatorname{conv}(A)$.
(2) *If $T : \mathcal{X} \to \mathcal{Y}$ is linear, then* $T(\operatorname{conv}(A)) = \operatorname{conv}(T(A))$.
(3) *If $r \in R \setminus (\ell d(0) \cup rd(0))$ and $B^+ r = rB^+$, then* $r\operatorname{conv}(A) = \operatorname{conv}(rA)$.

The final results of this subsection are aimed at studying the topological behaviour of convex sets.

LEMMA 2.38. *Let $\mathcal{X}$ be a topological module over a topological ring R. Let B^+ be a unit segment of R. If $A \subseteq \mathcal{X}$ is convex, then* $\operatorname{cl}(A)$ *is convex as well.*

PROOF. Take $x_1, \ldots, x_k \in \operatorname{cl}(A)$ and $r_1, \ldots, r_k \in B^+$ with $\sum_{i=1}^{k} r_i = 1$. Fix an arbitrary 0-neighborhood $U_0 \subseteq \mathcal{X}$. We can find another 0-neighborhood $U \subseteq \mathcal{X}$ such that $U + \overset{k}{\cdots} + U \subseteq U_0$. We can find a 0-neighborhood V of $\mathcal{X}$ in such a way that $r_i V \subseteq U$ for all $i \in \{1, \ldots, k\}$. For every $i \in \{1, \ldots, k\}$, there exists $y_i \in A$ with $y_i \in x_i + V$. Then $r_1 y_1 + \cdots + r_k y_k \in r_1 x_1 + \cdots + r_k x_k + (r_1 V + \cdots + r_k V) \subseteq r_1 x_1 + \cdots + r_k x_k + \left(U + \overset{k}{\cdots} + U\right) \subseteq r_1 x_1 + \cdots + r_k x_k + U_0$. This shows that $r_1 x_1 + \cdots + r_k x_k \in \operatorname{cl}(A)$. □

LEMMA 2.39. *Let $\mathcal{X}$ be a topological module over a topological ring R. Let B^+ be a unit segment of R satisfying that $B^+ \setminus \{0\} \subseteq \mathcal{U}(R)$. Let A be a convex subset of $\mathcal{X}$. If $x \in \operatorname{int}(A)$ and $y \in A$, then*

$$\Delta(x, y) \setminus \{y\} \subseteq \{rx + sy : r, s \in B^+, r + s = 1, r \neq 0\} \subseteq \operatorname{int}(A).$$

PROOF. Let U be a neighborhood of 0 in $\mathcal{X}$ such that $x+U \subseteq A$. Fix arbitrary elements $r, s \in B^+$ with $r+s=1$ and $r \neq 0$. Note that $B^+ \setminus \{0\} \subseteq \mathcal{U}(R)$ by hypothesis. Then rU is a neighborhood of 0 in $\mathcal{X}$ because r is invertible. We will prove that $(rx+sy)+rU \subseteq A$. If $u \in U$, then

$$(rx+sy)+ru = r(x+u)+sy \in \Delta(x+u, y) \subseteq A$$

because $x+u, y \in A$. □

Recall that the commutant of a subset A of a ring R is defined as $A' := \{r \in R : \forall a \in A \;\; ra = ar\}$. The center of R is precisely defined as R' and usually denoted by $\mathsf{Z}(R)$.

THEOREM 2.40. *Let $\mathcal{X}$ be a topological module over a topological ring R. Let B^+ be a unit segment of R satisfying that $B^+ \setminus \{0\} \subseteq \mathcal{U}(R)$ and $B^+ \subseteq (B^+)'$. Let A be a convex subset of $\mathcal{X}$. If $x \in \operatorname{int}(A)$ and $y \in \operatorname{cl}(A)$, then*

$$\Delta(x,y) \setminus \{y\} \subseteq \{rx+sy : r, s \in B^+, r+s=1, r \neq 0\} \subseteq \operatorname{int}(A).$$

PROOF. Fix arbitrary elements $r, s \in B^+$ with $r+s=1$ and $r \neq 0$. If $r=1$, then $rx+sy = x \in \operatorname{int}(A)$, so we will suppose that $r \neq 1$. Denote $z := rx+sy$. We will show that $z \in \operatorname{int}(A)$. Consider the function

$$\begin{array}{rccl} g: & \mathcal{X} & \to & \mathcal{X} \\ & p & \mapsto & g(p) := z - rs^{-1}(p-z) \end{array}$$

Notice that g is a homeomorphism because rs^{-1} is invertible. Also, $g(z) = z$. On the other hand,

$$g(x) = rx+sy-rs^{-1}(x-rx-sy) = rx+sy-rs^{-1}s(x-y) = rx+sy-r(x-y) = y.$$

Observe that for all $p \in \mathcal{X}$, since r and s commute, then so do r and s^{-1}, thus

$$rp+sg(p) = rp+s\left(z-rs^{-1}(p-z)\right) = rp+sz-srs^{-1}(p-z) = rp+sz-r(p-z) = z.$$

Next, since $x \in \operatorname{int}(A)$, there exists an open neighborhood U of x contained in A. Then $g(U)$ is an open neighborhood of y, therefore $g(U) \cap A \neq \varnothing$. Fix $u \in U$ with $g(u) \in A$. Observe that, by relying first on Lemma 2.39, we conclude that $z = ru + sg(u) \in \operatorname{int}(A)$. □

COROLLARY 2.41. *Let $\mathcal{X}$ be a module over a topological ring R. Let B^+ be a unit segment of R satisfying that $B^+ \setminus \{0\} \subseteq \mathcal{U}(R)$ and $0 \in \operatorname{cl}(B^+ \setminus \{0,1\})$. Let A be a convex subset of $\mathcal{X}$ with $\operatorname{int}(A) \neq \varnothing$. Then:*

1. $\operatorname{cl}(\operatorname{int}(A)) = \operatorname{cl}(A)$.
2. *If multiplicative inversion is continuous and $B^+ \subseteq (B^+)'$, then* $\operatorname{int}(\operatorname{cl}(A)) = \operatorname{int}(A)$.

PROOF.

1. Since $\operatorname{int}(A) \subseteq A$ we have that $\operatorname{cl}(\operatorname{int}(A)) \subseteq \operatorname{cl}(A)$. Fix an arbitrary $x \in \operatorname{cl}(A)$. Let U be an open neighborhood of x. Notice that $U \cap A \neq \varnothing$. Take $y \in U \cap A$. Since $\operatorname{int}(A) \neq \varnothing$, we can fix an element $x_0 \in \operatorname{int}(A)$. In accordance with Lemma 2.39, the function

   $$\begin{array}{rcl} B^+ & \to & \operatorname{int}(A) \cup \{y\} \\ r & \mapsto & (1-r)x_0 + ry \end{array}$$

   is well defined and continuous at 1. Then there exists a neighborhood W of 1 in R such that if $r \in W$, then $(1-r)x_0+ry \in U$. Since $1 \in \operatorname{cl}(B^+ \setminus \{1\})$, we can find $r_0 \in W \cap (B^+ \setminus \{1\})$. Hence, $(1 - r_0)x_0 + r_0 y \in \operatorname{int}(A) \cap U$. As a consequence, $x \in \operatorname{cl}(\operatorname{int}(A))$.

(2) Since $A \subseteq \operatorname{cl}(A)$ we have that $\operatorname{int}(A) \subseteq \operatorname{int}(\operatorname{cl}(A))$. Fix an arbitrary $x \in \operatorname{int}(\operatorname{cl}(A))$ and let U be an open neighborhood of 0 such that $x + U \subseteq \operatorname{cl}(A)$. Since $\operatorname{int}(A) \neq \varnothing$, we can fix an element $x_0 \in \operatorname{int}(A)$. There exists a neighborhood V of 0 in R such that $V(x_0 - x) \subseteq U$. According to Lemma 2.25, there exists $r_0 \in B^+ \setminus \{0, 1\}$ such that $v_0 := -(1 - r_0)^{-1} r_0 \in V$ and

$$x = r_0 x_0 + (1 - r_0)(x + v_0(x_0 - x)) \in \Delta(x_0, x + v_0(x_0 - x)).$$

Note that $x + v_0(x_0 - x) \in x + V(x_0 - x) \subseteq x + U \subseteq \operatorname{cl}(A)$. Since $r_0 \in B^+ \setminus \{0\}$, Theorem 2.40 assures that $x \in \operatorname{int}(A)$.

□

COROLLARY 2.42. *Let $\mathcal{X}$ be a topological module over a topological ring R. Let B^+ be a completely and totally full unit segment of R satisfying that $B^+ \subseteq (B^+)'$. Let A be a convex subset of $\mathcal{X}$ and $x, y \in \operatorname{cl}(A)$. If $\Delta(x, y) \cap \operatorname{int}(A) \neq \varnothing$, then $\Delta(x, y) \setminus \{x, y\} \subseteq \operatorname{int}(A)$.*

PROOF. Let $z \in \Delta(x, y) \cap \operatorname{int}(A)$. In accordance with Lemma 2.34(2),

$$\Delta(x, y) = \Delta(x, z) \cup \Delta(z, y).$$

By bearing in mind Theorem 2.40, we have that

$$\Delta(x, y) \setminus \{x, y\} \subseteq (\Delta(x, z) \setminus \{x\}) \cup (\Delta(z, y) \setminus \{y\}) \subseteq \operatorname{int}(A).$$

□

COROLLARY 2.43. *Let $\mathcal{X}$ be a topological module over a topological ring R such that $2 \in \mathcal{U}(R)$. Let B^+ be a unit segment of R satisfying that $2^{-1} \in B^+ \setminus \{0\} \subseteq \mathcal{U}(R)$. Let A be a convex subset of $\mathcal{X}$. If $A \subseteq \mathcal{X}$ is convex, additively symmetric, and $\operatorname{int}(A) \neq \varnothing$, then $0 \in \operatorname{int}(A)$.*

PROOF. Let $x \in \operatorname{int}(A)$. Notice that $-x \in \operatorname{int}(A)$. According to Lemma 2.39, we have that

$$0 = 2^{-1}x + 2^{-1}(-x) \in \Delta(x, -x) \subseteq \operatorname{int}(A).$$

□

2.4. Extremality. In [**34**, Definition 3.12] and [**36**, Definition 4.1], the notion of 2-extreme point for rings and modules was introduced.

DEFINITION 2.44 (2-Extreme point). Let $\mathcal{X}$ be a module over a ring R. Let $z \in A \subseteq \mathcal{X}$. We will say that z is a 2-extreme point of A provided that the condition $2z = x + y$, with $x, y \in A$, implies that $x = y = z$. The set of 2-extreme points of A is denoted by $\operatorname{ext}_2(A)$.

Next, we will make use of unit segments to introduce the concept of extreme point.

DEFINITION 2.45 (Extreme point). Let $\mathcal{X}$ be a module over a ring R. Let B^+ be a unit segment of R. Let $z \in A \subseteq \mathcal{X}$. We will say that z is an extreme point of A provided that, for all $x, y \in A$, if $z \in \Delta(x, y)$, then either $x = z$ or $y = z$. The set of extreme points of A is denoted by $\operatorname{ext}(A)$.

The following notion is the one of extremal set.

DEFINITION 2.46 (Extremal set). Let $\mathcal{X}$ be a module over a ring R. Let B^+ be a nontrivial unit segment of R. Let $E \subseteq A \subseteq \mathcal{X}$. We will say E is extremal in A if, for any $x, y \in A$ such that $E \cap \Theta(x, y) \neq \varnothing$, then $\{x, y\} \subseteq E$.

The notion of extremal set arises the one of extremal point.

DEFINITION 2.47 (Extremal point). Let $\mathcal{X}$ be a module over a ring R. Let B^+ be a nontrivial unit segment of R. Let $z \in A \subseteq \mathcal{X}$. We will say that z is an extremal point of A if $\{z\}$ is extremal in A. The set of extremal points of A is denoted by $\operatorname{extl}(A)$.

Next proposition unveils the inclusion relations between the above concepts.

PROPOSITION 2.48. *Let $\mathcal{X}$ be a module over a ring R. Let B^+ be a nontrivial unit segment of R. Take $A \subseteq \mathcal{X}$. Then:*

(1) $\operatorname{extl}(A) \subseteq \operatorname{ext}(A)$.
(2) *If $\mathcal{X}$ is torsionfree and R is an integral domain, then* $\operatorname{extl}(A) = \operatorname{ext}(A)$.
(3) *If $2 \in \mathcal{U}(R)$ and $2^{-1} \in B^+$, then* $\operatorname{ext}(A) \subseteq \operatorname{ext}_2(A)$.
(4) *If $\mathcal{X}$ is torsionfree, R is an integral domain, B^+ is strong, and A is convex, then* $\operatorname{ext}_2(A) \subseteq \operatorname{extl}(A)$.

PROOF.

(1) Fix any arbitrary $z \in \operatorname{extl}(A)$. Suppose to the contrary that there exist $x, y \in A$ satisfying that $z \in \Delta(x, y) \setminus \{x, y\}$. According to Remark 2.27, $\Delta(x, y) \setminus \{x, y\} \subseteq \Theta(x, y)$. Since $z \in \operatorname{extl}(A)$, we reach the contradiction that $\{x, y\} \subseteq \{z\}$.
(2) Take any $z \in \operatorname{ext}(A)$. Let $x, y \in A$ such that $z \in \Theta(x, y)$. We have to prove that $\{x, y\} \subseteq \{z\}$. If $x = y$, then $z \in \Theta(x, x) = \{x\}$, so we are done. Let us assume then that $x \neq y$. There are $r, s \in B^+ \setminus \{0, 1\}$ with $r + s = 1$ such that $z = rx + sy$. By hypothesis, either $x = z$ or $y = z$. Let us assume, without any loss of generality, that $x = z$. Then $x = rx + sy$, hence $s(x - y) = 0$. Since $\mathcal{X}$ is torsionfree and $x \neq y$, we conclude that s is a zero divisor. Since R is an integral domain, $s = 0$, which is a contradiction. As a consequence, $x = y = z$.
(3) Take any $z \in \operatorname{ext}(A)$. Suppose that $2z = x + y$ with $x, y \in A$. Then $z = 2^{-1}x + 2^{-1}y$. By hypothesis, $2^{-1} \in B^+$, therefore, $z \in \Theta(x, y) \subseteq \Delta(x, y)$. Since $z \in \operatorname{ext}(A)$, either $x = z$ or $y = z$. Suppose, without any loss of generality, that $x = z$. Then $2z = x + y = z + y$, hence $z = y$. This shows that $z \in \operatorname{ext}_2(A)$.
(4) Take $z \in \operatorname{ext}_2(A)$ and write $z = rx + sy$ with $x, y \in A$, $r, s \in B^+ \setminus \{0, 1\}$, and $r + s = 1$. We have to prove that $\{x, y\} \subseteq \{z\}$. If $x = y$, then $z = rx + sx = (r + s)x = x = y$. Thus, let us suppose that $x \neq y \in A$. By hypothesis, we may assume, without any loss of generality, that $2r \in B^+$. Then

$$2z = 2(rx + (1 - r)y) = y + ((2r)x + (1 - 2r)y).$$

Observe that $(2r)x + (1 - 2r)y \in \Delta(x, y) \subseteq A$. Since $z \in \operatorname{ext}_2(A)$, we conclude that $y = (2r)x + (1 - 2r)y$, which implies that $(2r)(x - y) = 0$. Since $\mathcal{X}$ is torsionfree and R is an integral domain, we have that $2r = 0$. Recall that the existence of a unit segment B^+ in R implies that $\operatorname{char}(R) \neq 2$ (see Proposition 2.10). Then $r = 0$ because R is an integral domain. This is a contradiction. $\square$

COROLLARY 2.49. *Let $\mathcal{X}$ be a torsionfree module over an integral domain ring R such that $2 \in \mathcal{U}(R)$. Let B^+ be a strong and centered unit segment of R. For every convex subset $A \subseteq \mathcal{X}$, $\operatorname{ext}_2(A) = \operatorname{extl}(A) = \operatorname{ext}(A)$.*

THEOREM 2.50. *Let $\mathcal{X}$ be a module over a ring R. Let B^+ be a completely full nontrivial unit segment of R. Suppose that $A \subseteq \mathcal{X}$ is convex and $E \subseteq A$ is extremal in A. For every $x_1, \dots, x_k \in A$, if $E \cap \Theta(x_1, \dots, x_k) \neq \varnothing$, then $\{x_1, \dots, x_k\} \subseteq E$.*

PROOF. Take $z \in E$ and $r_1, \dots, r_k \in B^+ \setminus \{0, 1\}$ with $r_1 + \cdots + r_k = 1$ and $z = r_1 x_1 + \cdots + r_k x_k$. Notice that $r_2 + \cdots + r_k = 1 - r_1 \in B^+ \setminus \{0\}$. Since B^+ is completely full, by Theorem 2.21, we have that $(1 - r_1)^{-1} r_j \in B^+$ for all $j = 2, \dots, k$. Observe that

$$\sum_{j=2}^{k} (1 - r_1)^{-1} r_j = (1 - r_1)^{-1} \sum_{j=2}^{k} r_j = (1 - r_1)^{-1}(1 - r_1) = 1,$$

meaning that

$$\sum_{j=2}^{k} (1 - r_1)^{-1} r_j x_j \in \Delta(x_2, \dots, x_k) \subseteq A.$$

Then

$$z = r_1 x_1 + (1 - r_1) \sum_{j=2}^{k} (1 - r_1)^{-1} r_j x_j \in \Theta\left(x_1, \sum_{j=2}^{k} (1 - r_1)^{-1} r_j x_j \right).$$

Finally, since E is extremal in A,

$$\left\{ x_1, \sum_{j=2}^{k} (1 - r_1)^{-1} r_j x_j \right\} \subseteq A.$$

By interchanging the roles of the x_i's, we end up having that $\{x_1, \dots, x_k\} \subseteq E$. □

Next theorem shows that, under not so restrictive hypotheses, affine internal points can never be 2-extreme points.

THEOREM 2.51. *Let $\mathcal{X}$ be a module over a topological ring R. Assume that $\mathcal{X}$ is torsionfree, R is nondiscrete, an integral domain, and $\operatorname{char}(R) \neq 2$. If $A \subseteq \mathcal{X}$ is not a singleton, then $\operatorname{ainter}(A) \cap \operatorname{ext}_2(A) = \varnothing$.*

PROOF. Suppose that there exists $x \in \operatorname{ainter}(A) \cap \operatorname{ext}_2(A)$. Since A is not a singleton, there exists $y \in \operatorname{aff}(A) \setminus \{x\}$. Since $x \in \operatorname{ainter}(A)$, there exists an additively symmetric neighborhood V of 0 in R such that $x + V(y - x) \subseteq A$. Since R is nondiscrete, we can find $v \in V \setminus \{0\}$. Observe that $2x = (x + v(y - x)) + (x - v(y - x))$. Since $x \in \operatorname{ext}_2(A)$, we conclude that $x + v(y - x) = x - v(y - x)$, which implies that $2v(y - x) = 0$. Since $y - x \neq 0$, $\mathcal{X}$ is torsionfree, and R is an integral domain, we have that $2v = 0$. Now, $v \neq 0$, so, we use again that R is an integral domain to obtain $2 = 0$. However, this contradicts the fact that $\operatorname{char}(R) \neq 2$. □

Next theorem is in some sense a generalization of Theorem 2.51.

THEOREM 2.52. *Let $\mathcal{X}$ be a module over a topological inversion ring R. Let B^+ be a unit segment of R such that $B^+ \setminus \{0\} \subseteq \mathcal{U}(R)$ and $0 \in \operatorname{cl}(B^+ \setminus \{0, 1\})$. Let $E \subseteq A \subseteq \mathcal{X}$. If E is extremal in A and $E \cap \operatorname{ainter}(A) \neq \varnothing$, then $E = A$.*

PROOF. We may assume that A is not a singleton. Let $x \in E \cap \operatorname{ainter}(A)$. Fix an arbitrary $y \in A \setminus \{x\}$. We will prove that $y \in E$. Since $y \in \operatorname{aff}(A)$, there exists a neighborhood V of 0 in R such that $x + V(y-x) \subseteq A$. According to Lemma 2.25, there exists $r_0 \in B^+ \setminus \{0,1\}$ such that $v_0 := -(1-r_0)^{-1} r_0 \in V$ and

$$x = r_0 y + (1-r_0)\,(x + v_0(y-x)) \in \Theta(y, x + v_0(y-x)).$$

Notice that $\{y, x+v_0(y-x)\} \subseteq A$ and $E \cap \Theta(y, x+v_0(y-x)) \neq \varnothing$. Since E is extremal in A, we conclude that $\{y, x+v_0(y-x)\} \subseteq E$. The arbitrariness of y implies that $A = E$. □

Extremality properties are preserved under translations, homotecies, and linear operators in a similar way like convexity properties were also preserved.

THEOREM 2.53. *Let $\mathcal{X}, \mathcal{Y}$ be modules over a ring R. Let B^+ be a nontrivial unit segment of R. Let $E \subseteq A \subseteq \mathcal{X}$. Then:*

(1) *If $\mathcal{X}$ is torsionfree and $r \in R \setminus (\ell d(0) \cup rd(0))$ verifies that $rB^+ = B^+ r$, then E is extremal in A if and only if rE is extremal in rA. In particular,* $\operatorname{extl}(rA) = r\operatorname{extl}(A)$.
(2) *E is extremal in A if and only if $x+E$ is extremal in $x+A$ for all $x \in \mathcal{X}$. In particular,* $\operatorname{extl}(x+A) = x + \operatorname{extl}(A)$.
(3) *If $T : \mathcal{X} \to \mathcal{Y}$ is injective and linear, then $T(E)$ is extremal in $T(A)$ if and only if E is extremal in A. In particular,* $T(\operatorname{extl}(A)) = \operatorname{extl}(T(A))$.

PROOF. We will only prove the first item. The others follow similar ideas.

$\Rightarrow$ Suppose first that E is extremal in A. Let $x, y \in A$ such that $rE \cap \Theta(rx, ry) \neq \varnothing$. Take $z \in E$ and $t, s \in B^+ \setminus \{0,1\}$ with $t+s=1$ such that $rz = t(rx) + s(ry)$. By hypothesis, $rz = r\,(t'x + s'y)$ for some $t', s' \in B^+$. In view of Lemma 2.29, $t'+s'=1$. Since $r \in R \setminus (\ell d(0) \cup rd(0))$, it can be easily shown that $t', s' \in B^+ \setminus \{0,1\}$. Then $t'x + s'y \in \Theta(x,y)$. Next, $\mathcal{X}$ is torsionfree by hypothesis and $r \in R \setminus (\ell d(0) \cup rd(0))$, therefore, $z = t'x + s'y \in \Theta(x,y)$. Since E is extremal in A, we conclude that $\{x,y\} \subseteq E$, meaning that $\{rx, ry\} \subseteq rE$.

$\Leftarrow$ Conversely, suppose that rE is extremal in rA. Let $x, y \in A$ such that $E \cap \Theta(x,y) \neq \varnothing$. Take $z \in E$ and $t, s \in B^+ \setminus \{0,1\}$ with $t+s=1$ such that $z = tx + sy$. Note that $rz = r(tx) + r(sy)$. By hypothesis, $rz = t'(rx) + s'(ry)$ for some $t', s' \in B^+$. In view of Lemma 2.29, $t'+s'=1$. Since $r \in R \setminus (\ell d(0) \cup rd(0))$, it can be easily shown that $t', s' \in B^+ \setminus \{0,1\}$. Then $rz = t'(rx) + s'(ry) \in \Theta(rx, ry)$. Since rE is extremal in rA, we conclude that $\{rx, ry\} \subseteq rE$. Next, $\mathcal{X}$ is torsionfree by hypothesis, therefore, $\{x,y\} \subseteq E$.

□

REMARK 2.54. Let $\mathcal{X}$ be a module over a ring R. Let B^+ be a nontrivial unit segment of R. Let $F \subseteq E \subseteq A \subseteq \mathcal{X}$. If F is extremal in E and E is extremal in A, then F is extremal in A. Indeed, take any $x, y \in A$ such that $F \cap \Theta(x,y) \neq \varnothing$. Then $E \cap \Theta(x,y) \neq \varnothing$, so $\{x,y\} \subseteq E$. Finally, $\{x,y\} \subseteq F$ because F is extremal in E.

The final results in this subsection are aimed at obtaining unit segments from closed unit neighborhoods of zero. For this, the next technical lemmas will be crucial.

LEMMA 2.55. *Let R be a ring and let $A \subseteq R$. Then $A \cap -A = \{0\}$ if and only if $0 \in \mathrm{ext}_2(A)$.*

PROOF. Suppose first that $0 \in \mathrm{ext}_2(A)$. Let $a_1, a_2 \in A$ such that $a_1 = -a_2$. Then $2 \cdot 0 = a_1 + a_2$. Since $0 \in \mathrm{ext}_2(1+B)$, we conclude that $0 = a_1 = a_2$. Conversely, suppose that $A \cap -A = \{0\}$ and let $a_1, a_2 \in A$ such that $2 \cdot 0 = a_1 + a_2$. Then $a_1 = -a_2 \in A \cap -A = \{0\}$, meaning that $a_1 = a_2 = 0$. □

LEMMA 2.56. *Let R be a topological ring. Let B be a closed unit neighborhood of 0 in R. Then:*

(1) $2(1+B) \subseteq (1+B)(1+B)$.
(2) $(1+B)(1+B) = 2(1+B)$ *if and only if* $(b-1)+B(b+1) \subseteq 2B$ *for all* $b \in B$.
(3) *If* $(b-1)+B(b+1) \subseteq 2B$ *for all* $b \in B$*, then* $1+B \subseteq 2B$.

PROOF.

(1) Fix any arbitrary $b \in B$. Since $1 \in B$, we clearly have that $2(1+b) = (1+1)(1+b) \in (1+B)(1+B)$.
(2) Suppose first that $(b-1)+B(b+1) \subseteq 2B$ for all $b \in B$. We only need to show that $(1+B)(1+B) \subseteq 2(1+B)$. Let $b, c \in B$. By hypothesis, there exists $a \in B$ such that $(c-1)+b(c+1) = 2a$. Then $b+c+bc = 1+2a$, so $(1+b)(1+c) = 2+2a = 2(1+a) \in 2(1+B)$. Conversely, assume that $(1+B)(1+B) = 2(1+B)$. Let $b, c \in B$. By hypothesis, there exists $a \in B$ such that $(1+b)(1+c) = 2(1+a)$. Then $(c-1)+b(c+1) = -2+(1+b)(1+c) = -2+2(1+a) = 2a \in 2B$.
(3) Fix an arbitrary $b \in B$. Then $-(1+b) = (-b-1)+0(-b+1) \in 2B$ by hypothesis, hence $1+b \in -2B = 2(-B) = 2B$.

□

After Lemma 2.56, and under its settings, it is expected to take $B^+ := 2^{-1}(1+B)$ provided that $2 \in \mathcal{U}(R)$.

THEOREM 2.57. *Let R be a topological ring such that $2 \in \mathcal{U}(R)$. Let B be a closed unit neighborhood of 0 in R. If $0 \in \mathrm{ext}_2(1+B)$ and $(b-1)+B(b+1) \subseteq 2B$ for all $b \in B$, then $B^+ := 2^{-1}(1+B)$ is a unit segment of R contained in B such that $2^{-1} \in B^+$. If, in addition, $1 \notin \mathrm{int}(B)$, then B^+ verifies that $0 \in \mathrm{cl}\left(B^+ \setminus \{0,1\}\right)$.*

PROOF. In the first place, $1-B^+ = 2^{-1}(2-(1+B)) = 2^{-1}(1-B) = 2^{-1}(1+B) = B^+$. Lemma 2.55 assures that $B^+ \cap B^- = \{0\}$. By lemma 2.56(2), $B^+B^+ \subseteq B^+$. Because of Lemma 2.56(3), $B^+ \subseteq B$. Next, if, in addition, $1 \notin \mathrm{int}(B)$, then $-1 \notin \mathrm{int}(B)$ either. Since $1 \in \mathrm{cl}(\mathrm{int}(B))$, we have that $1 = 2^{-1}(1+1) \in 2^{-1}(1+\mathrm{cl}(\mathrm{int}(B))) = \mathrm{cl}\left(2^{-1}(1+\mathrm{int}(B))\right) \subseteq \mathrm{cl}\left(2^{-1}(1+B \setminus \{-1,1\})\right) = \mathrm{cl}\left(B^+ \setminus \{0,1\}\right)$. Finally, Lemma 2.13(4) assures the fact that $0 \in \mathrm{cl}\left(B^+ \setminus \{0,1\}\right)$. □

Next example shows that not all closed unit zero-neighborhoods satisfy the hypotheses of previous theorem.

EXAMPLE 2.58. Let $\mathbb{D} := \{z \in \mathbb{C} : |z| \leq 1\}$ denote the closed unit disk, which is a closed unit neighborhood of 0 in $\mathbb{C}$ in view of [**34**, Theorem 3.7]. Denote $\mathbb{D}^+ := 2^{-1}(1+\mathbb{D}) \subseteq \mathbb{D}$. The additive symmetricity of $\mathbb{D}$ allows $1+\mathbb{D}^- = \mathbb{D}^+$. Also,

$0 \in \operatorname{ext}_2(\mathbb{D}^+)$, hence $\mathbb{D}^+ \cap \mathbb{D}^- = \{0\}$ in view of Lemma 2.55. However, $\mathbb{D}^+$ is not multiplicatively idempotent and thus it is not a unit segment of $\mathbb{C}$. Indeed,

$$\left(\frac{1+i}{2}\right)^2 = \frac{2i}{4} = \frac{i}{2}.$$

If $z \in \mathbb{C}$ is so that $\frac{1+z}{2} = \frac{i}{2}$, then $z = i - 1$ so $|z| = |i-1| = \sqrt{2} > 1$.

2.5. Locally convex module topologies. This final subsection is devoted to construct the finest locally convex module topology. For this, the notion of balancedness will be crucial. In [**36**, Definition 2.1], it was introduced the concept of B-balancedness, in the sense that a subset A of a topological module is B-balanced provided that $BA \subseteq A$, where B is a closed unit neighborhood of zero of the underlying ring. This definition requires that the underlying ring be endowed with a nontrivial closed unit neighborhood of zero. Here, we will generalize this notion to fit a larger class of topological rings.

Definition 2.59 (Balanced set). Let $\mathcal{X}$ be a module over a topological ring R. A subset A of $\mathcal{X}$ is said to be balanced provided that it is additively symmetric and there exists a 0-neighborhood $W \subseteq R$ such that $WA \subseteq A$.

The above notion not only generalizes [**36**, Definition 2.1] but it is also strictly more general than the classical notion of balancedness in real vector spaces. Indeed, the set $A := \{-3\} \cup [-1,1] \cup \{3\}$ is not balanced in the classical sense since $[-1,1]A \nsubseteq A$, however A is balanced in the sense of Definition 2.59 since $\left[\frac{-1}{3}, \frac{1}{3}\right] A \subseteq A$. From now on, whenever we talk about balancedness we will refer to Definition 2.59.

Lemma 2.60. *Let $\mathcal{X}$ be a topological module over a topological ring R. Then:*

(1) *If V is an additively symmetric bounded neighborhood of 0 in $\mathcal{X}$, then V is balanced.*
(2) *If $A \subseteq \mathcal{X}$ and $P \subseteq R$ is left-balanced in R, then PA is balanced.*
(3) *If $A \subseteq \mathcal{X}$ and $W \subseteq R$ is an additively symmetric left-bounded 0-neighborhood, then WA is balanced.*

Proof.

(1) There are $W_0 \subseteq R$ and $V_0 \subseteq \mathcal{X}$ neighborhoods of 0 in R and $\mathcal{X}$, respectively, in such a way that $W_0 V_0 \subseteq V$. By hypothesis, V is bounded, so there exists an invertible $u \in \mathcal{U}(R)$ satisfying that $V \subseteq uV_0$. Then $W_0 u^{-1}$ is a neighborhood of 0 in R and $\left(W_0 u^{-1}\right) V \subseteq W_0 u^{-1} u V_0 = W_0 V_0 \subseteq V$.
(2) Note that PA is additively symmetric because so is P. By hypothesis, there exists a 0-neighborhood W_1 in R such that $W_1 P \subseteq P$. Then $W_1(PA) \subseteq PA$.
(3) By Lemma 2.60(1), W is left-balanced in R. By Lemma 2.60(2), WA is balanced.

□

Lemma 2.61. *Let $\mathcal{X}$ be a topological module over a practical topological ring R. If W is a neighborhood of 0 in R and V is a neighborhood of 0 in $\mathcal{X}$, then WV is a neighborhood of 0 in $\mathcal{X}$.*

Proof. Simply note that there exists $u \in \mathcal{U}(R)$ with $u \in W$. Now $uV \subseteq WV$ and uV is a neighborhood of 0 in $\mathcal{X}$ because u is invertible. □

Lemma 2.61 does not hold if the topological ring is not practical. Indeed, take $R := \mathbb{Z}$ endowed with the inherited topology from $\mathbb{R}$ and $\mathcal{X} := \mathbb{R}$. If $W := \{0\}$ and $V := [-1, 1]$, then $WV = \{0\}$ is not a neighborhood of 0 in $\mathcal{X}$.

On the other hand, notice that, if $\mathcal{B}$ is a base of neighborhoods of 0 in a topological module $\mathcal{X}$, then $\mathcal{B}_s := \{A \cap -A : A \in \mathcal{B}\}$ and $\mathcal{B}_S := \{A \cup -A : A \in \mathcal{B}\}$ are both bases of additively symmetric neighborhoods of 0 in $\mathcal{X}$.

THEOREM 2.62. *Let $\mathcal{X}$ be a topological module over a practical topological ring R. Let W_0 be a right-bounded neighborhood of 0 in R and V_0 a bounded neighborhood of 0 in $\mathcal{X}$. Then:*

(1) *If $\mathcal{B}$ is a base of neighborhoods of 0 in $\mathcal{X}$, then $\mathcal{B}_{W_0} := \{W_0 V : V \in \mathcal{B}\}$ is a base of neighborhoods of 0 in $\mathcal{X}$.*
(2) *If $\mathcal{B}$ is a base of neighborhoods of 0 in R, then $\mathcal{B}_{V_0} := \{W V_0 : W \in \mathcal{B}\}$ is a base of neighborhoods of 0 in $\mathcal{X}$.*

PROOF.

(1) On the the hand, $\mathcal{B}_{W_0}$ is a filter base of $\mathcal{N}_0(\mathcal{X})$. Indeed, $\mathcal{B}_{W_0}$ is a nonempty subset of $\mathcal{N}_0(\mathcal{X})$ in view of Lemma 2.61, and if $V_1, V_2 \in \mathcal{B}$, there exists $V_3 \in \mathcal{B}$ such that $V_3 \subseteq V_1 \cap V_2$, meaning that $W_0 V_3 \subseteq W_0(V_1 \cap V_2) \subseteq W_0 V_1 \cap W_0 V_2$. On the other hand, let U be a neighborhood of 0 in $\mathcal{X}$. There exist a neighborhood W of 0 in R and a neighborhood V of 0 in $\mathcal{X}$ such that $WV \subseteq U$. Since W_0 is right-bounded, there exists $u \in \mathcal{U}(R)$ such that $W_0 \subseteq Wu$. Next, there exists $V_1 \in \mathcal{B}$ such that $V_1 \subseteq u^{-1}V$. Finally, $W_0 V_1 \subseteq Wuu^{-1}V = WV \subseteq U$ and $W_0 V_1 \in \mathcal{B}_{W_0}$.
(2) In the first place, following a similar reasoning as above, $\mathcal{B}_{V_0}$ is a filter base of $\mathcal{N}_0(\mathcal{X})$. Now, let U be a neighborhood of 0 in $\mathcal{X}$. There exist a neighborhood W of 0 in R and a neighborhood V of 0 in $\mathcal{X}$ such that $WV \subseteq U$. Since V_0 is bounded, there exists an invertible $u \in \mathcal{U}(R)$ such that $V_0 \subseteq uV$. By hypothesis, we can find $W_1 \in \mathcal{B}$ with $W_1 \subseteq Wu^{-1}$. As a consequence, $W_1 V_0 \subseteq Wu^{-1}uV = WV \subseteq U$ and $W_1 V_0 \in \mathcal{B}_{V_0}$.

□

COROLLARY 2.63. *Let $\mathcal{X}$ be a topological module over a practical topological ring R. If there exists a bounded neighborhood W_0 of 0 in R, then the collection of all balanced neighborhoods of 0 in $\mathcal{X}$ is a base of 0-neighborhoods in $\mathcal{X}$.*

PROOF. Let $\mathcal{B}$ denote the collection of all balanced neighborhoods of 0 in $\mathcal{X}$. Notice that $\mathcal{B}$ is a filter base of $\mathcal{N}_0(\mathcal{X})$. We may assume that W_0 is additively symmetric. Next, by Lemma 2.60(3), $\mathcal{B}$ contains $\{W_0 V : V \in \mathcal{N}_0(\mathcal{X})\}$, which is base of neighborhoods of 0 in $\mathcal{X}$ in virtue of Theorem 2.62(1). □

Corollary 2.63 motivates the following definition.

DEFINITION 2.64 (Locally convex module topology). Let $\mathcal{X}$ be a topological module over a topological ring R endowed with a unit segment B^+. We say that the module topology of $\mathcal{X}$ is locally convex if there exists a base of neighborhoods of 0 which are all balanced, convex, and absorbing.

Before constructing the finest locally convex module topology, we will make use of several technical lemmas.

LEMMA 2.65. *Let $\mathcal{X}$ be a module over a topological ring R such that R is an integral domain. Let B^+ be a commuting unit segment of R. If $A \subseteq \mathcal{X}$ is balanced, then* $\operatorname{conv}(A)$ *is balanced as well.*

PROOF. First off, A is additively symmetric and so is $\operatorname{conv}(A)$. Take W a neighborhood of 0 in R such that $WA \subseteq A$. We will show that $W\operatorname{conv}(A) \subseteq \operatorname{conv}(A)$. Fix an arbitrary $w \in W$. We may assume that $w \neq 0$ since $0 \in A \subseteq \operatorname{conv}(A)$. According to Theorem 2.36, a generic element of $\operatorname{conv}(A)$ has the form $\sum_{i=1}^k r_i a_i$ where $r_1, \ldots, r_k \in B^+$ with $\sum_{i=1}^k r_i = 1$ and $a_1, \ldots, a_k \in A$. By hypothesis, B^+ is commuting, so $wB^+ = B^+w$, meaning that for every $i \in \{1, \ldots, k\}$ there exists $s_i \in B^+$ with $wr_i = s_i w$. In view of Lemma 2.29, we have that $\sum_{i=1}^k s_i = 1$. Then

$$w \sum_{i=1}^k r_i a_i = \sum_{i=1}^k (wr_i) a_i = \sum_{i=1}^k s_i (wa_i) \in \operatorname{conv}(WA) \subseteq \operatorname{conv}(A).$$

□

According to Theorem 1.9(1), every zero-neighborhood in a topological module is absorbing.

LEMMA 2.66. *Let R be a practical topological ring and $\mathcal{X}$ a topological R-module. If $W \subseteq R$ is a 0-neighborhood and $A \subseteq \mathcal{X}$ is absorbing, then WA is absorbing.*

PROOF. Take $w \in \mathcal{U}(R)$ with $w \in W$. Fix an arbitrary $x \in \mathcal{X}$. There exists a 0-neighborhood $V \subseteq R$ with $Vx \subseteq A$. Finally, keep in mind that $(wV)x \subseteq WVx \subseteq WA$ and wV is a neighborhood of 0 in R because w is invertible. □

We are finally in the right position to prove the main result of this manuscript.

THEOREM 2.67 (Finest locally convex module topology). *Let $\mathcal{X}$ be a module over a practical topological ring R. Suppose that R is an integral domain, $2 \in \mathcal{U}(R)$, there exists a bounded neighborhood of* 0 *in R, and R is endowed with a commuting and centered full unit segment B^+. Then*

$$\mathcal{B}_{\mathcal{X}} := \{A \subseteq \mathcal{X} : A \textit{ is balanced, convex, and absorbing}\}$$

is base of neighborhoods of 0 *for a module topology $\tau_{\mathcal{X}}$ in $\mathcal{X}$ in such a way that if τ is a locally convex module topology in $\mathcal{X}$, then $\tau \subseteq \tau_{\mathcal{X}}$.*

PROOF. In the first place, we will check that $\mathcal{B}_{\mathcal{X}}$ verifies the requirements to be a filter base of $\mathcal{X}$:

- $\mathcal{B}_{\mathcal{X}}$ is clearly nonempty and $\varnothing \notin \mathcal{B}_{\mathcal{X}}$. Indeed, $\mathcal{X}$ is trivially convex, additively symmetric, absorbing, and $W\mathcal{X} \subseteq \mathcal{X}$ for every 0-neighborhood $W \subseteq R$. On the other hand, if $A \subseteq \mathcal{X}$ is balanced, convex, and absorbing, then $0 \in A$, hence $A \neq \varnothing$.
- If $C, D \in \mathcal{B}_{\mathcal{X}}$, then there exists $A \in \mathcal{B}_{\mathcal{X}}$ with $A \subseteq C \cap D$. Indeed, note that $C \cap D$ is convex and additively symmetric. Also, $0 \in \operatorname{inter}(C) \cap \operatorname{inter}(D) = \operatorname{inter}(C \cap D)$, meaning that $C \cap D$ is absorbing. Finally, if W_1, W_2 are 0-neighborhoods in R such that $W_1 C \subseteq C$ and $W_2 D \subseteq D$, then $(W_1 \cap W_2)(C \cap D) \subseteq (W_1 \cap W_2) C \cap (W_1 \cap W_2) D \subseteq C \cap D$. Therefore, $A := C \cap D$ verifies the desired properties.

Next, we will check that $\mathcal{B}_{\mathcal{X}}$ verifies Theorem 1.7(1,2,3,4,5):

(1) If $A \in \mathcal{B}_{\mathcal{X}}$, then there exists $C \in \mathcal{B}_{\mathcal{X}}$ with $C + C \subseteq A$. Indeed, if we take $C := 2^{-1}A$, then C is convex in view of Corollary 2.31(3), C is trivially additively symmetric, and C is absorbing because inter $(2^{-1}A) = 2^{-1}\text{inter}(A)$. Also, if W is a 0-neighborhood in R such that $WA \subseteq A$, then $WC = W2^{-1}A = 2^{-1}WA \subseteq 2^{-1}A = C$. As a consequence, $C \in \mathcal{B}_{\mathcal{X}}$. Finally, the convexity of A together with the fact that $2^{-1} \in B^+$ allow that $C + C = 2^{-1}A + 2^{-1}A \subseteq A$.

(2) If $A \in \mathcal{B}_{\mathcal{X}}$, then there exists $C \in \mathcal{B}_{\mathcal{X}}$ with $-C \subseteq A$. Indeed, since A is additively symmetric, it suffices to take $C := A$.

(3) If $A \in \mathcal{B}_{\mathcal{X}}$, then there exist $C \in \mathcal{B}_{\mathcal{X}}$ and $W \in \mathcal{N}_0(R)$ such that $WC \subseteq A$. Indeed, since A is balanced, there exists a 0-neighborhood W in R such that $WA \subseteq A$.

(4) If $A \in \mathcal{B}_{\mathcal{X}}$ and $r \in R$, then there exists $C \in \mathcal{B}_{\mathcal{X}}$ such that $rC \subseteq A$. Indeed, let W_0 be a neighborhood of 0 in R such that $W_0A \subseteq A$. There exists an additively symmetric neighborhood W_1 of 0 in R such that $rW_1 \subseteq W_0$. We may assume that W_1 is left-bounded since otherwise it can be intersected with an additively symmetric left-bounded neighborhood of 0 in R. Let us take $C := \text{conv}\,(W_1A)$. By taking into account Corollary 2.37, $rC = r\text{conv}\,(W_1A) = \text{conv}\,(rW_1A) \subseteq \text{conv}\,(W_0A) \subseteq \text{conv}(A) = A$. It remains to show that $C \in \mathcal{B}_{\mathcal{X}}$. On the one hand, C is clearly convex and additively symmetric. Next, A is absorbing and R is practical, hence W_1A is also absorbing in view of Lemma 2.66, meaning that C is absorbing since $W_1A \subseteq \text{conv}\,(W_1A) = C$. Finally, W_1A is balanced in view of Lemma 2.60(3), thus $C = \text{conv}(W_1A)$ is balanced in accordance with Lemma 2.65.

(5) If $A \in \mathcal{B}_{\mathcal{X}}$ and $x \in \mathcal{X}$, then there exists $W \in \mathcal{N}_0(R)$ such that $Wx \subseteq A$. Indeed, since A is absorbing, there exists $W \in \mathcal{N}_0(R)$ such that $Wx \subseteq A$.

□

The conclusion of this survey is a nontrivial example of a topological module over a topological ring satisfying the hypotheses of Theorem 2.67.

EXAMPLE 2.68. Take $R := \mathbb{Z}\left[\frac{1}{2}\right]$ with the inherited topology from $\mathbb{R}$. Notice that R is practical since $\frac{1}{2^n} \to 0$ as $n \to \infty$, an integral domain, and $2 \in \mathcal{U}(R)$. Observe that $B := [-1, 1] \cap R$ is a bounded neighborhood of 0 in R. Also, $B^+ := [0, 1] \cap R$ is a commuting and centered full unit segment of R. According to Theorem 2.67, if $\mathcal{X}$ is an R-module, then there exists the finest locally convex module topology in $\mathcal{X}$. A nontrivial example of R-module is, for instance, the module of bounded sequences with coefficients in R, $\ell_\infty\left(\mathbb{Z}\left[\frac{1}{2}\right]\right)$.

EXAMPLE 2.69. Take $R := \mathbb{Q}[\pi]$ with the inherited topology from $\mathbb{R}$. Notice that R is practical, an integral domain, and $2 \in \mathcal{U}(R)$. Also notice that π is trascendental over $\mathbb{Q}$, therefore π is not invertible in $\mathbb{Q}[\pi]$, hence $\mathbb{Q}[\pi]$ is not a field. Observe that $B := [-1, 1] \cap R$ is a bounded neighborhood of 0 in R. Also, $B^+ := [0, 1] \cap R$ is a commuting and centered full unit segment of R. According to Theorem 2.67, if $\mathcal{X}$ is an R-module, then there exists the finest locally convex module topology in $\mathcal{X}$. A nontrivial example of R-module is, for instance, the module of null sequences with coefficients in R, $c_0(\mathbb{Q}[\pi])$.

Future efforts will be paid into studying the finest locally convex module topology in modules of bounded operators, such as, for instance, $\mathcal{X} := \mathcal{B}(K, H)$, which is a (left) module over the ring $R := \mathcal{B}(H)$ by left composition, for H, K infinite-dimensional complex Hilbert spaces. Notice that R is not an integral domain, however $I + I \in \mathcal{U}(R)$.

Other future efforts will be paid into proving a version of the Krein-Milman Theorem in the topological module setting.

Acknowledgements

The author would like to express his deepest gratitude towards the Department of Mathematical Sciences at Kent State University.

References

[1] S. Abedi and M. S. Moslehian, *Power-norms based on Hilbert C^*-modules*, Rev. R. Acad. Cienc. Exactas Fís. Nat. Ser. A Mat. RACSAM **117** (2023), no. 1, Paper No. 7, 17, DOI 10.1007/s13398-022-01341-2. MR4496640

[2] S. Abedi and M. S. Moslehian, *Extensions of the Hilbert-multi-norm in Hilbert C^*-modules*, Positivity **27** (2023), no. 1, Paper No. 7, 17, DOI 10.1007/s11117-022-00960-8. MR4519196

[3] A. Aizpuru, S. Moreno-Pulido, and F. Rambla-Barreno, *Phillips lemma on effect algebras of sets*, Math. Slovaca **63** (2013), no. 3, 639–646, DOI 10.2478/s12175-013-0124-3. MR3071981

[4] A. Aizpuru and M. Tamayo, *Classical properties of measure theory on effect algebras*, Fuzzy Sets and Systems **157** (2006), no. 15, 2139–2143, DOI 10.1016/j.fss.2006.03.010. MR2236488

[5] V. I. Arnautov, *General theory of radicals of topological rings* (Russian, with English summary), Izv. Akad. Nauk Respub. Moldova Mat. **2** (1996), 5–45. MR1484009

[6] V. I. Arnautov, S. T. Glavatsky, and A. V. Mikhalev, *Introduction to the theory of topological rings and modules*, Monographs and Textbooks in Pure and Applied Mathematics, vol. 197, Marcel Dekker, Inc., New York, 1996. MR1368852

[7] A. Avallone and G. Barbieri, *Lyapunov measures on effect algebras*, Comment. Math. Univ. Carolin. **44** (2003), no. 3, 389–397. MR2025808

[8] A. Avallone, G. Barbieri, and P. Vitolo, *Hahn decomposition of modular measures and applications*, Comment. Math. (Prace Mat.) **43** (2003), no. 2, 149–168. MR2029889

[9] A. Avallone and A. Basile, *On a Marinacci uniqueness theorem for measures*, J. Math. Anal. Appl. **286** (2003), no. 2, 378–390, DOI 10.1016/S0022-247X(03)00274-9. MR2008838

[10] A. Avallone and P. Vitolo, *Decomposition and control theorems in effect algebras*, Sci. Math. Jpn. **58** (2003), no. 1, 1–14. MR1987813

[11] G. Barbieri, *Lyapunov's theorem for measures on D-posets*, Internat. J. Theoret. Phys. **43** (2004), no. 7-8, 1613–1623, DOI 10.1023/B:IJTP.0000048807.37145.cc. MR2108298

[12] G. Barbieri, F. J. García-Pacheco, and S. Moreno-Pulido, *Measures on effect algebras*, Math. Slovaca **69** (2019), no. 1, 159–170, DOI 10.1515/ms-2017-0211. MR3903635

[13] G. Barbieri, A. Valente, and H. Weber, *Decomposition of ℓ-group-valued measures*, Czechoslovak Math. J. **62(137)** (2012), no. 4, 1085–1100, DOI 10.1007/s10587-012-0065-y. MR3010258

[14] K. P. S. Bhaskara Rao and M. Bhaskara Rao, *Theory of charges: A study of finitely additive measures*, Pure and Applied Mathematics, vol. 109, Academic Press, Inc. [Harcourt Brace Jovanovich, Publishers], New York, 1983. With a foreword by D. M. Stone. MR751777

[15] N. Bourbaki, *Topological vector spaces. Chapters 1–5*, Elements of Mathematics (Berlin), Springer-Verlag, Berlin, 1987. Translated from the French by H. G. Eggleston and S. Madan, DOI 10.1007/978-3-642-61715-7. MR910295

[16] J. B. Conway, *A course in operator theory*, Graduate Studies in Mathematics, vol. 21, American Mathematical Society, Providence, RI, 2000, DOI 10.1090/gsm/021. MR1721402

[17] N. Dunford and J. T. Schwartz, *Linear operators. Part I: General theory*, Wiley Classics Library, John Wiley & Sons, Inc., New York, 1988. With the assistance of William G. Bade and Robert G. Bartle; Reprint of the 1958 original; A Wiley-Interscience Publication. MR1009162

[18] N. Dunford and J. T. Schwartz, *Linear operators. Part II: Spectral theory. Selfadjoint operators in Hilbert space*, Wiley Classics Library, John Wiley & Sons, Inc., New York, 1988. With the assistance of William G. Bade and Robert G. Bartle; Reprint of the 1963 original; A Wiley-Interscience Publication. MR1009163

[19] N. Dunford and J. T. Schwartz, *Linear operators. Part III: Spectral operators*, Wiley Classics Library, John Wiley & Sons, Inc., New York, 1988. With the assistance of William G. Bade and Robert G. Bartle; Reprint of the 1971 original; A Wiley-Interscience Publication. MR1009164

[20] A. Dvurečenskij and S. Pulmannová, *New trends in quantum structures*, Mathematics and its Applications, vol. 516, Kluwer Academic Publishers, Dordrecht; Ister Science, Bratislava, 2000, DOI 10.1007/978-94-017-2422-7. MR1861369

[21] J. Diestel and J. J. Uhl Jr., *Vector measures*, Mathematical Surveys, No. 15, American Mathematical Society, Providence, RI, 1977. With a foreword by B. J. Pettis. MR453964

[22] L. E. Dubins, *On extreme points of convex sets*, J. Math. Anal. Appl. **5** (1962), 237–244, DOI 10.1016/S0022-247X(62)80007-9. MR143106

[23] P. H. Enflo, M. S. Moslehian, and J. B. Seoane-Sepúlveda, *A history of solving some famous problems in mathematical analysis*, Br. J. Hist. Math. **37** (2022), no. 1, 64–80, DOI 10.1080/26375451.2022.2037358. MR4425919

[24] R. Eskandari, X. Fang, M. S. Moslehian, and Q. Xu, *Pedersen-Takesaki operator equation and operator equation $AX = B$ in Hilbert C^*-modules*, J. Math. Anal. Appl. **521** (2023), no. 1, Paper No. 126878, 14, DOI 10.1016/j.jmaa.2022.126878. MR4519337

[25] F. J. Freniche, *The Vitali-Hahn-Saks theorem for Boolean algebras with the subsequential interpolation property*, Proc. Amer. Math. Soc. **92** (1984), no. 3, 362–366, DOI 10.2307/2044835. MR759653

[26] F. J. García-Pacheco, *Complementation of the subspace of G-invariant vectors*, J. Algebra Appl. **16** (2017), no. 7, 1750124, 7, DOI 10.1142/S0219498817501249. MR3660407

[27] F. J. García-Pacheco, *Regularity in topological modules*, Mathematics **8** (2020), 1580, 17pp.

[28] F. J. García-Pacheco, *A solution to the faceless problem*, J. Geom. Anal. **30** (2020), no. 4, 3859–3871, DOI 10.1007/s12220-019-00220-4. MR4167267

[29] F. J. García-Pacheco, *Abstract calculus—a categorical approach*, Monographs and Research Notes in Mathematics, CRC Press, Boca Raton, FL, 2022. MR4492773

[30] F. J. García-Pacheco, *Convex functions on topological modules*, Linear Multilinear Algebra **70** (2022), no. 18, 3511–3524, DOI 10.1080/03081087.2020.1844615. MR4511243

[31] F. J. García-Pacheco, *The range of a module measure defined on an effect algebra*, Symmetry **14** (2022), no. 1819, 15 pp.

[32] F. J. García-Pacheco, M. A. Moreno-Frías, and M. Murillo-Arcila, *Topological ordered rings and measures*, Results Math. **78** (2023), no. 6, Paper No. 212, 16, DOI 10.1007/s00025-023-01988-2. MR4630560

[33] F. J. García-Pacheco and S. Sáez-Martínez, *Normalizing rings*, Banach J. Math. Anal. **14** (2020), no. 3, 1143–1176, DOI 10.1007/s43037-020-00055-0. MR4123326

[34] F. J. García-Pacheco and P. Piniella, *Unit neighborhoods in topological rings*, Banach J. Math. Anal. **9** (2015), no. 4, 234–242, DOI 10.15352/bjma/09-4-12. MR3336891

[35] F. J. García-Pacheco and P. Piniella, *Linear topologies and sequential compactness in topological modules*, Quaest. Math. **40** (2017), no. 7, 897–908, DOI 10.2989/16073606.2017.1334716. MR3740950

[36] F. J. García-Pacheco and P. Piniella, *Geometry of balanced and absorbing subsets of topological modules*, J. Algebra Appl. **18** (2019), no. 6, 1950119, 13, DOI 10.1142/S0219498819501196. MR3954673

[37] D. J. Foulis and M. K. Bennett, *Effect algebras and unsharp quantum logics*, Found. Phys. **24** (1994), no. 10, 1331–1352, DOI 10.1007/BF02283036. Special issue dedicated to Constantin Piron on the occasion of his sixtieth birthday. MR1304942

[38] P. R. Halmos, *Measure Theory*, D. Van Nostrand Co., Inc., New York, 1950. MR33869

[39] J. Hedlíková and S. Pulmannová, *Generalized difference posets and orthoalgebras*, Acta Math. Univ. Comenian. (N.S.) **65** (1996), no. 2, 247–279. MR1451174

[40] H. T. Hwang, L. Li, and H. Kim, *Bounded vector measures on effect algebras*, Bull. Austral. Math. Soc. **72** (2005), no. 2, 291–298, DOI 10.1017/S0004972700035085. MR2183410

[41] P. De Lucia and P. Morales, *Measures on effect algebras*, Atti Sem. Mat. Fis. Univ. Modena **51** (2003), no. 2, 273–293. MR2045074

[42] R. I. Loebl and V. I. Paulsen, *Some remarks on C^*-convexity*, Linear Algebra Appl. **35** (1981), 63–78, DOI 10.1016/0024-3795(81)90266-4. MR599846

[43] F. García Mazarío, *Convergence theorems for topological group valued measures on effect algebras*, Bull. Austral. Math. Soc. **64** (2001), no. 2, 213–231, DOI 10.1017/S0004972700039885. MR1860059

[44] G. J. Murphy, *C^*-algebras and operator theory*, Academic Press, Inc., Boston, MA, 1990. MR1074574

[45] A. Di Nola, A. Dvurečenskij, M. Hyčko, and C. Manara, *Entropy on effect algebras with the Riesz decomposition property. I. Basic properties*, Kybernetika (Prague) **41** (2005), no. 2, 143–160. MR2138765

[46] P. Piniella, *Existence of non-trivial complex unit neighborhoods*, Carpathian J. Math. **33** (2017), no. 1, 107–114. MR3727208

[47] W. Schachermayer, *On some classical measure-theoretic theorems for non-sigma-complete Boolean algebras*, Dissertationes Math. (Rozprawy Mat.) **214** (1982), 33. MR673286

[48] M. H. Stone, *The theory of representations for Boolean algebras*, Trans. Amer. Math. Soc. **40** (1936), no. 1, 37–111, DOI 10.2307/1989664. MR1501865

[49] M. Talagrand, *Propriété de Nikodým et propriété de Grothendieck* (French, with English summary), Studia Math. **78** (1984), no. 2, 165–171, DOI 10.4064/sm-78-2-165-171. MR766712

[50] S. Warner, *Topological rings that generalize Artinian rings*, J. Reine Angew. Math. **293(294)** (1977), 99–108, DOI 10.1515/crll.1977.293-294.99. MR447355

[51] S. Warner, *A new approach to Gel′fand-Mazur theory and the extension theorem*, Michigan Math. J. **26** (1979), no. 1, 13–17. MR514957

[52] S. Warner, *Topological fields*, North-Holland Mathematics Studies, vol. 157, North-Holland Publishing Co., Amsterdam, 1989. Notas de Matemática [Mathematical Notes], 126. MR1002951

[53] S. Warner, *Modern algebra*, 2nd ed., Dover Books on Advanced Mathematics, Dover Publications, Inc., New York, 1990. Two volumes bound as one. MR1068318

[54] S. Warner, *Topological rings*, North-Holland Mathematics Studies, vol. 178, North-Holland Publishing Co., Amsterdam, 1993. MR1240057

[55] S. Warner, *Linearly compact chain rings*, Forum Math. **6** (1994), no. 3, 391–397, DOI 10.1515/form.1994.6.391. MR1269848

[56] H. Weber, *Extension of modular functions and measures*, Ann. Mat. Pura Appl. (4) **197** (2018), no. 3, 761–774, DOI 10.1007/s10231-017-0703-y. MR3802693

[57] J.-D. Wu and Z.-H. Ma, *The Brooks-Jewett theorem on effect algebras with the sequential completeness property*, Czechoslovak J. Phys. **53** (2003), no. 5, 379–383, DOI 10.1023/A:1024046900156. MR2000495

Department of Mathematics, College of Engineering, University of Cadiz, Puerto Real 11519, Spain

Email address: garcia.pacheco@uca.es

Proceedings of Symposia in Pure Mathematics
Volume **106**, 2024
https://doi.org/10.1090/pspum/106/01931

Derivations with Values into Bimodules of τ-measurable Operators

Jinghao Huang and Fedor Sukochev

ABSTRACT. In this paper, we present a survey of recent developments concerning derivations with values in an order ideal of the $*$-algebra of all measurable operators affiliated with a semifinite von Neumann algebra.

1. Introduction

Let $\mathcal{A}$ be a C^*-algebra and let J be an $\mathcal{A}$-bimodule[**103**]. A derivation $\delta : \mathcal{A} \to J$ is a linear mapping satisfying $\delta(xy) = \delta(x)y + x\delta(y)$, $x, y \in \mathcal{A}$. In particular, if $a \in J$, then $\delta_a(x) := xa - ax$ is a derivation. A derivation $\delta : \mathcal{A} \to J$ is said to be inner, if $\delta = \delta_a$ for some $a \in J$. The theory of derivations is a fundamental tool in operator algebras. For instance, Connes[**34, 35**] characterized injective von Neumann algebras by the innerness of derivations into dual normal Banach bimodules.[1]

At a conference held in 1953, Kaplansky asked Singer if he had an idea of what the derivations of $C(X)$ (the algebra of continuous functions on a compact Hausdorff space X) might be. A day later, Singer gave Kaplansky a short, clever argument that such derivations must be all inner, in other words, taking into account that $C(X)$ is commutative C^*-algebra, this means that derivations map $C(X)$ to 0 [**68**] (see also [**104**]). Kaplansky [**78**] had extended Singer's result to AW^*-algebras of type I and his paper generated strong interest in derivations on C^* and W^* algebras with values in their bimodules.

In the course of his argument, Kaplansky proved that each such derivation is (norm-)continuous and conjectured that every derivation $\delta : \mathcal{A} \to \mathcal{A}$ is necessarily continuous for every C^*-algebra $\mathcal{A}$. This conjecture was proved a few years later by Sakai [**99**] and extended by Ringrose [**97**] to derivations of a C^*-algebra into a Banach bimodule. These were among the earliest automatic-continuity results. In

2020 *Mathematics Subject Classification.* Primary 46L52; Secondary 47B07, 46L10, 46B85.
Key words and phrases. Derivation, Murray–von Neumann algebra, symmetric space.
The first author was supported by the NNSF of China (No. 12301160 and 12031004).
The second author's research was supported by the Australian Research Council (FL170100052).

[1]Let $\mathcal{M}$ be a von Neumann algebra. An $\mathcal{M}$-bimodule X is said to be a dual normal X-bimodule if X is a dual space and the maps $m \mapsto mx$ and $m \mapsto xm$ are both ultraweak-weak* continuous from $\mathcal{M}$ into X for each fixed element $x \in X$. Any dual normal $\mathcal{M}$-bimodule can be represented as a weak* closed subspace of the $*$-algebra $B(K)$ of all bounded linear operator on some Hilbert space K[**52**, Theorem 4.1].

[**67, 100**], it was proved that each derivation of a C^*-algebra acting on $\mathcal{H}$ extends to a derivation of the closure (in the strong operator topology) of that algebra, a von Neumann algebra, and that each derivation of a von Neumann algebra is inner. However, as one can easily see, there exist C^*-algebras $\mathcal{A}$ with non-inner derivations $\delta : \mathcal{A} \to \mathcal{A}$, e.g., if $\mathcal{A}$ is the C^*-algebra $K(\mathcal{H})$ of all compact linear operators on $\mathcal{H}$, then for any $a \in B(\mathcal{H}) \setminus (K(\mathcal{H}) + \mathbb{C}\mathbf{1})$, δ_a is a non-inner derivation on $\mathcal{A}$ [**101**, Example 4.1.8]. We quote the following question from [**59**] (see also [**103**, p.60]):

> "*The derivation equation makes sense for linear maps D from the C^*-algebra $\mathfrak{A}$ to a two sided $\mathfrak{A}$-module J. Here again it can be asked if such derivations are inner; that is, are they induced by an element of J as above? In fancier language, the question asks if the cohomology group $H^1(\mathfrak{A}, J)$ is trivial.*"

A similar question was also asked by Johnson [**64**, Section 10.11]) in the setting of groups G:

> "*It would be desirable to identify those spaces $\mathfrak{X}$ with $H^1(G, \mathfrak{X}^*) = 0$ for all G* ".

During the past decades, a number of important special cases have been studied (see e.g. [**3, 9, 22, 64, 66, 69, 70, 85, 95, 98**]). We recall several important results in this direction:

(1) *Derivations from a C^*-algebra $\mathcal{A}$ into any reflexive $\mathcal{A}$-bimodule are inner* [**64**];
(2) *Derivations from a hyperfinite von Neumann algebra $\mathcal{A}$ into any dual normal $\mathcal{A}$-bimodule are inner* [**65**];
(3) *Derivations from a nuclear C^*-algebra $\mathcal{A}$ into a dual Banach $\mathcal{A}$-bimodule are inner* [**57**].

When one considers more general C^*-algebras $\mathcal{A}$ and $\mathcal{A}$-bimodules J, there are examples of non-inner derivations for some specific $\mathcal{A}$ and J (see e.g. [**15, 19, 49–51, 96, 101**]).

Due to the rapid development of noncommutative analysis and partly motivated by questions due to Johnson[**64**] and Hoover[**59**] (see above), there are also a substantial number of articles (see e.g. [**9, 13, 17, 29, 110**]) devoted to various versions of the following question:

QUESTION 1. Let $\mathcal{A}$ be a C^*-subalgebra of a semifinite von Neumann algebra $\mathcal{M}$ and J be an order ideal of the $*$-algebra $S(\mathcal{M})$ of all measurable operators affiliated with $\mathcal{M}$. Is every derivation from $\mathcal{A}$ into J necessarily inner?

When $J = B(\mathcal{H})$ (the algebra of all bounded linear operators on a Hilbert space $\mathcal{H}$), Question 1 above amounts to one of the most famous long-standing open problems (called the Derivation Problem) in the theory of operator algebras. There are results giving affirmative answers to this problem under some additional conditions on the subalgebra $\mathcal{A}$ [**90, 93, 103**]. For those who are interested in this special case, we refer to [**30, 32, 33, 37, 93, 103**]. In the present paper, we mainly consider Question 1 in the setting when J is a symmetric space of τ-compact operators affiliated with a semifinite von Neumann algebra $\mathcal{M}$ (see e.g. [**83, 84**]) and Section 2 below. Experts in the operator theory are probably more familiar with symmetrically normed ideals of compact operators on a separable Hilbert space $\mathcal{H}$[**55**], which are a special case of noncommutative symmetric spaces, when $\mathcal{M}$ coincides with the algebra $B(\mathcal{H})$ of all bounded linear operators on $\mathcal{H}$.

2. Preliminaries

In this section, we recall some notions of the theory of noncommutative integration.

In what follows, $\mathcal{H}$ is a Hilbert space and $B(\mathcal{H})$ is equipped with the uniform norm $\|\cdot\|_\infty$, and $\mathbf{1}$ is the identity operator on $\mathcal{H}$. Let $\mathcal{M}$ be a von Neumann algebra on $\mathcal{H}$. We denote by $\mathcal{P}(\mathcal{M})$ the collection of all projections in $\mathcal{M}$, by $\mathcal{M}'$ the commutant of $\mathcal{M}$ and by $Z(\mathcal{M})$ the center of $\mathcal{M}$. For details on von Neumann algebra theory, the reader is referred to e.g. [**27, 39**], [**71, 72**] or [**108**]. General facts concerning measurable operators may be found in [**87**], [**102**] (see also [**109**, Chapter IX] and the forthcoming book [**47**]). For convenience of the reader, some of the basic definitions are recalled.

2.1. τ-measurable operators and generalized singular value functions. A closed, densely defined operator $x : \mathfrak{D}(x) \to \mathcal{H}$ with the domain $\mathfrak{D}(x)$ is said to be *affiliated* with $\mathcal{M}$ if $yx \subseteq xy$ for all $y \in \mathcal{M}'$, where $\mathcal{M}'$ is the commutant of $\mathcal{M}$. A closed, densely defined operator $x : \mathfrak{D}(x) \to \mathcal{H}$ affiliated with $\mathcal{M}$ is said to be *measurable* if there exists a sequence $\{p_n\}_{n=1}^\infty \subset \mathcal{P}(\mathcal{M})$, such that $p_n \uparrow \mathbf{1}$, $p_n(\mathcal{H}) \subseteq \mathfrak{D}(x)$ and $\mathbf{1} - p_n$ is a finite projection (with respect to $\mathcal{M}$) for all n. The collection of all measurable operators with respect to $\mathcal{M}$ is denoted by $S(\mathcal{M})$, which is a unital $*$-algebra with respect to strong sums and products (denoted simply by $x + y$ and xy for all $x, y \in S(\mathcal{M})$).

Let x be a self-adjoint operator affiliated with $\mathcal{M}$. We denote its spectral measure by $\{e^x\}$. It is well known that if x is an operator affiliated with $\mathcal{M}$ with the polar decomposition $x = u|x|$, then $u \in \mathcal{M}$ and $e \in \mathcal{M}$ for all projections $e \in \{e^{|x|}\}$. Moreover, $x \in S(\mathcal{M})$ if and only if $e^{|x|}(\lambda, \infty)$ is a finite projection for some $\lambda > 0$. It follows immediately that in the case when $\mathcal{M}$ is a von Neumann algebra of type III or a type I factor, we have $S(\mathcal{M}) = \mathcal{M}$. For type II von Neumann algebras, this is no longer true. From now on, let $\mathcal{M}$ be a semifinite von Neumann algebra equipped with a faithful normal semifinite trace τ.

An operator $x \in S(\mathcal{M})$ is called τ-measurable if there exists a sequence $\{p_n\}_{n=1}^\infty$ in $P(\mathcal{M})$ such that $p_n \uparrow \mathbf{1}$, $p_n(\mathcal{H}) \subseteq \mathfrak{D}(x)$ and $\tau(\mathbf{1} - p_n) < \infty$ for all n. The collection $S(\mathcal{M}, \tau)$ of all τ-measurable operators is a unital $*$-subalgebra of $S(\mathcal{M})$. It is well known that a linear operator x belongs to $S(\mathcal{M}, \tau)$ if and only if $x \in S(\mathcal{M})$ and there exists $\lambda > 0$ such that $\tau(e^{|x|}(\lambda, \infty)) < \infty$. Alternatively, an unbounded operator x affiliated with $\mathcal{M}$ is τ-measurable (see [**53**]) if and only if

$$\tau\left(e^{|x|}(n, \infty)\right) \to 0, \quad n \to \infty.$$

A closed operator x affiliated with $\mathcal{M}$ is called *locally measurable* if there exists a sequence $\{z_n\}_{n=1}^\infty$ of central projections in $\mathcal{M}$ such that $z_n \uparrow \mathbf{1}$ and $xz_n \in S(\mathcal{M})$ for any $n \in \mathbb{N}$. The collection of all locally measurable operators with respect to $\mathcal{M}$ is denoted by $LS(\mathcal{M})$, which is a unital $*$-algebra with respect to strong sums and products (denoted simply by $x + y$ and xy for all $x, y \in LS(\mathcal{M})$)[**16, 17**].

For convenience of the reader, we also recall the definition of the measure topology t_τ on the algebra $S(\mathcal{M}, \tau)$. For every $\varepsilon, \delta > 0$, we define the set

$$V(\varepsilon, \delta) = \{x \in S(\mathcal{M}, \tau) : \ \exists p \in \mathcal{P}(\mathcal{M}) \text{ such that } \|x(\mathbf{1} - p)\|_\infty \leq \varepsilon, \ \tau(p) \leq \delta\}.$$

The topology generated by the sets $V(\varepsilon,\delta)$, $\varepsilon,\delta>0$, is called the *measure topology* t_τ on $S(\mathcal{M},\tau)$ [**47, 53, 87**]. It is well known that the algebra $S(\mathcal{M},\tau)$ equipped with the measure topology is a complete metrizable topological algebra. A sequence $\{x_n\}_{n=1}^\infty \subset S(\mathcal{M},\tau)$ converges to zero with respect to measure topology t_τ if and only if $\tau\big(e^{|x_n|}(\varepsilon,\infty)\big)\to 0$ as $n\to\infty$ for all $\varepsilon>0$ [**47**].

Another important vector topology on $S(\mathcal{M},\tau)$ is the *local measure topology.* For convenience we denote by $P_f(\mathcal{M})$ the collection of all τ-finite projections in $\mathcal{M}$, that is, the set of all $e\in\mathcal{P}(\mathcal{M})$ satisfying $\tau(e)<\infty$. A neighbourhood base for this topology is given by the sets $V(\varepsilon,\delta;e)$, $\varepsilon,\delta>0$, $e\in P_f(\mathcal{M})$, where

$$V(\varepsilon,\delta;e)=\{x\in S(\mathcal{M},\tau)\ :\ exe\in V(\varepsilon,\delta)\}.$$

Obviously, local measure topology is weaker than measure topology[**46**]. We note here, that the local measure topology used in the present paper differs from the local measure topology defined in e.g. [**16, 17**].

DEFINITION 2.1. Let $\mathcal{M}$ be a von Neumann algebra equipped with a faithful normal semi-finite trace τ and let $x\in S(\mathcal{M},\tau)$. The generalized singular value function $\mu(x): t\mapsto \mu(t;x)$ of the operator x is defined by setting

$$\mu(s;x)=\inf\{\|xp\|_\infty:\ p\in\mathcal{P}(\mathcal{M}) \text{ with } \tau(\mathbf{1}-p)\le s\}, \forall s\in(0,\infty).$$

An equivalent definition in terms of the distribution function of the operator $|x|$ is the following. for every operator $x\in S(\mathcal{M},\tau)$, setting $d_{|x|}(t)=\tau(e^{|x|}(t,\infty))$, $t>0$, we have (see e.g. [**53**])

$$\mu(t;x)=\inf\{s\ge 0:\ d_{|x|}(s)\le t\}. \tag{2.1}$$

Note that $\mu(x)$ is a function defined on $(0,\infty)$ even if the trace τ is finite. In particular, $\mu(s;x)=0$ when $s\ge\tau(\mathbf{1})$.

Consider the algebra $\mathcal{M}=L^\infty(0,\infty)$ of all Lebesgue measurable essentially bounded functions on $(0,\infty)$. The algebra $\mathcal{M}$ can be seen as an abelian von Neumann algebra acting via multiplication on the Hilbert space $\mathcal{H}=L^2(0,\infty)$, with the trace given by integration with respect to Lebesgue measure m. It is easy to see that the algebra of all τ-measurable operators affiliated with $\mathcal{M}$ can be identified with the subalgebra $S(0,\infty)$ of the algebra of Lebesgue measurable functions $L_0(0,\infty)$ which consists of all functions x such that $m(\{|x|>s\})$ is finite for some $s>0$. It should also be pointed out that the generalized singular value function $\mu(x)$ is precisely the decreasing rearrangement $\mu(x)$ of the function $|x|$ (see e.g. [**79**]) defined by

$$\mu(t;x)=\inf\{s\ge 0:\ m(\{|x|\ge s\})\le t\}.$$

If $\mathcal{M}=B(\mathcal{H})$ (respectively, ℓ_∞) and τ is the standard trace Tr (respectively, the counting measure on $\mathbb{N}$), then it is not difficult to see that $S(\mathcal{M})=S(\mathcal{M},\tau)=\mathcal{M}$. In this case, for $x\in S(\mathcal{M},\tau)$ we have

$$\mu(n;x)=\mu(t;x),\quad t\in[n,n+1),\quad n\ge 0.$$

The sequence $\{\mu(n;x)\}_{n\ge 0}$ is just the sequence of singular values of the operator x.

If $x,y\in S(\mathcal{M},\tau)$, then x is said to be submajorized by y, denoted by $x\prec\prec y$, if

$$\int_0^t \mu(s;x)ds\le\int_0^t\mu(s;y)ds \text{ for all } t\ge 0.$$

In particular, for $x, y \in S(0,\infty)$, $x \prec\prec y$ if and only if $\int_0^t \mu(s;x)ds \le \int_0^t \mu(s;y)ds$, $t \ge 0$.

2.2. Symmetric spaces. A special case of bimodules of τ-measurable operators is given by symmetric spaces in $S(\mathcal{M},\tau)$.

DEFINITION 2.2. A linear subspace E of $S(\mathcal{M},\tau)$ equipped with a complete norm $\|\cdot\|_E$, is called a symmetric space (of τ-measurable operators) if $x \in S(\mathcal{M},\tau)$, $y \in E$ and $\mu(x) \le \mu(y)$ imply that $x \in E$ and $\|x\|_E \le \|y\|_E$.

It is well-known that any symmetric space E is a normed $\mathcal{M}$-bimodule, that is, $axb \in E$ for any $x \in E$, $a, b \in \mathcal{M}$ and $\|axb\|_E \le \|a\|_\infty \|b\|_\infty \|x\|_E$ [**46, 47**].

A symmetric space $E(\mathcal{M},\tau) \subset S(\mathcal{M},\tau)$ is called *strongly symmetric* if its norm $\|\cdot\|_E$ has the additional property that $\|x\|_E \le \|y\|_E$ whenever $x, y \in E(\mathcal{M},\tau)$ satisfy $x \prec\prec y$. In addition, if $x \in S(\mathcal{M},\tau)$, $y \in E(\mathcal{M},\tau)$ and $x \prec\prec y$ imply that $x \in E(\mathcal{M},\tau)$ and $\|x\|_E \le \|y\|_E$, then $E(\mathcal{M},\tau)$ is called *fully symmetric space* (of τ-measurable operators).

If $E \subset S(\mathcal{M},\tau)$ is a symmetric space, then the norm $\|\cdot\|_E$ is called *order continuous* if $\|x_\alpha\|_E \to 0$ whenever $\{x_\alpha\}$ is a downwards directed net in E^+ satisfying $x_\alpha \downarrow 0$. A symmetric space $E(\mathcal{M},\tau)$ is said to have the *Fatou property* if for every upwards directed net $\{x_\beta\}$ in $E(\mathcal{M},\tau)^+$, satisfying $\sup_\beta \|x_\beta\|_E < \infty$, there exists an element $x \in E(\mathcal{M},\tau)^+$ such that $x_\beta \uparrow x$ in $E(\mathcal{M},\tau)$ and $\|x\|_E = \sup_\beta \|x_\beta\|_E$. Examples such as Schatten–von Neumann operator ideals, Lorentz operator ideals, etc. all have symmetric norms which have the Fatou property[**47, 79, 82**].

If $E(\mathcal{M},\tau)$ is a symmetric space, then the carrier projection $c_E \in \mathcal{P}(\mathcal{M})$ is defined by setting

$$c_E = \bigvee \{p : p \in \mathcal{P}(\mathcal{M}),\ p \in E(\mathcal{M},\tau)\}.$$

We remark that, replacing the von Neumann algebra $\mathcal{M}$ by the reduced von Neumann algebra $\mathcal{M}_{c_E}$ (note that $E(\mathcal{M},\tau) = c_E E(\mathcal{M},\tau) c_E$, see e.g. [**46**, Corollary 6]), it is often assumed that the carrier projection of $E(\mathcal{M},\tau)$ is equal to $\mathbf{1}$.

If $E(\mathcal{M},\tau)$ is a symmetric space, then the Köthe dual $E(\mathcal{M},\tau)^\times$ of $E(\mathcal{M},\tau)$ is defined by

$$E(\mathcal{M},\tau)^\times = \{x \in S(\mathcal{M},\tau) : \sup_{\|y\|_E \le 1, y \in E} \tau(|xy|) < \infty\},$$

and for every $x \in E(\mathcal{M},\tau)^\times$, we set $\|x\|_{E^\times} = \sup\{\tau(|yx|) : y \in E(\mathcal{M},\tau),\ \|y\|_E \le 1\}$ (see e.g. [**46**, Section 5.2], see also [**42, 83**]). It is well-known that $\|\cdot\|_{E^\times}$ is a norm on $E(\mathcal{M},\tau)^\times$ if and only if the carrier projection c_E of $E(\mathcal{M},\tau)$ is equal to $\mathbf{1}$. In this case, for a strongly symmetric space $E(\mathcal{M},\tau)$, the Köthe dual $E(\mathcal{M},\tau)^\times$ is a strongly symmetric space and the following statements are equivalent [**44–46**].

- $E(\mathcal{M},\tau)$ has the Fatou property.
- $E(\mathcal{M},\tau)^{\times\times} (:= (E(\mathcal{M},\tau)^\times)^\times) = E(\mathcal{M},\tau)$ and $\|x\|_E = \|x\|_{E^{\times\times}}$ for all $X \in E(\mathcal{M},\tau)$.
- The norm closed unit ball B_E of $E(\mathcal{M},\tau)$ is closed in $S(\mathcal{M},\tau)$ with respect to the local measure topology .

If $E(\mathcal{M},\tau)$ is a strongly symmetric space with $c_E = \mathbf{1}$ (or a symmetric space affiliated with a semifinite von Neumann algebra which is either atomless or atomic with all minimal projections having equal trace), then [**46**, Lemma 30] (see also

[**47**, Theorem IV.5.7] and [**79**])

$$\|x\|_E = \|x\|_{E^{\times\times}}, \; x \in (L_1 \cap L_\infty)(\mathcal{M}, \tau), \tag{2.2}$$

and

$$L_1 \cap L_\infty(\mathcal{M}, \tau) \subset E(\mathcal{M}, \tau) \subset (L_1 + L_\infty)(\mathcal{M}, \tau). \tag{2.3}$$

A wide class of symmetric operator spaces associated with the von Neumann algebra $\mathcal{M}$ can be constructed from concrete symmetric function spaces studied extensively in e.g. [**10, 79**]. Let $(E(0,\infty), \|\cdot\|_{E(0,\infty)})$ be a symmetric function space on the semi-axis $(0,\infty)$. The pair

$$E(\mathcal{M}, \tau) = \{x \in S(\mathcal{M}, \tau) : \mu(x) \in E(0,\infty)\}, \quad \|x\|_{E(\mathcal{M},\tau)} := \|\mu(x)\|_{E(0,\infty)}$$

is a symmetric operator space affiliated with $\mathcal{M}$ with $c_E = \mathbf{1}$ [**76**] (see also [**83, 106, 107**]). For brevity, we denote $\|\cdot\|_{E(\mathcal{M},\tau)}$ by $\|\cdot\|_E$.

2.3. The ideal of τ-compact operators. For a self-adjoint operator $x \in S(\mathcal{M}, \tau)$, we denote by $s(x)$ its support. The two-sided ideal $\mathcal{F}(\mathcal{M}, \tau)$ in $\mathcal{M}$ consisting of all elements of τ-finite range is defined by setting

$$\mathcal{F}(\mathcal{M}, \tau) = \{x \in \mathcal{M} \; : \; \tau(s(|x|)) < \infty\}.$$

The C^*-algebra $C_0(\mathcal{M}, \tau)$ of all τ-compact bounded operators can be described as the closure in the norm $\|\cdot\|_\infty$ of the linear span of all τ-finite projections [**83**, Definition 2.6.8]. Equivalently, $C_0(\mathcal{M}, \tau)$ the set of all elements $x \in \mathcal{M}$ such that $\tau(e^{|x|}(\lambda,\infty)) < \infty$ for every $\lambda > 0$ (see e.g. [**47**, Chapter II, Section 4], [**105**]). The space $C_0(\mathcal{M}, \tau)$ is associated to the ideal of essentially bounded functions vanishing at infinity (see [**83**, Lemma 2.6.9]), that is,

$$C_0(\mathcal{M}, \tau) = \Big\{a \in S(\mathcal{M}, \tau) : \mu(a) \in L_\infty(0,\infty), \; \mu(\infty; a) := \lim_{t\to\infty} \mu(t; a) = 0\Big\}.$$

In particular, if τ is finite, then $\mathcal{M} = C_0(\mathcal{M}, \tau)$ (see e.g. [**83**, Page 64]); if $\mathcal{M} = B(\mathcal{H})$ and τ is the standard trace Tr, then $C_0(\mathcal{M}, \tau) = K(\mathcal{H})$. The $*$-algebra $S_0(\mathcal{M}, \tau)$ of all τ-compact operators is the space associated to the algebra of functions from $S(0,\infty)$ vanishing at infinity, that is,

$$S_0(\mathcal{M}, \tau) = \{x \in S(\mathcal{M}, \tau) : \; \mu(\infty; x) = 0\}.$$

This is a two-sided ideal in $S(\mathcal{M}, \tau)$ and, clearly, $C_0(\mathcal{M}, \tau) = S_0(\mathcal{M}, \tau) \cap \mathcal{M}$.

3. Derivations with values in an ideal of a von Neumann algebra

In this section, we consider the following special version of Question 1.

QUESTION 2. Assume that $\mathcal{M}$ is a von Neumann algebra. Let $\mathcal{E}$ be an ideal in $\mathcal{M}$ and let $\mathcal{A}$ be an arbitrary C^*-subalgebra of $\mathcal{M}$. Is every derivation δ from $\mathcal{A}$ into $\mathcal{E}$ necessarily inner?

Various versions of Question 2 for derivations with values in ideals of a von Neumann algebra were stated and discussed in [**19, 20, 59, 60, 66, 75, 95, 96**].

3.1. The case when $\mathcal{A} = \mathcal{M}$. In 1966, Kadison [**67**] and Sakai [**100**] gave an affirmative answer to Question 2 the special case when the algebra $\mathcal{A}$ and the ideal $\mathcal{E}$ of $\mathcal{M}$ coincide with the algebra $\mathcal{M}$ itself. In 2012, Ber and Sukochev [**25, 26**] established commutator estimates in von Neumann algebras, which have been further improved in [**21**].

THEOREM 3.1. [**21**] *Let $\mathcal{M}$ be a von Neumann algebra and let $a = a^* \in LS(\mathcal{M})$. Then, for any $\varepsilon > 0$, there exists $u_\varepsilon = u_\varepsilon^* \in U(\mathcal{M})$ and $c_a \in Z(LS(\mathcal{M}))$ such that such that*

$$|[a, u_\varepsilon]| \geq (1 - \varepsilon)(|a - c_a| + u_\varepsilon|a - c_a|u_\varepsilon).$$

If $\mathcal{M}$ is a finite von Neumann algebra or else a purely infinite σ-finite von Neumann algebra, then there exists $u_0 = u_0^* \in U(\mathcal{M})$ such that

$$|[a, u_0]| = |a - c_a| + u_0|a - c_a|u_0.$$

As a direct consequence of Theorem 3.1, it is shown in [**26**] that when Question 2 has an affirmative answer in the special case when $\mathcal{A} = \mathcal{M}$, and for any (not necessarily Banach) ideal of $\mathcal{M}$, which generalized the Kadison–Sakai Theorem significantly.

THEOREM 3.2. [**26**] *Let $\mathcal{M}$ be a von Neumann algebra and J be an arbitrary ideal of $\mathcal{M}$. Every derivation from $\mathcal{M}$ into J is inner.*

3.2. The case when $\mathcal{A}$ is a von Neumann subalgebra of $\mathcal{M}$. In 1972, Johnson and Parrot[**66**] initiated the study of derivations with values in ideals of a von Neumann algebra by considering the special case of Question 2 when the range of δ is contained in $K(\mathcal{H})$, the ideal of all compact linear operators on $\mathcal{H}$. In that paper, the authors proved that if $\mathcal{A}$ is an abelian von Neumman subalgebra of $\mathcal{M}$, then every derivation δ from $\mathcal{A}$ to $K(\mathcal{H})$ is automatically inner. As an easy consequence, they were also able to treat the case when $\mathcal{A}$ has no certain type II_1 factors as direct summands. The remaining case, when $\mathcal{A}$ is a von Neumann algebra of type II_1 was later resolved by Popa [**95**]. The results in [**66**] and [**95**] can be therefore formulated as follows.

THEOREM 3.3 (Johnson–Parrott–Popa). *Let $\mathcal{A}$ be a von Neumann subalgebra of $B(\mathcal{H})$. Then, every derivation*

$$\delta : \mathcal{A} \to K(\mathcal{H})$$

is inner.

This result may be well seen as a precursor to Question 2 in the setting of symmetrically normed ideals in $\mathcal{M} = B(\mathcal{H})$. Note that the condition that $\mathcal{A}$ is a von Neumann algebra can not be relaxed to the setting of a C^*-algebra as shown in Section 1 and [**101**, Example 4.1.8].

A natural development of the Johnson–Parrott–Popa theorem is to establish a suitable semifinite version of the result. Kaftal[**73, 74**] introduced the $*$-ideal $J(\mathcal{M})$ of compact operators in a semifinite von Neumann algebra, where $J(\mathcal{M})$ is the ideal of $\mathcal{M}$ generated by all (algebraically) finite projections in a semifinite von Neumann algebra $\mathcal{M}$. In 1985, Kaftal and Weiss [**75**] proved that if $\mathcal{A}$ is an abelian (or properly infinite) von Neumann subalgebra of $\mathcal{M}$ containing the center $Z(\mathcal{M})$ of $\mathcal{M}$, then any derivation $\delta : \mathcal{A} \to J(\mathcal{M})$ is inner. This result was later extended to the case when the type I_{fin} direct sum of $\mathcal{A}$ is *locally compatible* with the center of $\mathcal{M}$ [**96**] (in particular, Popa and Radulescu resolved the case when $\mathcal{A}$ is a type II_1 von Neumann subalgebra of $\mathcal{M}$, which was left untreated in [**75**]). Recall that a $*$-subalgebra $\mathcal{N}$ of $\mathcal{M}$ is locally compatible with the center $Z(\mathcal{M})$ of $\mathcal{M}$, if there exists a partition of the unity $\{p_i\}_{i\in I}$ in the center $Z(\mathcal{N})$ of $\mathcal{N}$ such that for each i, we have either

$$Z(\mathcal{N})p_i \subset Z(\mathcal{M})p_i \text{ or } Z(\mathcal{M})p_i \subset Z(\mathcal{N})p_i.$$

THEOREM 3.4. [**96**] *Let $\mathcal{M}$ be a semifinite von Neumann algebra. Let $\mathcal{A} \subset \mathcal{M}$ be a weakly closed $*$-subalgebra of $\mathcal{M}$ and suppose the finite type I summand of $\mathcal{N}$ is locally compatible with $Z(\mathcal{M})$. Then, for any derivation $\delta : \mathcal{A} \to J(\mathcal{M})$, there exists $k \in J(\mathcal{M})$, $\|k\|_\infty \leq 2\,\|\delta\|_{\mathcal{A}\to J(\mathcal{M})}$ with $\delta = \delta_k$.*

In particular, every derivation $\delta : \mathcal{A} \to J(\mathcal{M})$ is inner if $\mathcal{A}$ is of type II_1 or properly infinite. Popa and Rădulescu also established the existence of non-inner derivations $\delta : \mathcal{A} \to J(\mathcal{M})$ for some specific semifinite von Neumann algebra $\mathcal{M}$ and abelian von Neumann subalgebra $\mathcal{A}$ of $\mathcal{M}$ (see [**96**, Theorem 1.2 and Section 8]), which is somewhat unexpected and yields the first example of non-inner derivations in von Neumann algebras.

EXAMPLE 3.5. let $\mathcal{M} = L_\infty([0,1],\lambda)\overline{\otimes}B(L_2(\mathbb{T},\mu))$, $\mathcal{A} = \mathbf{1} \otimes L_\infty(\mathbb{T},\mu) \subset \mathcal{M}$, where μ is the Lebesgue measure on the torus $\mathbb{T}$ and λ is the Lebesgue measure on the unit interval $[0,1]$. There exists an operator $a \in \mathcal{M}$ such that δ_a is a derivation from $\mathcal{A}$ into $J(\mathcal{M})$ which is not inner, i.e., there exists no elements $k \in J(\mathcal{M})$ such that $\delta_a = \delta_k$.

Another semifinite version of the Johnson–Parrott–Popa theorem was initiated by Christensen [**31**]. The notion of the so-called generalized compacts $C(\mathcal{M})$ associated with a von Neumann algebra $\mathcal{M}$ was introduced by Wright [**111**] in 1984. Let $x \in B(\mathcal{H})$ and let $\mathcal{M}$ be a von Neumann algebra. Then x is said to be a *weakly compact multiplier* of $\mathcal{M}$ if the operators r_x and l_x of $\mathcal{M}$ into $B(\mathcal{H})$, given by $r_x m = m$ and $l_x m = xm$, $m \in \mathcal{M}$, are weakly compact. The so-called generalized compacts $C(\mathcal{M})$ associated with $\mathcal{M}$ is the set of weakly compact multipliers of $\mathcal{M}$. If $\mathcal{M} = B(\mathcal{H})$, then the set $C(\mathcal{M})$ coincides with $K(\mathcal{H})$.

Christensen [**31**] showed that every derivation from $\mathcal{M}$ into $C(\mathcal{M})$ is inner if $\mathcal{M}$ is a properly inifinite von Neumann algebra. However, the case when $\mathcal{M}$ is a type II_1 von Neumann algebra was left open in [**31**] and was resolved by Galatan and Popa [**54**] only in 2017.

For a general semifinite von Neumann algebra $\mathcal{M}$, there is another natural notion of compactness, so-called τ-compactness, which comes from a semifinite trace τ defined on the algebra $\mathcal{M}$. Hence, it is natural to consider a problem, similar to that considered in [**96**], with the ideal $J(\mathcal{M})$ of all compact operators replaced by the ideal $C_0(\mathcal{M},\tau)$ of all τ-compact operators affiliated with $\mathcal{M}$ (see Section 2 for precise definitions).

THEOREM 3.6 ([**20**]). *Assume that $\mathcal{M}$ is a semifinite von Neumann algebra equipped with a faithful normal semifinite trace τ. Every derivation from an arbitrary von Neumann subalgebra $\mathcal{A}$ of $\mathcal{M}$ into $C_0(\mathcal{M},\tau)$ is inner.*

Although, the ideals $C_0(\mathcal{M},\tau)$ and $J(\mathcal{M})$ are quite similar in many respects (see [**20**]), the result of Theorem 3.6 is in strong contrast with Example 3.5, which is somewhat unexpected. Namely, the additional assumption on the type I summand of the von Neumann subalgebra $\mathcal{A}$ which plays an important role in [**96**] could be dispensed with in the setting of $C_0(\mathcal{M},\tau)$. We note that the assumption that $\mathcal{A}$ is von Neumann subalgebra of $\mathcal{M}$ is sharp, that is, this assumption can not be relaxed to the assumption that $\mathcal{A}$ is a C^*-subalgebra of $\mathcal{M}$ for derivations into $J(\mathcal{M})$ or into $C_0(\mathcal{M},\tau)$ as the Johnson–Parrott–Popa theorem can not be extended to the setting of a C^*-algebra.

Let $\mathcal{M}$ be a von Neumann algebra equipped with a semifinite faithful normal trace τ. Denote by $L_p(\mathcal{M},\tau)$, $p \geq 1$, the noncommutative L_p-space affiliated with $\mathcal{M}$[**94**], and denote $C_p(\mathcal{M},\tau) := L_p(\mathcal{M},\tau) \cap \mathcal{M}$. In general, $C_p(\mathcal{M},\tau)$ is not reflexive even for $1 < p < \infty$ and therefore the Ryll-Nardzewski fixed point theorem can not be applied directly (the method used in [**9**] or [**92**] is not applicable either because, in general, $C_p(\mathcal{M},\tau)$ is not an L-embedded space). In 1985, using Johnson and Parrott's trick [**66**], Kaftal and Weiss [**75**] showed that every derivation from an abelian (or properly infinite) von Neumann subalgebra of $\mathcal{M}$ into the C_p ideal of $\mathcal{M}$ is inner when $1 \leq p < \infty$. However, the cases when the $\mathcal{M}$ von Neumann subalgebra is of type I or type II_1 were left unresolved in [**75**], which led to the following question.

QUESTION 3. Let $\mathcal{M}$ be a semifinite von Neumann algebra equipped with a semifinite faithful normal trace τ and let $p \in [1,\infty)$. Assume that $\mathcal{A}$ is an arbitrary von Neumann subalgebra of $\mathcal{M}$. Is every derivation

$$\delta : \mathcal{A} \to C_p(\mathcal{M},\tau)$$

automatically inner?

As an application of Theorem 3.6, it was shown in [**20**] that Question 2 has a positive answer for a very extensive class of symmetric ideals $\mathcal{E}$ of $\mathcal{M}$ by using techniques from noncommutative integration, which, in particular, has allowed to fully resolve Question 3 (see Theorem 3.9 for an alternative resolution).

THEOREM 3.7. [**20**, Theorem 1.3] *Assume that $\mathcal{M}$ is a semifinite von Neumann algebra equipped with a faithful normal semifinite trace τ. If $E(\mathcal{M},\tau)$ is a strongly symmetric space of τ-measurable operators with the Fatou property, then every derivation δ from an arbitrary von Neumann subalgebra $\mathcal{A}$ of $\mathcal{M}$ into $E(\mathcal{M},\tau) \cap C_0(\mathcal{M},\tau)$ is necessarily inner.*

Theorem 3.7 above generalizes results due to Kaftal and Weiss [**75**] in two directions. Firstly, all additional conditions imposed on the von Neumann subalgebra $\mathcal{A}$ of $\mathcal{M}$ in [**75**] are removed. Secondly, Theorem 3.7 extends significantly the class of symmetric ideals associated with $\mathcal{M}$ for which the result is applicable. We point out that the Fatou property is a fascinating analogue of the dual normal property: let $E(\mathcal{M},\tau)$ be a strongly symmetric space having the Fatou property such that the carrier projection is the identity. Then, $E(\mathcal{M},\tau)^{\times\times} = E(\mathcal{M},\tau)$. In particular, $E(\mathcal{M},\tau)$ has a predual in the sense of Köthe duality. Let $\{m_i\}_i$ be a net ultraweakly converging to m in $\mathcal{M}$. Then, for any $x \in E(\mathcal{M},\tau)$ and $y \in E(\mathcal{M},\tau)^{\times}$, we have

$$\tau(m_i x y)$$

converges to $\tau(mxy)$. That is, in the $\sigma(E(\mathcal{M},\tau)^{\times\times}, E(\mathcal{M},\tau)^{\times})$-topology, $\{m_i x\}_i$ converges in $E(\mathcal{M},\tau)$.

It is well-known that every derivation from a hyperfinite von Neumann algebra $\mathcal{A}$ into a dual normal $\mathcal{A}$-bimodule is inner (see Section 1 or [**103**, Theorem 2.4.3]). However, Theorem 3.7 is applicable for any von Neumann subalgebra $\mathcal{A}$ of $\mathcal{M}$.

Actually, the majority of Banach symmetric function spaces used in analysis are strongly symmetric, and most of them have the Fatou property. Theorem 3.7 above give an affirmative answer to Question 2 for 'almost' every **proper** symmetric ideal $\mathcal{E}$ in $\mathcal{M}$, that is, for most proper symmetric ideals $\mathcal{E}$ in $\mathcal{M}$, derivations from an arbitrary von Neumann subalgebra of $\mathcal{M}$ into $\mathcal{E}$ are automatically inner. One should note that the class of symmetric ideals characterized in [**20**] covers almost every ideal $\mathcal{E}$ corresponding to a symmetric function space in the sense of Calkin (see [**76, 83**]). In the meantime, the ideal $\mathcal{J}(\mathcal{M})$ of all compact operators in $\mathcal{M}$ is not corresponding to any symmetric function space whenever $\mathcal{M} \neq \mathcal{J}(\mathcal{M}) \neq \mathcal{C}_0(\mathcal{M},\tau)$ (see e.g. [**20**] or [**96**, Section 8]).

QUESTION 4. Does there exist a semifinite von Neumann algebra $\mathcal{M}$ with a von Neumann subalgebra $\mathcal{A}$ and a symmetric ideal $\mathcal{E}$ in $\mathcal{M}$ with a non-inner derivation $\delta : \mathcal{A} \to \mathcal{E}$. Obviously, such an ideal should fail the Fatou property and $\mathcal{E} \neq C_0(\mathcal{M},\tau)$, see Theorem 3.7.

Note that any von Neumann algebra is a Grothendieck space[**91**], and any bounded linear operator from a Grothendieck space into a separable Banach space (more generally, a weakly compactly generated Banach space) is weakly compact[**38**]. Therefore, by Ringrose' theorem[**97**], any derivation from a von Neumann subalgebra $\mathcal{A}$ of a semifinite von Neumann subalgebra $\mathcal{M}$ into a separable symmetric space $E(\mathcal{M},\tau)$ is weakly compact. This observation may prove to be useful for future work on Question 4.

3.3. The case when $\mathcal{A}$ is a C^*-subalgebra of $\mathcal{M}$. There are some other extensions of the result by Johnson and Parrott [**66**]. In 1977, using the Ryll-Nardzewski fixed point theorem (as suggested by Johnson [**64, 66**], see Theorem 4.3 below) and the reflexivity of the Schatten–von Neumann p-class $C_p(\mathcal{H})$, $1 < p < \infty$, Hoover [**59**] obtained the following result where the algebra $K(\mathcal{H})$ in the Johnson–Parrott–Popa theorem was replaced by the Schatten p-class. Hoover also resolved the special case for the trace class $C_1(\mathcal{H})$ by a p-convexification technique [**59**], see e.g. [**43, 112**] for detailed expositions concerning p-convexification of a symmetric spaces.

THEOREM 3.8. [**59**] *Let $\mathcal{A}$ be a unital C^*-subalgebra of $B(\mathcal{H})$ and let $C_p(\mathcal{H})$ be the Schatten p-class, $1 \leq p < \infty$. If $\delta : \mathcal{A} \to C_p(\mathcal{H})$ is a derivation, then there exists a $a \in C_p(\mathcal{H})$ such that $\delta = \delta_a$.*

Ideas used by Hoover in the proof for the special case when $p = 1$ (the ideal coincides with the predual of $B(\mathcal{H})$) are subtle and the proof relies on the fact that $C_1(\mathcal{H}) \subset C_p(\mathcal{H})$ for any $p \geq 1$. However, when $B(\mathcal{H})$ is replaced by a general semifinite von Neumann algebra $\mathcal{M}$, the corresponding ideal $C_p(\mathcal{M},\tau)$ is not necessarily a reflexive space and therefore the Ryll-Nardzewski fixed point theorem can not be applied directly. Recall that Kaftal and Weiss [**75**] obtained the result concerning derivations from an abelian von Neumann subalgebra into an C_p-ideal $C_p(\mathcal{M},\tau)$, $1 \leq p < \infty$. In our recent paper co-authored with Ber and Levitina[**19**], we obtain the following extension of results due to Hoover[**59**], and Kaftal and Weiss[**75**]. This provides an alternative solution to Question 3.

Recall that, if a noncommutative symmetric space $E(\mathcal{M},\tau)$ has the Fatou property and order continuous norm, then it is said to be a *KB*-space (or Kantorovich–Banach space)[**46**].

THEOREM 3.9. [**19**] *Let $\mathcal{M}$ be a von Neumann algebra equipped with a semifinite faithful normal trace τ and let $\mathcal{A}$ be an arbitrary C^*-subalgebra of $\mathcal{M}$. Let $E(0,\infty)$ be a symmetric KB-function space. Then, every derivation δ from $\mathcal{A}$ into $E(\mathcal{M},\tau)\cap\mathcal{M}$ is automatically inner. Precisely speaking, there exists $a\in E(\mathcal{M},\tau)\cap\mathcal{M}$ such that $\delta=\delta_a$ with $\|a\|_\infty\le\|\delta\|_{\mathcal{A}\to\mathcal{M}}$ and $\|a\|_E\le\|\delta\|_{\mathcal{A}\to E}$.*

Let us briefly compare Theorem 3.7 with Theorem 3.9. In Theorem 3.9, the derivations act on a larger class of algebras (i.e. C^*-subalgebras of $\mathcal{M}$) but the result is only applicable for symmetric operator spaces $E(\mathcal{M},\tau)$ corresponding to Kantorovich–Banach symmetric function spaces $E(0,\infty)$, that is, the assumption in [**19**] that the norm of $E(0,\infty)$ is order continuous is not required in Theorem 3.7.

4. Derivations with values into symmetric spaces (of possibly unbounded operators)

4.1. Derivations into noncommutative L_p-spaces and symmetric spaces. In this section, we collect some results on derivations into more general bimodules.

The so-call *Ryll-Nardzewski Fixed Point Theorem* is paramount to this area of study of derivations. Before introducing the Ryll-Nardzewski fixed point theorem, we recall the definition of *noncontracting family of maps.*

DEFINITION 4.1 ([**36**, Definition 10.6]). Let $\mathfrak{X}$ be a locally convex space and let Q be a nonempty subset of $\mathfrak{X}$. If $\mathcal{G}$ is a family of maps (not necessarily linear) of Q into Q, then $\mathcal{G}$ is said to be a *noncontracting* family of maps if for any two distinct points x and y in Q, 0 is not in the closure of

$$\{T(x)-T(y):T\in\mathcal{G}\}.$$

THEOREM 4.2 (Ryll-Nardzewski Fixed Point Theorem [**36**, Theorem 10.8],[**88**]). *If $\mathfrak{X}$ is a locally convex space, Q is a weakly compact convex subset of $\mathfrak{X}$, and $\mathcal{G}$ is a noncontracting semigroup of weakly continuous affine maps of Q into Q, then there is a point x_0 in Q such that $T(x_0)=x_0$ for every $T\in\mathcal{G}$.*

The following theorem, due to Johnson [**64**] (see also [**19, 24, 59, 110**]), is a straightforward application of Theorem 4.2. We present the full proof for completeness of exposition. Our proof follows the line of thought from [**24, 59, 64, 110**] with some minor modifications.

THEOREM 4.3. *Let $\mathcal{A}$ be a C^*-algebra and let J be a reflexive $\mathcal{A}$-bimodule. Then, for every derivation $\delta:\mathcal{A}\to J$, there exists a $T\in J$ such that $\delta=\delta_a$ and $\|a\|_J\le\|\delta\|_{\mathcal{A}\to J}$. Moreover, $a\in\overline{\mathrm{co}\{\delta(u)u^*:\ u\in\mathcal{U}(\mathcal{A})\}}^{\|\cdot\|_J}$, where $\mathrm{co}(S)$ denotes the convex hull of a set S and $\mathcal{U}(\mathcal{A})$ is the group of all unitary elements in $\mathcal{A}$.*

PROOF. We firstly consider the case when $\mathcal{A}$ is unital.

By Ringrose's theorem [**97**], the derivation $\delta:(\mathcal{A},\|\cdot\|_\infty)\to(J,\|\cdot\|_J)$ is bounded. Let us define the sets $K_{00}:=\{\delta(u)u^*:\ u\in\mathcal{U}(\mathcal{A})\}\subset J$ and $K_0:=\mathrm{co}(K_{00})$. It is clear that K_{00}, and therefore K_0, lie in the ball of radius $\|\delta\|_{\mathcal{A}\to J}$ in J.

We set $K := \overline{K_0}^{\|\cdot\|_J}$. Note that K is weakly closed in J (see e.g. [**36**, Chapter V, Theorem 1.4]). Since J is reflexive, it follows that K is a convex weakly compact subset of J (see [**36**, Chapter V, Theorem 4.2]), contained in the ball of radius $\|\delta\|_{\mathcal{A}\to J}$.

For every $u \in \mathcal{U}(\mathcal{A})$, we have $\delta(u) \in J$, and therefore we can define the mapping $\alpha_U : J \longrightarrow J$, by setting

$$\alpha_u(x) := uxu^* + \delta(u)u^*.$$

For every $u, v \in \mathcal{U}(\mathcal{A})$, we have

$$\begin{aligned}\alpha_u(\alpha_v(x)) &= uvxv^*u^* + u\delta(v)v^*u^* + \delta(u)u^* \\ &= (uv)x(uv)^* + u\delta(v)v^*u^* + \delta(u)vv^*u^* \\ &= (uv)x(uv)^* + \delta(uv)(uv)^* = \alpha_{uv}(x).\end{aligned}$$

In addition, the equality $\delta(\mathbf{1}) = \delta(\mathbf{1}^2) = 2\delta(\mathbf{1})$ implies that $\delta(\mathbf{1}) = 0$, and therefore $\alpha_{\mathbf{1}}(x) = x$, $x \in J$. Thus, α is an action of the group $\mathcal{U}(\mathcal{A})$ on J.

We claim that the set K is invariant with respect to α. Since $\delta(u)u^* = \alpha_u(0)$, it follows that K_{00} is an orbit of 0 with respect to α, and therefore, is an invariant subset with respect to α. In addition, for any positive scalars s and t with $s+t = 1$, we have

$$\begin{aligned}\alpha_u(sx + ty) &= suxu^* + tuyu^* + (s+t)\delta(u)u^* \\ &= s\alpha_u(x) + t\alpha_u(y), \qquad \forall x, y \in J.\end{aligned}$$

Hence, for every $u \in \mathcal{U}(\mathcal{A})$ the mapping α_u is affine, which implies that $K_0 = \mathrm{co}(K_{00})$ is also an invariant subset with respect to α. Now, the equality $\alpha_u(x) - \alpha_u(y) = u(x-y)u^*$, $x, y \in J$ implies that every $\alpha_u, u \in \mathcal{U}(\mathcal{A})$, is an isometry on J. Hence, K is an invariant subset with respect to α.

Furthermore, the fact that α_U is an isometry implies that the family $\mathcal{G} := \{\alpha_u : u \in \mathcal{U}(\mathcal{A})\}$ is a noncontracting family of affine mappings (see e.g. [**36**, Chapter V, Lemma 10.7]). Clearly, α_u is weakly continuous for every $U \in \mathcal{U}(\mathcal{A})$. Thus, the set k and the family $\mathcal{G}$ satisfy the assumptions of Theorem 4.2. Hence, there exists a point $a \in K$ fixed with respect to α, that is, we have $t = \alpha_u(t) = utu^* + \delta(u)u^*$ for every $u \in \mathcal{U}(\mathcal{A})$ and $\|a\|_J \le \|\delta\|_{\mathcal{A}\to J}$. Therefore $tu = ut + \delta(u)$ for every $u \in \mathcal{U}(\mathcal{A})$. thus, $\delta(u) = [a, u]$ for every $u \in \mathcal{U}(\mathcal{A})$. Since every element $X \in \mathcal{A}$ is a linear combination of four elements from $\mathcal{U}(\mathcal{A})$, we obtain that $\delta = \delta_a$ on $\mathcal{A}$ and complete the proof for the case when $\mathcal{A}$ is unital.

Next, if $\mathcal{A}$ is not unital, then we define $\mathcal{A}_1 := \mathbb{C} \oplus \mathcal{A}$, which is a unital C^*-algebra equipped with the norm

$$\|(\lambda, a)\|_1 = |\lambda| + \|a\|_{\mathcal{A}}, \ \lambda \in \mathbb{C}, a \in \mathcal{A}.$$

For every $x \in J$ and $a_1 = (\alpha, a), b_1 = (\beta, b) \in \mathcal{A}$, we define

$$a_1 x = (\alpha, a)x = \alpha x + ax,$$

and

$$xb_1 = (\beta, b)x = \beta x + xb.$$

Define $\delta_1 : \mathcal{A}_1 \to J$ by $\delta_1((\alpha, a)) = \delta(a)$, $a \in \mathcal{A}$ and $\alpha \in \mathbb{C}$. It is clear that δ_1 is a derivation from a unital C^*-subalgebra $\mathcal{A}_1$ into J. Hence, the assertion follows from the proved above. $\square$

Throughout this section, we assume that $\mathcal{M}$ is a semifinite von Neumann algebra equipped with a semifinite faithful normal trace τ. As a corollary of Theorem 4.3, we immediately obtain the following result.

COROLLARY 4.1. *Let $\mathcal{A}$ be a C^*-subalgebra of $\mathcal{M}$ and let $E(\mathcal{M},\tau)$ be a reflexive symmetric space. Then, for every derivation $\delta : \mathcal{A} \to E(\mathcal{M},\tau)$, there exists a $a \in E(\mathcal{M},\tau)$ such that $\delta = \delta_T$ and $\|a\|_E \le \|\delta\|_{\mathcal{A}\to E}$. Moreover, $a \in \overline{\mathrm{co}\{\delta(u)u^* : u \in \mathcal{U}(\mathcal{A})\}}^{\|\cdot\|_E}$.*

Noting that $L_p(\mathcal{M},\tau)$ is reflexive when $p > 1$, we obtain the following corollary.

COROLLARY 4.2. *Let $\mathcal{A}$ be a unital C^*-subalgebra of $\mathcal{M}$ and $\delta : \mathcal{A} \to L_p(\mathcal{M},\tau)$, $p > 1$, be a derivation. Then, there exists an element $T \in L_p(\mathcal{M},\tau)$ such that $\delta = \delta_T$ and $\|T\|_{L_p} \le \|\delta\|_{\mathcal{A}\to L_p}$.*

One should note that Corollary 4.2 does not recover in full the result by Kaftal and Weiss[**75**] since the derivation δ in Corollary 4.2 takes the values in the symmetric space $L_p(\mathcal{M},\tau)$ (of possibly unbounded operators), while Kaftal and Weiss showed (under some additional assumptions on the von Neumann subalgebra $\mathcal{A}$) that every derivation into the symmetric ideal $C_p(\mathcal{M},\tau) := L_p(\mathcal{M},\tau)\cap\mathcal{M}$ is inner, and so there exists a uniformly bounded operator $T \in L_p(\mathcal{M},\tau)$, such that $\delta = \delta_T$.

It is well-known that $L_1(\mathcal{M},\tau)$ is not reflexive unless $\mathcal{M}$ is finite-dimensional. On the other hand, $L_1(\mathcal{M},\tau)$ is not necessarily a dual space. It was a long-standing open question whether every derivation a C^*-subalgebra of $\mathcal{M}$ into $L_1(\mathcal{M},\tau)$ is inner (see e.g. [**29**]). The special case when $\mathcal{M} = B(\mathcal{H})$ was resolved by Hoover (see Theorem 3.8). The case when $\mathcal{M}$ is a finite von Neumann algebra was resolved by Bunce and Paschke [**29**]. Using a noncommutative version of the Grothendieck inequality, Haagerup proved that the derivations from a von Neumann algebra into its predual are inner[**57**, Theorem 4.1]. Recently, the question was resolved completely by Bader, Gelander and Monod [**9**].

Since the unit ball of L_1 is not weakly compact, the Ryll-Nardzewski Fixed Point Theorem can not be applied directly. Bader, Gelander and Monod considered the so-called Chebyshev center[2] (which is weakly compact) in $L_1(\mathcal{M},\tau)$ rather than the $\|\cdot\|_1$-closure of convex hull of $\delta(u)u^*$ as in Theorem 4.3. Consequently, the Ryll-Nardzewski fixed point theorem can be applied to this Chebyshev center, and the long standing open question on derivations with values in $L_1(\mathcal{M},\tau)$ was fully resolved.

THEOREM 4.4. [**9**] *Let $\mathcal{N}$ be a von Neumann algebra and $\mathcal{A}$ be a C^*-subalgebra of $\mathcal{N}$. Every derivation from $\mathcal{A}$ into the predual of $\mathcal{N}$ is inner.*

REMARK 4.5. Pfitzner (see [**92**, Theorem 8.2]) indicated that the above theorem can also be proved by applying a variant of the Ryll-Nardzewski fixed point theorem obtained by Japón [**63**]. Moreover, the element implementing the derivation can be found in the closure (in the sense of the so-called *abstract measure topology*) of the convex hull of $\delta(u)u^*$.

In a recent paper[**61**], it is shown that the L-embeddedness carries over from symmetric function spaces to their noncommutative counterparts (see [**58**] for a

[2]Let A be a non-empty bounded subset of a metric space V. The *circumradius* of A in V is $\rho_V(A) := \inf\{r \ge 0 : \exists x \in V \text{ with } A \subset \overline{B}(x,r)\}$, where $\overline{B}(x,r)$ denotes the closed r-ball around x. The *Chebyshev center* of A in V is the (possibly empty) set $C_V(A) = \{c \in V : A \subset \overline{B}(c,\rho_V(A))\}$.

detailed exposition on L-embedded spaces). Consequently, a derivation theorem for L-embedded symmetric spaces is obtained in [**61**]. As one can see from the surveyed literature, very little is known for derivations from a C^*-subalgebra $\mathcal{A}$ of $\mathcal{M}$ into a symmetric spaces $E(\mathcal{M},\tau)$ of possibly unbounded τ-measurable operators.

Recall the derivation theorems for noncommutative L_1-spaces [**9**] and for symmetric KB-ideals[**19**]. It is natural to ask whether derivations into interpolation spaces between the noncommutative L_1-space and symmetric KB-ideals are inner. However, there are no available interpolation techniques in the study of innerness of derivations:

QUESTION 5. Assume that $\mathcal{M}$ is a von Neumann algebra equipped with a faithful normal semifinite trace τ. Let $E(\mathcal{M},\tau)$ be a symmetric space of τ-compact operators affiliated with $\mathcal{M}$. How can one identity those $E(\mathcal{M},\tau)$ such that derivations from an arbitrary C^*-subalgebra $\mathcal{A}$ of $\mathcal{M}$ into $E(\mathcal{M},\tau)$ are necessarily inner?

The method used in [**9**] (or [**92**]) does not have any chance to deliver the full answer to Question 5. This fact was emphaiszed in [**9**, Section 3], where the following points were raised:

a. In marked contrast to the classical fixed point theorems, there is no hope to find a fixed point inside a general bounded closed convex subset of L^1 $\cdots$ the weak compactness $\cdots$ seems almost unavoidable $\cdots$
b. $\cdots$ a canonical norm one projection $V^{**}\to V$ is not enough.
c. It would be interesting to find a purely geometric version of the proposition $\cdots$

The fact that the "fixed point" obtained in [**9**] is not inside a general bounded closed convex subset of $L_1(\mathcal{M},\tau)$ leads to extra difficulties in the general case.

The special case when $\mathcal{A}=\mathcal{M}$ has been studied by Ber, Chilin and the second author [**16, 17, 24**]. In particular, they proved the following result.

THEOREM 4.6 ([**17**, Corollary 6.11][**12**]). *Let $\mathcal{M}$ be a semifinite von Neumann algebra with a faithful semifinite normal trace τ and $E(\mathcal{M},\tau)$ be a symmetric space. Every derivation from $\mathcal{M}$ into $E(\mathcal{M},\tau)$ is inner.*

A symmetric space $E(\mathcal{M},\tau)$ is said to have the *Levi property*[3], if for every upwards directed net $\{x_\beta\}$ in $E(\mathcal{M},\tau)^+$, satisfying $\sup_\beta\|x_\beta\|_E<\infty$, there exists an element $x\in E(\mathcal{M},\tau)^+$ such that $x_\beta\uparrow x$ in $E(\mathcal{M},\tau)$. It is well known that if a norm is Levi, then necessarily it is also weak Fatou [**1**, p.89] (see also [**5**, Lemma]), i.e., there exists a constant $K\geq 1$ such that

$$0\leq x_\beta\uparrow x\Rightarrow\|x\|_E\leq K\lim_\beta\|x_\beta\|_E\,.$$

In our recent paper[**62**], we completely resolve Question 5 above (i.e., questions due to Johnson and Hoover et al. in the special setting of symmetric spaces), see Theorem 4.8 below. The main ingredients used in the proof of Theorem 4.8 are the following noncommutative version of Grothendieck's theorem[**56**] obtained by

[3]In the case when $\mathcal{M}=L_\infty(0,\infty)$, this definition coincides with that in [**1**, Definition 7]. This property appears in the literature under many different names, e.g., a monotone complete norm or property (B)[**77**, Chapter X.4].

Akemann, Dodds and Gamlen[**2**, Theorem 4.2], weak compactness criteria in symmetric spaces obtained in [**40**, **48**], and a noncommutative analogue of the reflexive gate theorem[4].

THEOREM 4.7 ([**2**]). *A bounded linear map from a C^*-algebra into a weakly sequentially complete Banach space is weakly compact.*

THEOREM 4.8 ([**62**]). *Let $\mathcal{M}$ be a von Neumann algebra with a faithful normal semifinite trace τ and let $\mathcal{A}$ be a C^*-subalgebra of $\mathcal{M}$. If $E(\mathcal{M},\tau)$ is a strongly symmetric space of τ-compact operators (i.e., $E(\mathcal{M},\tau) \subset S_0(\mathcal{M},\tau)$[5]) having the Fatou property (resp., the Levi property), then every derivation $\delta : \mathcal{A} \to E(\mathcal{M},\tau)$ is inner. That is, there exists an element $a \in \overline{\mathrm{conv}\,\{\delta(u)u^* \mid u \in \mathcal{U}(\mathcal{A})\}}^{t_\tau} \subset E(\mathcal{M},\tau)$ with $\|a\|_E \le \|\delta\|_{\mathcal{A}\to E}$ (resp., $\|a\|_E \le c\,\|\delta\|_{\mathcal{A}\to E}$ for some constant c depending on $E(\mathcal{M},\tau)$ only) such that $\delta = \delta_a$ on $\mathcal{A}$.*

Observe that Theorem 4.8 above is (in some ideological sense) an interpolation type result between [**9**] and [**19**], which shows that Question 5 has an affirmative answer for any symmetric space having the Fatou property sitting between noncommutative L_1-space $L_1(\mathcal{M},\tau)$ (affiliated with a semifinite von Neumann algebra $\mathcal{M}$) and a symmetric KB-ideal in $\mathcal{M}$. Let us emphasize that the main results of [**9**, **92**] (for derivations into the predual of a semifinite von Neumann algebra) and [**19**, **20**, **24**, **29**, **59**, **61**, **75**, **110**] are immediate consequences of our Theorem 4.8.

The new approach devised in [**62**] answers to points (a), (b) and (c) above raised in [**9**], which provides an alternative proof for the resolution of the question raised by Bunce and Paschke [**29**], without involving weak compactness of a subset in aN L-embedded space as [**9**] and [**92**] did. This enables us to find a "fixed point" (implementing the derivation) from a not necessarily weakly compact closed convex subset of a noncommutative symmetric space which is 1-complemented subspace of its bidual but not necessarily an L-embedded Banach space (i.e., a symmetric space having the Levi property[**44**, **45**][6]), see Theorem 4.8. On the other hand, the Levi property of the space $E(\mathcal{M},\tau)$ means that $E(\mathcal{M},\tau)$ coincides with its second Köthe dual and this geometrical condition is the only one required in Theorem 4.8 thus delivering (at least spiritually) an answer to the question suggested in [**9**, Comment c] above.

It is important to emphasize that the Fatou/Levi property was hatched in the theory of Banach lattices[**1**, **28**], and was even included into the original definition of Banach function spaces over σ-finite measure spaces (see [**10**, **86**]). The property is somewhat analogous to the so-called "dual normal" property. The importance of the Fatou/Levi property in the theory of Banach function spaces and symmetric operator spaces is hard to overestimate [**42**, **44**, **45**]. It seems appropriate to recall here that every derivation from a hyperfinite von Neumann algebra $\mathcal{A}$ into a dual normal $\mathcal{A}$-bimodule is inner (see e.g. [**103**, Theorem 2.4.3], [**34**] and [**65**]). Recall also, that derivations from a nuclear C^*-algebra $\mathcal{A}$ into a dual Banach $\mathcal{A}$-module are inner[**35**, **57**]. However, Theorem 4.8 below holds for arbitrary C^*-subalgebras $\mathcal{A}$ of $\mathcal{M}$ and for symmetric spaces which may not have a predual space.

[4]Recall the reflexive gate type result for symmetric function spaces $E(0,1)$[**89**, Corollary 3.2.3], i.e., if $E(0,1) \neq L_1(0,1)$ and $E(0,1) \neq L_\infty(0,1)$, then there exist two reflexive symmetric spaces $F_1(0,1)$ and $F_2(0,1)$ such that $F_2(0,1) \subset E(0,1) \subset F_1(0,1)$.

[5]If $\tau(\mathbf{1}) < \infty$, then $E(\mathcal{M},\tau) \subset S_0(\mathcal{M},\tau)$ holds for any symmetric space $E(\mathcal{M},\tau)$ affiliated with $\mathcal{M}$.

[6]Indeed, Theorem 4.8 holds for the case when the projection constant is not necessarily 1.

We also demonstrate why the Fatou/Levi property of the space $E(\mathcal{M},\tau)$ is a necessary condition for an affirmative answer to Question 5 in the class of all semifinite von Neumann algebras.

COROLLARY 4.3. *For a given symmetric function space $E(0,\infty)\subset S_0(0,\infty)$, the following two statements are equivalent:*

(1) *for any von Neumann algebra $\mathcal{M}$ equipped with a semifinite faithful normal trace τ and any C^*-subalgebra $\mathcal{A}$ of $\mathcal{M}$, derivations $\delta:\mathcal{A}\to E(\mathcal{M},\tau)$ are necessarily inner;*
(2) *$E(0,\infty)$ has the Levi property.*

We feel that the following two questions are now within the reach.

QUESTION 6. To characterize those semifinite von Neumann algebras $\mathcal{M}$ such that for a symmetric function space $E(0,\infty)\subset S_0(0,\infty)$, the following two statements are equivalent:

(1) for any C^*-subalgebra $\mathcal{A}$ of $\mathcal{M}$, derivations $\delta:\mathcal{A}\to E(\mathcal{M},\tau)$ are necessarily inner;
(2) $E(0,\infty)$ has the Levi property.

QUESTION 7. The case for von Neumann subalgebras of $\mathcal{M}$ is still open: how to characterize those symmetric function spaces $E(0,\infty)$ such that for any von Neumann algebra $\mathcal{M}$ equipped with a semifinite faithful normal trace τ and for any von Neumann subalgebra $\mathcal{A}$ of $\mathcal{M}$, derivations $\delta:\mathcal{A}\to E(\mathcal{M},\tau)$ are necessarily inner. This question is a generalization of Question 4.

It is of interest to emphasize that the direction of derivations in quasi-normed ideals of compact operators (and more generally, quasi-normed ideals of symmetric spaces) is almost completely unexplored. For example, to the best of our knowledge, the following question is still open.

QUESTION 8. Let $\mathcal{M}$ be a von Neumann algebra equipped with a semifinite faithful normal trace τ. It is an open question whether derivations from a C^*-/von Neumann subalgebra $\mathcal{A}$ into $L_p(\mathcal{M},\tau)$ is inner when $0<p<1$.

4.2. Derivations with values in the algebra $S(\mathcal{M})$. In 2000, Ayupov suggested a problem concerning the description of derivations on algebras of measurable operators affiliated with von Neumann algebras [**6**].

The answer to Ayupov's problem for commutative von Neumann algebras was independently obtained by Ber, Chilin and the second author [**11, 14, 15**] (see also [**18**]), and Kusraev [**80, 81**]: for an abelian von Neumann algebra $\mathcal{M}$, the algebra $S(\mathcal{M})$ admits non-trivial derivations if and only if the Boolean algebra of projections $P(\mathcal{M})$ of $\mathcal{M}$ is non-atomic.

The situation for von Neumann algebras $\mathcal{M}$ without finite type I direct summands differs dramatically. It was proved in 2009 that for infinite von Neumann algebras $\mathcal{M}$ of type I, all derivations on $S(\mathcal{M})$ are inner [**4, 23**]. Later, it was shown that any continuous (in the local measure topology) derivation on $S(\mathcal{M})$ is automatically inner [**8, 17**], which was an important step towards the resolution of Ayupov's problem. In particular, it was shown that derivations are automatically continuous in the case of properly infinite von Neumann algebras [**7, 16**].

Let $\mathcal{M}$ be a von Neumann algebra of type II_1. The set $S(\mathcal{M})$ is the so-called Murray–von Neumann algebra. Kadison and Liu noted in [**69**, p. 210-211] (see also

[**70**, p. 2090]) that

> "the complete cohomological result would say that each derivation of $S(\mathcal{M})$ is inner .... The authors strongly feel that this is true; but it is still open."

In [**22**], Ber, Kudaybergenov and the second author answered the above conjecture due to Kadison and Liu in the affirmative, showing that all derivations of a Murray–von Neumann algebra are inner, which provides a complete answer to Ayupov's problem.

THEOREM 4.9. *Let $\mathcal{M}$ be an arbitrary von Neumann. The following assertion are equivalent:*

(1) *any derivation of $S(\mathcal{M})$ (respectively, $LS(\mathcal{M})$ or $S(\mathcal{M},\tau)$) is inner;*
(2) *the type $I_{\rm fin}$ direct summand of $\mathcal{M}$ is atomic.*

QUESTION 9. To describe derivations from a C^*/von Neumann subalgebra $\mathcal{A}$ of a von Neumann algebra $\mathcal{M}$ into $S(\mathcal{M},\tau)/S(\mathcal{M})/LS(\mathcal{M})$, which should be considered as a generalization of the famous derivation problem[**90**] and the Ayupov–Kadison–Liu problem[**6, 69, 70**].

References

[1] Y. A. Abramovich and C. D. Aliprantis, *Positive operators*, Handbook of the geometry of Banach spaces, Vol. I, North-Holland, Amsterdam, 2001, pp. 85–122, DOI 10.1016/S1874-5849(01)80004-8. MR1863690

[2] C. A. Akemann, P. G. Dodds, and J. L. B. Gamlen, *Weak compactness in the dual space of C^*-algebra*, J. Functional Analysis **10** (1972), 446–450, DOI 10.1016/0022-1236(72)90040-7. MR344898

[3] S. Albeverio, Sh. A. Ayupov, and K. K. Kudaybergenov, *Non-commutative Arens algebras and their derivations*, J. Funct. Anal. **253** (2007), no. 1, 287–302, DOI 10.1016/j.jfa.2007.04.010. MR2362425

[4] S. Albeverio, Sh. A. Ayupov, and K. K. Kudaybergenov, *Structure of derivations on various algebras of measurable operators for type I von Neumann algebras*, J. Funct. Anal. **256** (2009), no. 9, 2917–2943, DOI 10.1016/j.jfa.2008.11.003. MR2502428

[5] I. Amemiya, *A generalization of Reisz-Fischer's theorem*, J. Math. Soc. Japan **5** (1953), 353–354, DOI 10.2969/jmsj/00530353. MR61769

[6] Sh.A. Ayupov, *Derivations in algebras of measurable operators*, Dokl. Uzbek Akad. Nauk **3** (2000), 14–17.

[7] Sh. A. Ayupov and K. K. Kudaybergenov, *Additive derivations on algebras of measurable operators*, J. Operator Theory **67** (2012), no. 2, 495–510, DOI 10.2478/s11533-011-0136-6. MR2928326

[8] S. Ayupov and K. Kudaybergenov, *Innerness of continuous derivations on algebras of measurable operators affiliated with finite von Neumann algebras*, J. Math. Anal. Appl. **408** (2013), no. 1, 256–267, DOI 10.1016/j.jmaa.2013.06.005. MR3079963

[9] U. Bader, T. Gelander, and N. Monod, *A fixed point theorem for L^1 spaces*, Invent. Math. **189** (2012), no. 1, 143–148, DOI 10.1007/s00222-011-0363-2. MR2929085

[10] C. Bennett and R. Sharpley, *Interpolation of operators*, Pure and Applied Mathematics, vol. 129, Academic Press, Inc., Boston, MA, 1988. MR928802

[11] A. F. Ber, *Derivations in commutative regular algebras* (Russian, with Russian summary), Mat. Tr. **13** (2010), no. 1, 3–14, DOI 10.3103/s1055134411030011; English transl., Siberian Adv. Math. **21** (2011), no. 3, 161–169. MR2682765

[12] A. F. Ber, G. B. Levitina, and V. I. Chilin, *Derivations with values in quasinormable bimodules of locally measurable operators* (Russian, with Russian summary), Mat. Tr. **17** (2014), no. 1, 3–18, DOI 10.3103/s1055134415030025; English transl., Siberian Adv. Math. **25** (2015), no. 3, 169–178. MR3236359

[13] A. Ber, V. Chilin, G. Levitina, F. Sukochev, and D. Zanin, *Derivations on algebras of unbounded operators*, unpublished manuscript.

[14] A. F. Ber, F. A. Sukochev, and V. I. Chilin, *Derivations in commutative regular algebras* (Russian), Mat. Zametki **75** (2004), no. 3, 453–454, DOI 10.1023/B:MATN.0000023321.64184.5a; English transl., Math. Notes **75** (2004), no. 3-4, 418–419. MR2068808

[15] A. F. Ber, V. I. Chilin, and F. A. Sukochev, *Non-trivial derivations on commutative regular algebras*, Extracta Math. **21** (2006), no. 2, 107–147. MR2292743

[16] A. F. Ber, V. I. Chilin, and F. A. Sukochev, *Continuity of derivations of algebras of locally measurable operators*, Integral Equations Operator Theory **75** (2013), no. 4, 527–557, DOI 10.1007/s00020-013-2039-3. MR3032667

[17] A. F. Ber, V. I. Chilin, and F. A. Sukochev, *Continuous derivations on algebras of locally measurable operators are inner*, Proc. Lond. Math. Soc. (3) **109** (2014), no. 1, 65–89, DOI 10.1112/plms/pdt070. MR3237736

[18] A. F. Ber, V. I. Chilin, and F. A. Sukochev, *Derivations in disjointly complete commutative regular algebras*, Quaest. Math (2022) (to appear).

[19] A. Ber, J. Huang, G. Levitina, and F. Sukochev, *Derivations with values in ideals of semifinite von Neumann algebras*, J. Funct. Anal. **272** (2017), no. 12, 4984–4997, DOI 10.1016/j.jfa.2017.02.010. MR3639520

[20] A. Ber, J. Huang, G. Levitina, and F. Sukochev, *Derivations with values in the ideal of τ-compact operators affiliated with a semifinite von Neumann algebra*, Comm. Math. Phys. **390** (2022), no. 2, 577–616, DOI 10.1007/s00220-022-04313-0. MR4384716

[21] A. Ber, J. Huang, and F. Sukochev, *Norms of skew-adjoint derivations with values in the predual of a semifinite von Neumann algebra*, J. Funct. Anal. **285** (2023), no. 8, Paper No. 110072, 23, DOI 10.1016/j.jfa.2023.110072. MR4610072

[22] A. Ber, K. Kudaybergenov, and F. Sukochev, *Derivations of Murray–von Neumann algebras*, J. Reine Angew. Math. **791** (2022), 283–301, DOI 10.1515/crelle-2022-0051. MR4489632

[23] A. F. Ber, B. de Pagter, and F. A. Sukochev, *Derivations in algebras of operator-valued functions*, J. Operator Theory **66** (2011), no. 2, 261–300. MR2844466

[24] A. Ber and F. Sukochev, *Derivations in the Banach ideals of τ-compact operators*, `arXiv:1204.4052v1`, 2012.

[25] A. F. Ber and F. A. Sukochev, *Commutator estimates in W^*-factors*, Trans. Amer. Math. Soc. **364** (2012), no. 10, 5571–5587, DOI 10.1090/S0002-9947-2012-05568-1. MR2931339

[26] A. F. Ber and F. A. Sukochev, *Commutator estimates in W^*-algebras*, J. Funct. Anal. **262** (2012), no. 2, 537–568, DOI 10.1016/j.jfa.2011.09.018. MR2854712

[27] B. Blackadar, *Operator algebras: Theory of C^*-algebras and von Neumann algebras*, Encyclopaedia of Mathematical Sciences, vol. 122, Springer-Verlag, Berlin, 2006. Operator Algebras and Non-commutative Geometry, III, DOI 10.1007/3-540-28517-2. MR2188261

[28] A. V. Buhvalov, A. I. Veksler, and V. A. Geĭler, *Normed lattices* (Russian), Mathematical analysis, Vol. 18 (Russian), Itogi Nauki i Tehniki, Akad. Nauk SSSR, Vsesoyuz. Inst. Nauchn. i Tekhn. Inform., Moscow, 1980, pp. 125–184. MR597904

[29] J. W. Bunce and W. L. Paschke, *Derivations on a C^*-algebra and its double dual*, J. Functional Analysis **37** (1980), no. 2, 235–247, DOI 10.1016/0022-1236(80)90043-9. MR578934

[30] E. Christensen, *Extension of derivations*, J. Functional Analysis **27** (1978), no. 2, 234–247, DOI 10.1016/0022-1236(78)90029-0. MR481217

[31] E. Christensen, *The C^*-algebra of generalized compacts associated with a von Neumann algebra*, Operators in indefinite metric spaces, scattering theory and other topics (Bucharest, 1985), Oper. Theory Adv. Appl., vol. 24, Birkhäuser, Basel, 1987, pp. 51–58. MR903063

[32] E. Christensen, E. G. Effros, and A. Sinclair, *Completely bounded multilinear maps and C^*-algebraic cohomology*, Invent. Math. **90** (1987), no. 2, 279–296, DOI 10.1007/BF01388706. MR910202

[33] E. Christensen, F. Pop, A. M. Sinclair, and R. R. Smith, *Hochschild cohomology of factors with property Γ*, Ann. of Math. (2) **158** (2003), no. 2, 635–659, DOI 10.4007/annals.2003.158.635. MR2018931

[34] A. Connes, *Classification of injective factors. Cases II_1, II_∞, III_λ, $\lambda \neq 1$*, Ann. of Math. (2) **104** (1976), no. 1, 73–115, DOI 10.2307/1971057. MR454659

[35] A. Connes, *On the cohomology of operator algebras*, J. Functional Analysis **28** (1978), no. 2, 248–253, DOI 10.1016/0022-1236(78)90088-5. MR493383

[36] J. B. Conway, *A course in functional analysis*, Graduate Texts in Mathematics, vol. 96, Springer-Verlag, New York, 1985, DOI 10.1007/978-1-4757-3828-5. MR768926

[37] K. R. Davidson, *Nest algebras: Triangular forms for operator algebras on Hilbert space*, Pitman Research Notes in Mathematics Series, vol. 191, Longman Scientific & Technical, Harlow; copublished in the United States with John Wiley & Sons, Inc., New York, 1988. MR972978

[38] J. Diestel, *Grothendieck spaces and vector measures*, Vector and operator valued measures and applications (Proc. Sympos., Alta, Utah, 1972), Academic Press, New York-London, 1973, pp. 97–108. MR338774

[39] J. Dixmier, *Les algèbres d'opérateurs dans l'espace hilbertien (algèbres de von Neumann)* (French), Cahiers Scientifiques [Scientific Reports], Fasc. XXV, Gauthier-Villars Éditeur, Paris, 1969. Deuxième édition, revue et augmentée. MR352996

[40] P. G. Dodds, T. K. Dodds, and B. de Pagter, *Weakly compact subsets of symmetric operator spaces*, Math. Proc. Cambridge Philos. Soc. **110** (1991), no. 1, 169–182, DOI 10.1017/S0305004100070225. MR1104612

[41] P. G. Dodds, T. K. Dodds, and B. de Pagter, *Fully symmetric operator spaces*, Integral Equations Operator Theory **15** (1992), no. 6, 942–972, DOI 10.1007/BF01203122. MR1188788

[42] P. G. Dodds, T. K.-Y. Dodds, and B. de Pagter, *Noncommutative Köthe duality*, Trans. Amer. Math. Soc. **339** (1993), no. 2, 717–750, DOI 10.2307/2154295. MR1113694

[43] P. G. Dodds, T. K. Dodds, and F. A. Sukochev, *On p-convexity and q-concavity in non-commutative symmetric spaces*, Integral Equations Operator Theory **78** (2014), no. 1, 91–114, DOI 10.1007/s00020-013-2082-0. MR3147405

[44] P. G. Dodds, T. K. Dodds, F. A. Sukochev, and O. Ye. Tikhonov, *A non-commutative Yosida-Hewitt theorem and convex sets of measurable operators closed locally in measure*, Positivity **9** (2005), no. 3, 457–484, DOI 10.1007/s11117-005-1384-0. MR2188531

[45] P. G. Dodds and B. de Pagter, *The non-commutative Yosida-Hewitt decomposition revisited*, Trans. Amer. Math. Soc. **364** (2012), no. 12, 6425–6457, DOI 10.1090/S0002-9947-2012-05569-3. MR2958942

[46] P. G. Dodds and B. de Pagter, *Normed Köthe spaces: a non-commutative viewpoint*, Indag. Math. (N.S.) **25** (2014), no. 2, 206–249, DOI 10.1016/j.indag.2013.01.009. MR3151815

[47] P. Dodds, B. de Pagter, and F. Sukochev, *Theory of noncommutative integration*, unpublished manuscript.

[48] P. G. Dodds, F. A. Sukochev, and G. Schlüchtermann, *Weak compactness criteria in symmetric spaces of measurable operators*, Math. Proc. Cambridge Philos. Soc. **131** (2001), no. 2, 363–384, DOI 10.1017/S0305004101005114. MR1857125

[49] G. A. Elliott, *Some C^*-algebras with outer derivations*, Rocky Mountain J. Math. **3** (1973), 501–506, DOI 10.1216/RMJ-1973-3-3-501. MR365158

[50] G. A. Elliott, *Some C^*-algebras with outer derivations. II*, Canadian J. Math. **26** (1974), 185–189, DOI 10.4153/CJM-1974-018-6. MR365159

[51] G. A. Elliott, *Some C^*-algebras with outer derivations. III*, Ann. of Math. (2) **106** (1977), no. 1, 121–143, DOI 10.2307/1971162. MR448093

[52] E. G. Effros and Z.-J. Ruan, *Representations of operator bimodules and their applications*, J. Operator Theory **19** (1988), no. 1, 137–158. MR950830

[53] T. Fack and H. Kosaki, *Generalized s-numbers of τ-measurable operators*, Pacific J. Math. **123** (1986), no. 2, 269–300. MR840845

[54] A. Galatan and S. Popa, *Smooth bimodules and cohomology of* II_1 *factors*, J. Inst. Math. Jussieu **16** (2017), no. 1, 155–187, DOI 10.1017/S1474748015000122. MR3591964

[55] I. C. Gohberg and M. G. Kreĭn, *Introduction to the theory of linear nonselfadjoint operators*, Translations of Mathematical Monographs, Vol. 18, American Mathematical Society, Providence, RI, 1969. Translated from the Russian by A. Feinstein. MR246142

[56] A. Grothendieck, *Sur les applications linéaires faiblement compactes d'espaces du type $C(K)$* (French), Canad. J. Math. **5** (1953), 129–173, DOI 10.4153/cjm-1953-017-4. MR58866

[57] U. Haagerup, *All nuclear C^*-algebras are amenable*, Invent. Math. **74** (1983), no. 2, 305–319, DOI 10.1007/BF01394319. MR723220

[58] P. Harmand, D. Werner, W. Werner, *M-ideals in Banach spaces and Banach algebras*, Springer-Verlag, Berlin/New York, 1993.

[59] T. B. Hoover, *Derivations, homomorphisms, and operator ideals*, Proc. Amer. Math. Soc. **62** (1977), no. 2, 293–298, DOI 10.2307/2041031. MR435860

[60] J. Huang, *Derivations wuth values into ideals of a semifinite von Neumann algebra*, Ph.D. thesis, UNSW, 2019.

[61] J. Huang, G. Levitina, and F. Sukochev, *M-embedded symmetric operator spaces and the derivation problem*, Math. Proc. Cambridge Philos. Soc. **169** (2020), no. 3, 607–622, DOI 10.1017/s030500411900029x. MR4170617

[62] J. Huang and F. Sukochev, *Innerness of derivations into noncommutative symmetric spaces is determined commutatively*, arXiv:2301.03874, 2023.

[63] M. A. Japón Pineda, *Some fixed point results on L-embedded Banach spaces*, J. Math. Anal. Appl. **272** (2002), no. 1, 380–391, DOI 10.1016/S0022-247X(02)00107-5. MR1930722

[64] B. E. Johnson, *Cohomology in Banach algebras*, Memoirs of the American Mathematical Society, No. 127, American Mathematical Society, Providence, RI, 1972. MR374934

[65] B. E. Johnson, R. V. Kadison, and J. R. Ringrose, *Cohomology of operator algebras. III. Reduction to normal cohomology*, Bull. Soc. Math. France **100** (1972), 73–96. MR318908

[66] B. E. Johnson and S. K. Parrott, *Operators commuting with a von Neumann algebra modulo the set of compact operators*, J. Functional Analysis **11** (1972), 39–61, DOI 10.1016/0022-1236(72)90078-x. MR341119

[67] R. V. Kadison, *Derivations of operator algebras*, Ann. of Math. (2) **83** (1966), 280–293, DOI 10.2307/1970433. MR193527

[68] R. V. Kadison, *Which Singer is that?*, Surveys in differential geometry, Surv. Differ. Geom., vol. 7, Int. Press, Somerville, MA, 2000, pp. 347–373, DOI 10.4310/SDG.2002.v7.n1.a12. MR1919431

[69] R. V. Kadison and Z. Liu, *A note on derivations of Murray–von Neumann algebras*, Proc. Natl. Acad. Sci. USA **111** (2014), no. 6, 2087–2093, DOI 10.1073/pnas.1321358111. MR3200173

[70] R. V. Kadison and Z. Liu, *Derivations of Murray–von Neumann algebras*, Math. Scand. **115** (2014), no. 2, 206–228, DOI 10.7146/math.scand.a-19223. MR3291726

[71] R. V. Kadison and J. R. Ringrose, *Fundamentals of the theory of operator algebras. Vol. I: Elementary theory*, Pure and Applied Mathematics, vol. 100, Academic Press, Inc. [Harcourt Brace Jovanovich, Publishers], New York, 1983. MR719020

[72] R. V. Kadison and J. R. Ringrose, *Fundamentals of the theory of operator algebras. Vol. II: Advanced theory*, Pure and Applied Mathematics, vol. 100, Academic Press, Inc., Orlando, FL, 1986, DOI 10.1016/S0079-8169(08)60611-X. MR859186

[73] V. Kaftal, *On the theory of compact operators in von Neumann algebras.* I, Indiana Univ. Math. J. **26** (1977), no. 3, 447–457, DOI 10.1512/iumj.1977.26.26035. MR463970

[74] V. Kaftal, *On the theory of compact operators in von Neumann algebras. II*, Pacific J. Math. **79** (1978), no. 1, 129–137. MR526672

[75] V. Kaftal and G. Weiss, *Compact derivations relative to semifinite von Neumann algebras*, J. Funct. Anal. **62** (1985), no. 2, 202–220, DOI 10.1016/0022-1236(85)90003-5. MR791847

[76] N. J. Kalton and F. A. Sukochev, *Symmetric norms and spaces of operators*, J. Reine Angew. Math. **621** (2008), 81–121, DOI 10.1515/CRELLE.2008.059. MR2431251

[77] L. V. Kantorovich and G. P. Akilov, *Funktsional′nyĭ analiz* (Russian), 3rd ed., "Nauka", Moscow, 1984. MR788496

[78] I. Kaplansky, *Modules over operator algebras*, Amer. J. Math. **75** (1953), 839–858, DOI 10.2307/2372552. MR58137

[79] S. G. Kreĭn, Yu. Ī. Petunīn, and E. M. Semënov, *Interpolation of linear operators*, Translations of Mathematical Monographs, vol. 54, American Mathematical Society, Providence, RI, 1982. Translated from the Russian by J. Szűcs. MR649411

[80] A. G. Kusraev, *Derivations and automorphisms in the algebra of measurable complex-valued functions* (Russian, with Russian summary), Vladikavkaz. Mat. Zh. **7** (2005), no. 3, 45–49. MR2193468

[81] A. G. Kusraev, *Automorphisms and derivations in an extended complex f-algebra* (Russian, with Russian summary), Sibirsk. Mat. Zh. **47** (2006), no. 1, 97–107, DOI 10.1007/s11202-006-0010-0; English transl., Siberian Math. J. **47** (2006), no. 1, 77–85. MR2215299

[82] J. Lindenstrauss and L. Tzafriri, *Classical Banach spaces. II: Function spaces*, Ergebnisse der Mathematik und ihrer Grenzgebiete [Results in Mathematics and Related Areas], vol. 97, Springer-Verlag, Berlin-New York, 1979. MR540367

[83] S. Lord, F. Sukochev, and D. Zanin, *Singular traces: Theory and applications*, De Gruyter Studies in Mathematics, vol. 46, De Gruyter, Berlin, 2013. MR3099777

[84] S. Lord, F. Sukochev, and D. Zanin, *Singular traces. Vol. 1. Theory*, De Gruyter Studies in Mathematics, vol. 46/1, De Gruyter, Berlin, [2021] ©2021. 2nd edition [of 3099777], DOI 10.1515/9783110378054-201. MR4304074

[85] V. Losert, *The derivation problem for group algebras*, Ann. of Math. (2) **168** (2008), no. 1, 221–246, DOI 10.4007/annals.2008.168.221. MR2415402

[86] W. A. J. Luxemburg, *Banach function spaces*, Technische Hogeschool te Delft, Delft, 1955. Thesis. MR72440

[87] E. Nelson, *Notes on non-commutative integration*, J. Functional Analysis **15** (1974), 103–116, DOI 10.1016/0022-1236(74)90014-7. MR355628

[88] I. Namioka and E. Asplund, *A geometric proof of Ryll-Nardzewski's fixed point theorem*, Bull. Amer. Math. Soc. **73** (1967), 443–445, DOI 10.1090/S0002-9904-1967-11779-8. MR209904

[89] S.Ya. Novikov, *Sequences of functions in symmetric spaces*, Publishing house "Samara University", Samara, 2008 (in Russian).

[90] N. Ozawa, *An invitation to the similarity problems (after Pisier)*, Surikaisekikenkyusho Kokyuroku, 1486 (2006), 27–40.

[91] H. Pfitzner, *Weak compactness in the dual of a C^*-algebra is determined commutatively*, Math. Ann. **298** (1994), no. 2, 349–371, DOI 10.1007/BF01459739. MR1256621

[92] H. Pfitzner, *L-embedded Banach spaces and measure topology*, Israel J. Math. **205** (2015), no. 1, 421–451, DOI 10.1007/s11856-014-1136-6. MR3314594

[93] G. Pisier, *Similarity problems and completely bounded maps*, Second, expanded edition, Lecture Notes in Mathematics, vol. 1618, Springer-Verlag, Berlin, 2001. Includes the solution to "The Halmos problem", DOI 10.1007/b55674. MR1818047

[94] G. Pisier and Q. Xu, *Non-commutative L^p-spaces*, Handbook of the geometry of Banach spaces, Vol. 2, North-Holland, Amsterdam, 2003, pp. 1459–1517, DOI 10.1016/S1874-5849(03)80041-4. MR1999201

[95] S. Popa, *The commutant modulo the set of compact operators of a von Neumann algebra*, J. Funct. Anal. **71** (1987), no. 2, 393–408, DOI 10.1016/0022-1236(87)90011-5. MR880987

[96] S. Popa and F. Rădulescu, *Derivations of von Neumann algebras into the compact ideal space of a semifinite algebra*, Duke Math. J. **57** (1988), no. 2, 485–518, DOI 10.1215/S0012-7094-88-05722-5. MR962517

[97] J. R. Ringrose, *Automatic continuity of derivations of operator algebras*, J. London Math. Soc. (2) **5** (1972), 432–438, DOI 10.1112/jlms/s2-5.3.432. MR374927

[98] V. Runde, *$\mathcal{B}(\ell^p)$ is never amenable*, J. Amer. Math. Soc. **23** (2010), no. 4, 1175–1185, DOI 10.1090/S0894-0347-10-00668-5. MR2669711

[99] S. Sakai, *On a conjecture of Kaplansky*, Tohoku Math. J. (2) **12** (1960), 31–33, DOI 10.2748/tmj/1178244484. MR112055

[100] S. Sakai, *Derivations of W^*-algebras*, Ann. of Math. (2) **83** (1966), 273–279, DOI 10.2307/1970432. MR193528

[101] S. Sakai, *C^*-algebras and W^*-algebras*, Classics in Mathematics, Springer-Verlag, Berlin, 1998. Reprint of the 1971 edition, DOI 10.1007/978-3-642-61993-9. MR1490835

[102] I. E. Segal, *A non-commutative extension of abstract integration*, Ann. of Math. (2) **57** (1953), 401–457, DOI 10.2307/1969729. MR54864

[103] A. M. Sinclair and R. R. Smith, *Hochschild cohomology of von Neumann algebras*, London Mathematical Society Lecture Note Series, vol. 203, Cambridge University Press, Cambridge, 1995, DOI 10.1017/CBO9780511526190. MR1336825

[104] I. M. Singer and J. Wermer, *Derivations on commutative normed algebras*, Math. Ann. **129** (1955), 260–264, DOI 10.1007/BF01362370. MR70061

[105] A. Ströh and G. P. West, *τ-compact operators affiliated to a semifinite von Neumann algebra*, Proc. Roy. Irish Acad. Sect. A **93** (1993), no. 1, 73–86. MR1241841

[106] F. Sukochev, *Symmetric spaces of measurable operators on finite von Neumann algebras*, Ph.D. Thesis (Russian), Tashkent 1987.

[107] F. Sukochev, *Completeness of quasi-normed symmetric operator spaces*, Indag. Math. (N.S.) **25** (2014), no. 2, 376–388, DOI 10.1016/j.indag.2012.05.007. MR3151823

[108] M. Takesaki, *Theory of operator algebras. I*, Springer-Verlag, New York-Heidelberg, 1979. MR548728

[109] M. Takesaki, *Theory of operator algebras. II: Operator Algebras and Non-commutative Geometry, 6*, Encyclopaedia of Mathematical Sciences, vol. 125, Springer-Verlag, Berlin, 2003, DOI 10.1007/978-3-662-10451-4. MR1943006

[110] M. Weigt, *Derivations of τ-measurable operators*, Operator algebras, operator theory and applications, Oper. Theory Adv. Appl., vol. 195, Birkhäuser Verlag, Basel, 2010, pp. 273–286. MR2683202

[111] S. Wright, *Weakly compact, operator-valued derivations of type* I *von Neumann algebras*, Canad. J. Math. **36** (1984), no. 3, 436–457, DOI 10.4153/CJM-1984-027-x. MR752977

[112] Q. H. Xu, *Analytic functions with values in lattices and symmetric spaces of measurable operators*, Math. Proc. Cambridge Philos. Soc. **109** (1991), no. 3, 541–563, DOI 10.1017/S030500410006998X. MR1094753

Institute for Advanced Study in Mathematics of HIT, Harbin Institute of Technology, Harbin, 150001, People's Republic of China
Email address: `jinghao.huang@hit.edu.cn`

School of Mathematics and Statistics, University of New South Wales, Kensington, 2052, Australia
Email address: `f.sukochev@unsw.edu.au`

Proceedings of Symposia in Pure Mathematics
Volume **106**, 2024
https://doi.org/10.1090/pspum/106/01932

Singular Integrals in Uniformly Convex Spaces

Tuomas Hytönen

This paper is dedicated to Professor Per H. Enflo

Abstract. We consider the action of finitely truncated singular integral operators on functions taking values in a Banach space. Such operators are bounded for any Banach space, but we show a quantitative improvement over the trivial bound in any space with an equivalent uniformly convex norm. This answers a question asked by Naor and the author, who previously proved the result in the important special case of the finite Hilbert transforms.

The proof, which splits the operator into a cancellative part and two paraproducts, follows the broad outline of similar results for genuinely singular (non-truncated) operators in the narrower class of UMD spaces. Thus we revisit and survey the recent techniques behind such results, but our precise setting, the main theorem, and some aspects of its proof, are new.

Curiously, it turns out that the paraproducts admit somewhat better bounds than the full operator. In a large class of spaces other than UMD, they remain bounded even without the finite truncations.

1. Introduction

The theory of singular integrals of Banach space -valued functions is well developed in the setting of so-called UMD (unconditional martingale differences) spaces. In fact, this is known to be the maximal generality in which the boundedness of several basic singular operators, like the Hilbert transform, can be extended [**3**, **4**].

In view of this, the title of this paper might seem outrageous. After all, the questions of boundedness of singular integrals depend only on the isomorphic structure of the target Banach space, and the class of spaces that admit an equivalent uniformly convex norm is known to be strictly larger than the class of UMD spaces [**3**, **18**].

However, our point of departure is the observation that there are many singular integral type objects of finite nature, whose boundedness is qualitatively trivial but where one can still hope to beat the trivial estimates by more clever quantitative

2020 *Mathematics Subject Classification.* Primary 46E40; Secondary 42B20, 60G46.

Key words and phrases. Singular integral, uniformly convex space, martingale type.

The author was supported by the Academy of Finland / Research Council of Finland through project No. 346314 (Finnish Centre of Excellence in Randomness and Structures "FiRST").

considerations. A prototype example consists of the finite Hilbert transforms

$$Hf(i) = \sum_{\substack{j=1 \\ j \neq i}}^{N} \frac{f(j)}{i-j}, \quad i \in \{1, \dots, N\}.$$

It is not difficult to check that $\|Hf\|_{\ell_s^N(\mathcal{X})} \lesssim \log N \cdot \|f\|_{\ell_s^N(\mathcal{X})}$ for every Banach space $\mathcal{X}$, while the uniform bound $\|Hf\|_{\ell_s^N(\mathcal{X})} \lesssim \|f\|_{\ell_s^N(\mathcal{X})}$ is equivalent to the boundedness of the usual Hilbert transform on $L_s(\mathbb{R}; \mathcal{X})$, and hence to the UMD property of $\mathcal{X}$. However, an intermediate behaviour $\|Hf\|_{\ell_s^N(\mathcal{X})} \lesssim (\log N)^\theta \|f\|_{\ell_s^N(\mathcal{X})}$ is also possible; this phenomenon was found, and its geometric implications explored, by Naor and the author [**11**].

Motivated by this, in the present work, our aim is to obtain similar improvements over the trivial bound for a general class of singular integrals of finite type when acting on functions taking values in a space that admits a uniformly convex norm. Actually, we will not make direct use of uniform convexity as such, but rather resort to the rich theory of its equivalent formulations. Recall that the spaces with an equivalent uniformly convex norm are precisely those with an equivalent uniformly smooth norm, and further the same as the super-reflexive ones [**6**]. They can further be given an equivalent p-uniformly smooth or q-uniformly convex norm [**17**], or, what is most relevant for the present needs, they satisfy related martingale inequalities known as martingale type and cotype [**17**, **19**]:

Definition 1.1. A Banach space $\mathcal{X}$ is said to have *martingale type* $p \in (1,2]$, if for some (equivalently, all) $s \in (1, \infty)$ and for every probability space (equivalently, every σ-finite measure space) $(E, \mathcal{E}, \mu)$, every martingale $(f_n)_{n=0}^N$ of arbitrary finite length in $L_s(\mu; \mathcal{X})$ satisfies

$$\|f_N\|_{L_s(\mu;\mathcal{X})} \lesssim \Big\| \Big(\|f_0\|_{\mathcal{X}}^p + \sum_{n=1}^{N} \|f_n - f_{n-1}\|_{\mathcal{X}}^p \Big)^{1/p} \Big\|_{L_s(\mu)},$$

where the implied constant may depend at most on $\mathcal{X}$, p, and s.

A Banach space $\mathcal{X}$ is said to have *martingale cotype* $q \in [2, \infty)$ if, for the same quantities as above, there holds

$$\Big\| \Big(\|f_0\|_{\mathcal{X}}^q + \sum_{n=1}^{N} \|f_n - f_{n-1}\|_{\mathcal{X}}^q \Big)^{1/q} \Big\|_{L_s(\mu)} \lesssim \|f_N\|_{L_s(\mu;\mathcal{X})}.$$

In [**19**, p. 221], the case $s = p$ is taken as the definition for martingale type p, and the case $s = q$ for martingale cotype q, but the equivalence with other values of s is observed immediately afterwards. See also [**20**, Chapter 10] for more information on these and related conditions. The definition is usually formulated for probability spaces, but the equivalence with any σ-finite measure space follows easily (approximating arbitrary martingales by ones supported on a set of finite measure, and multiplying the measure by a constant to achieve a probability space). See [**13**, Chapter 3] for a systematic development of martingale theory in the σ-finite setting, and in particular [**13**, Section 3.5.d] for martingale type and cotype.

Representations of singular integrals with the help of martingales are known in various contexts, and this provides the link to our present aims. Our main result can be formulated as follows. We refer the reader to Section 2 for relevant definitions.

THEOREM 1.2. *Let (E, d, μ) be a doubling metric measure space, and let $Tf(u) = \int_E K(u,v)f(v)d\mu(v)$ be an integral operator whose kernel K satisfies the Calderón–Zygmund standard estimates and the additional finiteness property*

$$K(u,v) = 0 \quad \textit{unless} \quad r \le d(u,v) < R.$$

Let $n := 1 + \log(R/r)$.

(1) *If $\mathcal{X}$ is any Banach space and $s \in [1,\infty]$, then $\|Tf\|_{L_s(\mu;\mathcal{X})} \lesssim n\|f\|_{L_s(\mu;\mathcal{X})}$.*
(2) *If $s \in (1,\infty)$ and T is bounded on the scalar-valued $L_s(\mu)$ with $\|Tf\|_{L_s(\mu)} \lesssim \|f\|_{L_s(\mu)}$, and if $\mathcal{X}$ has an equivalent uniformly convex norm, then for some $\theta \in [0,1)$, we have $\|Tf\|_{L_s(\mu;\mathcal{X})} \lesssim n^\theta \|f\|_{L_s(\mu;\mathcal{X})}$.*
(3) *If $1 < p \le s \le q < \infty$ and $\mathcal{X}$ has martingale type p and martingale cotype q, then one can take $\theta = 1/p - 1/q$.*

Theorem 1.2 applies in particular to the case when $E = \{1, \dots, N\}$, d is the usual distance, μ is the counting measure, and T is the finite Hilbert transform. Then one can take $r = 1$ and $R = N$. This important case of Theorem 1.2 was obtained in [**11**], where it is also shown that (3) is the best possible statement in general. The question, answered by Theorem 1.2, of extending this particular case to general Calderón–Zygmund operators was also raised in [**11**].

The proof of this special case used a transference between the finite Hilbert transform on $\{1, \dots, N\}$ and truncations of the Hilbert transform on $\mathbb{R}$, as well as Petermichl's dyadic martingale representation of the latter [**16**]. The possibility of working directly with the original operator, making use of the more general representation theorems of singular integrals developed in [**14**] and, in the context of abstract spaces, in [**15**], was observed there as an alternative. The present paper implements this programme in detail. By now, several variants of the dyadic representation theorem of [**14**] are available in the literature. The present approach is perhaps closest to the one designed in [**8**].

2. Set-up

A metric measure space (E, d, μ) is said to be doubling μ is a Borel measure on E and the balls satisfy

$$\mu(B(u, 2t)) \lesssim \mu(B(u,t)).$$

We denote $V(u,t) := \mu(B(u,t))$ and $V(u,v) := V(u, d(u,v))$. It follows from doubling that $V(u,v) \eqsim V(v,u)$.

Let $\dot{E}^2 := \{(u,v) \in E \times E : u \neq v\}$. Let $\omega : [0,1] \to [0,\infty)$ be continuous, non-decreasing and doubling in the sense that $\omega(2t) \lesssim \omega(t)$. A function $K : \dot{E}^2 \to \mathbb{C}$ is called an ω-standard kernel if

$$|K(u,v)| \lesssim \frac{1}{V(u,v)}$$

for all $(u,v) \in \dot{E}^2$, and

$$|K(u,v) - K(u,w)| + |K(v,u) - K(w,u)| \lesssim \omega\Big(\frac{d(v,w)}{d(u,v)}\Big)\frac{1}{V(u,v)}$$

for all $(u,v) \in \dot{E}^2$ and $w \in E$ with $d(v,w) \le \frac{1}{2}d(u,v)$ (hence also $(u,w) \in \dot{E}^2$).

Some control of ω is usually required, and we define

$$\|\omega\|_{\mathrm{Dini}^\nu} := \int_0^1 \omega(t)\Big(1+\log\frac{1}{t}\Big)^\nu \frac{dt}{t}.$$

Much of the classical theory of singular integrals is valid under the standard Dini condition with $\nu = 0$, but slightly stronger conditions on this scale are also often required. In Theorem 1.2 (and many other results in the area), it suffices to take $\nu = 1$. In the course of the proof, we will see that somewhat less is actually enough, but we will not insist too much in this at this point. Many results in the literature are formulated for $\omega(t) = t^\delta$, which are easily seen to satisfy the Dini conditions with any $\nu \geq 0$.

In this paper, we will impose the following additional assumption: for some $0 < r < R < \infty$,

$$K(u,v) = 0 \quad \text{unless} \quad r \leq d(u,v) < R. \tag{2.1}$$

This assumption effectively kills the singularity of the kernel $K(u,v)$, making the $L_s(\mu)$ boundedness of the related integral operator

$$Tf(u) = \int_E K(u,v)f(v)d\mu(v)$$

qualitatively trivial:

PROPOSITION 2.1. *Under the above assumptions,*

$$\int_E |K(u,v)|d\mu(u) + \int_E |K(v,u)|d\mu(u) \lesssim 1 + \log\frac{R}{r} =: n,$$

hence

$$\|Tf\|_{L_s(\mu;\mathcal{X})} \lesssim n \cdot \|f\|_{L_s(\mu;\mathcal{X})} \quad \textit{for all} \quad s \in [1,\infty], \quad f \in L_s(\mu;\mathcal{X}), \tag{2.2}$$

where $\mathcal{X}$ is an arbitrary Banach space.

PROOF.

$$\int_E |K(u,v)|d\mu(u) \leq \sum_{j: r \leq 2^j \leq \frac{1}{2}R} \int_{2^j \leq |u-v| < 2^{j+1}} |K(u,v)|d\mu(u),$$

where

$$|K(u,v)| \leq \frac{1}{\lambda(u,d(u,v))} \lesssim \frac{1}{\lambda(v,d(u,v))} \leq \frac{1}{\lambda(v,2^j)},$$

thus

$$\int_{2^j \leq |u-v| < 2^{j+1}} |K(u,v)|d\mu(u) \lesssim \frac{\mu(B(v,2^{j+1}))}{\lambda(v,2^j)} \leq \frac{\lambda(v,2^{j+1})}{\lambda(v,2^j)} \lesssim 1,$$

and hence

$$\int_E |K(u,v)|d\mu(u) \lesssim \sum_{j: r \leq 2^j \leq \frac{1}{2}R} 1 \lesssim 1 + \log\frac{R}{r}.$$

The estimate of $K(v,u)$ follows by symmetry of the assumptions, and the estimate of Tf is then standard. □

But the point we wish to address is conditions under which we can beat the trivial bound (2.2).

The notation

$$n := 1 + \log\frac{R}{r}$$

is motivated by the key example of $E = \{1, \dots, 2^n\}$. Since $1 \le d(u,v) < 2^n$ for all $(u,v) \in \dot{E}^2$, the assumption (2.1) is automatically satisfied with $r = 1$, $R = 2^n$, hence $1 + \log R/r = 1 + n \log 2 \asymp n$. Another basic example arises from usual (smooth) truncations of a standard kernel.

3. Dyadic cubes

A construction of dyadic cubes in general doubling metric spaces is first due to Christ [**5**]. Rather than just one system, we will require an ensemble of dyadic systems equipped with a suitable probability measure, leading to a notion of a random dyadic system. In the generality of doubling metric (even quasi-metric) spaces, a first such construction is from [**10**] with subsequent variants in [**1**, **9**, **12**, **15**]. We will quote some results from [**1**], specialising them to a metric space (thus taking $A_0 = 1$ for the quasi-triangle constant featuring in [**1**].)

There are reference points x_α^k, where $k \in \mathbb{Z}$ and $\alpha \in \mathcal{A}_k$, some countable index sets. There is parameter space $\Omega = \prod_{k \in \mathbb{Z}} \Omega_k$, where each Ω_k is a copy of the same finite set Ω_0. Thus, there is a natural probability measure on Ω.

For every $\omega \in \Omega$, there is a partial order relation $\le_\omega$ among pairs (k, α) and (ℓ, γ). Within a fixed level, we declare that $(k, \alpha) \le_\omega (k, \beta)$ if and only if $\alpha = \beta$. The restriction of the relation $\le_\omega$ between levels k and $k+1$ depends only on the component ω_k. The relation between arbitrary levels $\ell \ge k$ is obtained via extension by transitivity.

There are also new random reference points $z_\alpha^k = z_\alpha^k(\omega_k)$ such that $d(z_\alpha^k(\omega), x_\alpha^k) < 2\delta^k$. (This follows from [**1**, Eq. (2.1)] and the definition of z_α^k further down the same page.)

By [**1**, Theorem 2.11], there are open and closed sets $\tilde{Q}_\alpha^k(\omega)$ and $\bar{Q}_\alpha^k(\omega)$ that satisfy

$$B(z_\alpha^k, \tfrac{1}{6}\delta^k) \subseteq \tilde{Q}_\alpha^k(\omega) \subseteq \bar{Q}_\alpha^k(\omega) \subseteq B(z_\alpha^k, 6\delta^k).$$

We refer to these sets as "(dyadic) cubes". Here $\delta \in (0,1)$ is a small parameter that indicates the ratio of the scales of two consecutive generations of the dyadic system. In $\mathbb{R}^d$, one typically takes $\delta = \frac{1}{2}$. These cubes satisfy the disjointness

$$\tilde{Q}_\alpha^k(\omega) \cap \bar{Q}_\beta^k(\omega) = \varnothing \quad (\alpha \ne \beta)$$

and the covering properties

$$E = \bigcup_\alpha \bar{Q}_\alpha^k(\omega), \quad \bar{Q}_\alpha^k(\omega) = \bigcup_{\beta:(k+1,\beta) \le_\omega (k,\alpha)} \bar{Q}_\beta^k(\omega).$$

While these properties are valid for each fixed $\omega \in \Omega$, we additionally have the following probabilistic properties:

$$\begin{aligned} &\mathbb{P}_\omega\Big(u \in \bigcup_\alpha \partial_\varepsilon Q_\alpha^k(\omega)\Big) \le C\varepsilon^\eta, \\ &\qquad\qquad \partial_\varepsilon Q_\alpha^k(\omega) := \{u \in \bar{Q}_\alpha^k(\omega) : d(u, (\tilde{Q}_\alpha^k(\omega))^c) < \varepsilon\delta^k\}, \end{aligned} \tag{3.1}$$

for some fixed $0 < \eta \le 1 \le C < \infty$ and all $\varepsilon > 0$, and in particular the negligible boundary:

$$\mathbb{P}_\omega\Big(x \in \bigcup_{k,\alpha} \partial Q_\alpha^k(\omega)\Big) = 0 \quad \partial Q_\alpha^k(\omega) := \bar{Q}_\alpha^k(\omega) \setminus \tilde{Q}_\alpha^k(\omega).$$

The sets $\tilde{Q}^k_\alpha(\omega)$ and $\bar{Q}^k_\alpha(\omega)$ (and hence also $\partial_\varepsilon Q^k_\alpha(\omega)$ and $\partial Q^k_\alpha(\omega)$) depend only on $(\omega_i)_{i=k}^\infty$. We will also need to understand the following event:

LEMMA 3.1. *Let us fix $\varepsilon > 0$ so small that the bound in (3.1) satisfies $C\varepsilon^\eta \leq \frac{1}{2}$. Then there exists an $m_0 \in \mathbb{Z}_+$ such that we have the following: For some $k \in \mathbb{Z}$ and $m \geq m_0$, let $x^{k+m}_\beta, x^{k+m}_\gamma$ be two reference points at level $k+m$ with $d(x^{k+m}_\beta, x^{k+m}_\gamma) \leq \frac{1}{2}\varepsilon\delta^k$. Consider the random event*

$$\begin{aligned} A &:= \{\omega \in \Omega : \exists \alpha \text{ such that } (k+m,\beta) \leq_\omega (k,\alpha) \text{ and } (k+m,\gamma) \leq_\omega (k,\alpha)\} \\ &= \{\omega \in \Omega : \exists \alpha \text{ such that } \bar{Q}^{k+m}_\beta(\omega) \subseteq \bar{Q}^k_\alpha(\omega) \text{ and } \bar{Q}^{k+m}_\gamma(\omega) \subseteq \bar{Q}^k_\alpha(\omega)\} \end{aligned}$$

Then

(1) *A depends only on the components $(\omega_{k+i})_{i=0}^{m-1}$, and*
(2) *$\mathbb{P}(A) \geq \frac{1}{2}$.*

PROOF. Claim (1) is immediate from the facts that the restriction of the relation $\leq_\omega$ between levels k and $k+1$ depends only on the component ω_k, and that in general it is defined via extension by transitivity.

When $\varepsilon > 0$ is chosen as stated, with probability at least $\frac{1}{2}$, the point x^{k+m}_β is not contained in any $\partial_\varepsilon Q^k_\alpha(\omega)$. In particular, choosing α so that $x^{k+m}_\beta \in \bar{Q}^k_\alpha(\omega)$ (which must exist by the covering property), it follows that $d(x^{k+m}_\beta, (\tilde{Q}^k_\alpha(\omega))^c) \geq \varepsilon\delta^k$. Now suppose that $d(x^{k+m}_\gamma, x^{k+m}_\beta) \leq \frac{1}{2}\varepsilon\delta^k$. Then

$$\begin{aligned} d(z^{k+m}_\gamma(\omega), (\tilde{Q}^k_\alpha(\omega))^c) &\geq d(x^{k+m}_\beta, (\tilde{Q}^k_\alpha(\omega))^c) - d(z^{k+m}_\gamma(\omega), x^{k+m}_\gamma) \\ &\quad - d(x^{k+m}_\gamma, x^{k+m}_\beta) \\ &> \varepsilon\delta^k - 2\delta^{k+m} - \frac{1}{2}\varepsilon\delta^k \geq (\frac{1}{2}\varepsilon - \delta^{m_0})\delta^k > 0 \end{aligned}$$

provided that m_0 is large enough. Hence $z^{k+m}_\gamma(\omega) \in \tilde{Q}^k_\alpha(\omega)$, and thus $(k+m,\gamma) \leq_\omega (k,\alpha)$. □

In the sequel, we will drop the somewhat heavy notation above, and denote a typical $Q^k_\alpha(\omega)$ simply by Q. We let $\mathcal{D}_k = \mathcal{D}_k(\omega) := \{Q^k_\alpha(\omega) : \alpha \in \mathcal{A}_k\}$ be the dyadic cubes of generation k. We also write $\ell(Q) := \delta^k$ if $Q \in \mathcal{D}_k$. We denote by $B_Q := B(z^k_\alpha, 6\delta^k)$ the (non-random!) ball that contains $Q = Q^k_\alpha(\omega)$.

4. A dyadic decomposition of finite kernels

With the dyadic cubes defined above, we have the corresponding averaging (or conditional expectation; hence the notation) operators

$$\langle f\rangle_Q := \frac{1}{\mu(Q)}\int_Q f d\mu, \quad \mathbb{E}_Q f := 1_Q\langle f\rangle_Q, \quad \mathbb{E}_i f := \sum_{Q\in\mathcal{D}_i} \mathbb{E}_Q f$$

and their differences

$$\mathbb{D}_i f := \mathbb{E}_{i+1} f - \mathbb{E}_i f = \sum_{Q\in\mathcal{D}_i} \mathbb{D}_Q f, \quad \mathbb{D}_Q f = \sum_{P\in\operatorname{ch}(Q)} \mathbb{E}_P f - \mathbb{E}_Q f,$$

where, for $Q \in \mathcal{D}_i$, we denote by

$$\operatorname{ch}(Q) := \{P \in \mathcal{D}_{i+1} : P \subseteq Q\}$$

the collection of its children, which has some cardinality between 1 and a fixed finite M depending only on the geometry of (E, d).

For any $-\infty < \sigma < \tau < \infty$, we can decompose

$$f = \mathbb{E}_\sigma f + \sum_{i=\sigma}^{\tau} \mathbb{D}_i f + \mathbb{F}_\tau f, \quad \mathbb{F}_\tau f := (f - \mathbb{E}_\tau)f = \sum_{Q \in \mathcal{D}_\tau} 1_Q (f - \langle f \rangle_Q)$$

LEMMA 4.1. *If τ is large enough so that $\delta^\tau \ll r$, then*

$$\|T\mathbb{F}_\tau f\|_{L_s(\mu)} \lesssim \|\omega\|_{\mathrm{Dini}} \|f\|_{L_s(\mu)}, \qquad \|\omega\|_{\mathrm{Dini}} := \int_0^1 \omega(t) \frac{dt}{t}.$$

PROOF. Since $\int_Q (f(v) - \langle f \rangle_Q) d\mu(v) = 0$, we have

$$\begin{aligned} T\mathbb{F}_\tau f(u) &= \sum_{Q \in \mathcal{D}_\tau} \int_Q K(u, v)(f(v) - \langle f \rangle_Q) d\mu(v) \\ &= \sum_{Q \in \mathcal{D}_\tau} \int_Q (K(u, v) - K(u, z_Q))(f(v) - \langle f \rangle_Q) d\mu(v). \end{aligned}$$

If $d(u, v) \leq r$ for all $v \in Q$ (thus also for $v = z_Q$), this Q does not contribute to the sum above. If $d(u, v) > r$ for some $v \in Q$, then $d(u, z_Q) \geq d(u, v) - d(v, z_Q) \gtrsim r$ if $\ell(Q) = \delta^\tau \ll r$. Then

$$\begin{aligned} &\int_Q |(K(u, v) - K(u, z_Q))(f(v) - \langle f \rangle_Q)| d\mu(v) \\ &\quad \lesssim \int_Q \omega\Big(\frac{d(v, z_Q)}{d(u, z_Q)}\Big) \frac{1}{V(u, d(u, z_Q))} |f(v) - \langle f \rangle_Q | d\mu(v) \\ &\quad \lesssim \omega\Big(\frac{r}{d(u, z_Q)}\Big) \frac{1}{V(u, d(u, z_Q))} \int_Q |f(v)| d\mu(v) \\ &\quad \lesssim \int_Q \omega\Big(\frac{r}{d(u, v) + r}\Big) \frac{1}{V(u, d(u, v) + r)} |f(v)| d\mu(v) \\ &\quad =: \int_Q K_r(u, v) |f(v)| d\mu(v) \end{aligned}$$

Hence, summing over $Q \in \mathcal{D}_\tau$,

$$|T\mathbb{F}_\tau f(u)| \lesssim \int_E K_r(u, v) |f(v)| d\mu(v).$$

The kernel here satisfies

$$\begin{aligned} &\int_E K_r(u, v) d\mu(v) \\ &\leq \int_{B(u,r)} \omega(1) \frac{1}{V(u, r)} d\mu(v) + \sum_{j=0}^{\infty} \int_{2^j r \leq |u-v| < 2^{j+1} r} \omega(2^{-j}) \frac{1}{V(u, 2^j r)} d\mu(v) \\ &\leq \omega(1) \frac{1}{V(u, r)} \mu(B(u, r)) + \sum_{j=0}^{\infty} \omega(2^{-j}) \frac{1}{V(u, 2^j r)} \mu(B(u, 2^{j+1} r)) \\ &\lesssim \sum_{j=0}^{\infty} \omega(2^{-j}) \eqsim \int_0^1 \omega(t) \frac{dt}{t}, \end{aligned}$$

and a similar bound for $\int_E K_r(v,u)d\mu(v)$ by symmetry; hence

$$\Big\| u \mapsto \int_E K_r(u,v)|f(v)|d\mu(v)\Big\|_{L_s(\mu)} \lesssim \|\omega\|_{\mathrm{Dini}}\|f\|_{L_s(\mu)}$$

by standard considerations. □

LEMMA 4.2. *If σ is small enough so that $\delta^\sigma \gtrsim R$, then*

$$\|T\mathbb{E}_\sigma f\|_{L_s(\mu;\mathcal{X})} \lesssim \sup_{P\in\mathcal{D}_\sigma} \frac{\|T(1_P)\|_{L_s(\mu)}}{\mu(P)^{1/s}}\|f\|_{L_s(\mu;\mathcal{X})}.$$

Thus $T\mathbb{E}_\sigma$ is bounded on $L_s(\mu;\mathcal{X})$ assuming the cube testing condition *that the supremum above is finite.*

PROOF. If $v \in Q \in \mathcal{D}_\sigma$ and $u \in E$, then $K(u,v)$ is non-zero only if $d(u,v) < R \lesssim \ell(Q)$, and hence only if $u \in cB_Q$ for some constant c. Thus

$$T\mathbb{E}_\sigma f(u) = \sum_{Q\in\mathcal{D}_\sigma}\int_Q K(u,v)d\mu(v)\langle f\rangle_Q = \sum_{Q\in\mathcal{D}_\sigma} 1_{cB_Q}(u)T(1_Q)(u)\langle f\rangle_Q.$$

By the geometric doubling property, the balls cB_Q have bounded overlap, and hence

$$\begin{aligned}\|T\mathbb{E}_\sigma f\|_{L_s(\mu)} &\lesssim \Big(\sum_{Q\in\mathcal{D}_\sigma}\|1_{cB_Q}T(1_Q)\langle f\rangle_Q\|^s_{L_s(\mu;\mathcal{X})}\Big)^{1/s}\\ &\leq \sup_{P\in\mathcal{D}_\sigma}\frac{\|T(1_P)\|_{L_s(\mu)}}{\mu(P)^{1/s}}\Big(\sum_{Q\in\mathcal{D}_\sigma}\mu(Q)\|\langle f\rangle_Q\|^s_{\mathcal{X}}\Big)^{1/s}\\ &\leq \sup_{P\in\mathcal{D}_\sigma}\frac{\|T(1_P)\|_{L_s(\mu)}}{\mu(P)^{1/s}}\|f\|_{L_s(\mu;\mathcal{X})}.\end{aligned}$$

□

To prove the boundedness of T on $L_s(\mu;\mathcal{X})$, we consider the pairing

$$\langle Tf, g\rangle, \quad f\in L_s(\mu;\mathcal{X}), \quad g \in L_{s'}(\mu;\mathcal{X}').$$

This can be expanded as

$$\langle Tf,g\rangle = \langle T\mathbb{E}_\sigma f, g\rangle + \sum_{i=\sigma}^{\tau}\langle T\mathbb{D}_i f, g\rangle + \langle T\mathbb{F}_{\tau+1}f,g\rangle.$$

We already dealt with the first and the last terms in Lemmas 4.1 and 4.2. In the sum, expanding also g, we obtain

$$\sum_{i=\sigma}^{\tau}\langle T\mathbb{D}_i f,g\rangle = \sum_{i=\sigma}^{\tau}\langle T\mathbb{D}_i f,\mathbb{E}_\sigma g\rangle + \sum_{i,j=\sigma}^{\tau}\langle T\mathbb{D}_i f,\mathbb{D}_j g\rangle + \sum_{i=\sigma}^{\tau}\langle T\mathbb{D}_i f,\mathbb{F}_{\tau+1}g\rangle \tag{4.1}$$

The double sum in (4.1) can be reorganised as

$$\sum_{i,j=\sigma}^{\tau} \langle T\mathbb{D}_i f, \mathbb{D}_j g\rangle = \sum_{\substack{i,j=\sigma \\ i=j}}^{\tau} \langle T\mathbb{D}_i f, \mathbb{D}_j g\rangle + \sum_{\substack{i,j=\sigma \\ i<j}}^{\tau} \langle T\mathbb{D}_i f, \mathbb{D}_j g\rangle + \sum_{\substack{i,j=\sigma \\ i>j}}^{\tau} \langle T\mathbb{D}_i f, \mathbb{D}_j g\rangle$$

$$= \sum_{i=\sigma}^{\tau}\langle T\mathbb{D}_i f, \mathbb{D}_i g\rangle + \sum_{j=\sigma}^{\tau}\langle T(\mathbb{E}_j f - \mathbb{E}_\sigma f), \mathbb{D}_j g\rangle + \sum_{i=\sigma}^{\tau}\langle T\mathbb{D}_i f, \mathbb{E}_i g - \mathbb{E}_\sigma g\rangle$$

$$= \sum_{i=\sigma}^{\tau}\Big(\langle T\mathbb{D}_i f, \mathbb{D}_i g\rangle + \langle T\mathbb{E}_i f, \mathbb{D}_i g\rangle + \langle T\mathbb{D}_i f, \mathbb{E}_i g\rangle\Big)$$

$$- \sum_{i=\sigma}^{\tau}\langle T\mathbb{E}_\sigma f, \mathbb{D}_j g\rangle - \sum_{i=\sigma}^{\tau}\langle T\mathbb{D}_i, \mathbb{E}_\sigma g\rangle.$$

Substituting back to (4.1), the last term right above cancels out with the first term in (4.1), leaving

$$\sum_{i=\sigma}^{\tau}\langle T\mathbb{D}_i f, g\rangle = \sum_{i=\sigma}^{\tau}\Big(\langle T\mathbb{D}_i f, \mathbb{D}_i g\rangle + \langle T\mathbb{E}_i f, \mathbb{D}_i g\rangle + \langle T\mathbb{D}_i f, \mathbb{E}_i g\rangle\Big)$$
$$- \sum_{i=\sigma}^{\tau}\langle T\mathbb{E}_\sigma f, \mathbb{D}_j g\rangle + \sum_{i=\sigma}^{\tau}\langle T\mathbb{D}_i f, \mathbb{F}_{\tau+1} g\rangle.$$

Here

$$\sum_{i=\sigma}^{\tau}\langle T\mathbb{E}_\sigma f, \mathbb{D}_j g\rangle = \langle T\mathbb{E}_\sigma f, \mathbb{E}_\tau g - \mathbb{E}_\sigma g\rangle$$

is controlled by Lemma 4.2 and the contractivity of $\mathbb{E}_i$, while

$$\sum_{i=\sigma}^{\tau}\langle T\mathbb{D}_i f, \mathbb{F}_\tau g\rangle = \langle \mathbb{E}_\tau - \mathbb{E}_\sigma f, T^*\mathbb{F}_\tau g\rangle$$

is controlled by an application of Lemma 4.1 on the dual side, observing the symmetry of the assumptions.

Altogether, we find that

$$\Big|\langle Tf, g\rangle - \sum_{i=\sigma}^{\tau}\Big(\langle T\mathbb{D}_i f, \mathbb{D}_i g\rangle + \langle T\mathbb{E}_i f, \mathbb{D}_i g\rangle + \langle T\mathbb{D}_i f, \mathbb{E}_i g\rangle\Big)\Big|$$
$$\lesssim \Big(\sup_{P\in\mathscr{D}_\sigma} \frac{\|T(1_P)\|_{L_s(\mu)}}{\mu(P)^{1/s}} + \|\omega\|_{\mathrm{Dini}}\Big)\|f\|_{L_s(\mu;\mathcal{X})}\|g\|_{L_{s'}(\mu;\mathcal{X}')}.$$

This is valid for any Banach space $\mathcal{X}$, and hence the core is estimating the sum on the left. For the previous estimate, we need $\delta^\tau \ll r$ and $\delta^\sigma \gtrsim R$, where the implied constants are absolute. Thus we may take $\tau \asymp \log\frac{1}{r}$ and $\sigma \asymp \log\frac{1}{R}$ so that the length of the sum is

$$1 + (\tau - \sigma) \asymp 1 + \log\frac{R}{r} = n.$$

Proving a uniform bound for each term and applying the triangle inequality would recover the trivial estimate, which is linear in n, and which we already obtained by more elementary considerations. We wish to make use of the above decomposition to beat this trivial bound.

5. Estimates for Haar coefficients

The difference operators $\mathbb{D}_i$ have the piecewise local expansion

$$\mathbb{D}_i f = \sum_{Q\in\mathscr{D}_i} \mathbb{D}_Q f,$$

where each $\mathbb{D}_Q$, in turn, can be expanded in terms of a bounded number of "Haar" functions associated with Q; they are constant on the dyadic children of Q:

$$\mathbb{D}_Q f = \sum_{\alpha=1}^{m_Q-1} \langle f, h_Q^\alpha\rangle h_Q^\alpha.$$

Here $m_Q \in \{1,\ldots,M\}$ is the number of dyadic children of Q. In general it can happen that $m_Q = 1$, in which case the sum above is empty, and $\mathbb{D}_Q = 0$. Denoting $h_Q^0 := \mu(Q)^{-1/2} 1_Q$, we obtain similar formulas for the averaging operators

$$\mathbb{E}_i f = \sum_{Q\in\mathscr{D}_i} \mathbb{E}_Q f, \quad \mathbb{E}_Q f = \langle f\rangle_Q 1_Q = \langle f, h_Q^0\rangle h_Q^0.$$

Thus each of the three types of terms $\langle T\mathbb{D}_i f, \mathbb{D}_i g\rangle$, $\langle T\mathbb{E}_i f, \mathbb{D}_i g\rangle$, $\langle T\mathbb{D}_i f, \mathbb{E}_i g\rangle$ can be expanded in terms of a bounded number of sums of the type

$$\sum_{P,Q\in\mathscr{D}_i} \chi_{\alpha,\beta} \langle T h_P^\alpha, h_Q^\beta\rangle \langle f, h_P^\alpha\rangle \langle g, h_Q^\beta\rangle,$$

where $\chi_{\alpha,\beta} \in \{0,1\}$. Since there is at most one $\mathbb{E}_i$, at most one of α and β can be 0 in any given sum of this type.

Estimates for the Haar coefficients $\langle T h_P^\alpha, h_Q^\beta\rangle$ are well known in the Euclidean setting. The extension to doubling metric spaces depends on the fact that a certain Hardy inequality remains valid for the dyadic cubes in this situation, as observed by Auscher and Routin [**2**]. We only need the following special case of their [**2**, Lemma 2.4]: If $P, Q \in \mathscr{D}$ are disjoint, then

$$\int_P \int_Q \frac{1}{V(u,v)} d\mu(u) d\mu(v) \lesssim \mu(P)^{1/s} \mu(Q)^{1/s'} \tag{5.1}$$

for $s \in (1,\infty)$. Then

$$\langle T h_P^\alpha, h_Q^\beta\rangle = \sum_{\substack{P'\in \mathrm{ch}(P)\\ Q'\in\mathrm{ch}(Q)}} \langle T 1_{P'}, 1_{Q'}\rangle \langle h_P^\alpha\rangle_{P'} \langle h_Q^\beta\rangle_{Q'}.$$

If $\ell(P) = \ell(Q)$ and $P' \neq Q'$, then $P' \cap Q' = \varnothing$, and (5.1) applies to give

$$\begin{aligned} |\langle T 1_{P'}, 1_{Q'}\rangle| &= \Big| \int_{P'} \int_{Q'} K(u,v) du dv \Big| \\ &\lesssim \int_{P'} \int_{Q'} \frac{1}{V(u,v)} du dv \lesssim \mu(P')^{1/2} \mu(Q')^{1/2}. \end{aligned}$$

Thus

$$\begin{aligned} &\Big| \sum_{\substack{P'\in \mathrm{ch}(P)\\ Q'\in\mathrm{ch}(Q)\\ P'\neq Q'}} \langle T 1_{P'}, 1_{Q'}\rangle \langle h_P^\alpha\rangle_{P'} \langle h_Q^\beta\rangle_{Q'} \Big| \\ &\lesssim \sum_{P'\in\mathrm{ch}(P)} \mu(P')^{1/2} |\langle h_P^\alpha\rangle_{P'}| \sum_{Q'\in\mathrm{ch}(Q)} \mu(Q')^{1/2} |\langle h_Q^\beta\rangle_{Q'}|, \end{aligned} \tag{5.2}$$

where

$$\sum_{P'\in\mathrm{ch}(P)}\mu(P')^{1/2}|\langle h_P^\alpha\rangle_{P'}|\leq M^{1/2}\Big(\sum_{P'\in\mathrm{ch}(P)}\mu(P')|\langle h_P^\alpha\rangle_{P'}|^2\Big)^{1/2}$$
$$=M^{1/2}\|h_P^\alpha\|_{L_2(\mu)}=M^{1/2},$$

and similarly for the sum over $Q'\in\mathrm{ch}(Q)$. Thus (5.2) $\lesssim 1$.

If $P,Q\in\mathcal{D}_i$ are different, then all their children P',Q' satisfy $P'\neq Q'$, and we have proved that

$$|\langle Th_P^\alpha,h_Q^\beta\rangle|\lesssim 1,\quad P,Q\in\mathcal{D}_i,\quad P\neq Q.$$

If $P=Q$, then we need to deal in addition with the sum

$$\Big|\sum_{P'\in\mathrm{ch}(P)}\langle T1_{P'},1_{P'}\rangle\langle h_P^\alpha\rangle_{P'}\langle h_P^\beta\rangle_{P'}\Big|$$
$$\leq\sup_{Q\in\mathcal{D}}\frac{|\langle T1_Q,1_Q\rangle|}{\mu(Q)}\sum_{P'\in\mathrm{ch}(P)}\mu(P')|\langle h_P^\alpha\rangle_{P'}\langle h_P^\beta\rangle_{P'}|,$$

where

$$\sum_{P'\in\mathrm{ch}(P)}\mu(P')|\langle h_P^\alpha\rangle_{P'}\langle h_P^\beta\rangle_{P'}|$$
$$\leq\Big(\sum_{P'\in\mathrm{ch}(P)}\mu(P')|\langle h_P^\alpha\rangle_{P'}|^2\Big)^{1/2}\Big(\sum_{P'\in\mathrm{ch}(P)}\mu(P')|\langle h_P^\beta\rangle_{P'}|^2\Big)^{1/2}$$
$$=\|h_P^\alpha\|_{L_2(\mu)}\|h_P^\beta\|_{L_2(\mu)}=1.$$

The finiteness of

$$\|T\|_{\mathrm{wbp}(\mathcal{D})}:=\sup_{Q\in\mathcal{D}}\frac{|\langle T1_Q,1_Q\rangle|}{\mu(Q)}$$

is the so-called weak boundedness property. It follows from the finiteness of the testing quantity

$$\|T\|_{\mathrm{test}^s(\mathcal{D})}:=\sup_{Q\in\mathcal{D}}\frac{\|T1_Q\|_{L_s(\mu)}}{\mu(Q)^{1/s}},$$

which in turn is clearly dominated by the operator norm of T on $L_s(\mu)$.

If P and Q are well separated, we need and have a better estimate. Assuming for instance that $\alpha\neq 0$ (the case $\beta\neq 0$ being symmetric), we have

$$\langle Th_P^\alpha,h_Q^\beta\rangle=\int_P\int_Q K(u,v)h_P^\alpha(v)h_Q^\beta(u)d\mu(u)d\mu(v)$$
$$=\int_P\int_Q[K(u,v)-K(u,z_P)]h_P^\alpha(v)h_Q^\beta(u)d\mu(u)d\mu(v).$$

If $d(P,Q)\gg\ell(P)=\ell(Q)$, then $v\in P$ and $u\in Q$ satisfy $d(v,z_P)\lesssim\ell(P)\ll d(u,v)$, and hence

$$|\langle Th_P^\alpha,h_Q^\beta\rangle|\lesssim\int_P\int_Q\omega\Big(\frac{\ell(P)}{d(u,z_P)}\Big)\frac{1}{V(u,z_P)}|h_P^\alpha(v)h_Q^\beta(u)|d\mu(u)d\mu(v)$$
$$\lesssim\omega\Big(\frac{\ell(P)}{d(z_Q,z_P)+\ell(P)}\Big)\frac{\|h_P^\alpha\|_{L_1(\mu)}\|h_Q^\beta\|_{L_1(\mu)}}{V(z_Q,d(z_Q,z_P)+\ell(P))}\tag{5.3}$$
$$\lesssim\omega\Big(\frac{\ell(P)}{d(z_Q,z_P)+\ell(P)}\Big)\frac{\sqrt{\mu(P)\mu(Q)}}{V(z_Q,d(z_Q,z_P)+\ell(P))}.$$

Actually, this final bound is valid also for $d(P,Q) \lesssim \ell(P) = \ell(Q)$. In this case, $d(z_Q, z_P) + \ell(P) \eqsim \ell(P)$, so that the argument of ω is roughly 1, and

$$V(z_Q, d(z_Q, z_P) + \ell(P)) \eqsim \mu(Q) \eqsim \mu(P).$$

Thus, when $d(P,Q) \lesssim \ell(P) = \ell(Q)$, the bound (5.3) reduces to the uniform bound obtained earlier.

6. Extracting paraproducts

In the two sums

$$\sum_{i=\sigma}^{\tau} \langle T\mathbb{E}_i f, \mathbb{D}_i g\rangle, \quad \sum_{i=\sigma}^{\tau} \langle T\mathbb{D}_i f, \mathbb{E}_i g\rangle,$$

we need to force some cancellation as follows. Let us deal with the first one, the second being symmetric. A typical term in this sum is

$$\langle T\mathbb{E}_i f, \mathbb{D}_i g\rangle = \sum_{P,Q\in\mathcal{D}_i} \langle T1_P, \mathbb{D}_Q g\rangle \langle f\rangle_P.$$

Writing

$$\langle f\rangle_P = (\langle f\rangle_P - \langle f\rangle_Q) + \langle f\rangle_Q,$$

we have introduced some cancellation into the first term. On the other hand, in the sum involving the second term, the only factor depending on P is the indicator 1_P in the pairing $\langle T1_P, \mathbb{D}_Q g\rangle$. Thus, summing over $P \in \mathcal{D}_i$ first, we obtain by linearity

$$\sum_{P,Q\in\mathcal{D}_i} \langle T1_P, \mathbb{D}_Q g\rangle \langle f\rangle_Q = \sum_{Q\in\mathcal{D}_i} \Big\langle T \sum_{P\in\mathcal{D}_i} 1_P, \mathbb{D}_Q g\Big\rangle \langle f\rangle_Q = \sum_{Q\in\mathcal{D}_i} \langle T1, \mathbb{D}_Q g\rangle \langle f\rangle_Q.$$

For the finite singular integrals that we consider, there is no trouble in making sense of the expression "$T1$" above, as the action of the integral operator T on bounded function is well defined. Denoting (as usual) $b := T1$, and using the self-adjointness of $\mathbb{D}_Q$, we find that

$$\sum_{i=\sigma}^{\tau} \sum_{Q\in\mathcal{D}_i} \langle b, \mathbb{D}_Q g\rangle \langle f\rangle_Q = \Big\langle \sum_{i=\sigma}^{\tau} \sum_{Q\in\mathcal{D}_i} \mathbb{D}_Q b \langle f\rangle_Q, g\Big\rangle =: \Big\langle \Pi_b^{\sigma,\tau} f, g\Big\rangle.$$

Here

$$\Pi_b^{\sigma,\tau} f := \sum_{i=\sigma}^{\tau} \sum_{Q\in\mathcal{D}_i} \mathbb{D}_Q b \langle f\rangle_Q = \sum_{i=\sigma}^{\tau} \mathbb{D}_i b \mathbb{E}_i f$$

is a truncated version of a *dyadic paraproduct*. We can estimate it as follows. Noting that $\mathbb{D}_i$ are martingale differences, the assumption that $\mathcal{X}$ has martingale type $p \in [1,2]$ implies that

$$\begin{aligned}
\|\Pi_b^{\sigma,\tau} f\|_{L_s(\mu;\mathcal{X})} &\lesssim \Big\| \Big(\sum_{i=\sigma}^{\tau} |\mathbb{D}_i b|^p \|\mathbb{E}_i f\|_{\mathcal{X}}^p \Big)^{1/p} \Big\|_{L_s(\mu;\mathcal{X})} \\
&\le n^{1/p-1/2} \Big\| \Big(\sum_{i=\sigma}^{\tau} |\mathbb{D}_i b|^2 \|\mathbb{E}_i f\|_{\mathcal{X}}^2 \Big)^{1/2} \Big\|_{L_s(\mu;\mathcal{X})} \\
&\le n^{1/p-1/2} \Big\| \Big(\sum_{i=\sigma}^{\tau} \sum_{Q\in\mathcal{D}_i} |\mathbb{D}_Q b|^2 \langle \|f\|_{\mathcal{X}}\rangle_Q^2 \Big)^{1/2} \Big\|_{L_s(\mu;\mathcal{X})}.
\end{aligned}$$

Next, we make a stopping time construction as follows. The initial layer of stopping cubes consist of all $Q \in \mathcal{D}_\sigma$. Assuming that some stopping cube $S \in \mathcal{D}$ has been picked, we look for its maximal dyadic subcubes $S' \subsetneq S$ such that either

$$\langle \|f\|_{\mathcal{X}} \rangle_{S'} > 4 \langle \|f\|_{\mathcal{X}} \rangle_S,$$

or

$$\sum_{\substack{Q \in \mathcal{D} \\ S' \subsetneq Q \subseteq S}} |D_Q b(u)|^2 > \lambda^2 \quad \text{for all} \quad u \in S',$$

where $\lambda > 0$ will be specified shortly. Note that the left-hand side of the stopping criterion is constant on S', since $D_Q b$ is constant on the dyadic children of Q, so the condition "for all $u \in S'$" could be equivalently replaced by "for some $u \in S'$". Let us refer to the stopping cubes arising from the first or the second criterion as being of the first or the second kind, respectively.

For the stopping cubes of the first kind (pairwise disjoint by their maximality), it is immediate from the stopping criterion that

$$\sum_{S'} \mu(S') \leq \sum_{S'} \frac{1}{4\langle \|f\|_{\mathcal{X}} \rangle_S} \int_{S'} \|f\|_{\mathcal{X}} d\mu \leq \frac{1}{4\langle \|f\|_{\mathcal{X}} \rangle_S} \int_S \|f\|_{\mathcal{X}} d\mu = \frac{1}{4}\mu(S).$$

To derive a similar estimate for the stopping cubes of the second kind, let us consider the dyadic square function and its truncations

$$\mathfrak{S}b := \Big(\sum_{Q \in \mathcal{D}} |D_Q b|^2 \Big)^{1/2}, \quad \mathfrak{S}_P b := \Big(\sum_{\substack{Q \in \mathcal{D} \\ Q \subseteq P}} |D_Q b|^2 \Big)^{1/2} = \mathfrak{S}(1_P (b - \langle b \rangle_P)).$$

If $S' \subsetneq S$ is one of the stopping cubes of the second kind, then

$$\mathfrak{S}_S b(u)^2 \geq \sum_{\substack{Q \in \mathcal{D} \\ S' \subsetneq Q \subseteq S}} |D_Q b(u)|^2 > \lambda^2 \quad \text{for all} \quad u \in S'.$$

Thus

$$\begin{aligned} \mu\Big(\bigcup S'\Big) \leq \mu(\mathfrak{S}_S b > \lambda) &\leq \frac{1}{\lambda^2} \int_E (\mathfrak{S}_S b)^2 d\mu \\ &= \frac{1}{\lambda^2} \int_S |b - \langle b \rangle_S|^2 d\mu \leq \frac{1}{\lambda^2} \|b\|^2_{\mathrm{BMO}^2(\mathcal{D})} \mu(S), \end{aligned}$$

where

$$\|b\|_{\mathrm{BMO}^2(\mathcal{D})} := \sup_{Q \in \mathcal{D}} \Big(\frac{1}{\mu(Q)} \int_Q |b - \langle b \rangle_Q|^2 d\mu \Big)^{1/2}$$

is the dyadic BMO norm based on L_2 averages. (The different BMO norms are equivalent by the John–Nirenberg inequality, which remains valid in this generality.)

If we choose some $\lambda \gg \|b\|_{\mathrm{BMO}^2(\mathcal{D})}$, then $\mu(\bigcup S') \ll \mu(S)$. Denoting by $\mathcal{S}$ the collection of all stopping cubes of both kinds, and

$$E_S := S \setminus \bigcup_{\substack{S' \in \mathcal{S} \\ S' \subsetneq S}} S',$$

these subsets $E_S \subseteq S$ are pairwise disjoint, and

$$\mu(E_S) \gtrsim \mu(S).$$

The existence of such subsets is referred to as the *sparseness* of the collection $\mathcal{S}$.

Let us now return to the estimation of $\Pi_b^{\sigma,\tau} f$. For each $Q \in \mathcal{D}_i$ for some $i \in [\sigma, \tau]$, let $\pi Q \in \mathcal{S}$ denote the minimal stopping cube that contains Q. Regrouping the summation under these stopping parents, we have

$$\begin{aligned}\sum_{i=\sigma}^{\tau} \sum_{Q \in \mathcal{D}_i} |\mathbb{D}_Q b|^2 \langle \|f\|_{\mathcal{X}} \rangle_Q^2 &\leq \sum_{S \in \mathcal{S}} \sum_{\substack{Q \in \mathcal{D} \\ \pi Q = S}} |\mathbb{D}_Q b|^2 \langle \|f\|_{\mathcal{X}} \rangle_Q^2 \\ &\lesssim \sum_{S \in \mathcal{S}} \Big(\sum_{\substack{Q \in \mathcal{D} \\ \pi Q = S}} |\mathbb{D}_Q b|^2 \Big) \langle \|f\|_{\mathcal{X}} \rangle_S^2 \\ &\lesssim \sum_{S \in \mathcal{S}} \|b\|_{\mathrm{BMO}^2(\mathcal{D})}^2 1_S \langle \|f\|_{\mathcal{X}} \rangle_S^2.\end{aligned}$$

Hence

$$\begin{aligned}\Big\| \Big(\sum_{i=\sigma}^{\tau} \sum_{Q \in \mathcal{D}_i} |\mathbb{D}_Q b|^2 \langle \|f\|_{\mathcal{X}} \rangle_Q^2 \Big)^{1/2} \Big\|_{L_s(\mu)} & \\ \lesssim \|b\|_{\mathrm{BMO}^2(\mathcal{D})} \Big\| \Big(\sum_{S \in \mathcal{S}} 1_S \langle \|f\|_{\mathcal{X}} \rangle_S^2 \Big)^{1/2} \Big\|_{L_s(\mu)}&\end{aligned}$$

Here

$$\Big\| \Big(\sum_{S \in \mathcal{S}} 1_S \langle \|f\|_{\mathcal{X}} \rangle_S^2 \Big)^{1/2} \Big\|_{L_s(\mu)} \lesssim \Big\| \sum_{S \in \mathcal{S}} 1_S \langle \|f\|_{\mathcal{X}} \rangle_S \Big\|_{L_s(\mu)}.$$

Dualising with $h \in L_{s'}(\mu)$ and using sparseness and the boundedness of the dyadic maximal operator

$$M_{\mathcal{D}} \phi(u) := \sup_{Q \in \mathcal{D}} 1_Q(u) \langle |\phi| \rangle_Q,$$

we find that

$$\begin{aligned}\int_E \Big(\sum_{S \in \mathcal{S}} 1_S \langle \|f\|_{\mathcal{X}} \rangle_S \Big) h d\mu &= \sum_{S \in \mathcal{S}} \langle \|f\|_{\mathcal{X}} \rangle_S \langle h \rangle_S \mu(S) \lesssim \sum_{S \in \mathcal{S}} \langle \|f\|_{\mathcal{X}} \rangle_S \langle h \rangle_S \mu(E_S) \\ &\leq \sum_{S \in \mathcal{S}} \int_{E_S} M_{\mathcal{D}}(\|f\|_{\mathcal{X}})(u) M_{\mathcal{D}} h(u) d\mu(u) \\ &\leq \int_E M_{\mathcal{D}}(\|f\|_{\mathcal{X}})(u) M_{\mathcal{D}} h(u) d\mu(u) \\ &\leq \big\| M_{\mathcal{D}}(\|f\|_{\mathcal{X}}) \big\|_{L_s(\mu)} \|M_{\mathcal{D}} h\|_{L_{s'}(\mu)} \\ &\lesssim \big\| \|f\|_{\mathcal{X}} \big\|_{L_s(\mu)} \|h\|_{L_{s'}(\mu)} = \|f\|_{L_s(\mu;\mathcal{X})} \|h\|_{L_{s'}(\mu)}.\end{aligned}$$

Thus

$$\Big\| \sum_{S \in \mathcal{S}} 1_S \langle \|f\|_{\mathcal{X}} \rangle_S \Big\|_{L_s(\mu)} \lesssim \|f\|_{L_s(\mu;\mathcal{X})},$$

hence

$$\Big\|\Big(\sum_{i=\sigma}^{\tau}\sum_{Q\in\mathcal{D}_i}|\mathbb{D}_Q b|^2\langle\|f\|_{\mathcal{X}}\rangle_Q^2\Big)^{1/2}\Big\|_{L_s(\mu)}\lesssim\|b\|_{\mathrm{BMO}^2(\mathcal{D})}\|f\|_{L_s(\mu;\mathcal{X})}$$

and therefore

$$\|\Pi_b^{\sigma,\tau}f\|_{L_s(\mu;\mathcal{X})}\lesssim n^{1/p-1/2}\|b\|_{\mathrm{BMO}^2(\mathcal{D})}\|f\|_{L_s(\mu;\mathcal{X})},$$

provided that $\mathcal{X}$ has martingale type $p\in[1,2]$.

Recall that $b=T1$. That this belongs to BMO is a well known necessary condition for the boundedness of T on $L_s(\mu)$. The standard argument becomes particularly simple in the present finite case, since $T1$ is a pointwise well-defined function (not an equivalence class modulo constants as in the general theory). We have

$$T1=T(1_{cB_Q})+T(1_{(cB_Q)^c})$$

and hence

$$\begin{aligned}\Big(\int_Q|T1-\langle T1\rangle_Q|^s d\mu\Big)^{1/s}&\leq 2\inf_z\Big(\int_Q|T1-z|^s d\mu\Big)^{1/s}\\
&\leq 2\Big(\int_Q|T1-T(1_{(cB_Q)^c})(z_Q)|^s d\mu\Big)^{1/s}\\
&\leq 2\Big(\int_Q|T(1_{cB_Q})|^s d\mu\Big)^{1/s}+2\Big(\int_Q|T(1_{(cB_Q)^c})-T(1_{(cB_Q)^c})(z_Q)|^s d\mu\Big)^{1/s}.\end{aligned}$$

The first term is dominated by the ball testing condition

$$\|T\|_{\mathrm{test}^s(\mathcal{B})}:=\sup_{B\in\mathcal{B}}\frac{\|T1_B\|_{L_s(\mu)}}{\mu(B)^{1/s}},$$

where $\mathcal{B}$ is the family of all balls $B(u,t)\subset E$; namely

$$\Big(\int_Q|T(1_{cB_Q})|^s d\mu\Big)^{1/s}\leq\|T\|_{\mathrm{test}^s(\mathcal{B})}\mu(cB_Q)^{1/s}\lesssim\|T\|_{\mathrm{test}^s(\mathcal{B})}\mu(Q)^{1/s}.$$

For the remaining integral, we observe that

$$\begin{aligned}&\int_Q|T(1_{(cB_Q)^c})(u)-T(1_{(cB_Q)^c})(z_Q)|^s d\mu(u)\\
&\quad\leq\int_Q\int_{(cB_Q)^c}|K(u,v)-K(z_Q,v)|d\mu(v)d\mu(u)\\
&\quad\lesssim\int_Q\int_{(cB_Q)^c}\omega\Big(\frac{d(u,z_Q)}{d(v,z_Q)}\Big)\frac{1}{V(v,z_Q)}d\mu(v)d\mu(u).\end{aligned}$$

The inner integral is bounded by a constant by computations like those in the proof of Lemma 4.1, and the outer integral over Q then gives simply $\mu(Q)$.

7. Reorganisation of the remaining parts

After the extraction of the paraproducts, we are left with estimating sums of the form

$$\sum_{i=a}^{b}\sum_{P,Q\in\mathcal{D}_i} T(P,Q),\quad T(P,Q)\in\begin{cases}\langle T\mathbb{D}_P f,\mathbb{D}_Q g\rangle,\\ (\langle f\rangle_P-\langle f\rangle_Q)\langle T1_P,\mathbb{D}_Q g\rangle,\\ \langle T\mathbb{D}_P f,1_Q\rangle(\langle g\rangle_Q-\langle g\rangle_P),\end{cases}\tag{7.1}$$

where we note that the last two "products" are actually duality pairings between e.g. $(\langle f\rangle_P-\langle f\rangle_Q)\in\mathcal{X}$ and $\langle T1_P,\mathbb{D}_Q g\rangle\in\mathcal{X}'$.

Dropping for the moment the summands, and recalling the parameter m_0 guaranteed by Lemma 3.1, we further reorganise the summation as

$$\sum_{P,Q\in\mathcal{D}_i}=\sum_{\substack{P,Q\in\mathcal{D}_i\\ d(x_P,x_Q)\leq\frac{1}{2}\varepsilon\delta^{-m_0}\ell(P)}}+\sum_{m=m_0+1}^{\infty}\sum_{\substack{P,Q\in\mathcal{D}_i\\ \frac{1}{2}\varepsilon\delta^{1-m}\ell(P)<d(x_P,x_Q)\leq\frac{1}{2}\varepsilon\delta^{-m}\ell(P)}}.$$

In the m-th sum, the Haar coefficients (appearing when expanding the summands $T(P,Q)$) can be estimated by

$$|\langle Th_P^{\alpha},h_Q^{\beta}\rangle|\lesssim\omega(\delta^m)\frac{\sqrt{\mu(P)\mu(Q)}}{V(z_P,\delta^{-m}\ell(P))},\tag{7.2}$$

which depends on the lower bound on $d(x_P,x_Q)$. The case $m=m_0$ is included, in which case this bound reduces to just $O(1)$, regarding m_0 as fixed.

Let us observe that we can also truncate the summation over m from above. Due to the property that $K(u,v)\neq 0$ only if $d(u,v)<R$, a pairing like $\langle T\mathbb{D}_P f,\mathbb{D}_Q g\rangle$ (or one with $\mathbb{D}_P f$ replaced by 1_P, or $\mathbb{D}_Q g$ replaced by 1_Q) can only be non-zero if $d(P,Q)<R$ and hence also $d(x_P,x_Q)\lesssim R$. But in the m-th term we also have the lower bound $d(x_P,x_Q)>\frac{1}{2}\varepsilon\delta^{1-m}\ell(P)\gtrsim\frac{1}{2}\varepsilon\delta^{1-m}r$. Hence this term gives no contribution unless $\delta^{-m}\lesssim R/r$, thus $m\lesssim n$.

To proceed further, we will need to make use of a random selection of our dyadic systems $\mathcal{D}$. We can write the decomposition of the pairing $\langle Tf,g\rangle$ relative to any such dyadic system, and then take the expectation over the choice of $\mathcal{D}$. By linearity, we can move the expectation inside the summations. If we assume that f and g are boundedly supported (such functions are in any case dense in the spaces that we consider), we entirely avoid any issues of convergence: The summation over i is finite already, and for each i, the bounded support of f can only intersect finitely many of the cube $P,Q\in\mathcal{D}_i$.

Let $A_m(P,Q)$ denote the random event—as in Lemma 3.1, only with a slightly different notation—that the two (random) cubes $P,Q\in\mathcal{D}_i$ share a common ancestor in the (random) dyadic system $\mathcal{D}_{i-m}$. By the summation condition that $d(x_P,x_Q)\leq\frac{1}{2}\varepsilon\delta^{-m}\ell(P)$, Lemma 3.1 guarantees that $\mathbb{P}(A_m(P,Q))\in[\frac{1}{2},1]$. Moreover, Lemma 3.1 also says that $A_m(P,Q)$ depends only on $(\omega_j)_{i-m\leq j<i}$ (observe the difference in indexing compared to Lemma 3.1). On the other hand, the cubes P,Q, as well as their dyadic children, and hence quantities like $\mathbb{D}_P f$ and eventually $T(P,Q)$, depend only on $(\omega_j)_{j\geq i}$. Thus

$$T(P,Q)\text{ and }A_m(P,Q)\text{ are independent.}$$

We can then manipulate

$$
\begin{aligned}
\mathbb{E} & \sum_{\substack{P,Q\in\mathcal{D}_i \\ \frac{1}{2}\varepsilon\delta^{1-m}\ell(P)<d(x_P,x_Q) \\ \leq\frac{1}{2}\varepsilon\delta^{-m}\ell(P)}} T(P,Q) \\
&= \sum_{\substack{P,Q\in\mathcal{D}_i \\ \frac{1}{2}\varepsilon\delta^{1-m}\ell(P)<d(x_P,x_Q) \\ \leq\frac{1}{2}\varepsilon\delta^{-m}\ell(P)}} \mathbb{E}T(P,Q)\frac{\mathbb{E}1_{A_m(P,Q)}}{\mathbb{P}(A_m(P,Q))} \\
&= \sum_{\substack{P,Q\in\mathcal{D}_i \\ \frac{1}{2}\varepsilon\delta^{1-m}\ell(P)<d(x_P,x_Q) \\ \leq\frac{1}{2}\varepsilon\delta^{-m}\ell(P)}} \mathbb{E}[T(P,Q)1_{A_m(P,Q)}]\frac{1}{\mathbb{P}(A_m(P,Q))} \\
&= \mathbb{E} \sum_{\substack{P,Q\in\mathcal{D}_i \\ \frac{1}{2}\varepsilon\delta^{1-m}\ell(P)<d(x_P,x_Q) \\ \leq\frac{1}{2}\varepsilon\delta^{-m}\ell(P)}} T(P,Q)1_{A_m(P,Q)}\frac{1}{\mathbb{P}(A_m(P,Q))} \\
&=: \mathbb{E} \sum_{\substack{P,Q\in\mathcal{D}_i \\ \frac{1}{2}\varepsilon\delta^{1-m}\ell(P)<d(x_P,x_Q) \\ \leq\frac{1}{2}\varepsilon\delta^{-m}\ell(P)}} T_m(P,Q).
\end{aligned}
$$

Thus, at the small cost of multiplying the summands by numerical factors in the range $[1,2]$, we could insert the extra summation condition $A_m(P,Q)$ that $P,Q\in\mathcal{D}_i$ share a common ancestor $S\in\mathcal{D}_{i-m}$.

We can now reorganise the sums under these common ancestors. Thus, abbreviating the summation condition

$$
\begin{cases} d(x_P,x_Q)\leq\frac{1}{2}\varepsilon\delta^{-m_0}, & \text{if } m=m_0, \\ \frac{1}{2}\varepsilon\delta^{1-m}<d(x_P,x_Q)\leq\frac{1}{2}\varepsilon\delta^{-m}, & \text{if } m>m_0, \end{cases}
$$

simply as $\sum^m$, we have

$$
\begin{aligned}
\mathbb{E}\sum_{i=\sigma}^{\tau}\sum_{P,Q\in\mathcal{D}_i} T(P,Q) &= \mathbb{E}\sum_{m=m_0}^{\infty}\sum_{i=\sigma}^{\tau}\sum_{P,Q\in\mathcal{D}_i}^{m} \tilde{T}_m(P,Q) \\
&= \mathbb{E}\sum_{m=m_0}^{\infty}\sum_{i=\sigma}^{\tau}\sum_{S\in\mathcal{D}_{i-m}}\sum_{\substack{P,Q\in\mathcal{D}_i \\ P,Q\subseteq S}}^{m} \tilde{T}_m(P,Q).
\end{aligned}
$$

Let us investigate

$$
A_S^m(f,g) := \sum_{\substack{P,Q\in\mathcal{D}_i \\ P,Q\subseteq S}}^{m} \tilde{T}_m(P,Q),
$$

where $T_m(P,Q)$ is just $T(P,Q)$ multiplied by a number in $\{0\}\cup[1,2]$, and $T(P,Q)$ is as in (7.1).

7.1. Cancellation properties. We see that both f and g are acted on by operators that annihilate constants, only observe these functions on $P,Q\subseteq S$.

Thus, we may replace f by $1_S(f - \langle f\rangle_S)$, and similarly with g. Moreover, in the case of $\mathbb{D}_P f$, we may replace f by

$$\mathbb{D}_S^m f := \sum_{\substack{S'\in\mathcal{D}\\ \ell(S')=\delta^m\ell(S)}} \mathbb{D}_{S'} f,$$

and in the case of $\langle f\rangle_P - \langle f\rangle_Q$, by

$$\mathbb{E}_S^m f := \sum_{\substack{S'\in\mathcal{D}\\ \ell(S')=\delta^m\ell(S)}} \mathbb{D}_{S'} f,$$

as this operator only observed the averages of f on the of the cubes P and Q. Combining these observations, we can actually replace f, in the latter case, by

$$1_S(\mathbb{E}_S^m f - \langle \mathbb{E}_S^m f\rangle_S) = \sum_{\substack{S'\in\mathcal{D}\\ \ell(S')>\delta^m\ell(S)}} \mathbb{D}_{S'} f =: \mathbb{D}_S^{[0,m)} f.$$

Similar observations apply on the g side. This describes the cancellation properties present in $A_S^m(f,g)$.

7.2. Size properties. Noting that $\ell(S) = \delta^{-m}\ell(P)$ and $z_P \in S$, we can simplify the denominator in (7.2) as

$$V(z_P, \delta^{-m}\ell(P)) = V(z_P, \ell(S)) \asymp \mu(S).$$

In the case of $\mathbb{D}_P$ and $\mathbb{D}_Q$ on both sides, we have

$$A_S^m(f,g) = \iint_S a_S^m(u,v) f(v) g(u)\, d\mu(v)\, d\mu(u),$$

where

$$a_S^m(u,v) = \sum_{\alpha,\beta} \sum_{\substack{P,Q\in\mathcal{D}_i\\ P,Q\subseteq S}}^{m} \chi_{\alpha,\beta}^m \langle T h_P^\alpha, h_Q^\beta\rangle h_P^\alpha(v) h_Q^\beta(u)$$

satisfies

$$\begin{aligned}|a_S^m(u,v)| &\lesssim \sum_{\substack{P,Q\in\mathcal{D}_i\\ P,Q\subseteq S}}^{m} \omega(\delta^m) \frac{\sqrt{\mu(P)\mu(Q)}}{\mu(S)} \frac{1_P(v)}{\mu(P)^{1/2}} \frac{1_Q(u)}{\mu(Q)^{1/2}}\\ &= \omega(\delta^m)\frac{1}{\mu(S)} \sum_{\substack{P,Q\in\mathcal{D}_i\\ P,Q\subseteq S}}^{m} 1_P(v) 1_Q(u) \leq \omega(\delta^m)\frac{1}{\mu(S)} 1_{S\times S}(u,v).\end{aligned}$$

In the case of $\langle f\rangle_P - \langle f\rangle_Q$, we get

$$a_S^m(u,v) = \sum_{\beta} \sum_{\substack{P,Q\in\mathcal{D}_i\\ P,Q\subseteq S}}^{m} \chi_{\alpha,\beta}^m \langle T1_P, h_Q^\beta\rangle \Big(\frac{1_P(v)}{\mu(P)} - \frac{1_Q(v)}{\mu(Q)}\Big) h_Q^\beta(u),$$

which satisfies

$$
\begin{aligned}
|a_S^m(u,v)| &\lesssim \sum_{\substack{P,Q\in\mathcal{D}_i\\ P,Q\subseteq S}}^{m} \omega(\delta^m)\frac{\mu(P)\sqrt{\mu(Q)}}{\mu(S)}\Big(\frac{1_P(v)}{\mu(P)}+\frac{1_Q(v)}{\mu(Q)}\Big)\frac{1_Q(u)}{\mu(Q)^{1/2}} \\
&= \omega(\delta^m)\frac{1}{\mu(S)}\sum_{\substack{P,Q\in\mathcal{D}_i\\ P,Q\subseteq S}}^{m}\Big(1_P(v)1_Q(u)+\mu(P)\frac{1_Q(v)1_Q(u)}{\mu(Q)}\Big) \\
&\le \omega(\delta^m)\Big(\frac{1}{\mu(S)}1_{S\times S}(u,v)+\sum_{\substack{Q\subseteq S\\ \ell(Q)=\delta^m\ell(S)}}\frac{1_{Q\times Q}(u,v)}{\mu(Q)}\Big)
\end{aligned}
$$

The case of $\langle g\rangle_Q-\langle g\rangle_P$ is of course symmetric.

Note that $1_{Q\times Q}(u,v)/\mu(Q)$ is the kernel of the averaging operator $\mathbb{E}_Q$.

Altogether, we find that A_S^m takes one of the following forms:

$$
A_S^m=\omega(\delta^m)\times\begin{cases}\mathbb{D}_S^m\dot{A}_S^m\mathbb{D}_S^m,\\ \mathbb{D}_S^{[0,m)}\dot{A}_S^m\mathbb{D}_S^m,\\ \mathbb{D}_S^m\dot{A}_S^m\mathbb{D}_S^{[0,m)},\end{cases}
$$

where

$$
\|\dot{A}_S^m f\|_{\mathcal{X}}\lesssim \mathbb{E}_S\|f\|_{\mathcal{X}}+\mathbb{E}_S^m\|f\|_{\mathcal{X}}.
$$

8. Final estimates

We are left with estimating, e.g.,

$$
\sum_{m=m_0}^{\infty}\omega(\delta^m)\sum_{i=\sigma}^{\tau}\langle\mathbb{D}_{i-m}^{[0,m)}\dot{A}_{i-m}^m\mathbb{D}_{i-m}^m f,g\rangle, \tag{8.1}
$$

where

$$
G_i=\sum_{S\in\mathcal{D}_i}G_S,\quad G\in\{D^{[0,m)},\dot{A}^m,\mathbb{D}^m\},
$$

as well as two similar sums with either the positions of $\mathbb{D}_S^{[0,m)}$ and $\mathbb{D}_S^m$ interchanged, or both replaced by $\mathbb{D}_S^m$.

Let us estimate the term that we wrote down above.

LEMMA 8.1. *Let $1<p\le s\le q<\infty$ and suppose that $\mathcal{X}$ has martingale type p and martingale cotype q. Then*

$$
\Big|\sum_{i=\sigma}^{\tau}\langle\mathbb{D}_{i-m}^{[0,m)}\dot{A}_{i-m}^m\mathbb{D}_{i-m}^m f,g\rangle\Big|\lesssim n^{1/p-1/q}m^{1/p'}\|f\|_{L_s(\mu;\mathcal{X})}\|g\|_{L_s(\mu;\mathcal{X}')}.
$$

PROOF. By self-adjointness of the difference operators,

$$
\begin{aligned}
|\langle\mathbb{D}_i^{[0,m)}\dot{A}_i^m\mathbb{D}_i^m f,g\rangle| &= |\langle\dot{A}_i^m\mathbb{D}_i^m f,\mathbb{D}_i^{[0,m)}g\rangle| \\
&\le\langle(\mathbb{E}_i+\mathbb{E}_{i+m})\|\mathbb{D}_i^m f\|_{\mathcal{X}},\|\mathbb{D}_i^{[0,m)}g\|_{\mathcal{X}'}\rangle.
\end{aligned}
$$

Hence

$$\begin{aligned}\sum_{i=\sigma}^{\tau}|\langle\mathbb{D}_{i-m}^{[0,m)}\dot{A}_{i-m}^{m}\mathbb{D}_{i-m}^{m}f,g\rangle| &= \sum_{i=\sigma-m}^{\tau-m}|\langle\mathbb{D}_{i}^{[0,m)}\dot{A}_{i}^{m}\mathbb{D}_{i}^{m}f,g\rangle| \\ &\leq \Big\|\Big(\sum_{i=\sigma-m}^{\tau-m}[(\mathbb{E}_i+\mathbb{E}_{i+m})\|\mathbb{D}_i^m f\|_{\mathcal{X}}]^q\Big)^{1/q}\Big\|_{L_s(\mu)} \\ &\quad\times\Big\|\Big(\sum_{i=\sigma-m}^{\tau-m}[\|\mathbb{D}_i^{[0,m)}g\|_{\mathcal{X}}]^{q'}\Big)^{1/q'}\Big\|_{L_{s'}(\mu)}\end{aligned}$$

for any $q,s\in(1,\infty)$. In the first term, we can pull out the conditional expectations (even after a crude estimate $(\mathbb{E}_i+\mathbb{E}_{i+m})\phi\leq\sup_{j\in\mathbb{Z}}\mathbb{E}_j\phi$ if we wish) by the vector-valued Doob maximal inequality [**13**, Theorem 3.2.7]:

$$\Big\|\Big(\sum_{i=\sigma-m}^{\tau-m}[(\mathbb{E}_i+\mathbb{E}_{i+m})\|\mathbb{D}_i^m f\|_{\mathcal{X}}]^q\Big)^{1/q}\Big\|_{L_s(\mu)}\lesssim\Big\|\Big(\sum_{i=\sigma-m}^{\tau-m}\|\mathbb{D}_i^m f\|_{\mathcal{X}}^q\Big)^{1/q}\Big\|_{L_s(\mu)}.$$

If $\mathcal{X}$ has martingale cotype q, then

$$\Big\|\Big(\sum_{i=\sigma-m}^{\tau-m}\|\mathbb{D}_i^m f\|_{\mathcal{X}}^q\Big)^{1/q}\Big\|_{L_s(\mu)}\lesssim\|f\|_{L_s(\mu;\mathcal{X})},$$

essentially by definition, since $\mathbb{D}_i^m f$ are martingale differences.

The dual side is slightly trickier, since the compound differences

$$\mathbb{D}_i^{[0,m)}=\sum_{k=0}^{m-1}\mathbb{D}_{i-k}$$

involve overlapping scales. For the optimal estimate, we need to be somewhat clever with the use of the triangle inequality. This is accomplished as follows: We now assume that $p\leq s\leq q$; hence $q'\leq s'\leq p'$, where $\mathcal{X}$ has martingale type p and martingale cotype q. Then

$$\begin{aligned}&\Big\|\Big(\sum_{i=\sigma-m}^{\tau-m}\|\mathbb{D}_i^{[0,m)}g\|_{\mathcal{X}}^{q'}\Big)^{1/q'}\Big\|_{L_{s'}(\mu)} \\ &\quad\leq n^{1/q'-1/s'}\Big\|\Big(\sum_{i=\sigma-m}^{\tau-m}\|\mathbb{D}_i^{[0,m)}g\|_{\mathcal{X}'}^{s'}\Big)^{1/s'}\Big\|_{L_{s'}(\mu)} \\ &\quad= n^{1/q'-1/s'}\Big(\sum_{j=0}^{m-1}\Big\|\Big(\sum_{\substack{i=\sigma-m\\ i\equiv j \mod m}}^{\tau-m}\|\mathbb{D}_i^{[0,m)}g\|_{\mathcal{X}'}^{s'}\Big)^{1/s'}\Big\|_{L_{s'}(\mu)}^{s'}\Big)^{1/s'}.\end{aligned}$$

The inner summation contains every m-th term among a sequence of length n. Recalling that $m\lesssim n$, the number of such terms is $O(n/m)$. (There could of course be a single term even if $m\gg n$, in which case the estimate would no longer be valid.) Thus, by Hölder's inequality, we can continue with

$$\lesssim n^{1/q'-1/s'}(n/m)^{1/s'-1/p'}\Big(\sum_{j=0}^{m-1}\Big\|\Big(\sum_{\substack{i=\sigma-m\\ i\equiv j \mod m}}^{\tau-m}\|\mathbb{D}_i^{[0,m)}g\|_{\mathcal{X}'}^{p'}\Big)^{1/p'}\Big\|_{L_{s'}(\mu)}^{s'}\Big)^{1/s'}.$$

For each fixed j, the sequence $(\mathbb{D}_i^{[0,m)})_{i\equiv j \mod m}$ consists of martingale differences, If $\mathcal{X}$ has martingale type p, its dual $\mathcal{X}'$ has martingale cotype p', and we can further continue the estimate

$$\begin{aligned}
&\lesssim n^{1/q'-1/s'}(n/m)^{1/s'-1/p'}\Big(\sum_{j=0}^{m-1}\|g\|_{L_{s'}(\mu;\mathcal{X}')}^{s'}\Big)^{1/s'}\\
&= n^{1/q'-1/s'}(n/m)^{1/s'-1/p'}(n/m)^{1/s'-1/p'}m^{1/s'}\|g\|_{L_{s'}(\mu;\mathcal{X}')}\\
&= n^{1/q'-1/p'}m^{1/p'}\|g\|_{L_{s'}(\mu;\mathcal{X}')} = n^{1/p-1/q}m^{1/p'}\|g\|_{L_{s'}(\mu;\mathcal{X}')}.
\end{aligned}$$

□

By symmetry, we infer that

$$\Big|\sum_{i=\sigma}^{\tau}\langle\mathbb{D}_{i-m}^m\dot{A}_{i-m}^m\mathbb{D}_{i-m}^{[0,m)}f,g\rangle\Big| \lesssim n^{1/p-1/q}m^{1/q}\|f\|_{L_s(\mu;\mathcal{X})}\|g\|_{L_s(\mu;\mathcal{X}')}.$$

In the case of $\mathbb{D}_{i-m}^m$ on both sides, the estimation is more straightforward, and we simply estimate

$$\begin{aligned}
&\Big\|\Big(\sum_{i=\sigma-m}^{\tau-m}\|\mathbb{D}_i^m g\|_{\mathcal{X}}^{q'}\Big)^{1/q'}\Big\|_{L_{s'}(\mu)}\\
&\qquad\le n^{1/q'-1/p'}\Big\|\Big(\sum_{i=\sigma-m}^{\tau-m}\|\mathbb{D}_i g\|_{\mathcal{X}'}^{p'}\Big)^{1/p'}\Big\|_{L_{s'}(\mu)} \lesssim n^{1/q'-1/p'}\|g\|_{L_{s'}(\mu;\mathcal{X}')}
\end{aligned}$$

and hence

$$\Big|\sum_{i=\sigma}^{\tau}\langle\mathbb{D}_{i-m}^m\dot{A}_{i-m}^m\mathbb{D}_{i-m}^m f,g\rangle\Big| \lesssim n^{1/p-1/q}\|f\|_{L_s(\mu;\mathcal{X})}\|g\|_{L_s(\mu;\mathcal{X}')}.$$

The largest of the different bounds that we have obtained for the m-th term is hence

$$n^{1/p-1/q}m^{\max(1/p',1/q)}\|f\|_{L_s(\mu;\mathcal{X})}\|g\|_{L_s(\mu;\mathcal{X}')}.$$

Recalling from (8.1) that we still need to sum over m, we finally obtain the bound

$$\begin{aligned}
n^{1/p-1/q}\sum_{m=m_0}^{\infty}\omega(\delta^m)m^{\max(1/p,1/q')} &\eqsim n^{1/p-1/q}\int_0^1\omega(t)(1+\log\frac{1}{t})^{\max(1/p,1/q')}\frac{dt}{t}\\
&=: n^{1/p-1/q}\|\omega\|_{\mathrm{Dini}^{\max(1/p,1/q')}}.
\end{aligned}$$

9. Concluding remarks

We have now concluded the (unavoidably somewhat lengthy) proof of Theorem 1.2. In the course of the proof, after some preliminary reductions, the main part of the operator was decomposed into two paraproducts and the cancellative part. Recall that we assume that our target Banach space has martingale type p and martingale cotype q.

For the paraproduct, we found the estimate $O(n^{1/p-1/2})$. By symmetry, the dual paraproduct then satisfies the bound $O(n^{1/q'-1/2}) = O(n^{1/2-1/q})$. Finally, the cancellative part of the operator had the bound $O(n^{1/p-1/q})$, provided that the kernel satisfies a Dini condition of order $\max(1/p, 1/q')$.

It might at first seem counterintuitive that the order of the required Dini condition increases with improving martingale type or cotype. On the other hand, the dependence on the finiteness parameter n of the kernel decreases at the same time. A way to think of this is as follows: If the space is better, then we are able to make better use of the properties of the kernel, whereas a poor space only observes a Dini condition of low order, even if more would be available.

If (and only if) $\mathcal{X}$ is (isomorphic to) a Hilbert space, then one can take $p = q = 2$, and all bounds become $O(1)$ in terms of n, allowing one to dispense with the truncations and deal with genuine singular integrals. In this case, one needs a Dini condition of order $1/2$ to run the argument. It seems to be open whether this can be relaxed, even in the scalar case.

In fact, with the present argument, the only way to get $O(1)$ in terms of n in the cancellative term is to have $p = q = 2$. This is due to the fact that necessarily $p \leq 2$ and $q \geq 2$; hence $1/p - 1/q = (1/p - 1/2) + (1/2 - 1/q)$ can only be zero if both terms vanish. On the other hand, for the paraproducts, it suffices to have just one of p or q equal to 2, which provides a much larger class of examples.

The boundedness of the dyadic paraproduct Π_b with a scalar-valued symbol $b \in \mathrm{BMO}$, is well known on $L_s(\mu; \mathcal{X})$, in the case that $\mathcal{X}$ is a UMD space. The result is attributed to Bourgain, and written down in [**7**]. Our considerations in Section 6 show that they are also bounded on $L_s(\mu; \mathcal{X})$, if $\mathcal{X}$ has martingale type 2. These two classes (UMD and martingale type 2) are not mutually comparable. This raises the interesting question about the maximal class of spaces $\mathcal{X}$ such that the dyadic paraproduct induces a bounded operator on $L_s(\mu; \mathcal{X})$.

Acknowledgments

I would like to thank the anonymous referees for their constructive comments on the manuscript.

References

[1] P. Auscher and T. Hytönen, *Orthonormal bases of regular wavelets in spaces of homogeneous type*, Appl. Comput. Harmon. Anal. **34** (2013), no. 2, 266–296, DOI 10.1016/j.acha.2012.05.002. MR3008566

[2] P. Auscher and E. Routin, *Local Tb theorems and Hardy inequalities*, J. Geom. Anal. **23** (2013), no. 1, 303–374, DOI 10.1007/s12220-011-9249-1. MR3010282

[3] J. Bourgain, *Some remarks on Banach spaces in which martingale difference sequences are unconditional*, Ark. Mat. **21** (1983), no. 2, 163–168, DOI 10.1007/BF02384306. MR727340

[4] D. L. Burkholder, *A geometric condition that implies the existence of certain singular integrals of Banach-space-valued functions*, Conference on harmonic analysis in honor of Antoni Zygmund, Vol. I, II (Chicago, Ill., 1981), Wadsworth Math. Ser., Wadsworth, Belmont, CA, 1983, pp. 270–286. MR730072

[5] M. Christ, *A $T(b)$ theorem with remarks on analytic capacity and the Cauchy integral*, Colloq. Math. **60/61** (1990), no. 2, 601–628, DOI 10.4064/cm-60-61-2-601-628. MR1096400

[6] P. Enflo, *Banach spaces which can be given an equivalent uniformly convex norm*, Israel J. Math. **13** (1972), 281–288 (1973), DOI 10.1007/BF02762802. MR336297

[7] T. Figiel and P. Wojtaszczyk, *Special bases in function spaces*, Handbook of the geometry of Banach spaces, Vol. I, North-Holland, Amsterdam, 2001, pp. 561–597, DOI 10.1016/S1874-5849(01)80016-4. MR1863702

[8] A. G. de la Herrán and T. Hytönen, *Dyadic representation and boundedness of nonhomogeneous Calderón-Zygmund operators with mild kernel regularity*, Michigan Math. J. **67** (2018), no. 4, 757–786, DOI 10.1307/mmj/1531447374. MR3877436

[9] T. Hytönen and A. Kairema, *Systems of dyadic cubes in a doubling metric space*, Colloq. Math. **126** (2012), no. 1, 1–33, DOI 10.4064/cm126-1-1. MR2901199

[10] T. Hytönen and H. Martikainen, *Non-homogeneous Tb theorem and random dyadic cubes on metric measure spaces*, J. Geom. Anal. **22** (2012), no. 4, 1071–1107, DOI 10.1007/s12220-011-9230-z. MR2965363

[11] T. Hytönen and A. Naor, *A geometric dichotomy for the discrete Hilbert transform*, Manuscript, 2022.

[12] T. Hytönen and O. Tapiola, *Almost Lipschitz-continuous wavelets in metric spaces via a new randomization of dyadic cubes*, J. Approx. Theory **185** (2014), 12–30, DOI 10.1016/j.jat.2014.05.017. MR3233063

[13] T. Hytönen, J. van Neerven, M. Veraar, and L. Weis, *Analysis in Banach spaces. Vol. I. Martingales and Littlewood-Paley theory*, Ergebnisse der Mathematik und ihrer Grenzgebiete. 3. Folge. A Series of Modern Surveys in Mathematics [Results in Mathematics and Related Areas. 3rd Series. A Series of Modern Surveys in Mathematics], vol. 63, Springer, Cham, 2016. MR3617205

[14] T. P. Hytönen, *The sharp weighted bound for general Calderón-Zygmund operators*, Ann. of Math. (2) **175** (2012), no. 3, 1473–1506, DOI 10.4007/annals.2012.175.3.9. MR2912709

[15] F. Nazarov, A. Reznikov, and A. Volberg, *The proof of A_2 conjecture in a geometrically doubling metric space*, Indiana Univ. Math. J. **62** (2013), no. 5, 1503–1533, DOI 10.1512/iumj.2013.62.5098. MR3188553

[16] S. Petermichl, *Dyadic shifts and a logarithmic estimate for Hankel operators with matrix symbol* (English, with English and French summaries), C. R. Acad. Sci. Paris Sér. I Math. **330** (2000), no. 6, 455–460, DOI 10.1016/S0764-4442(00)00162-2. MR1756958

[17] G. Pisier, *Martingales with values in uniformly convex spaces*, Israel J. Math. **20** (1975), no. 3-4, 326–350, DOI 10.1007/BF02760337. MR394135

[18] G. Pisier, *Un exemple concernant la super-réflexivité* (French), Séminaire Maurey-Schwartz 1974–1975: Espaces L^p, applications radonifiantes et géométrie des espaces de Banach, École Polytech., Paris, 1975, pp. Annexe No. 2, 12. MR410340

[19] G. Pisier, *Probabilistic methods in the geometry of Banach spaces*, Probability and analysis (Varenna, 1985), Lecture Notes in Math., vol. 1206, Springer, Berlin, 1986, pp. 167–241, DOI 10.1007/BFb0076302. MR864714

[20] G. Pisier, *Martingales in Banach spaces*, Cambridge Studies in Advanced Mathematics, vol. 155, Cambridge University Press, Cambridge, 2016. MR3617459

Tuomas Hytönen: Department of Mathematics and Statistics, P.O.Box 68 (Pietari Kalmin katu 5), FI-00014 University of Helsinki, Finland

Email address: tuomas.hytonen@helsinki.fi

Proceedings of Symposia in Pure Mathematics
Volume **106**, 2024
https://doi.org/10.1090/pspum/106/01933

Some Easily Formulated Open Problems in Banach Space Theory

Vicente Montesinos and Václav Zizler

This paper is dedicated to the golden memory of the second author's dear daughter Jarmila Zizlerová

ABSTRACT. This chapter—that has a survey character—contains a collection of problems in analysis and geometry of Banach spaces followed by some others from operator theory. It claims to show that a huge amount of basic classical problems in analysis remains open in the modern theory of Banach spaces. Most of them are important for future developments, and many of them also for practical use in the area. We mention the sources of the problems, together with some literature for their proper understanding.

1. Introduction

This chapter contains a list of a few selected commented and easily formulated open problems in Banach space theory. Given the authors' area of expertise, most problems come from the renorming and smoothness areas. Some of them can be found in [**38**] and [**40**, Chapter 52], where many others are included. Some others appear here for the first time.

The chapter contains also some personal memories to show that the Banach space community forms a large friendly family of workers.

When several norms appear on the same space, we understand that they are equivalent. Banach spaces are considered on the real field. The word "space" means, if nothing is said to the contrary, a Banach space. If $\mathcal{X}$ is a Banach space equipped with a norm $\|\cdot\|$, the unit sphere of $\mathcal{X}$ is $S_{\mathcal{X}} := \{x \in \mathcal{X} : \|x\| = 1\}$ and the closed unit ball is $B_{\mathcal{X}} := \{x \in \mathcal{X} : \|x\| \leq 1\}$. Unless stated otherwise, by a subspace of a Banach space we mean a closed one. C^1-smoothness is meant in the Fréchet sense ([**40**]). If nothing is said otherwise, by a function on a Banach space we mean a real-valued function. By the word "operator" we mean a bounded linear operator. The term "smoothness" has the same meaning as differentiability and "rotundity" has the same meaning as strict convexity.

2020 *Mathematics Subject Classification.* Primary 46B20; Secondary 46B04, 46B10, 46B03, 46B22.

Key words and phrases. Banach spaces, renorming, structure of Banach spaces, open problems.

The first author was supported in part by Project PID2021-122126NB-C33/MCIN/AEI/ 10.13039/501100011033 (FEDER), Spain. 2020.

The weak topology on $\mathcal{X}$ associated to a dual pair $\langle \mathcal{X}, \mathcal{Y} \rangle$ will be denoted $w(\mathcal{X}, \mathcal{Y})$. As is usual, $w(\mathcal{X}, \mathcal{X}^*)$ will be just written w, and w^* will stand for $w(\mathcal{X}^*, \mathcal{X})$. Alternatively, we shall use the terms "weak" and "weak-star" topologies, respectively.

If we assign a problem to some author we sometimes mean that we met it first in connection with this author.

Most of the required prerequisites for following this chapter are, e.g., in [**6**], [**70**], [**63**], [**22**], [**40**], [**21**], or [**44**].

2. Asplund spaces; C^1-Fréchet differentiability

PROBLEM 2.1. If $\mathcal{X}$ is an Asplund space, does $\mathcal{X}$ admit a Fréchet C^1-smooth bump?

An **Asplund space** is a Banach space $\mathcal{X}$ each separable subspace of which has a separable dual space. It is equivalent to the fact that on $\mathcal{X}$ every convex continuous function is Fréchet differentiable at each point of some dense G_δ subset (see, e.g., [**22**, Definition 7.38 and Theorem 11.8] or [**40**, Definition 58, and Section 11.6 and 46.1]).

By a **bump** we mean a real-valued function with bounded nonempty support.

By the Smooth Variational Principle (see, e.g., [**22**, Theorem 7.43]), the existence of a C^1-smooth bump implies that the space is Asplund (J. Kurzweil, I. Ekeland, J. Borwein, D. Preiss, R. Deville, G. Godefroy, and V. Zizler —chronologically, see also [**40**, Theorem 289]).

In separable spaces $\mathcal{X}$, the existence of a C^1-smooth bump on $\mathcal{X}$ implies the existence of a Fréchet smooth equivalent norm there (M. I. Kadets, see [**15**, page 58]), which is not true in general (R. Haydon, see, e.g., [**40**, Theorems 817 and 818]). For extensions of the previously mentioned result of Kadets to nonseparable spaces see e.g. [**40**].

The existence of a uniformly Fréchet or a uniformly Gâteaux differentiable bump —in this case, the space is superreflexive, see, e.g., [**40**, Theorem 447]— already implies the existence of a norm with the given differentiability property (M. Fabian, W. Tang, see the aforementioned reference).

PROBLEM 2.2 (G. Godefroy). If all the w^*-convergent sequences on the dual unit sphere $S_{\mathcal{X}^*}$ of a nonseparable space $\mathcal{X}$ are norm-convergent, is $\mathcal{X}$ necessarily Asplund?

This problem was first presented in the Frolík-Lukeš winter schools. It may help to get an alternative proof of the Josefson–Nissenzweig theorem (see, e.g., [**22**, Exercise 3.39]) by using just differentiability concepts and the magical Šmulyan's lemma [**22**, Exercise 3.39 and Section 7.2]; see also, e.g., [**40**, Remark 583].

PROBLEM 2.3 (D. Preiss). Assume that the norm of a separable Banach space $(\mathcal{X}, \|\cdot\|)$ has the property that its restriction to each subspace is somewhere Fréchet differentiable. Is $\mathcal{X}$ necessarily Asplund?

In connection with this, it is good to mention that the dual of Day's norm on c_0 is Fréchet differentiable at points of a dense subset of ℓ_1 (by Phelps' and Šmulyan's theorems, see [**40**, Remark 242]) and yet, the dual of ℓ_1 is not separable.

PROBLEM 2.4. Assume that a separable Banach space $(\mathcal{X}, \|\cdot\|)$ is saturated with copies of ℓ_2. Is the bidual norm $\|\cdot\|^{**}$ on $\mathcal{X}^{**}$ Fréchet differentiable somewhere?

"Saturated with copies of ℓ_2" means that each infinite-dimensional subspace contains an isomorphic copy of ℓ_2. See [**40**, Remark after Theorem 517] and the reference therein.

Related to this, we mention W. Schachermayer's result that this holds true for the James JT space (see e.g. [**45**] and [**38**, Problem 150]).

PROBLEM 2.5. If f_1, f_2, f_3 are real-valued Lipschitz functions on ℓ_2, do they have a common point of Fréchet differentiability?

This is mentioned, e.g., in G. Godefroy's report [**31**] on the book [**61**]. See also [**38**, Problem 124] and the related comments there, and [**40**, paragraph preceding Remark 438]. For two functions, the corresponding problem has a positive solution (see this deep result in [**61**, Corollary 13.1.2]).

It is important to mention here the following result of D. Preiss [**73**]: *Every real-valued Lipschitz function on an Asplund space has points of Fréchet differentiability* (see e.g. [**40**, Theorem 60 and Section 22.2], where this result is discussed). It is also worth mentioning that the question in Problem 2.5 for any finite number of real-valued Lipschitz functions defined on a nonempty open subset of a separable asymptotically smooth Banach space *has a positive answer for the concept of ε-Fréchet differentiability (for any $\varepsilon > 0$)* (check this result and the associated details in [**61**, Theorem 4.3.2]).

PROBLEM 2.6. Assume that a separable Banach space $\mathcal{X}$ has the property that each Gâteaux differentiable norm on $\mathcal{X}$ is Fréchet differentiable somewhere. Is the same true for all Lipschitz real-valued functions on $\mathcal{X}$?

The second named author could not solve it with Wee-Kee Tang when Wee-Kee Tang prepared his PhD thesis in Edmonton in 1995. See also [**38**, Problem 125].

PROBLEM 2.7. Let $\mathcal{X}$ be a reflexive Banach space and $\|\cdot\|$ be a Fréchet differentiable norm on a subspace of $\mathcal{X}$. Can $\|\cdot\|$ be extended to a Fréchet differentiable norm on $\mathcal{X}$?

For more in this direction see, e.g, [**40**, Problem 6 in Section 52.1.2], where it is mentioned that Problem 2.7 above is open even in the separable setting (see also [**22**, Open Problem 8.6.1]). A similar question for Gâteaux differentiability in separable spaces has a negative answer (see [**82**], see also [**15**, Theorem II.8.3]). For the uniform Fréchet case the question is also open. The corresponding questions for rotundity and related concepts (uniform rotundity, local uniform rotundity, weakly uniform rotundity, uniform rotundity in every direction, among others) are now completely solved by the result of Wee-Kee Tang mentioned in [**40**, Theorem 122], where more results in these directions are quoted (as, for example, [**54**], the first result in this area).

PROBLEM 2.8. Does the three-space problem hold for Fréchet smooth renorming? *In other words,* if for a subspace $\mathcal{Y}$ of a Banach space $\mathcal{X}$, both $\mathcal{Y}$ and $\mathcal{X}/\mathcal{Y}$ do admit a Fréchet differentiable norm, does $\mathcal{X}$ necessarily admit a Fréchet differentiable norm?

Three-space problems are the subject of the monograph [**13**]. Problem 2.8 is mentioned in [**13**, Section 5.3], where the authors refer to [**15**, Problem VII.5]. The question is also open for Fréchet differentiable bumps. For this and more see, e.g., [**40**, Theorems 229, 648, and Chapter 41].

Problem 2.9. Does a Banach space $\mathcal{X}$ have the Mazur intersection property if $\mathcal{X}$ admits a Fréchet smooth bump?

A Banach space has the **Mazur intersection property** whenever every convex closed bounded set in $\mathcal{X}$ is an intersection of balls in $\mathcal{X}$; see, in particular, [**34**].

Problem 2.9 is mentioned in [**40**, Section 52.8, Problem 1] and [**38**, Problem 187]. The answer is yes if, moreover, $\mathcal{X}$ has the Radon–Nikodým property (see [**14**], see also [**15**, page 88]).

A Banach space $\mathcal{X}$ has the **Radon–Nikodým property** (**RNP**, for short) if every closed convex bounded set C in $\mathcal{X}$ is the closed convex hull of its strongly exposed points. A point $x_0 \in C$ is a **strongly** (or **weakly**) **exposed point** of C if there is $f \in \mathcal{X}^*$ with $\sup_C f = f(x_0) = 1$ and $x_n \to x_0$ in norm (or weakly) whenever $f(x_n) \to 1$.

M. Bačák and P. Hájek proved in [**4**] that the question of whether the Asplund property directly implies the Mazur intersection property is not decidable in ZFC.

Let us mention that the notion of the Mazur intersection property is of interest to people from operator theory, as was the case with the second named author's late Professor V. Pták.

Problem 2.10. Assume that a Banach space $\mathcal{X}$ can be renormed by a Fréchet differentiable norm and also by a norm with the LUR property. Can $\mathcal{X}$ be renormed by a norm that is simultaneously LUR and Fréchet differentiable?

The **LUR property** of the norm $\|\cdot\|$ means that for a sequence (x_n) and a point x, both in the unit sphere $S_{\mathcal{X}}$, $\|x_n - x\| \to 0$ whenever $\|x_n + x\| \to 2$.

Problem 2.10 is a question related to the so-called "Asplund averaging technique" (see, e.g., [**40**, Chapter 8 and Section 10.8]). For more on this question see, e.g., [**15**], [**40**, pages 129 and 189] and [**46**]. Notice that the problem whether a Banach space with a Fréchet smooth norm admits an equivalent LUR norm is open (see, e.g., [**65**, Question 6.14]).

Problem 2.11. Assume that $\mathcal{X}$ admits a Fréchet differentiable bump. Does $\mathcal{X}$ allow for C^1-smooth approximation for all continuous functions on $\mathcal{X}$?

Problem 2.11 is mentioned in [**15**]: We refer the reader to [**15**, Section VIII.3, in particular Theorem VIII.3.2 and Lemma VIII.3.18] for a detailed study of this and related questions. It appears also in [**40**, Problem 3 in the Introduction]. See also [**43**, Problem 108].

Problem 2.12. Are the set of all C^1-smooth norms dense in the set of all norms on $C[0, \omega_1]$?

Problem 2.12 is mentioned in [**40**, Comments after Theorem 247 and Problem 4 in Section 52.1.7]. Some related results are discussed there.

Problem 2.13 (M. Fabian). If f is a continuous convex function on c_0, does there exist $x \in c_0$ such that

$$f(x+h) + f(x-h) - 2f(x) = O(\|h\|^2)$$

as $\|h\| \to 0$?

In other words, this problem asks whether c_0 is a "Lipschitz Asplund space" (M. Fabian used the term "LD-space"). It is discussed in [**40**, Chapter 36]. We refer to the works of M. Fabian quoted there, in particular to [**19**].

It is in a sense unbelievable that the following problem seems to be open.

PROBLEM 2.14. If a Banach space admits a C^1-smooth bump, does it admit also a Lipschitz C^1-smooth bump?

We refer to [**15**, Problem II.1].

3. C^k-Fréchet differentiability

PROBLEM 3.1. Let a separable space $\mathcal{X}$ admit a C^k-smooth bump. Does $\mathcal{X}$ admit a C^k-smooth norm?

This problem is mentioned in [**40**, Introduction and Section 52.1.3] . For a discussion on this long-standing problem see, e.g., [**43**, Problem 154]. Notice that a Banach space admitting a C^2-smooth bump either contains c_0 or it is superreflexive (see, e.g., [**22**], or [**40**]).

PROBLEM 3.2. Assume that every separable subspace of a Banach space $\mathcal{X}$ admits a C^2-smooth bump. Does $\mathcal{X}$ necessarily admit a C^2-smooth bump?

Notice that there is a nonseparable Banach space $\mathcal{X}$, in fact a $C(K)$ space, such that $\mathcal{X}$ does not admit any Gâteaux differentiable norm and yet, every separable subspace of $\mathcal{X}$ admits a real-analytic norm ([**15**, Problem V.3]).

PROBLEM 3.3. Does there exist a nonseparable Banach space $\mathcal{X}$ such that $\mathcal{X}$ admits a C^k-smooth norm ($k = 1, 2, \ldots$ or ∞) and another norm that cannot be approximated, uniformly on bounded sets by C^k-smooth norms?

It is a result of P. Hájek and J. Talponen in [**49**] that, for separable Banach spaces, the above question has a negative answer. This result and its proof can be found in [**43**, Theorem 103]. In this direction there is a result of V. Bible and R.J. Smith [**9**] that the answer is "no" also for $c_0(\Gamma)$. For more information see [**48**]. Related to this problem see [**15**, Theorem V.1.7].

PROBLEM 3.4 (R. Deville). Assume that a nonseparable Banach space $\mathcal{X}$ admits a norm that locally depends on finitely many coordinates. Does $\mathcal{X}$ admit a C^∞-smooth norm?

The norm $\|\cdot\|$ on $\mathcal{X}$ **locally depends on finitely many functionals (or coordinates)** if for every nonzero $x \in \mathcal{X}$ there is a neighborhood $\mathcal{V}$ of x in $\mathcal{X}$, functionals $f_1, f_2, \ldots, f_n \in \mathcal{X}^*$ and a function ψ on $\mathbb{R}^n$ such that, for all $y \in \mathcal{V}$, we have

$$\|y\| = \psi\Big(f_1(y), f_2(y), \ldots f_n(y)\Big).$$

A related concept is polyhedrality. A Banach space $\mathcal{X}$ is said to be **polyhedral** if for every finite-dimensional subspace $F \subset \mathcal{X}$, the closed unit ball B_F of F is the convex hull of a finite set of points. It is a result of V. Fonf [**26**] and P. Hájek [**39**] that $\mathcal{X}$ is separable and isomorphic to a polyhedral space if, and only if, $\mathcal{X}$ admits a norm that locally depends on finitely many coordinates.

Problem 3.4 is mentioned in [**83**]. For separable spaces it has a positive solution (P. Hájek's PhD thesis in Edmonton, see [**22**, Theorem 10 12]).

PROBLEM 3.5. Can every norm on ℓ_2 be approximated (uniformly on bounded sets) by real-analytic norms?

A function is **real-analytic** if it can be locally expressed as a Taylor series. We refer, e.g., to [**43**, Section 7.3]. See also [**50**] and [**77**].

PROBLEM 3.6. Does every $C(K)$ space, where K is a scattered compact space, admit a C^∞-smooth bump?

A compact space K is **scattered** whenever every closed subset L has an isolated point in L. See, e.g., [**40**, Problem 136] and the comments therein. See also, e.g., [**83**], [**51**], and [**52**].

PROBLEM 3.7. Can every continuous function on c_0 be uniformly approximated by real-analytic functions?

See, e.g., [**43**, Problem 106] and the comments therein. Problem 3.7 formulates one of the main questions in this area.

PROBLEM 3.8. If Γ is uncountable, can every continuous function on $\ell_2(\Gamma)$ be uniformly approximated by real-analytic functions?

For Γ countable the answer is positive. This is an important result of J. Kurzweil [**60**], see also [**43**].

PROBLEM 3.9. If $C(K)$ admits a Fréchet smooth norm, does it admit a C^∞-smooth norm?

See, e.g., [**83**], [**38**, Problem 136], and [**40**, Chapter 27]. Observe that, under the assumption on $C(K)$ in Problem 3.9, the compact space K should be scattered, since $C(K)$ is then Asplund.

PROBLEM 3.10. If $\mathcal{X}$ admits a C^2-smooth norm, does it admit a LUR norm?

See, e.g., [**65**]. We mentioned above that [**65**, Question 6.14] asks whether every Banach space with a Fréchet smooth norm admits an equivalent LUR norm.

PROBLEM 3.11. Does c_0 admit a C^∞-smooth norm the dual of which is LUR?

See, e.g., [**24**].

PROBLEM 3.12 (A. Pełczyński). If a separable space $\mathcal{X}$ admits C^k-smooth norms for every $k \in \mathbb{N}$, does $\mathcal{X}$ admit a C^∞-smooth norm?

PROBLEM 3.13. Let f be a convex continuous function on ℓ_2. Do there exist $x, y \in \ell_2$ and a continuous symmetric bilinear form T_2 on ℓ_2 such that for all $h \in \ell_2$ we have

$$f(x + th) - f(x) - (y, th) - t^2 T_2(h, h) = o(t^2)$$

as $t \to 0$?

For finite-dimensional spaces this is the celebrated Alexandrov theorem. This problem appears as [**15**, Problem IV.4]. See the references there. It has a negative answer for nonseparable spaces [**10**].

PROBLEM 3.14. Does the space of compact operators from ℓ_2 into ℓ_2 admit a real-analytic norm?

This space admits a C^∞-smooth norm by the result of N. Tomczak-Jaegermann in [**80**].

PROBLEM 3.15. Let K be a scattered compact. Does $C(K)$ admit a Fréchet smooth norm if, and only if, $C(K)$ admits a LUR norm?

R. Haydon proved in [Ha99] that the answer is yes when K is a tree. For more in this direction see, e.g., [**40**] or [**15**]. See also [**65**, Question 6.19].

4. Gâteaux differentiability

PROBLEM 4.1. Assume that a Banach space $\mathcal{X}$ admits a Gâteaux differentiable norm and also that $\mathcal{X}$ admits an SSD norm. Must $\mathcal{X}$ admit a Fréchet differentiable norm?

The norm $\|\cdot\|$ is **SSD (strongly subdifferentiable)** at x if the one sided limit

$$\lim_{t\to 0+}\frac{\|x+th\|-\|x\|}{t}$$

exists uniformly for h in the unit sphere of $\mathcal{X}$. It is a result of G. Godefroy [**28**] that every space with an SSD norm is Asplund. For some other results related to this concept see, e.g., [**33**].

PROBLEM 4.2. If a Banach space $\mathcal{X}$ is WLD, is every convex continuous function on $\mathcal{X}$ Gâteaux differentiable somewhere?

A Banach space $\mathcal{X}$ is **weakly Lindelöf determined** (**WLD**, for short) if the dual ball of $\mathcal{X}^*$ is a Corson compact in its weak-star topology.

A compact space K is said to be **Corson compact** if K is homeomorphic to a subset S of $[0,1]^\Gamma$ (endowed with its pointwise topology) and such that for every $x \in S$ we have card $\{\alpha\in\Gamma : x(\alpha)\neq 0\} \le \aleph_0$.

Regarding Problem 4.2, see [**83**] and [**2**].

PROBLEM 4.3. If a nonseparable $\mathcal{X}$ has an unconditional basis and admits a Gâteaux differentiable norm, does $\mathcal{X}$ admit a norm whose dual is strictly convex?

The norm $\|\cdot\|$ of a Banach space is said to be **strictly convex** (or **rotund**) whenever its unit sphere does not contain non-degenerate segments. In other words, all points $x \in S_\mathcal{X}$ are extreme points of the closed unit ball. Still a third equivalent definition: If $x, y \in S_\mathcal{X}$, then $\|x+y\| < 2$. For more information on Problem 4.3 see [**76**]. For the definition of an unconditional basis see Section 7 below.

PROBLEM 4.4 (M. Fabian). Is every subspace of a weak Asplund space again a weak Asplund space?

A Banach space $\mathcal{X}$ is a **weak Asplund space** if every continuous convex function on $\mathcal{X}$ is Gâteaux differentiable at points of a dense G_δ-set in $\mathcal{X}$.

PROBLEM 4.5 (M. Fabian). If $\mathcal{X}$ is a weak Asplund space, is $\mathcal{X}\oplus\mathbb{R}$ a weak Asplund space?

PROBLEM 4.6 (M. Fabian). If $\mathcal{X}$ is a weak Asplund space and K is the dual closed unit ball of $\mathcal{X}^*$ in the w^*-topology, is $C(K)$ necessarily a weak Asplund space?

For all these three problems we refer to [**20**].

PROBLEM 4.7. If a Banach space $\mathcal{X}$ is weak Asplund, does $\mathcal{X}^*$ admit a strictly convex (not necessarily dual) norm?

For Problems 4.4, 4.5, 4.6, and 4.7 above, we refer to [**20**].

5. Rotund renormings

PROBLEM 5.1. Does every nonseparable reflexive Banach space admit a *2R* norm?

The norm $\|\cdot\|$ is **2R** (i.e., **2-rotund**) if whenever

$$\lim_{n,m\to\infty} \|x_m + x_n\| = 2$$

for some sequence (x_n) in the unit sphere, then (x_n) is norm convergent.

See, e.g., [**69**], [**29**], and [**40**]

Note that *every separable reflexive Banach space admits a 2R norm* [**69**], and that *every space with a 2R norm is reflexive*, see, e.g., [**22**, Exercise 8.6].

Thus a positive solution to Problem 5.1 would provide for a nice renorming characterization of reflexive spaces.

PROBLEM 5.2. Does a renorming by a WUR norm have the three-space property?

The norm $\|\cdot\|$ is **WUR (weakly uniformly rotund)** if $x_n - y_n \to 0$ weakly whenever (x_n) and (y_n) are sequences in the unit sphere such that $\|x_n + y_n\| \to 2$. See [**83**], [**40**]. This problem is also mentioned after [**40**, Theorem 648 and Section 52.3]. For the meaning of a three-space property see the statement of Problem 2.8 above.

PROBLEM 5.3. Assume that in a Banach space $\mathcal{X}$ every closed convex bounded set C is the closed convex hull of its weakly exposed points. Does $\mathcal{X}$ have necessarily the RNP property?

See [Pi09]. The concepts involved were defined after Problem 2.9. It is known that *a space has the RNP if, and only if, it is dentable.* For the definition of dentability, see the comments after Problem 5.6 below. For other details see, e.g., [**22**, Chapter 11].

We believe that the next problem is still open:

PROBLEM 5.4. Assume that every separable subspace of a Banach space $\mathcal{X}$ has the Krein–Milman property. Does $\mathcal{X}$ necessarily have the Krein–Milman property?

See [**53**]. A Banach space $\mathcal{X}$ has the **Krein–Milman property** if each closed convex bounded set in $\mathcal{X}$ is the closed convex hull of its extreme points. In [**75**] is a sharpening of a result of R. Huff and P. Morris: *A dual space has the RNP if, and only if, all separable subspaces have the Krein–Milman property.* Thus, *for dual Banach spaces, the RNP and the Krein–Milman property are equivalent and separably determined.*

PROBLEM 5.5. If a Banach space $\mathcal{X}$ can be renormed by a norm with modulus of convexity p and also renormed with a norm with modulus of smoothness q, can $\mathcal{X}$ be renormed by a norm that has both these properties simultaneously?

See [**6**]. See also [**38**, Problem 198]. Problem 5.5 has a positive answer for $p = q = 2$ (see, e.g., [**15**, page 175]).

PROBLEM 5.6. Let $\mathcal{X}$ be a separable Banach space that does not contain a copy of ℓ_1.

(i) Can $\mathcal{X}^*$ be renormed by a (not necessarily dual) Gâteaux differentiable norm?

(ii) Is every w*-compact set in $\mathcal{X}^*$ norm-dentable?

(iii) Can $\mathcal{X}^*$ be renormed by a not necessarily dual LUR norm?

See [**38**] and [**45**]. A set C in a Banach space $\mathcal{X}$ is **dentable** if for every $\varepsilon > 0$ there is an open half-space in $\mathcal{X}$ whose nonempty intersection with C has diameter $< \varepsilon$.

It has been recently shown [**47**] that *there is a C^∞-smooth norm on c_0 whose dual norm is not rotund.* See also [**35**].

6. General structure

PROBLEM 6.1 (S. Banach, A. Pełczyński). Does every infinite-dimensional Banach space have an infinite-dimensional separable quotient?

An equivalent formulation of Problem 6.1 is the following: Given an infinite-dimensional Banach space $\mathcal{X}$, does there exist a separable infinite-dimensional Banach space $\mathcal{Z}$ and a bounded linear operator T from $\mathcal{X}$ onto $\mathcal{Z}$?

Below are two more equivalent formulations of Problem 6.1 (the equivalence with (ii) is due to H.P. Rosenthal):

(i) If $\mathcal{X}$ is any nonseparable Banach space, does there exist an infinite increasing sequence of distinct subspaces $(\mathcal{E}_n)_{n=1}^\infty$ such that $\overline{\bigcup_{n=1}^\infty \mathcal{E}_n} = \mathcal{X}$?

(ii) Does every nonseparable Banach space $\mathcal{X}$ contain an infinite-dimensional separable subspace $\mathcal{Y}$ that is **quasicomplemented in** $\mathcal{X}$, i.e., there is a subspace $\mathcal{Z}$ of $\mathcal{X}$ such that $\mathcal{Y} \cap \mathcal{Z} = \{0\}$ and $\overline{\mathcal{Y} + \mathcal{Z}} = \mathcal{X}$?

In this direction we refer to an excellent paper by J. Mujica [**68**]. See also [**72**], [**38**, Problem 32], [**38**, Problem 32], and the references therein.

PROBLEM 6.2 (G. Godefroy). If $\mathcal{X}$ is a subspace of ℓ_∞, does the class of spaces isomorphic to $\mathcal{X}$ have infinite diameter in the Banach–Mazur distance?

The **Banach–Mazur distance** $D(\mathcal{Y}, Z)$ of two isomorphic spaces $\mathcal{Y}$ and $\mathcal{Z}$ is

$$D(\mathcal{Y}, \mathcal{Z}) = \inf\{\|T\| \cdot \|T^{-1}\| : T \text{ isomorphism from } \mathcal{Y} \text{ onto } \mathcal{Z}\}.$$

See [**30**]. The answer is yes for all separable spaces $\mathcal{X}$, a result due to W.B. Johnson and E.W. Odell [**55**]. See also [**79**].

PROBLEM 6.3 (E.W. Odell). If $\mathcal{X}$ is a separable Banach space, do there exist two spaces $\mathcal{Y}$ and $\mathcal{Z}$, both isomorphic to $\mathcal{X}$, such that $\mathcal{Y}$ is not isometric to $\mathcal{Z}$ and yet, the Banach–Mazur distance of $\mathcal{Y}$ to $\mathcal{Z}$ is 1?

See [**38**, Problem 96]. Problem 6.3 was solved positively in c_0 by C. Bessaga and A. Pełczyński [**5**], see also [**22**].

PROBLEM 6.4 (S. Todorčević). Assume that the density of a Banach space $\mathcal{X}$ is $> c$. Does $\mathcal{X}$ necessarily contain a separable infinite-dimensional subspace with unconditional basis?

PROBLEM 6.5 (R. Deville). Does there exist a HI space with a C^2-smooth norm?

A **hereditarily indecomposable** (**HI**, for short) space is a space such that no subspace of it can be written as the direct sum of two infinite-dimensional subspaces. See, e.g., [**64**], [**40**].

PROBLEM 6.6. Is a Banach space $\mathcal{X}$ WCG if $\mathcal{X}^{**}$ is WCG?

Recall that a Banach space $\mathcal{X}$ is **weakly compactly generated** (**WCG**, for short) if $\mathcal{X}$ contains a weakly compact subset K whose closed linear hull equals $\mathcal{X}$.

For this problem see, e.g., [**83**] and [**3**].

PROBLEM 6.7. Let $\mathcal{X}$ be a separable Banach space with transitive norm. Is $\mathcal{X}$ linearly isometric to a Hilbert space?

The norm of a Banach space $\mathcal{X}$ is called **transitive** if for $x, y \in S_{\mathcal{X}}$ there is a linear isometry of $\mathcal{X}$ onto $\mathcal{X}$ such that $T(x) = T(y)$.

See, e.g., [**5**] and [**16**]. See also Problem 6.8 below.

PROBLEM 6.8. If $\mathcal{X}$ is a separable Banach space with a quasitransitive norm, is $\mathcal{X}$ isomorphic to a Hilbert space?

The norm of a Banach space $\mathcal{X}$ is called **quasitransitive** if for $x, y \in S_{\mathcal{X}}$ and for every $\varepsilon > 0$ there is a linear isometry T from $\mathcal{X}$ onto $\mathcal{X}$ such that $\|T(x) - T(y)\| < \varepsilon$.

See, e.g., [**16**] and [**11**]. See also Problem 6.7 above.

PROBLEM 6.9. Assume that a compact topological space K has the strong property $\mathcal{P}$. Is K necessarily Eberlein?

A compact topological space K is an **Eberlein compact** whenever it is homeomorphic to a weakly compact subset of a Banach space endowed with the restriction of the weak topology.

A compact topological space K has the **strong property** $\mathcal{P}$ (see [**23**]) whenever there is a norm $\|\cdot\|$ of $C(K)$ such that for every dense subset D of K, the D-lower semicontinuous envelope of $\|\cdot\|$ (i.e., the norm $\|x\|_D := \sup\{\langle x, x^*\rangle : x^* \in D,\ \|x^*\| \leq 1\}$, $x \in \mathcal{X}$) is Gâteaux differentiable.

Problem 6.9 pertains to the area of applications of smoothness to the analysis of the structure of nonseparable Banach spaces and to topological problems related to Eberlein compacta. See [**23**].

PROBLEM 6.10. Is every Chebyshev subset of ℓ_2 convex?

A subset C of a Banach space $\mathcal{X}$ is **Chebyshev** whenever each point of $\mathcal{X}$ has a unique closest point in C. We refer to the nice survey [**27**]. See also [**38**, Problem 79] and the references therein.

PROBLEM 6.11. Let $\mathcal{X}$ and $\mathcal{Y}$ be separable Banach spaces that are Lipschitz homeomorphic. Are they necessarily linearly isomorphic?

See [**6**]. There is a nonseparable $C(K)$ space that admits a C^∞-smooth norm that is Lipschitz homeomorphic but not linearly isomorphic to $c_0(\Gamma)$ (see, e.g., [**40**, Theorem 816]).

PROBLEM 6.12 (G. Godefroy). Does there exist a separable HI Banach space Lipschitz homeomorphic to one of its proper subspaces?

See [**25**]. For the definition of a HI space see Problem 6.5 above.

PROBLEM 6.13. If $\mathcal{X}$ is Lipschitz homeomorphic to $C[0,1]$ (or to ℓ_1), is $\mathcal{X}$ linearly isomorphic to $C[0,1]$ (or to ℓ_1, respectively)?

See [**6**].

PROBLEM 6.14. If $\mathcal{X}$ is strictly convex nonseparable C^2-smooth Banach space, must $\mathcal{X}$ necessarily linearly inject into a superreflexive space?

See, e.g., [**41**], [**42**], and [**40**]. A Banach space $\mathcal{X}$ is **superreflexive** whenever it admits a uniformly convex norm.

PROBLEM 6.15. If $\mathcal{X}$ has density $\leq c$, does $\mathcal{X}$ Lipschitz-embed into ℓ_∞?

See [**32**]. It is worth mentioning that if we add to the requirements on $\mathcal{X}$ to have an unconditional basis, then the answer is affirmative [**56**].

PROBLEM 6.16. Does there exist a separable Banach space that is not a Lipschitz retract of its bidual? *This means,* does there exist a separable or even WCG Banach space $\mathcal{X}$ such that there is no Lipschitz map T from $\mathcal{X}^{**}$ onto $\mathcal{X}$ such that $T(x) = x$ for $x \in \mathcal{X}$?

See [**38**], or [**32**].

The following problem is chosen to document the depth of the roots of Banach spaces in the Kingdom of Mathematics.

A closed convex set C in a Banach space $\mathcal{X}$ is called a **support set** if every point $x_0 \in C$ is a **proper support point** of C —meaning that there is $f \in \mathcal{X}^*$ such that

$$f(x_0) = \inf\{f(x) : x \in C\} < \sup\{f(x) : x \in C\}.$$

No separable Banach space contains a support set. This is a result of S. Rolewicz [**74**] (see, e.g., [**22**, Lemma 12.26] and [**44**, Theorem 8.2]). It was a question in [**74**] whether every nonseparable space contains a support set. It turned out that it cannot be answered in ZFC. See, e.g., [**44**, Section 8.1].

The following problem is open:

PROBLEM 6.17 (S. Todorčević). Assume that the density character of the space $\mathcal{X}$ is greater than $\aleph_1$. Must $\mathcal{X}$ necessarily contain a support set?

S. Todorčević [**78**] proved that this is the case for $C(K)$ spaces of density greater than $\aleph_1$. See Section 8.1 and, in particular, Theorems 8.11 and 8.12 of [**44**]. See also [**38**, Problem 290] and the references therein.

PROBLEM 6.18. Let $\mathcal{X}$ be a nonreflexive Banach space. Does there exist in $\mathcal{X}$ a closed convex bounded set without farthest points?

A point x_0 in a set $C \subset \mathcal{X}$ is a **farthest point in** C if there is a point $x \in \mathcal{X}$ such that $\|x_0 - x\| \geq \|y - x\|$ for all $y \in C$. We refer to [**15**, page 122], [**22**, Section 8.3] and [**66**]. Every nonempty weakly compact subset of a Banach space has farthest points. For a more precise statement, see, in particular [**22**, Theorems 8.19 and 8.20].

PROBLEM 6.19 (P. Morris). Characterize the Banach spaces $\mathcal{X}$ that can be renormed by an equivalent strictly convex norm $\|\cdot\|$ such that no point of $S_{(\mathcal{X},\|\cdot\|)}$ is an extreme point of $S_{(\mathcal{X}^{**},\|\cdot\|^{**})}$.

We refer to [**67**], [**36**], and [**37**].

7. Schauder bases

Recall that a sequence (x_n) in a Banach space $\mathcal{X}$ is a **Schauder basis of** $\mathcal{X}$ if for every $x \in \mathcal{X}$ there is a unique sequence of scalars (a_n) such that $x = \sum a_n x_n$. If the convergence of the series is unconditional for all $x \in \mathcal{X}$, the basis is called **unconditional**.

PROBLEM 7.1. Is a separable infinite-dimensional Banach space $\mathcal{X}$ isomorphic to ℓ_2 if each subspace of $\mathcal{X}$ has a subspace with unconditional basis?

See, e.g., [**12**], [**57**], and [**58**]. See also [**38**, Problem 1 (v)] and the references therein.

PROBLEM 7.2 (A. Pełczyński). Let $\mathcal{X}$ be an infinite-dimensional separable Banach space not isomorphic to ℓ_2. Does $\mathcal{X}$ contain a subspace with a Schauder basis that is not complemented in $\mathcal{X}$?

J. Lindenstrauss and L. Tzafriri proved in [**62**] that *if all subspaces of a Banach space $\mathcal{X}$ are complemented in $\mathcal{X}$, then $\mathcal{X}$ is isomorphic to ℓ_2*. For more, see [**38**, Problem 1 (i)] and the references therein.

PROBLEM 7.3 (A. Pełczyński). If in a separable Banach space $\mathcal{X}$ all infinite-dimensional subspaces with a Schauder basis are isomorphic, is the space $\mathcal{X}$ isomorphic to ℓ_2?

See [**38**, Problem 1 (ii)] and the references therein. R. Komorowski and N. Tomczak-Jaegermann proved in [**58**] that *the answer is positive if, in the preceding statement, the words "Schauder basis" are deleted.*

PROBLEM 7.4 (A. Pełczyński). Let $\mathcal{X}$ be a separable Banach space. Does $\mathcal{X}$ admit an Auerbach basis? Does $C[0,1]$ admit an Auerbach basis?

An **Auerbach basis** for a Banach space $\mathcal{X}$ is a biorthogonal system $\{x_i, f_i\}_{i \in I}$ such that the norm-closed linear hull of $\{x_i : i \in I\}$ is $\mathcal{X}$, the weak-star-closed linear hull of $\{f_i : i \in I\}$ is $\mathcal{X}^*$ and, moreover, $\|x_i\| = \|f_i\| = 1$ for all $i \in I$. See [**44**, Sections 1.2 and 1.4, and Problem 1.4]. See also [**38**, Problem 113] and the references therein.

PROBLEM 7.5 (G. Godefroy). Let $\mathcal{X}$ be a separable superreflexive Banach space. Does there exist a superreflexive separable Banach space $\mathcal{Z}$ with a Schauder basis, such that $\mathcal{X}$ is isomorphic to a subspace of $\mathcal{Z}$?

8. Operators

PROBLEM 8.1. Does there exist a reflexive separable space $\mathcal{X}$ such that every operator T from $\mathcal{X}$ into $\mathcal{X}$ is of the form $T = \lambda I +$compact operator for some scalar λ? (here, I is the identity operator)

See [**1**]. The famous "scalar-plus-compact problem" was open for a long time. S. Argyros and R. Haydon [**1**] provided the solution, showing that there exists a hereditarily indecomposable space $\mathcal{X}$ whose dual is isomorphic to ℓ_1 and such that every operator from $\mathcal{X}$ into $\mathcal{X}$ is a multiple of the identity plus a compact operator. See also, e.g., [**40**, Theorem 467] and [**71**]. Problem 8.1 also appears as [**38**, Problem 48].

P. Koszmider and N.J. Laustsen showed in [**59**] that there exists a closed subspace of ℓ_∞ generated by c_0 and the characteristic functions of elements of an uncountable, almost disjoint family $\mathcal{A}$ of infinite subsets on $\mathbb{N}$, that is of the form $C_0(K_{\mathcal{A}})$, for $K_{\mathcal{A}}$ a locally compact space known as Ψ-space or Isbell-Mrówka space, such that every bounded linear operator on $C_0(K_{\mathcal{A}})$ is the sum of a scalar multiple of the identity and an operator that factors through c_0 (which, in this case, is equivalent to having a separable range).

PROBLEM 8.2 (G. Pisier). Is the nuclear version of the scalar-plus-compact problem true? *This means:* Does there exist a separable infinite-dimensional Banach space $\mathcal{X}$ such that every operator T from $\mathcal{X}$ into $\mathcal{X}$ is of the form $T = \lambda I$ + a nuclear operator from $\mathcal{X}$ into $\mathcal{X}$.

A **nuclear operator** T from a Banach space $\mathcal{X}$ into $\mathcal{X}$ is an operator such that there exist sequences (x_i^*) in $\mathcal{X}^*$ and (y_i) in $\mathcal{X}$ with $\sum_{i=1}^{\infty} \|x_i^*\|\|y_i\| < \infty$ and $T(x) = \sum_{i=1}^{\infty} x_i^*(x) y_i$ for $x \in \mathcal{X}$.

We note that a positive answer to Problem 8.2 would solve Problem 8.3 below in the negative, see again [**71**].

PROBLEM 8.3. Let $\mathcal{X}$ be a Banach space such that every compact operator from $\mathcal{X}$ into $\mathcal{X}$ is the limit of finite-rank operators from $\mathcal{X}$ into $\mathcal{X}$ in the norm topology. Does $\mathcal{X}$ have the approximation property?

A Banach space $\mathcal{X}$ has the **approximation property** if for every compact set K in $\mathcal{X}$ and every $\varepsilon > 0$ there is a finite rank operator T from $\mathcal{X}$ into $\mathcal{X}$ such that $\|Tx - x\| \leq \varepsilon$ for every $x \in K$.

Problem 8.3 is a long-standing problem. See, e.g., [**63**], [**22**, page 711], [**12**], and [**38**, Problem 52]. See the references therein.

PROBLEM 8.4. Does the space H_∞ of all bounded analytic functions on the set $\{z : |z| < 1\}$ endowed with the supremum norm have the approximation property?

See, e.g., [**63**], [**38**, Problem 53], and [**12**].

PROBLEM 8.5. Does there exist an infinite-dimensional space where all compact operators are nuclear?

See, e.g., [**38**] and [**71**].

9. Nonlinear analysis

PROBLEM 9.1. If $\mathcal{X}$ is an infinite-dimensional Banach space, is the unit sphere uniformly homeomorphic to the unit ball?

See [**6**]. It is true if one removes the word "uniformly" above, see [**5**, pages 278-279], see also comments to Problem 9.3.

PROBLEM 9.2. Is it true that every normed linear space is homeomorphic to an inner product space?

PROBLEM 9.3. Is it true that every separable infinite-dimensional metric linear space is homeomorphic to ℓ_2?

It is a basic result that in infinite-dimensional Banach spaces, balls and spheres are homeomorphic to the whole space. Notice, too, that no incomplete normed space can be homeomorphic to a Banach space (see, for the problem and the remarks, [**8**], [**5**], and [**22**]).

PROBLEM 9.4. Can Bessaga's method of proof of the topological equivalence of infinite dimensional reflexive spaces of the same density *(in* [**7**]*)* be extended to WCG spaces?

Acknowledgments

We thank Marián Fabian and Petr Hájek for their help in preparing this paper. We thank also two referees for their precise work that highly improved the final version of the manuscript. The referencce to [**59**] after Problem 8.1 was suggested by one of them.

References

[1] S. A. Argyros and R. G. Haydon, *A hereditarily indecomposable $\mathcal{L}_\infty$-space that solves the scalar-plus-compact problem*, Acta Math. **206** (2011), no. 1, 1–54, DOI 10.1007/s11511-011-0058-y. MR2784662

[2] S. Argyros and S. Mercourakis, *On weakly Lindelöf Banach spaces*, Rocky Mountain J. Math. **23** (1993), no. 2, 395–446, DOI 10.1216/rmjm/1181072569. MR1226181

[3] S. A. Argyros and S. Mercourakis, *Examples concerning heredity problems of WCG Banach spaces*, Proc. Amer. Math. Soc. **133** (2005), no. 3, 773–785, DOI 10.1090/S0002-9939-04-07532-X. MR2113927

[4] M. Bačák and P. Hájek, *Mazur intersection property for Asplund spaces*, J. Funct. Anal. **255** (2008), no. 8, 2090–2094, DOI 10.1016/j.jfa.2008.05.016. MR2462587

[5] S. Banach, *Œuvres. Vol. II: Travaux sur l'analyse fonctionnelle*, PWN—Éditions Scientifiques de Pologne, Warsaw, 1979. Edited by Czesław Bessaga, Stanisław Mazur, Władysław Orlicz, Aleksander Pełczyński, Stefan Rolewicz and Wiesław Żelazko; With an article by Pełczyński and Bessaga. MR563126

[6] Y. Benyamini and J. Lindenstrauss, *Geometric nonlinear functional analysis. Vol. 1*, American Mathematical Society Colloquium Publications, vol. 48, American Mathematical Society, Providence, RI, 2000, DOI 10.1090/coll/048. MR1727673

[7] C. Bessaga, *Topological equivalence of non-separable reflexive Banach spaces. Ordinal resolutions of identity and monotone bases*, Symposium on Infinite-Dimensional Topology (Louisiana State Univ., Baton Rouge, La., 1967), Ann. of Math. Stud., No. 69, Princeton Univ. Press, Princeton, NJ, 1972, pp. 3–14. MR417764

[8] C. Bessaga and A. Pełczyński, *Selected topics in infinite-dimensional topology*, Monografie Matematyczne [Mathematical Monographs], Tom 58, PWN—Polish Scientific Publishers, Warsaw, 1975. MR478168

[9] V. Bible and R. J. Smith, *Smooth and polyhedral approximation in Banach spaces*, J. Math. Anal. Appl. **435** (2016), no. 2, 1262–1272, DOI 10.1016/j.jmaa.2015.11.018. MR3429641

[10] J. M. Borwein and D. Noll, *Second order differentiability of convex functions in Banach spaces*, Trans. Amer. Math. Soc. **342** (1994), no. 1, 43–81, DOI 10.2307/2154684. MR1145959

[11] F. Cabello Sánchez, V. Ferenczi, and B. Randrianantoanina, *On Mazur rotations problem and its multidimensional versions*, São Paulo J. Math. Sci. **16** (2022), no. 1, 406–458, DOI 10.1007/s40863-021-00234-3. MR4426402

[12] P. G. Casazza, *Approximation properties*, Handbook of the geometry of Banach spaces, Vol. I, North-Holland, Amsterdam, 2001, pp. 271–316, DOI 10.1016/S1874-5849(01)80009-7. MR1863695

[13] J. M. F. Castillo and M. González, *Three-space problems in Banach space theory*, Lecture Notes in Mathematics, vol. 1667, Springer-Verlag, Berlin, 1997, DOI 10.1007/BFb0112511. MR1482801

[14] R. Deville, G. Godefroy, and V. Zizler, *Smooth bump functions and geometry of Banach spaces*, Mathematika **40** (1993), no. 2, 305–321, DOI 10.1112/S0025579300007075. MR1260894

[15] R. Deville, G. Godefroy, and V. Zizler, *Smoothness and renormings in Banach spaces*, Pitman Monographs and Surveys in Pure and Applied Mathematics, vol. 64, Longman Scientific & Technical, Harlow; copublished in the United States with John Wiley & Sons, Inc., New York, 1993. MR1211634

[16] S. J. Dilworth and B. Randrianantoanina, *On an isomorphic Banach-Mazur rotation problem and maximal norms in Banach spaces*, J. Funct. Anal. **268** (2015), no. 6, 1587–1611, DOI 10.1016/j.jfa.2014.11.021. MR3306357

[17] P. Enflo, *On the invariant subspace problem for Banach spaces*, Acta Math. **158** (1987), no. 3-4, 213–313, DOI 10.1007/BF02392260. MR892591

[18] P. Enflo and V. Lomonosov, *Some aspects of the invariant subspace problem*, Handbook of the geometry of Banach spaces, Vol. I, North-Holland, Amsterdam, 2001, pp. 533–559, DOI 10.1016/S1874-5849(01)80015-2. MR1863701

[19] M. Fabián, *Lipschitz smooth points of convex functions and isomorphic characterizations of Hilbert spaces*, Proc. London Math. Soc. (3) **51** (1985), no. 1, 113–126, DOI 10.1112/plms/s3-51.1.113. MR788852

[20] M. J. Fabian, *Gâteaux differentiability of convex functions and topology: Weak Asplund spaces*, Canadian Mathematical Society Series of Monographs and Advanced Texts, John Wiley & Sons, Inc., New York, 1997. A Wiley-Interscience Publication. MR1461271

[21] M. Fabian, P. Habala, P. Hájek, V. Montesinos Santalucía, J. Pelant, and V. Zizler, *Functional analysis and infinite-dimensional geometry*, CMS Books in Mathematics/Ouvrages de Mathématiques de la SMC, vol. 8, Springer-Verlag, New York, 2001, DOI 10.1007/978-1-4757-3480-5. MR1831176

[22] M. Fabian, P. Habala, P. Hájek, V. Montesinos, and V. Zizler, *Banach space theory: The basis for linear and nonlinear analysis*, CMS Books in Mathematics/Ouvrages de Mathématiques de la SMC, Springer, New York, 2011, DOI 10.1007/978-1-4419-7515-7. MR2766381

[23] M. Fabian, V. Montesinos, and V. Zizler, *Pointwise semicontinuous smooth norms*, Arch. Math. (Basel) **78** (2002), no. 6, 459–464, DOI 10.1007/s00013-002-8272-7. MR1921736

[24] M. Fabian and V. Zizler, *On uniformly Gâteaux smooth $C^{(n)}$-smooth norms on separable Banach spaces*, Czechoslovak Math. J. **49(124)** (1999), no. 3, 657–672, DOI 10.1023/A:1022487822852. MR1708330

[25] V. Ferenczi, *Lipschitz homogeneous Banach spaces*, Q. J. Math. **54** (2003), no. 4, 415–419, DOI 10.1093/qjmath/54.4.415. MR2031174

[26] V. P. Fonf, *Three characterizations of polyhedral Banach spaces* (Russian, with Ukrainian summary), Ukrain. Mat. Zh. **42** (1990), no. 9, 1286–1290, DOI 10.1007/BF01056615; English transl., Ukrainian Math. J. **42** (1990), no. 9, 1145–1148 (1991). MR1093646

[27] J. Fletcher and W. B. Moors, *Chebyshev sets*, J. Aust. Math. Soc. **98** (2015), no. 2, 161–231, DOI 10.1017/S1446788714000561. MR3312270

[28] G. Godefroy, *Some applications of Simons' inequality*, Serdica Math. J. **26** (2000), no. 1, 59–78. MR1767034

[29] G. Godefroy, *Renormings of Banach spaces*, Handbook of the geometry of Banach spaces, Vol. I, North-Holland, Amsterdam, 2001, pp. 781–835, DOI 10.1016/S1874-5849(01)80020-6. MR1863706

[30] G. Godefroy, *On the diameter of the Banach-Mazur set*, Czechoslovak Math. J. **60(135)** (2010), no. 1, 95–100, DOI 10.1007/s10587-010-0021-7. MR2595073

[31] G. Godefroy, *Fréchet differentiability of Lipschitz functions and porous sets in Banach spaces [book review of refcno 2884141]*, Bull. Amer. Math. Soc. (N.S.) **51** (2014), no. 3, 515–517, DOI 10.1090/S0273-0979-2014-01431-4. MR3196801

[32] G. Godefroy, G. Lancien, and V. Zizler, *The non-linear geometry of Banach spaces after Nigel Kalton*, Rocky Mountain J. Math. **44** (2014), no. 5, 1529–1583, DOI 10.1216/RMJ-2014-44-5-1529. MR3295641

[33] G. Godefroy, V. Montesinos, and V. Zizler, *Strong subdifferentiability of norms and geometry of Banach spaces*, Comment. Math. Univ. Carolin. **36** (1995), no. 3, 493–502. MR1364490

[34] A. S. Granero, M. Jiménez-Sevilla, and J. P. Moreno, *Intersections of closed balls and geometry of Banach spaces*, Extracta Math. **19** (2004), no. 1, 55–92. MR2072538

[35] A. J. Guirao, V. Montesinos, and V. Zizler, *On a classical renorming construction of V. Klee*, J. Math. Anal. Appl. **385** (2012), no. 1, 458–465, DOI 10.1016/j.jmaa.2011.06.059. MR2834271

[36] A. J. Guirao, V. Montesinos, and V. Zizler, *On preserved and unpreserved extreme points*, Descriptive topology and functional analysis, Springer Proc. Math. Stat., vol. 80, Springer, Cham, 2014, pp. 163–193, DOI 10.1007/978-3-319-05224-3_9. MR3238208

[37] A. J. Guirao, V. Montesinos, and V. Zizler, *A note on extreme points of C^∞-smooth balls in polyhedral spaces*, Proc. Amer. Math. Soc. **143** (2015), no. 8, 3413–3420, DOI 10.1090/S0002-9939-2015-12617-2. MR3348784

[38] A. J. Guirao, V. Montesinos, and V. Zizler, *Open problems in the geometry and analysis of Banach spaces*, Springer, [Cham], 2016, DOI 10.1007/978-3-319-33572-8. MR3524558

[39] P. Hájek, *Smooth norms that depend locally on finitely many coordinates*, Proc. Amer. Math. Soc. **123** (1995), no. 12, 3817–3821, DOI 10.2307/2161911. MR1285993

[40] A. J. Guirao, V. Montesinos, and V. Zizler, *Renormings in Banach spaces—a toolbox*, Monografie Matematyczne [Mathematical Monographs], vol. 75, Birkhäuser/Springer, Cham, [2022] ©2022, DOI 10.1007/978-3-031-08655-7. MR4501243

[41] P. Hájek, *Smooth functions on $C(K)$*, Israel J. Math. **107** (1998), 237–252, DOI 10.1007/BF02764011. MR1658563

[42] P. Hájek, *Smooth norms on certain $C(K)$ spaces*, Proc. Amer. Math. Soc. **131** (2003), no. 7, 2049–2051, DOI 10.1090/S0002-9939-03-06819-9. MR1963749

[43] P. Hájek and M. Johanis, *Smooth analysis in Banach spaces*, De Gruyter Series in Nonlinear Analysis and Applications, vol. 19, De Gruyter, Berlin, 2014, DOI 10.1515/9783110258998. MR3244144

[44] P. Hájek, V. Montesinos Santalucía, J. Vanderwerff, and V. Zizler, *Biorthogonal systems in Banach spaces*, CMS Books in Mathematics/Ouvrages de Mathématiques de la SMC, vol. 26, Springer, New York, 2008. MR2359536

[45] P. Hájek, V. Montesinos, and V. Zizler, *Geometry and Gâteaux smoothness in separable Banach spaces*, Oper. Matrices **6** (2012), no. 2, 201–232, DOI 10.7153/oam-06-15. MR2976113

[46] P. Hájek and A. Procházka, *C^k-smooth approximations of LUR norms*, Trans. Amer. Math. Soc. **366** (2014), no. 4, 1973–1992, DOI 10.1090/S0002-9947-2013-05899-0. MR3152719

[47] P. Hájek and A. Quilis, *Counterexamples in rotundity of norms in Banach spaces*, to appear.

[48] P. Hájek and T. Russo, *Some remarks on smooth renormings of Banach spaces*, J. Math. Anal. Appl. **455** (2017), no. 2, 1272–1284, DOI 10.1016/j.jmaa.2017.05.077. MR3671225

[49] P. Hájek and J. Talponen, *Smooth approximations of norms in separable Banach spaces*, Q. J. Math. **65** (2014), no. 3, 957–969, DOI 10.1093/qmath/hat053. MR3261976

[50] P. Hájek and S. Troyanski, *Analytic norms in Orlicz spaces*, Proc. Amer. Math. Soc. **129** (2001), no. 3, 713–717, DOI 10.1090/S0002-9939-00-05773-7. MR1801996

[51] R. Haydon, *A counterexample to several questions about scattered compact spaces*, Bull. London Math. Soc. **22** (1990), no. 3, 261–268, DOI 10.1112/blms/22.3.261. MR1041141

[52] R. Haydon, *Trees in renorming theory*, Proc. London Math. Soc. (3) **78** (1999), no. 3, 541–584, DOI 10.1112/S0024611599001768. MR1674838

[53] R. E. Huff and P. D. Morris, *Geometric characterizations of the Radon-Nikodým property in Banach spaces*, Studia Math. **56** (1976), no. 2, 157–164, DOI 10.4064/sm-56-2-157-164. MR412776

[54] K. John and V. Zizler, *On extension of rotund norms*, Bull. Acad. Polon. Sci. Sér. Sci. Math. Astronom. Phys. **24** (1976), no. 9, 705–707. MR428011

[55] W. B. Johnson and E. Odell, *The diameter of the isomorphism class of a Banach space*, Ann. of Math. (2) **162** (2005), no. 1, 423–437, DOI 10.4007/annals.2005.162.423. MR2178965

[56] N. J. Kalton, *Lipschitz and uniform embeddings into ℓ_∞*, Fund. Math. **212** (2011), no. 1, 53–69, DOI 10.4064/fm212-1-4. MR2771588

[57] R. A. Komorowski and N. Tomczak-Jaegermann, *Banach spaces without local unconditional structure*, Israel J. Math. **89** (1995), no. 1-3, 205–226, DOI 10.1007/BF02808201. MR1324462

[58] R. A. Komorowski and N. Tomczak-Jaegermann, *Banach spaces without local unconditional structure*, Israel J. Math. **89** (1995), no. 1-3, 205–226, DOI 10.1007/BF02808201. MR1324462

[59] P. Koszmider and N. J. Laustsen, *A Banach space induced by an almost disjoint family, admitting only few operators and decompositions*, Adv. Math. **381** (2021), Paper No. 107613, 39, DOI 10.1016/j.aim.2021.107613. MR4206788

[60] J. Kurzweil, *On approximation in real Banach spaces by analytic operations*, Studia Math. **16** (1957), 124–129, DOI 10.4064/sm-16-2-124-129. MR93697

[61] J. Lindenstrauss, D. Preiss, and J. Tišer, *Fréchet differentiability of Lipschitz functions and porous sets in Banach spaces*, Annals of Mathematics Studies, vol. 179, Princeton University Press, Princeton, NJ, 2012. MR2884141

[62] J. Lindenstrauss and L. Tzafriri, *On the complemented subspaces problem*, Israel J. Math. **9** (1971), 263–269, DOI 10.1007/BF02771592. MR276734

[63] J. Lindenstrauss and L. Tzafriri, *Classical Banach spaces. I: Sequence spaces*, Ergebnisse der Mathematik und ihrer Grenzgebiete [Results in Mathematics and Related Areas], Band 92, Springer-Verlag, Berlin-New York, 1977. MR500056

[64] B. Maurey, *Banach spaces with few operators*, Handbook of the geometry of Banach spaces, Vol. 2, North-Holland, Amsterdam, 2003, pp. 1247–1297, DOI 10.1016/S1874-5849(03)80036-0. MR1999196

[65] A. Moltó, J. Orihuela, S. Troyanski, and M. Valdivia, *A nonlinear transfer technique for renorming*, Lecture Notes in Mathematics, vol. 1951, Springer-Verlag, Berlin, 2009, DOI 10.1007/978-3-540-85031-1. MR2462399

[66] V. Montesinos, P. Zizler, and V. Zizler, *Some remarks on farthest points*, Rev. R. Acad. Cienc. Exactas Fís. Nat. Ser. A Mat. RACSAM **105** (2011), no. 1, 119–131, DOI 10.1007/s13398-011-0012-z. MR2783800

[67] P. Morris, *Disappearance of extreme points*, Proc. Amer. Math. Soc. **88** (1983), no. 2, 244–246, DOI 10.2307/2044709. MR695251

[68] J. Mujica, *Separable quotients of Banach spaces*, Rev. Mat. Univ. Complut. Madrid **10** (1997), no. 2, 299–330. MR1605658

[69] E. Odell and Th. Schlumprecht, *Asymptotic properties of Banach spaces under renormings*, J. Amer. Math. Soc. **11** (1998), no. 1, 175–188, DOI 10.1090/S0894-0347-98-00251-3. MR1469118

[70] R. R. Phelps, *Convex functions, monotone operators and differentiability*, Lecture Notes in Mathematics, vol. 1364, Springer-Verlag, Berlin, 1989, DOI 10.1007/BFb0089089. MR984602

[71] A. Pietsch, *Long-standing open problems of Banach space theory: my personal top ten*, Quaest. Math. **32** (2009), no. 3, 321–337, DOI 10.2989/QM.2009.32.3.4.905. MR2569102

[72] G. Pisier, *Factorization of linear operators and geometry of Banach spaces*, CBMS Regional Conference Series in Mathematics, vol. 60, Published for the Conference Board of the Mathematical Sciences, Washington, DC; by the American Mathematical Society, Providence, RI, 1986, DOI 10.1090/cbms/060. MR829919

[73] D. Preiss, *Differentiability of Lipschitz functions on Banach spaces*, J. Funct. Anal. **91** (1990), no. 2, 312–345, DOI 10.1016/0022-1236(90)90147-D. MR1058975

[74] S. Rolewicz, *On convex sets containing only points of support*, Comment. Math. Spec. Issue **1** (1978), 279–281. MR504173

[75] W. Schachermayer, *The Radon-Nikodým property and the Kreĭn-Milman property are equivalent for strongly regular sets*, Trans. Amer. Math. Soc. **303** (1987), no. 2, 673–687, DOI 10.2307/2000690. MR902791

[76] R. J. Smith and S. Troyanski, *Unconditional bases and strictly convex dual renormings*, Bull. Lond. Math. Soc. **41** (2009), no. 5, 831–840, DOI 10.1112/blms/bdp059. MR2557464

[77] R. J. Smith and S. Troyanski, *Approximation of norms on Banach spaces*, J. Funct. Anal. **276** (2019), no. 10, 3226–3260, DOI 10.1016/j.jfa.2018.07.007. MR3944292

[78] S. Todorcevic, *Biorthogonal systems and quotient spaces via Baire category methods*, Math. Ann. **335** (2006), no. 3, 687–715, DOI 10.1007/s00208-006-0762-7. MR2221128

[79] N. Tomczak-Jaegermann, *Banach-Mazur distances and finite-dimensional operator ideals*, Pitman Monographs and Surveys in Pure and Applied Mathematics, vol. 38, Longman Scientific & Technical, Harlow; copublished in the United States with John Wiley & Sons, Inc., New York, 1989. MR993774

[80] N. Tomczak-Jaegermann, *On the differentiability of the norm in trace classes S_p*, Séminaire Maurey-Schwartz 1974–1975: Espaces L^p, applications radonifiantes et géométrie des espaces de Banach, École Polytech., Paris, 1975, pp. Exp. No. XXII, 9. MR410443

[81] H. M. Wark, *A non-separable uniformly convex Banach space on which there are few operators*, Studia Math. **241** (2018), no. 3, 241–256, DOI 10.4064/sm8740-5-2017. MR3756104

[82] V. E. Zizler, *Smooth extension of norms and complementability of subspaces*, Arch. Math. (Basel) **53** (1989), no. 6, 585–589, DOI 10.1007/BF01199818. MR1023974

[83] V. Zizler, *Nonseparable Banach spaces*, Handbook of the geometry of Banach spaces, Vol. 2, North-Holland, Amsterdam, 2003, pp. 1743–1816, DOI 10.1016/S1874-5849(03)80048-7. MR1999608

Instituto de Matemática Pura y Aplicada, Universitat Politécnica de Valéncia, Camino de Vera s/n, 46022 Valencia, Spain
Email address: vmontesinos@mat.upv.es

Mathematical Institute of the Czech Academy of Sciences. Žitná 25, Praha 1, 11567, the Czech Republic
Email address: vasekzizler@gmail.com

Proceedings of Symposia in Pure Mathematics
Volume **106**, 2024
https://doi.org/10.1090/pspum/106/01934

Geometry of Trinomials Revisited

Gustavo A. Muñoz-Fernández, Sara Ruiz, and Juan B. Seoane-Sepúlveda

Dedicated to our friend Per H. Enflo

Abstract. In this work we pursue a double objective. First we present a survey about the most relevant results on the geometry of Banach spaces of polynomials of low dimension appeared in the last 25 years, giving the reader some hints about the importance and relevance of such results. To accomplish our second objective we focus on one specific polynomial space. In particular we consider the space $\mathcal{P}_{m,n}$ with $m, n \in \mathbb{N}$ and $m > n$ consisting of the trinomials $ax^m + bx^n + c$ endowed with the supremum norm over the unit interval $[-1, 1]$. We provide a parametrization of the unit sphere $\mathsf{S}_{m,n}$ of $\mathcal{P}_{m,n}$ whenever m and n have different parity by projecting $\mathsf{S}_{m,n}$ over the plane $a = 0$ when m is odd and n is even, and over the plane $b = 0$ when m is even and n is odd. These parametrizations allow us to obtain the extreme points of the closed unit ball $\mathsf{B}_{m,n}$ of $\mathcal{P}_{m,n}$.

1. Introduction

The study of the extreme points of the closed unit ball of Banach spaces of polynomials has inspired a number of publications since the 1960's. Recall that if C is a convex subset of $\mathbb{R}^n$ then $e \in C$ is a extreme point of C ($e \in \text{ext}(C)$ for short) if e is not an interior point of any segment with endpoints in C. Konheim and Rivlin [**33**] provided in 1966 a characterization of the extreme points of the unit ball of the space consisting of the real polynomials of degree at most n, represented by $\mathcal{P}_n(\mathbb{R})$, endowed with the norm

$$\|P\|_{[-1,1]} = \sup\{|P(x)| : x \in [-1, 1]\}.$$

The study of the geometry of polynomial spaces has intensified during the last 25 years, giving special attention to spaces of low dimension, and in particular, of dimension 3. The monograph [**13**] is entirely devoted to represent the unit sphere of several 3-dimensional polynomial spaces, providing an explicit characterization of the extreme points of their unit balls. The following references represent a complete, although not exhaustive, list of works where the characterization of the geometry of the unit ball of a low dimensional space of polynonials is the main topic [**5**, **6**, **8**–**11**, **14**–**19**, **21**, **22**, **26**, **29**–**32**, **34**–**36**, **39**, **40**, **45**]. A combination of the Krein-Milman theorem and the geometrical studies mentioned above may provide a number of sharp polynomial inequalities by means of the so called Krein-Milman approach.

2020 *Mathematics Subject Classification.* Primary 52A21; Secondary 46B04.
Key words and phrases. Convexity, extreme points, polynomial norms, trinomials.

Several examples where the Krein-Milman approach has been successfully applied to obtain optimal polynomial inequalities can be found here [**1**–**4**, **12**, **24**, **25**, **37**, **38**, **40**–**43**].

Section 2 contains a brief revision of some of the most significant results on the the geometry of Banach spaces of polynomials of low dimension obtained specially during the last 25 years. In Section 3 we present one specific example, namely the space $\mathcal{P}_n(\mathbb{R})$ of the polynomials on the real line with real coefficients of degree at most n endowed with the supremum norm over $[-1,1]$. One application will also be provided in this section.

In Section 4 we study the geometry of the space of real tinomials on $\mathbb{R}$ endowed with the supremum norm over the interval $[-1,1]$. In fact this space was considered first in [**44**], however here we provide an original perspective.

2. Geometry of polynomial spaces of low dimension

As mentioned in the introduction a number of publications have been devoted to characterize the extreme points of certain low dimensional Banach spaces of polynomials in the last 25 years. In recent times a completed monography on this topic has been published [**13**], where the interested reader will find many other references. We provide below a representative list of polynomial spaces whose extreme points have been characterized. The list is not exhaustive:

(1) The space $\mathcal{P}_n(\mathbb{R})$ of real polynomials of degree at most n on the real line with the sup norm on $[-1,1]$. As mentioned above this space was studied by Konheim and Rivling [**33**] in 1966. The extreme points of the unit ball in this case are not explicitly described.

(2) The spaces $\mathcal{P}(^2\ell_p^2(\mathbb{R}))$ with $p=1,2,\infty$ consisting of the quadratic forms $ax^2+bxy+cy^2$, with $(x,y)\in\mathbb{R}^2$, endowed with the norm

$$\|ax^2+bxy+cy^2\| = \sup\{|ax^2+bxy+cy^2| : \|(x,y)\|_p \leq 1\}.$$

Choi, Ki and Kim studied these spaces in four publications [**8**–**11**] appeared between 1998 and 2004. Here, and in the rest of the paper, $\|\cdot\|_p$ with $p\in[1,\infty]$ denotes the classical ℓ_p-norm.

(3) The space $\mathcal{P}(^3H)$ of the homogeneous polynomials of degree 3 on the Hilbert space H endowed with the supremum norm over the closed unit ball of H. Recall that P is an n homogeneous polynomial on a vector space F if there exists an n linear form L such that $P(x)=L(x,\dots,x)$ for all $x\in E$. The extreme points of the closed unit ball of $\mathcal{P}(^3H)$ were characterized by Grecu [**20**] in 2000.

(4) The spaces $\mathcal{P}_2(\mathbb{R})$ and $\mathcal{P}_2(\mathbb{C})$ of real polynomials of degree at most two with the sup norm on $[-1,1]$ and the closed unit disk in $\mathbb{C}$, denoted by $\mathbb{D}$, respectively. These spaces were solved by Aron and Klimek [**5**] in 2001.

(5) The spaces $\mathcal{P}(^2\ell_p^2(\mathbb{R}))$ with $p\in(0,2)$ and $p\in(2,\infty)$ were considered by Grecu [**18**] in 2002.

(6) The space $\mathcal{P}_{m,n}(\mathbb{R})$ of real trinomials of the form ax^m+bx^n+c with $a,b,c\in\mathbb{R}$ and $m>n$ $(m,n\in\mathbb{N})$ endowed with the sup norm on $[-1,1]$. The geometry of this space was characterized by Muñoz and Seoane [**44**] in 2008.

(7) The space $\mathcal{P}_{m,n,k}(\mathbb{C})$ of the real trinomials of the form $az^m+bz^n+cz^k$ with $a,b,c\in\mathbb{R}$ and $m,n,k\in\mathbb{N}$ with $m>n>k$ endowed with the sup norm on

$\mathbb{D}$. Observe that the maximum modulus principle shows that $\mathcal{P}_{m,n,k}(\mathbb{C})$ is the same as the space $\mathcal{P}_{m-k,n-k}(\mathbb{C})$ of the trinomials $az^{m-k} + bz^{n-k} + c$ endowed with the sup norm over $\mathbb{D}$. This space was studied by Neuwirth [**45**] in 2008.

(8) The space $\mathcal{P}_2(\Delta)$ of the polynomials of degree at most 2 on $\mathbb{R}^2$ endowed with the supremum norm over the simplex Δ, i.e., the triangle of vertices $(0,0)$, $(1,0)$ and $(0,1)$. Observe that the elements of $\mathcal{P}_2(\Delta)$ are of the form $ax^2 + by^2 + cxy + dx + ey + f$ where $a, b, c, d, e, f \in \mathbb{R}$ and therefore $\mathcal{P}_2(\Delta)$ is a 6-dimensional space. The extreme points of the closed unit ball of $\mathcal{P}_2(\Delta)$ were characterized by Milev and Naidenov in several papers ranging from 2008 to 2013 (see [**34**–**36**]).

(9) The space $\mathcal{P}(^2\Delta)$ of the quadratic forms $ax^2 + bxy + cy^2$ with $a, b, c \in \mathbb{R}$ endowed with the sup norm on the simplex Δ. Muñoz, Seoane and Révész provided the extreme points of the unit ball of this space [**40**] in 2009 .

(10) The space $\mathcal{P}(^2\mathcal{O}_w^2)$ of the quadratic forms $ax^2 + bxy + cy^2$ with $a, b, c \in \mathbb{R}$ endowed with the norm

$$\|ax^2 + bxy + cy^2\|_{\mathcal{O}_w^2} = \sup\left\{|ax^2 + bxy + cy^2| : \|(x,y)\|_{\mathrm{oct}(w)} \leq 1\right\}$$

where

$$\|(x,y)\|_{\mathrm{oct}(w)} = \max\left\{|x|, |y|, \frac{|x| + |y|}{1+w}\right\}$$

for a fixed $w \in [0,1]$. Notice that the unit ball of $\mathcal{O}_w^2 = (\mathbb{R}^2, \|\cdot\|_{\mathrm{oct}(w)})$ is an octagon whenever $w \in (0,1)$ whereas $\mathcal{O}_0^2 = \ell_1^2$ and $\mathcal{O}_1^2 = \ell_\infty^2$. The extreme points of this space were characterized by Kim [**27**] in 2011.

(11) The space $\mathcal{P}(^2\square)$ of the quadratic forms $ax^2 + bxy + cy^2$ with $a, b, c \in \mathbb{R}$ endowed with the sup norm on the square $\square = [0,1]^2$. This space was studied by Gámez, Muñoz, Sánchez and Seoane [**14**] in 2013.

(12) The space $\mathcal{P}(^2\mathcal{H}_w^2)$ of the quadratic forms $ax^2 + bxy + cy^2$ with $a, b, c \in \mathbb{R}$ endowed with the norm

$$\|ax^2 + bxy + cy^2\|_{\mathcal{H}_w^2} = \sup\left\{|ax^2 + bxy + cy^2| : \|(x,y)\|_{\mathrm{hex}(w)} \leq 1\right\}$$

where

$$\|(x,y)\|_{\mathrm{hex}(w)} = \max\left\{|y|, (|x| + (1-w)|y|)\right\}$$

for a fixed $w \in [0,1]$. Notice that the unit ball of $\mathcal{H}_w^2 = (\mathbb{R}^2, \|\cdot\|_{\mathrm{hex}(w)})$ is an hexagon whenever $w \in (0,1)$ whereas $\mathcal{H}_0^2 = \ell_1^2$ and $\mathcal{H}_1^2 = \ell_\infty^2$. The extreme points of $\mathcal{P}(^2\mathcal{H}_{1/2}^2)$ were characterized by Kim [**28**] in 2017.

(13) The space $\mathcal{P}(^2D(\alpha,\beta))$ of the quadratic forms $ax^2+bxy+cy^2$ with $a, b, c \in \mathbb{R}$ endowed with the sup norm on the sector

$$D(\alpha,\beta) = \{re^{i\theta} : r \in [0,1], \theta \in [\alpha,\beta]\}.$$

Here $\alpha, \beta \in [0,\infty)$ with $\alpha \leq \beta$. This problem was investigated by Muñoz, Pellegrino, Seoane and Weber [**39**] in 2014 for $\beta - \alpha = \frac{\pi}{4}, \frac{\pi}{2}, \frac{3\pi}{4}$. The rest of the cases have been studied by Bernal, Muñoz, Rodríguez and Seoane [**6**] in 2020.

(14) The space $\mathcal{P}_{m,n}^h(\mathbb{R})$ of the homogeneous trinomials of the form $ax^m + bx^{m-n}y^n + cy^m$ with $a, b, c \in \mathbb{R}$ and $m > n$ $(m, n \in \mathbb{N})$ endowed with the sup norm on the square $[-1,1]^2$. The space $\mathcal{P}_{m,n}^h(\mathbb{R})$ has been studied in three different works: First by Jiménez, Muñoz and Rodríguez [**26**] in

2021. These authors considered the case where m is odd. In 2024 García, Muñoz and Seaone [15] considered the case where m and n are both even. The remaining case was under study at the time this manuscript was finished.

3. One specific example: The space $\mathcal{P}_n(\mathbb{R})$ with $n = 2$ and $n = 3$

In this section we consider the space $\mathcal{P}_n(\mathbb{R})$ of the polynomials with real coefficients of degree at most n endowed with the norm

$$\|P\|_{[-1,1]} = \max\{|P(x)| : x \in [-1,1]\}.$$

Let us denote the closed unit ball of this space by B_n, i.e.,

$$\mathsf{B}_n = \{P \in \mathcal{P}_n(\mathbb{R}) : \|P\|_{[-1,1]} \leq 1\}.$$

The extreme points of B_n were described by Konheim and Rivlin [33] in 1966 providing a characteristic property based on the total multiplicity $N(P)$ of the roots of the equations $P(x) = \pm 1$. Let us examine in detail Konheim and Rivlin characterization, for which a formal definition of $N(P)$ is needed.

DEFINITION 3.1. Let P be a real polynomial of degree at most n. We denote by $N(P, y)$ the total multiplicity with which the value y is assumed by P and, in particular, let us define the multiplicity of P by the number $N(P) := N(P, 1) + N(P, -1)$.

THEOREM 3.2 (Konheim and Rivlin [33]). *Let $P \in \mathcal{P}_n(\mathbb{R})$ with $\|P\|_{[-1,1]} \leq 1$. We have that P is an extreme polynomial if, and only if, $N(P) > n$.*

Although the characterization given in Theorem 3.2 is not explicit, it can be very useful to obtain more accurate descriptions of the extreme points of B_n. As a matter of fact the extreme points of B_3 can be deduced explicitly using Theorem 3.2 (see Section 3.2). First let us consider the geometry of B_2.

3.1. The space $\mathcal{P}_2(\mathbb{R})$.

Aron and Klimek [5] provided an explicit description of the extreme points of B_2 in 2001. Moreover, Aron and Klimek parametrized and sketched the unit sphere S_2 of $\mathcal{P}_2(\mathbb{R})$. Interestingly, Aron and Klimek did not use Konheim and Rivlin's characterization, on the contrary, their achievements are based on an explicit formula for $\|ax^2 + bx + c\|_{[-1,1]}$. For simplicity $ax^2 + bx + c$ will be represented from now on as (a, b, c).

THEOREM 3.3 (Aron and Klimek [5]). *Let $P(x) = ax^2 + bx + c$. We have*

$$\|(a,b,c)\|_{[-1,1]} = \begin{cases} \left|\frac{b^2}{4a} - c\right| & \text{if } |b| < 2|a| \text{ and } \frac{c}{a} + 1 < \frac{1}{2}\left(\left|\frac{b}{2a}\right| - 1\right)^2, \\ |a + c| + |b| & \text{otherwise.} \end{cases}$$

In order to sketch S_2 Aron and Klimek realized that it was essential to find the projection of S_n over the plane $c = 0$, also referred to as ab-plane from now on.

THEOREM 3.4 (Aron and Klimek [**5**]). *The projection of* B_2 *onto the ab-plane is the set* $U \cup V \cup W$ *(see Figure 1 for a representation of* $U \cup V \cup W$*) where*

$$U = \left\{(a,b) \in \mathbb{R}^2 \colon a \leq 0 \text{ and } |b| \leq \min\left\{2|a|, 2\left(\sqrt{2|a|} - |a|\right)\right\}\right\},$$

$$V = \left\{(a,b) \in \left[-\frac{1}{2}, \frac{1}{2}\right] \times [-1,1] \colon |b| \geq 2|a|\right\},$$

$$W = \left\{(a,b) \in \mathbb{R}^2 \colon a \geq 0 \text{ and } |b| \leq \min\left\{2|a|, 2\left(\sqrt{2|a|} - |a|\right)\right\}\right\}.$$

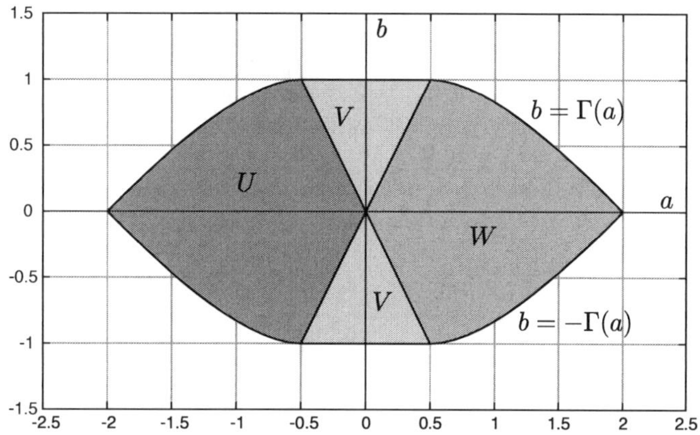

FIGURE 1. Projection of B_2 onto the *ab*-plane. Here $\Gamma(a) = 2\left(\sqrt{2|a|} - |a|\right)$.

THEOREM 3.5 (Aron and Klimek [**5**]). *Let us define the functions*

$$f_+(a,b) = 1 - |b| - |a|,$$

$$g_+(a,b) = \frac{b^2}{4a} - 1,$$

and also the functions $f_-(a,b) = -f_+(-a,b)$ *and* $g_-(a,b) = -g_+(-a,b)$*. We have*

(i) $\mathsf{S}_2 = \operatorname{graph}\left(f_+|_{(V\cup W)}\right) \cup \operatorname{graph}\left(f_-|_{(U\cup V)}\right) \cup \operatorname{graph}\left(g_+|_W\right) \cup \operatorname{graph}\left(g_-|_U\right)$ *(see Figure 2 for a representation of* B_2*).*

(ii) *The set of extreme points of* B_2*, denoted by* $\operatorname{ext}(\mathsf{B}_2)$*, is*

$$\operatorname{ext}(\mathsf{B}_2) = \left\{\pm\left(t, \pm 2(\sqrt{2t} - t), 1 + t - 2\sqrt{2t}\right) : t \in [1/2, 2]\right\} \bigcup \{\pm(0,0,1)\}.$$

3.2. The space $\mathcal{P}_3(\mathbb{R})$. In 2020 Araújo, Muñoz, Rodríguez and Seoane [**3**] provided an explicit characterization of the extreme points of the closed unit ball B_3 of $\mathcal{P}_3(\mathbb{R})$. Their results are based on Theorem 3.2 and are summarized below.

THEOREM 3.6 (Araújo, Muñoz, Rodríguez and Seoane [**3**]). *The extreme polynomials of the closed unit ball* B_3 *of* $\mathcal{P}_3(\mathbb{R})$ *are given by*

(i) $P_1(x) = \pm 1$;
(ii) $P_2(x) = \pm\left[1 - \frac{1}{4}(\pm x + 1)^3\right]$;
(iii) $P_3(x) = \pm(2x^2 - 1)$;
(iv) $P_4(x) = \pm\left[1 - \frac{1}{(1-q^2)^2}(x-q)^2(4qx + 2 + 2q^2)\right]$ *and*
$P_5(x) = \pm\left[1 + \frac{1}{(1-q^2)^2}(x+q)^2(4qx - 2 - 2q^2)\right]$*, for every* $q \in \left(-\frac{1}{3}, 0\right)$;

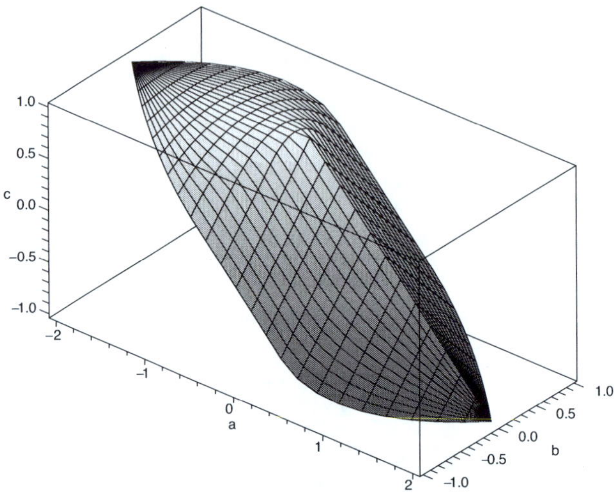

FIGURE 2. Unit sphere of $\mathcal{P}_2(\mathbb{R})$.

(v) $P_6(x) = \pm\left[1 + \frac{1}{(1+t)^2}(x-t)^2(x-1)\right]$ *and*
$P_7(x) = \pm\left[1 - \frac{1}{(1+t)^2}(x+t)^2(x+1)\right]$, *for every* $t \in \left(-\frac{1}{2}, 1\right)$;

(vi) $P_8(x) = \pm\left[1 + \frac{4}{(s-r)^3}(x-r)^2\left(x - \frac{3s-r}{2}\right)\right]$ *and*
$P_9(x) = \pm\left[1 - \frac{4}{(s-r)^3}(x+r)^2\left(x + \frac{3s-r}{2}\right)\right]$, *for every* $-1 \leq r < s \leq 1$ *such that* $s \geq \min\left\{3r+2, \frac{r+2}{3}\right\}$.

3.3. One application. Some of the most significant applications of the geometry of Banach spaces of polynomials rely on the so called Krein-Milman approach, used commonly to maximize a convex target function on a convex set. We will consider the following formulation of the Krein-Milman approach:

> If $C \subset \mathbb{R}^n$ is a nonempty, compact set and $f : C \to \mathbb{R}$ is a convex function that attains its maximum in C, then there exists an extreme point e in C such that $f(e) = \max\{f(x) : x \in C\}$.

The previous well known principle, in combination with the extreme points of a given polynomial space, can be applied to obtain a number of sharp polynomial inequalities. In this section we will focus our attention on the Bernstein-Markov inequalities, which provide sharp estimates on the derivatives of polynomials.

The fist problem related to estimating the derivative of polynomials was stated by the chemist D. Medeleev at the end of the 19th century. Mendeleev was interested in estimating the derivative of quadratic polynomials on the real line. A precise and general formulation of this problem is given below:

> If $P(x) = ax^2 + bx + c$ is a quadratic polynomial, $\alpha, \beta \in \mathbb{R}$, $\alpha < \beta$ and
> $$\|P\|_{[\alpha,\beta]} = \max\{|P(x)| : x \in [\alpha, \beta]\},$$
> what is the optimal (in the sense of smallest) constant $\mathcal{M}(\alpha, \beta)$ in
> $$|P'(x)| \leq \mathcal{M}(\alpha, \beta)\|P\|_{[\alpha,\beta]}$$
> for every $x \in [\alpha, \beta]$?

The previous question can be normalized using the interval $[-1,1]$ by means of the change of variables $x \to \phi(x) = (\alpha+\beta-(\alpha-\beta)x)/2$. As a matter of fact

$$\mathcal{M}_2(\alpha,\beta) = \frac{2}{\beta-\alpha}\mathcal{M}_2(-1,1).$$

For short we will use $\mathcal{M}_2$ instead of $\mathcal{M}_2(-1,1)$. It can be proved that $\mathcal{M}_2 = 4$. Medeleev communicated this problem to the brothers Markov, who generalized it in several forms. Let $n,k \in \mathbb{N}$ with $n \geq k$ and assume that $\mathcal{M}_{n,k}$ denotes the best (smallest) constant in

$$|P^{(k)}(x)| \leq \mathcal{M}_{n,k}\|P\|_{[-1,1]} \tag{3.1}$$

for every $P \in \mathcal{P}_n(\mathbb{R})$ and every $x \in [-1,1]$, where $P^{(k)}$ represents the kth derivative of P. We will use $\mathcal{M}_n$ instead of $\mathcal{M}_{n,1}$ for simplicity. The inequality (3.1) is usually known as Markov's inequality in the real line. The following classical results are of common use nowadays:

(1) A. Markov proved in 1889 that $\mathcal{M}_n = n^2$.
(2) V. Markov proved in 1892 that $\mathcal{M}_{n,k} = \frac{n^2(n^2-1^2)\cdots(n^2-(k-1)^2)}{1\cdot 3\cdots(2k-1)}$.

In both cases equality is attained in (3.1) at the end points of $[-1,1]$ for the n-th Chebyshev polynomial of the first kind, defined by $T_n(x) = \cos(n\arccos x)$ for $x \in [-1,1]$.

The uniform estimates $\mathcal{M}_{n,k}$ can be refined in the interior of $[-1,1]$. Indeed, for a fixed $x \in [-1,1]$, let us denote by $\mathcal{B}_{n,k}(x)$ the optimal (smallest) constant in the inequality

$$|P^{(k)}(x)| \leq \mathcal{B}_{n,k}(x)\|P\|_{[-1,1]} \tag{3.2}$$

for every $P \in \mathcal{P}_n(\mathbb{R})$. Here we will use $\mathcal{B}_n(x)$ in place of $\mathcal{B}_{n,1}(x)$ for short. According to another classical result due to Bernstein (1916) we have

$$\mathcal{B}_n(x) \leq \frac{n}{\sqrt{1-x^2}}$$

for every $x \in (-1,1)$. Although Bernstein's estimate is not optimal, specially if x is close to ± 1, it has important consequences. The interested reader is referred to [**7**] for an interesting revision on this topic.

The problem of finding the functions $\mathcal{B}_{n,k}(x)$ is extraordinarily complicated. V. Voronovskaja developed the theory required to solve this question and, as a matter of fact, she obtained the exact value of $\mathcal{B}_n(x)$ [**46**]. Also, Gusev obtained $\mathcal{B}_{n,k}(x)$ [**23**]. In all cases the solution depends strongly on the so called Zolotarev polynomials, which are difficult to be obtained explicitly. Alternatively we can apply the Krein-Milman Approach. Indeed, for fixed $n,k \in \mathbb{N}$ with $n \geq k$ and $x \in [-1,1]$, let $\Phi_{n,k,x} : \mathsf{B}_n \to \mathbb{R}$ be the convex function

$$\Phi_{n,k,x}(P) = |P^{(k)}(x)|.$$

Then we have

$$\mathcal{B}_{n,k}(x) = \max\{\Phi_{n,k,x}(P) : P \in \mathsf{B}_n\} = \max\{\Phi_{n,k,x}(P) : P \in \operatorname{ext}(\mathsf{B}_n)\}. \tag{3.3}$$

The previous identity can be used to determine $\mathcal{B}_{n,k}$ provided that $\operatorname{ext}(\mathsf{B}_n)$ is known.

If $n = 2$, according to Theorem 3.5 we have

$$\text{ext}\,(\mathsf{B}_2) = \left\{\pm 1, \pm\left(tx^2 \pm 2(\sqrt{2t} - t)x + 1 + t - 2\sqrt{2t}\right) : t \in [1/2, 2]\right\}.$$

Hence,

$$\mathcal{B}_2(x) = 2 \max_{t\in[1/2,2]} \left\{|tx \pm (\sqrt{2t} - t)|\right\} = 2 \max_{t\in[1/2,2]} \left\{t|x| + \sqrt{2t} - t\right\}.$$

If we call $R(t) = t|x| + \sqrt{2t} - t$, the maximum is attained either at $t = 1/2$, $t = 2$ or at the critical points of $R(t)$ that lie inside the interval $(1/2, 2)$. The only point that satisfies $R'(t) = 0$ is $t_0 = \frac{1}{2(1-|x|)^2}$. But notice that $\frac{1}{2} \le t_0 \le 2$ happens only when $|x| \le \frac{1}{2}$. Some calculations lead to

$$\begin{aligned} R(1/2) &= \frac{|x| + 1}{2}, \\ R(2) &= 2|x|, \\ R(t_0) &= \frac{1}{2(1 - |x|)}. \end{aligned}$$

And hence,

$$\mathcal{B}_2(x) = 2 \max_{t\in[1/2,2]} R(t) = 2 \begin{cases} \max\{R(1/2), R(2), R(t_0)\} \text{ if } |x| \le \frac{1}{2}, \\ \max\{R(1/2), R(2), \} \text{ if } \frac{1}{2} \le |x| \le 1, \end{cases}$$

which is computed to obtain

$$\mathcal{B}_2(x) = \begin{cases} \frac{1}{1-|x|} \text{ if } |x| \le \frac{1}{2}, \\ 4|x| \text{ if } \frac{1}{2} \le |x| \le 1. \end{cases}$$

Now it is easy to see that $\mathcal{M}_2 = 4$.

On the other hand, using the characterization of $\text{ext}(\mathsf{B}_3)$ given in Theorem 3.6 combined with (3.3) it is possible to show that (see [**3**])

$$\mathcal{B}_3(x) = \begin{cases} 3(1 - 4x^2) & \text{if } |x| \le \frac{\sqrt{7}-2}{6}, \\ \frac{7\sqrt{7}+10}{9(|x|+1)} & \text{if } \frac{\sqrt{7}-2}{6} \le |x| \le \frac{2\sqrt{7}-1}{9}, \\ \frac{-16x^3}{(1-9x^2)(1-x^2)} & \text{if } \frac{2\sqrt{7}-1}{9} \le |x| \le \frac{1+2\sqrt{7}}{9}, \\ \frac{7\sqrt{7}-10}{9(1-|x|)} & \text{if } \frac{1+2\sqrt{7}}{9} \le |x| \le \frac{\sqrt{7}+2}{6}, \\ 3(4x^2 - 1) & \text{if } |x| \ge \frac{\sqrt{7}+2}{6}. \end{cases} \tag{3.4}$$

A representation of (3.4) can be found in Figure 3. Observe that $\mathcal{M}_3 = 9$. Similarly

$$\mathcal{B}_{3,2}(x) = \begin{cases} \frac{4}{1-9x^2} & \text{if } |x| \le \frac{1}{9}, \\ \frac{32}{9(|x|-1)^2} & \text{if } \frac{1}{9} \le |x| \le \frac{1}{3}, \\ 24|x| & \text{if } |x| \ge \frac{1}{3}. \end{cases}$$

The calculation of $\mathcal{B}_3(x)$ and $\mathcal{B}_{3,2}(x)$ is highly nontrivial in any case. Both functions were known to Voronovskaja and Gusev in the 1960's. The interested reader can find in [**3**] how to obtain $\mathcal{B}_3(x)$ and $\mathcal{B}_{3,2}(x)$ using the Krein-Milman approach.

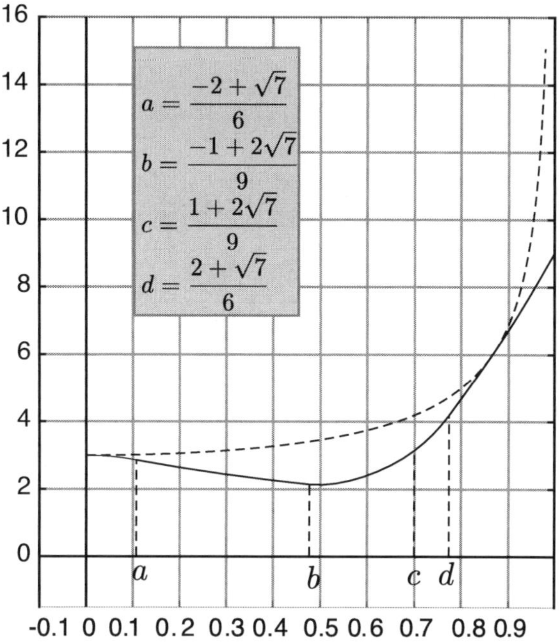

FIGURE 3. The functions $\mathcal{B}_{n,k}$ are even and hence it is enough to represent them in $[0,1]$. Here we have sketched B_3 (solid line) together with the Bernstein estimate $3/\sqrt{1-x^2}$ (dashed line).

4. The space $\mathcal{P}_{m,n}(\mathbb{R})$

In the present section we consider the space of trinomials on the real line of the form $ax^m + bx^n + c$ with $m, n \in \mathbb{N}$, $m > n$ and $a, b, c \in \mathbb{R}$ endowed with the norm

$$\|ax^m + bx^n + c\|_{m,n} = \sup\{|ax^m + bx^n + c| : x \in [-1,1]\}.$$

This space is represented by $\mathcal{P}_{m,n}(\mathbb{R})$. Observe that $\mathcal{P}_{m,n}(\mathbb{R})$ is in fact $\mathbb{R}^3$ since every polynomial $ax^m + bx^n + c$ can be identified with (a,b,c). From now on this identification will be frequently used.

The notations $\mathsf{B}_{m,n}$ and $\mathsf{S}_{m,n}$ stand for the closed unit ball and the unit sphere of $\mathcal{P}_{m,n}(\mathbb{R})$, respectively, that is

$$\begin{aligned} \mathsf{B}_{m,n} &= \{P \in \mathcal{P}_{m,n}(\mathbb{R}) : \|P\|_{m,n} \leq 1\}, \\ \mathsf{S}_{m,n} &= \{P \in \mathcal{P}_{m,n}(\mathbb{R}) : \|P\|_{m,n} = 1\}. \end{aligned}$$

Notice that $\mathsf{B}_{2,1}$ and $\mathsf{S}_{2,1}$ are nothing but B_2 and S_2 respectively, whereas $\|\cdot\|_{2,1}$ is $\|\cdot\|_{[-1,1]}$. As metioned earlier, the geometry of $\mathsf{B}_{2,1}$ was studied by Aron and Klimek in 2001 (see [**5**]), providing all its extreme points. Recall that Aron and Klimek proceeded as follows:

(1) They found a formula to calculate $\|ax^2 + bx + c\|_{2,1}$.
(2) Then they calculated the projection of $\mathsf{B}_{2,1}$ over the ab plane.
(3) Using the previous two steps they obtained a parametrization of $\mathsf{S}_{2,1}$.
(4) The parametrization of $\mathsf{S}_{2,1}$ allowed the authors to identify easily all the extreme points of $\mathsf{B}_{2,1}$.

In 2008 Muñoz and Seoane [**44**] generalized Aron and Klimek results for any choice of m and n using the previous scheme. Interestingly, the space $\mathcal{P}_{m,n}(\mathbb{C})$ of the trinomials $az^m + bz^n + c$ with $m, n \in \mathbb{N}$, $m > n$ and $a, b, c \in \mathbb{R}$ endowed with the norm

$$\|az^m + bz^n + c\|_{m,n,\mathbb{C}} = \sup\{|az^m + bz^n + c| : z \in \mathbb{D}\}$$

was studied in [**45**]. Notice that here $\mathbb{D}$ stands for the closed unit disk in $\mathbb{C}$.

The results obtained in [**44**] depend strongly on whether m and n are even or odd. In this note we will focus on the two cases where m and n have different parity, that is, either m is odd and n is even or m is even and n is odd. Although these two cases are solved satisfactorily in [**44**], the authors did not take advantage of all the symmetries present in the above mentioned cases. Indeed, observe that, if m is odd and n even, then

$$\begin{aligned}\|(-a, b, c)\|_{m,n} &= \max_{x\in[-1,1]} |-ax^m + bx^n + c| \\ &= \max_{x\in[-1,1]} |a(-x)^m + b(-x)^n + c| = \|(a, b, c)\|_{m,n},\end{aligned}$$

which indicates a symmetry with respect to the bc-plane. Similarly, if m is even and n odd, then

$$\begin{aligned}\|(a, -b, c)\|_{m,n} &= \max_{x\in[-1,1]} |ax^m - bx^n + c| \\ &= \max_{x\in[-1,1]} |a(-x)^m + b(-x)^n + c| = \|(a, b, c)\|_{m,n},\end{aligned}$$

which indicates a symmetry over the ac-plane. The use of these two symmetries does not only simplify some of the calculations employed in [**44**], but also provides a different perspective of the unit sphere $\mathsf{S}_{m,n}$ with m and n having different parity.

4.1. The case where m is odd and n is even. A formula to calculate $\|\cdot\|_{m,n}$ will be needed and can be obtained from [**44**]:

THEOREM 4.1 (Muñoz and Seoane [**44**]). *Let $m, n \in \mathbb{N}$ with m odd, n even and $m > n$. Then for every $(a, b, c) \in \mathbb{R}^3$, an explicit formula for the norm $\|(a, b, c)\|_{m,n}$ is given by*

$$\|(a, b, c)\|_{m,n} = \max\{|c|, |a| + |b + c|\},$$

that is

$$\|(a, b, c)\|_{m,n} = \begin{cases} |c| & \text{if } |c| \geq |a| + |b + c|, \\ |a| + |b + c| & \text{otherwise.} \end{cases} \tag{4.1}$$

Let us define the set

$$V = \{(b, c) \in \mathbb{R}^2 : |c| \leq 1 \text{ and } |b + c| \leq 1\}.$$

The reader can find a representation of V in Figure 4. The importance of the set V is revealed in the following result, where π_{bc} stands for the projection over the plane $a = 0$.

THEOREM 4.2. *For any $m, n \in \mathbb{N}$ with m odd, n even and $m > n$ we have that $\pi_{bc}(\mathsf{B}_{m,n}) = V$.*

PROOF. Since $\mathsf{B}_{m,n}$ is symmetric with respect to the bc-plane, $\pi_{bc}(\mathsf{B}_{m,n})$ coincides with the intersection of $\mathsf{B}_{m,n}$ with the plane $a = 0$, that is

$$\pi_{bc}(\mathsf{B}_{m,n}) = \{(b, c) \in \mathbb{R}^2 : \|(0, b, c)\|_{m,n} \leq 1\},$$

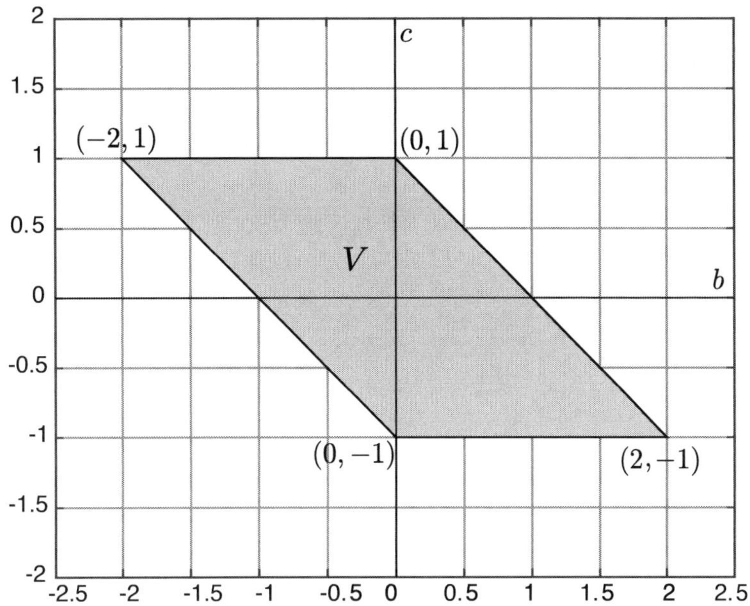

FIGURE 4. $\pi_{bc}(\mathsf{B}_{m,n})$ for m odd and n even.

which is the set bounded by the curve $\|(0,b,c)\|_{m,n} = 1$. By formula (4.1), we obtain

$$\|(0,b,c)\|_{m,n} = \begin{cases} |c| & \text{if } c \neq 0 \text{ and } |\frac{b}{c}+1| \leq 1, \\ |b+c| & \text{otherwise,} \end{cases}$$

$$= \begin{cases} |c| & \text{if } |b+c| \leq |c|, \\ |b+c| & \text{otherwise.} \end{cases} \tag{4.2}$$

So let us study when both parts of the formula are equal to 1. In the case where $|b+c| \leq |c|$, we have $\|(0,b,c)\|_{m,n} = |c| = 1$. Then c can take only the values $c = 1$, in which case $|b+1| \leq 1$ and hence $b \in [-2, 0]$, or $c = -1$, from which $|b-1| \leq 1$ and therefore $b \in [0, 2]$. In the other case, $\|(0,b,c)\|_{m,n} = |b+c| = 1$ with $|b+c| > |c|$, which means $c = 1-b$ or $c = -1-b$. In both cases we would have, from the condition $|b+c| > |c|$, that $|c| < 1$. That concludes the study of the boundary of $\pi_{bc}(\mathsf{B}_{m,n})$ and therefore also of the set $\pi_{bc}(\mathsf{B}_{m,n})$ itself. □

THEOREM 4.3. *If for every $m, n \in \mathbb{N}$ $(m > n)$ with m odd and n even and every $(b,c) \in \mathbb{R}^2$ we define*

$$f_+(b,c) = 1 - |b+c|, \quad f_-(b,c) = -f_+(b,c) = -1 + |b+c|,$$

and

$$W \;=\; \{(a,b,c) \in \mathbb{R}^3 : |c| = 1 \text{ and } f_-(b,c) \leq a \leq f_+(b,c)\},$$

then

(a) $\mathsf{S}_{m,n} = \operatorname{graph}(f_+|_V) \cup \operatorname{graph}(f_-|_V) \cup W$.
(b) $\operatorname{ext}(\mathsf{B}_{m,n}) = \{\pm(2,-1,0), \pm(1,-1,1), \pm(1,-1,-1), \pm(0,-1,0)\}$.

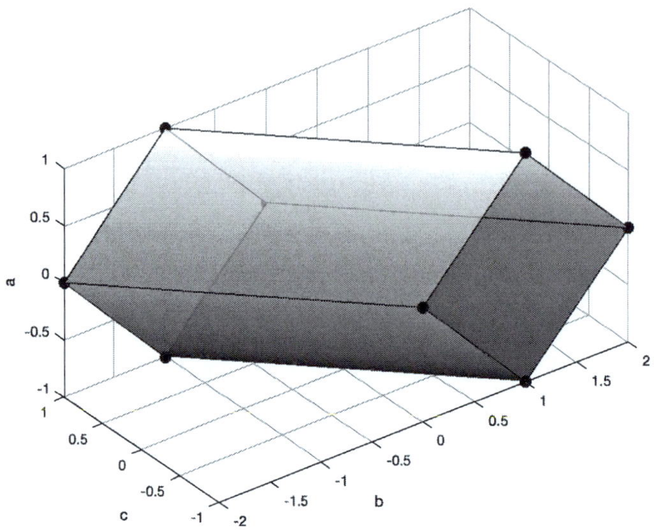

FIGURE 5. Unit sphere of $(\mathbb{R}^3, \|\cdot\|_{m,n})$ for m odd, n even and $m > n$, parameterized over the bc-plane. The extreme points have been drawn with isolated dots.

PROOF. To prove the first part of the theorem, it suffices to isolate the variable a in the formula $\|(a,b,c)\|_{m,n} = 1$, for which (4.1) will be essential. Using the second line of (4.1) we obtain $a = \pm(1 - |b+c|)$, whose domains turn out to be the entire projection V, and lastly, when $\|(a,b,c)\|_{m,n} = |c| = 1$, we encounter two vertical flat surfaces at $c = 1$ and $c = -1$, named W, closing the unit ball as seen in figure 5.

The calculation of the extreme points comes once again from $\mathsf{S}_{m,n}$ being a polytope, so its extreme points are located at its vertices. $\square$

4.2. The case where m is even and n is odd. Having a formula to calculate $\|\cdot\|_{m,n}$ when m is even and n is odd will be essential here too. For the sake of completeness we present below a formula taken from [**44**].

THEOREM 4.4 (Muñoz and Seaone [**44**]). *If $m, n \in \mathbb{N}$ are such that m is even, n is odd and $m > n$, defining $\mathcal{I}_{m,n}$ as the set of triples $(a,b,c) \in \mathbb{R}^3$ such that*

$$a \neq 0,\ \left|\frac{nb}{ma}\right| < 1 \text{ and } 1 + \frac{c}{a} < \frac{1}{2}\left[\frac{m-n}{n}\left(\frac{nb}{ma}\right)^{\frac{m}{m-n}} - \left|\frac{b}{a}\right| + 1\right],$$

we have

$$\|(a,b,c)\|_{m,n} = \begin{cases} \left|\frac{(m-n)a}{n} \cdot \left|\frac{nb}{ma}\right|^{\frac{m}{m-n}} - c\right| & \text{if } (a,b,c) \in \mathcal{I}_{m,n}, \\ |a+c| + |b| & \text{otherwise.} \end{cases} \tag{4.3}$$

The following technical result will also be required. This result is mentioned in [**44**] but no complete proof is given, for which reason we provide one here.

LEMMA 4.5. *Let $m, n \in \mathbb{N}$ with $m > n$. Then the equation*

$$\frac{(m-n)a}{n}\left|\frac{nb}{ma}\right|^{\frac{m}{m-n}} - 2 + a + b = 0, \tag{4.4}$$

defines implicitly a unique differentiable curve $b = \Gamma_{m,n}(a)$ defined on $(0,\infty)$ such that $\Gamma_{m,n}(2) = 0$ and $\Gamma_{m,n}(n/m) = 1$.

PROOF. The existence of a local solution to (4.5) follows immediately from the Implicit Function Theorem. To find a global solution we can use ODE's theory. To do so, given equation (4.4), we can derive both sides with respect to a (considering $b = b(a)$) to obtain

$$\frac{(m-n)}{n}\left|\frac{nb}{ma}\right|^{\frac{m}{m-n}} + \left|\frac{nb}{ma}\right|^{\frac{n}{m-n}}\frac{b'(a)a - |b|}{a} + 1 + b'(a) = 0,$$

from which we obtain the quivalent ODE

$$b'(a) = \frac{\left|\frac{nb}{ma}\right|^{\frac{n}{m-n}}\left|\frac{b}{a}\right| - \frac{(m-n)}{n}\left|\frac{nb}{ma}\right|^{\frac{m}{m-n}} - 1}{1 + \left|\frac{nb}{ma}\right|^{\frac{n}{m-n}}}.$$

Observe that the above ODE is of the form $b'(a) = \phi\left(\frac{b(a)}{a}\right)$. The initial value problem

$$\begin{cases} b'(a) &= \phi\left(\frac{b(a)}{a}\right), \\ b(2) &= 0, \end{cases} \tag{4.5}$$

has a unique local solution around 2. Moveover, since ϕ does not have any blow up, the local solution of the initial value problem (4.5) can be extended to $(0,\infty)$. If we call the function $b(a) = \Gamma_{m,n}(a)$, then the only thing left to prove is $\Gamma_{m,n}(\frac{n}{m}) = 1$, but this is immediate, as the point verifies the equation (4.4). □

For the rest of the section, let us define $L_{m,n} = \frac{m}{m-n}\left(\frac{m-n}{n}\right)^{\frac{n}{m}}$, as it will simplify most of the expressions. One more technical lemma will be needed.

LEMMA 4.6. *Let $m, n \in \mathbb{N}$ with $m > n$. Then the function $\Phi_{m,n}(a)$ defined on $(0,\infty)$ as $\Phi_{m,n}(a) = 1 - a - \Gamma_{m,n}(a)$ satisfies the equation*

$$L_{m,n}\,(1+c)^{\frac{m-n}{m}}\,a^{\frac{n}{m}} - 1 + a + c = 0. \tag{4.6}$$

Also, $\Phi_{m,n}(2) = -1$ and $\Phi_{m,n}(n/m) = -n/m$.

PROOF. It is a simple application of the information we have of $\Gamma_{m,n}(a)$ from the lemma 4.5. □

REMARK 4.7. In general $\Phi_{m,n}$ cannot be obtained explicitly. However, whenever n is odd and we choose $m = 2n$, its expression is

$$\Phi_{2n,n}(a) = 1 + a - 2\sqrt{2a}.$$

Our mission now is to have a parametrization of $\mathsf{S}_{m,n}$, which will be done by projecting $\mathsf{B}_{m,n}$ onto the plane that presents the symmetries, that is, the ac-plane.

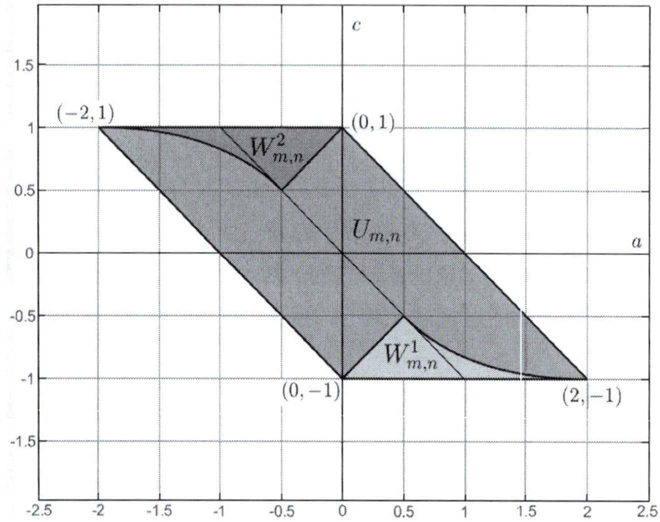

FIGURE 6. $\pi_{ac}(\mathsf{B}_{m,n})$ for n odd and $m = 2n$.

Let us define

$$\begin{aligned}
U_{m,n} &= \left\{|a+c|\leq 1, \min\left\{\frac{m-n}{n}a-1, \Phi_{m,n}(a)\right\}\leq c\leq \max\left\{\frac{m-n}{n}a+1, -\Phi_{m,n}|a|\right\}\right\},\\
W^1_{m,n} &= \left\{0\leq a\leq 2, -1\leq c\leq \min\left\{\frac{m-n}{n}a-1, \Phi_{m,n}(a)\right\}\right\},\\
W^2_{m,n} &= \left\{-2\leq a\leq 0, \max\left\{\frac{m-n}{n}a+1, -\Phi_{m,n}|a|\right\}\leq c\leq 1\right\},\\
W_{m,n} &= W^1_{m,n}\cup W^2_{m,n},\\
V &= U_{m,n}\cup W_{m,n}.
\end{aligned}$$

THEOREM 4.8. *For any $m, n \in \mathbb{N}$ with m even, n odd and $m > n$ we have that $\pi_{ac}(\mathsf{B}_{m,n}) = U_{m,n} \cup W_{m,n}$.*

PROOF. Due to the symmetry of $\mathsf{B}_{m,n}$ with respect to the plane $b = 0$ we have that $\pi_{ac}(\mathsf{B}_{m,n})$ is the intersection of $\mathsf{B}_{m,n}$ and the plane $b = 0$, or in other words,

$$\pi_{ac}(\mathsf{B}_{m,n}) = \{(a, c) \in \mathbb{R}^2 : \|(a, 0, c)\|_{m,n} \leq 1\},$$

which is a set bounded by the curve defined by $\|(a, 0, c)\|_{m,n} = 1$. By formula (4.3), we obtain

$$\|(a,0,c)\|_{m,n} = \begin{cases} |c| & \text{if } a\neq 0 \text{ and } \frac{c}{a} < -\frac{1}{2},\\ |a+c| & \text{otherwise.}\end{cases} \tag{4.7}$$

So let us study when both parts of the formula are equal to 1. In the case where $a \neq 0$ and $\frac{c}{a} < -\frac{1}{2}$, we have $\|(a, 0, c)\|_{m,n} = |c| = 1$. Hence either $c = 1$ or $c = -1$. If $c = 1$ then the inequality $\frac{1}{a} < -\frac{1}{2}$ with $a \neq 0$ is equivalento to $-2 < a < 0$. If $c = -1$ then $\frac{-1}{a} < -\frac{1}{2}$ $(a \neq 0)$ is equivalent to $0 < a < 2$. In the other case, $\|(a, 0, c)\|_{m,n} = |a + c| = 1$, which means $c = \pm 1 - a$. That occurs when either $a = 0$ or $\frac{c}{a} \geq -\frac{1}{2}$ with $a \neq 0$. Since the case $a = 0$ is trivial, we will assume for the rest of the proof that $a \neq 0$. If $c = 1 - a$ then $\frac{c}{a} \geq -\frac{1}{2}$ is equivalen to $\frac{1}{a} \geq \frac{1}{2}$, that is $0 < a \leq 2$. If now $c = -1 - a$ then $\frac{c}{a} \geq -\frac{1}{2}$ is equivalent to $-\frac{1}{a} \geq \frac{1}{2}$, or $-2 \leq a < 0$. That concludes the study of the boundary of $\pi_{ac}(\mathsf{B}_{m,n})$ and therefore also of the set $\pi_{ac}(\mathsf{B}_{m,n})$ itself. □

REMARK 4.9. Notice that the projection of $\mathsf{B}_{m,n}$ over the ac-plane when m is even and n is odd is exactly a copy of the projection of $\mathsf{B}_{m,n}$ over the bc-plane when m is odd and n is even. That is due to the similarity between formulas (4.2) and (4.7).

THEOREM 4.10. *Let $m, n \in \mathbb{N}$ with m even, n odd and $m > n$. If for every $(a, c) \in U_{m,n} \cup W_{m,n}$ we define*

$$\begin{aligned} f(a,c) &= 1 - |a+c|, \\ g(a,c) &= L_{m,n}\,(1+c)^{\frac{m-n}{m}}\;|a|^{\frac{n}{m}}, \\ h(a,c) &= g(a,-c) = L_{m,n}\,(1-c)^{\frac{m-n}{m}}\;|a|^{\frac{n}{m}}, \end{aligned}$$

then

(a) $\mathsf{S}_{m,n} = \operatorname{graph}(\pm f|_{U_{m,n}}) \cup \operatorname{graph}(\pm g|_{W^1_{m,n}}) \cup \operatorname{graph}(\pm h|_{W^2_{m,n}})$.

(b) *The extreme points of $\mathsf{B}_{m,n}$ are*

$$\pm(0,1,0) \quad \text{and} \quad \pm\left(t, \Phi_{m,n}(t), \pm(\Phi_{m,n}(t) + t - 1)\right),$$

where $t \in [n/m, 2]$.

PROOF. Since in this case $\mathsf{B}_{m,n}$ is symmetric with respect to $b = 0$ we can assume that $b > 0$ for the entire proof. To obtain the expressions of f, g and h we will isolate the variable b in $\|(a,b,c)\| = 1$, for which the formula (4.3) will be needed. The next step is to prove that the domains of f, g and h are, respectively, $U_{m,n}$, $W^1_{m,n}$ and $W^2_{m,n}$. The first observation is that the function f vanishes on the segments $|a + c| = 1$ that are contained in V, g vanishes on the segment $c = -1$ in V and h in the segment $c = 1$ in V. Let us now compute the intersection of the graph of the functions. By symmetries, let us assume $a > 0$ and $c < 0$.

In this case scenario, the only possible functions are f and g. For their intersection, we will consider two separate cases. When $a + c < 0$ and otherwise. If $a + c < 0$, then we are looking for a solution to the equation

$$1 + a + c = L_{m,n}\,(1+c)^{\frac{m-n}{m}}\,(|a|)^{\frac{n}{m}}\,.$$

This can be solved by a change of variables $1 + c = d$. The equation now is $d + a = L_{m,n} d^{\frac{m-n}{m}} |a|^{\frac{n}{m}}$, or equivalently if we divide by d, $L_{m,n}\left(\frac{a}{d}\right)^{\frac{n}{m}} = 1 + \frac{a}{d}$. By observing that the function $\tilde{f}(x) = L_{m,n}x^{\frac{n}{m}} - x - 1$ has exactly one root, located at its maximum, we can calculate such point and obtain c as a function of a, exactly what we are looking for. That is, $\tilde{f}'(x) = \frac{n}{m}L_{m,n}x^{\frac{n-m}{m}} - 1 = 0$ if and only if $x = \left(\frac{n}{m}L_{m,n}\right)^{\frac{m}{m-n}} = \frac{n}{m-n}$. If we call this point $\gamma_0 = \frac{n}{m-n}$, notice that $\tilde{f}(\gamma_0) = L_{m,n}\gamma_0^{\frac{n}{m}} - \gamma_0 - 1 = 0$ and it is the only root because to the left and right of the point the function is strictly monotone. An example graph of this function when $m = 2n$ and n is odd is shown in figure 7. So, in conclusion, the equation $L_{m,n}\left(\frac{a}{d}\right)^{\frac{n}{m}} = 1 + \frac{a}{d}$ has for unique solution $\frac{a}{d} = \gamma_0 = \frac{n}{m-n}$, that is $c = \frac{m-n}{n}a - 1$, and it intersects the line $a + c = 0$ at the point $\gamma_1 = \frac{n}{m} < 1$. Now, the intersection

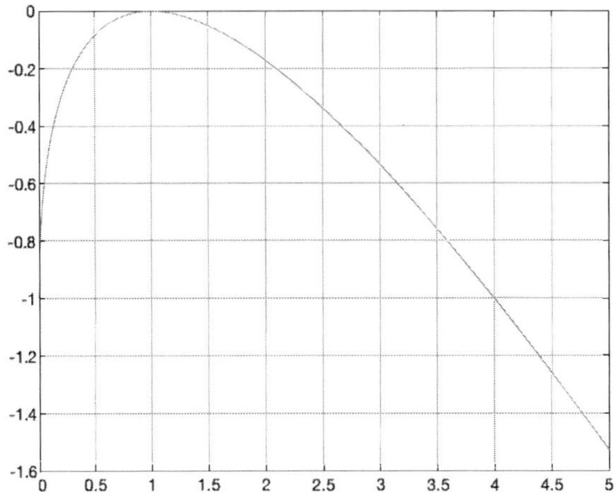

FIGURE 7. Function $\tilde{f}(x)$ for $m = 2n$ and n odd.

of the graphs of the functions f and g when $a + c \geq 0$ and $a > 0$, is the solution of the equation

$$1 - a - c = L_{m,n}\,(1 + c)^{\frac{m-n}{m}}\,a^{\frac{n}{m}},$$

which is exactly equation (4.6) and that means it implicitly defines the function $c = \Phi_{m,n}(a)$ that passes through the points $(\frac{n}{m}, -\frac{n}{m})$ and $(2, -1)$.

To prove (b), notice that if we fix $\lambda > 0$, then

$$f(a, \lambda a) = 1 - |1 + \lambda||a|,$$
$$g(a, -1 + \lambda a) = L_{m,n}\lambda^{\frac{m-n}{m}}|a| = h(a, 1 + \lambda a),$$

so the functions are affine on half-lines. Hence, all the extreme points of the ball are necessarily either in places where the graphs of these functions form a vertex, namely the points $\pm(0, 1, 0)$ or meet along non-affine curves, which has already been studied and are the curves $\pm(t, \Phi(t), \pm(1 - t - \Phi(t)))$ when $t \in [\frac{n}{m}, 2]$. To prove that all of them are indeed extreme points, it is enough to follow the argument in [**44**, theorem 4.6]. □

Having in mind that $\Phi_{2n,n}$ with $n \in \mathbb{N}$ odd can be obtained explicitly (see Remark 4.7), from the previous result we derive the following:

COROLLARY 4.11. *Let $n \in \mathbb{N}$ be odd. If for every $(a, c) \in U_{2n,n} \cup W_{2n,n}$ we define*

$$f(a, c) = 1 - |a + c|,$$
$$g(a, c) = 2\sqrt{(1 + c)\,|a|},$$
$$h(a, c) = g(a, -c) = 2\sqrt{(1 - c)\,|a|},$$

then

(a) $\mathsf{S}_{2n,n} = \text{graph}\big(\pm f|_{U_{2n,n}}\big) \cup \text{graph}\big(\pm g|_{W^1_{2n,n}}\big) \cup \text{graph}\big(\pm h|_{W^2_{2n,n}}\big)$.

(b) *The extreme points of $\mathsf{B}_{2n,n}$ are*

$$\pm(0, 1, 0) \quad \textit{and} \quad \pm\,(t, 1 + t - 2\sqrt{2t}, \pm 2(\sqrt{2t} - t)),$$

where $t \in [1/2, 2]$.

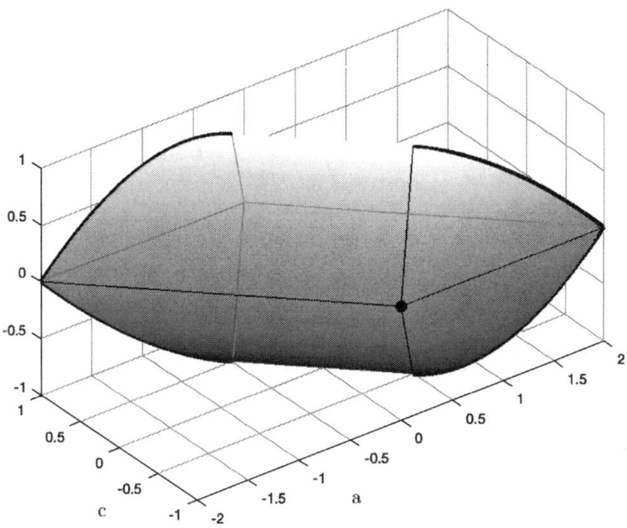

FIGURE 8. Unit sphere of $(\mathbb{R}^3, \|\cdot\|_{2n,n})$ with $n \in \mathbb{N}$ odd. The extreme points appear with a thicker line or big dots. The surfaces that form $\mathsf{S}_{2n,n}$ are delimited by thin lines.

Figure 8 shows what $\mathsf{S}_{2n,n}$ looks like with $n \in \mathbb{N}$ odd.

References

[1] G. Araújo, P. Jiménez-Rodríguez, G. A. Muñoz-Fernández, and J. B. Seoane-Sepúlveda, *Polynomial inequalities on the $\pi/4$-circle sector*, J. Convex Anal. **24** (2017), no. 3, 927–953. MR3684809

[2] G. Araújo, P. Jiménez-Rodríguez, G. A. Muñoz-Fernández, and J. B. Seoane-Sepúlveda, *Equivalent norms in polynomial spaces and applications*, J. Math. Anal. Appl. **445** (2017), no. 2, 1200–1220, DOI 10.1016/j.jmaa.2016.03.039. MR3545233

[3] G. Araújo, G. A. Muñoz-Fernández, D. L. Rodríguez-Vidanes, and J. B. Seoane-Sepúlveda, *Sharp Bernstein inequalities using convex analysis techniques*, Math. Inequal. Appl. **23** (2020), no. 2, 725–750, DOI 10.7153/mia-2020-23-61. MR4089483

[4] G. Araújo, P. H. Enflo, G. A. Muñoz-Fernández, D. L. Rodríguez-Vidanes, and J. B. Seoane-Sepúlveda, *Quantitative and qualitative estimates on the norm of products of polynomials*, Israel J. Math. **236** (2020), no. 2, 727–745, DOI 10.1007/s11856-020-1987-y. MR4093902

[5] R. M. Aron and M. Klimek, *Supremum norms for quadratic polynomials*, Arch. Math. (Basel) **76** (2001), no. 1, 73–80, DOI 10.1007/s000130050544. MR1808745

[6] L. Bernal-González, G. A. Muñoz-Fernández, D. L. Rodríguez-Vidanes, and J. B. Seoane-Sepúlveda, *A complete study of the geometry of 2-homogeneous polynomials on circle sectors*, Rev. R. Acad. Cienc. Exactas Fís. Nat. Ser. A Mat. RACSAM **114** (2020), no. 3, Paper No. 160, 31, DOI 10.1007/s13398-020-00892-6. MR4118482

[7] R. P. Boas Jr., *Inequalities for the derivatives of polynomials*, Math. Mag. **42** (1969), 165–174, DOI 10.2307/2688534. MR259429

[8] Y. S. Choi and S. G. Kim, *The unit ball of $\mathcal{P}(^2l_2^2)$*, Arch. Math. (Basel) **71** (1998), no. 6, 472–480, DOI 10.1007/s000130050292. MR1653404

[9] Y. S. Choi and S. G. Kim, *Smooth points of the unit ball of the space $\mathcal{P}(^2l_1)$*, Results Math. **36** (1999), no. 1-2, 26–33, DOI 10.1007/BF03322099. MR1706485

[10] Y. S. Choi and S. G. Kim, *Exposed points of the unit balls of the spaces $\mathcal{P}(^2l_p^2)$ $(p = 1, 2, \infty)$*, Indian J. Pure Appl. Math. **35** (2004), no. 1, 37–41. MR2032962

[11] Y. S. Choi, S. G. Kim, and H. Ki, *Extreme polynomials and multilinear forms on l_1*, J. Math. Anal. Appl. **228** (1998), no. 2, 467–482, DOI 10.1006/jmaa.1998.6161. MR1663577

[12] D. Diniz, G. A. Muñoz-Fernández, D. Pellegrino, and J. B. Seoane-Sepúlveda, *The asymptotic growth of the constants in the Bohnenblust-Hille inequality is optimal*, J. Funct. Anal. **263** (2012), no. 2, 415–428, DOI 10.1016/j.jfa.2012.04.014. MR2923418

[13] J. Ferrer, D. García, M. Maestre, G. A. Muñoz-Fernández, D. L. Rodríguez-Vidanes, and J. B. Seoane-Sepúlveda, *Geometry of the unit sphere in polynomial spaces*, SpringerBriefs in Mathematics, Springer, 2023.

[14] J. L. Gámez-Merino, G. A. Muñoz-Fernández, V. M. Sánchez, and J. B. Seoane-Sepúlveda, *Inequalities for polynomials on the unit square via the Krein-Milman theorem*, J. Convex Anal. **20** (2013), no. 1, 125–142. MR3086444

[15] D. García, G. A. Muñoz-Fernández, and J. B. Seoane-Sepúlveda, *Extreme homogeneous trinomials on the unit square*, J. Math. Anal. Appl. **529** (2024), no. 2, Paper No. 127134, DOI 10.1016/j.jmaa.2023.127134. MR4650778

[16] B. C. Grecu, *Geometry of homogeneous polynomials on two-dimensional real Hilbert spaces*, J. Math. Anal. Appl. **293** (2004), no. 2, 578–588, DOI 10.1016/j.jmaa.2004.01.020. MR2053900

[17] B. C. Grecu, *Extreme 2-homogeneous polynomials on Hilbert spaces*, Quaest. Math. **25** (2002), no. 4, 421–435, DOI 10.2989/16073600209486027. MR1946920

[18] B. C. Grecu, *Geometry of 2-homogeneous polynomials on l_p spaces,* $1 < p < \infty$, J. Math. Anal. Appl. **273** (2002), no. 2, 262–282, DOI 10.1016/S0022-247X(02)00217-2. MR1932488

[19] B. C. Grecu, *Smooth 2-homogeneous polynomials on Hilbert spaces*, Arch. Math. (Basel) **76** (2001), no. 6, 445–454, DOI 10.1007/PL00000456. MR1831501

[20] B. C. Grecu, *Geometry of three-homogeneous polynomials on real Hilbert spaces*, J. Math. Anal. Appl. **246** (2000), no. 1, 217–229, DOI 10.1006/jmaa.2000.6783. MR1761159

[21] B. C. Grecu, G. A. Muñoz-Fernández, and J. B. Seoane-Sepúlveda, *The unit ball of the complex $\mathcal{P}(^3H)$*, Math. Z. **263** (2009), no. 4, 775–785, DOI 10.1007/s00209-008-0438-y. MR2551598

[22] B. C. Grecu, G. A. Muñoz-Fernández, and J. B. Seoane-Sepúlveda, *Unconditional constants and polynomial inequalities*, J. Approx. Theory **161** (2009), no. 2, 706–722, DOI 10.1016/j.jat.2008.12.001. MR2563077

[23] V. A. Gusev, *Functionals of the derivatives of an algebraic polynomial and a theorem of V. A. Markov* (Russian), Izv. Akad. Nauk SSSR Ser. Mat. **25** (1961), 367–384. MR130943

[24] P. Jiménez-Rodríguez, G. A. Muñoz-Fernández, M. Murillo-Arcila, and J. B. Seoane-Sepúlveda, *Sharp values for the constants in the polynomial Bohnenblust-Hille inequality*, Linear Multilinear Algebra **64** (2016), no. 9, 1731–1749, DOI 10.1080/03081087.2015.1115810. MR3509496

[25] P. Jiménez-Rodríguez, G. A. Muñoz-Fernández, D. Pellegrino, and J. B. Seoane-Sepúlveda, *Bernstein-Markov type inequalities and other interesting estimates for polynomials on circle sectors*, Math. Inequal. Appl. **20** (2017), no. 1, 285–300, DOI 10.7153/mia-20-21. MR3610901

[26] P. Jiménez-Rodríguez, G. A. Muñoz-Fernández, and D. L. Rodríguez-Vidanes, *Geometry of spaces of homogeneous trinomials on $\mathbb{R}^2$*, Banach J. Math. Anal. **15** (2021), no. 4, Paper No. 61, 22, DOI 10.1007/s43037-021-00144-8. MR4292847

[27] S. G. Kim, *The unit ball of $\mathcal{P}(^2D_*(1,W)^2)$*, Math. Proc. R. Ir. Acad. **111A** (2011), no. 2, 79–94, DOI 10.3318/pria.2011.111.1.9. MR2852918

[28] S. G. Kim, *Extreme 2-homogeneous polynomials on the plane with a hexagonal norm and applications to the polarization and unconditional constants*, Studia Sci. Math. Hungar. **54** (2017), no. 3, 362–393, DOI 10.1556/012.2017.54.3.1371. MR3733217

[29] S. G. Kim, *Exposed polynomials of $\mathcal{P}(^2\mathbb{R}^2_{h(\frac{1}{2})})$*, Extracta Math. **33** (2018), no. 2, 127–143. MR3897208

[30] S. G. Kim, *Extreme 2-homogeneous polynomials on the plane with a hexagonal norm and applications to the polarization and unconditional constants*, Studia Sci. Math. Hungar. **54** (2017), no. 3, 362–393, DOI 10.1556/012.2017.54.3.1371. MR3733217

[31] S. G. Kim, *Exposed 2-homogeneous polynomials on the two-dimensional real predual of Lorentz sequence space*, Mediterr. J. Math. **13** (2016), no. 5, 2827–2839, DOI 10.1007/s00009-015-0658-4. MR3554281

[32] S. G. Kim, *Polarization and unconditional constants of $\mathcal{P}(^2d_*(1,w)^2)$*, Commun. Korean Math. Soc. **29** (2014), no. 3, 421–428, DOI 10.4134/CKMS.2014.29.3.421. MR3245197

[33] A. G. Konheim and T. J. Rivlin, *Extreme points of the unit ball in a space of real polynomials*, Amer. Math. Monthly **73** (1966), 505–507, DOI 10.2307/2315472. MR191891

[34] L. Milev and N. Naidenov, *Strictly definite extreme points of the unit ball in a polynomial space*, C. R. Acad. Bulgare Sci. **61** (2008), no. 11, 1393–1400. MR2490668

[35] L. Milev and N. Naidenov, *Indefinite extreme points of the unit ball in a polynomial space*, Acta Sci. Math. (Szeged) **77** (2011), no. 3-4, 409–424. MR2906511

[36] L. Milev and N. Naidenov, *Semidefinite extreme points of the unit ball in a polynomial space*, J. Math. Anal. Appl. **405** (2013), no. 2, 631–641, DOI 10.1016/j.jmaa.2013.04.026. MR3061039

[37] L. Milev and N. Naidenov, *An inequality for polynomials on the standard simplex*, Numerical methods and applications, Lecture Notes in Comput. Sci., vol. 13858, Springer, Cham, 2023, pp. 221–232, DOI 10.1007/978-3-031-32412-3_20. MR4596953

[38] G. A. Muñoz-Fernández, D. Pellegrino, and J. B. Seoane-Sepúlveda, *Estimates for the asymptotic behaviour of the constants in the Bohnenblust-Hille inequality*, Linear Multilinear Algebra **60** (2012), no. 5, 573–582, DOI 10.1080/03081087.2011.613833. MR2916842

[39] G. A. Muñoz-Fernández, D. Pellegrino, J. B. Seoane-Sepúlveda, and A. Weber, *Supremum norms for 2-homogeneous polynomials on circle sectors*, J. Convex Anal. **21** (2014), no. 3, 745–764. MR3243817

[40] G. A. Muñoz-Fernández, S. Gy. Révész, and J. B. Seoane-Sepúlveda, *Geometry of homogeneous polynomials on non symmetric convex bodies*, Math. Scand. **105** (2009), no. 1, 147–160, DOI 10.7146/math.scand.a-15111. MR2549803

[41] G. A. Muñoz-Fernández, V. M. Sánchez, and J. B. Seoane-Sepúlveda, *Estimates on the derivative of a polynomial with a curved majorant using convex techniques*, J. Convex Anal. **17** (2010), no. 1, 241–252. MR2642728

[42] G. A. Muñoz-Fernández, V. M. Sánchez, and J. B. Seoane-Sepúlveda, *L^p-analogues of Bernstein and Markov inequalities*, Math. Inequal. Appl. **14** (2011), no. 1, 135–145, DOI 10.7153/mia-14-11. MR2796983

[43] G. A. Muñoz-Fernández, Y. Sarantopoulos, and J. B. Seoane-Sepúlveda, *An application of the Krein-Milman theorem to Bernstein and Markov inequalities*, J. Convex Anal. **15** (2008), no. 2, 299–312. MR2422991

[44] G. A. Muñoz-Fernández and J. B. Seoane-Sepúlveda, *Geometry of Banach spaces of trinomials*, J. Math. Anal. Appl. **340** (2008), no. 2, 1069–1087, DOI 10.1016/j.jmaa.2007.09.010. MR2390911

[45] S. Neuwirth, *The maximum modulus of a trigonometric trinomial*, J. Anal. Math. **104** (2008), 371–396, DOI 10.1007/s11854-008-0028-2. MR2403441

[46] E. V. Voronovskaja, *The functional method and its applications*, Translations of Mathematical Monographs, Vol. 28, American Mathematical Society, Providence, RI, 1970. With an appendix by V. A. Gusev; Translated from the Russian by R. P. Boas. MR267056

Instituto de Matemática Interdisciplinar (IMI) and Departamento de Análisis Matemático y Matemática Aplicada, Facultad de Ciencias Matemáticas, Plaza de Ciencias 3, Universidad Complutense de Madrid, Madrid, 28040 (Spain)

Email address: `gustavo_fernandez@mat.ucm.es`

Facultad de Ciencias Matemáticas, Plaza de Ciencias 3, Universidad Complutense de Madrid, Madrid, 28040 (Spain)

Email address: `sarrui09@ucm.es`

Instituto de Matemática Interdisciplinar (IMI) and Departamento de Análisis Matemático y Matemática Aplicada, Facultad de Ciencias Matemáticas, Plaza de Ciencias 3, Universidad Complutense de Madrid, Madrid, 28040 (Spain)

Email address: `jseoane@ucm.es`

Proceedings of Symposia in Pure Mathematics
Volume **106**, 2024
https://doi.org/10.1090/pspum/106/01935

Metric Invariants in Banach and Jordan–Banach Algebras

Antonio M. Peralta

This paper is dedicated to Professor Per H. Enflo

Abstract. In this note we collect some significant contributions on metric invariants for complex Banach algebras and Jordan-Banach algebras established during the last fifteen years. This note is mainly expository, but it also contains complete proofs and arguments, which in many cases are new or have been simplified. We have also included several new results. The common goal in the results is to seek for "natural" subsets, $\mathfrak{S}_A$, associated with each complex Banach or Jordan-Banach algebra A, sets which when equipped with a certain metric, d_A, enjoy the property that each surjective isometry from $(\mathfrak{S}_A, d_A)$ to a similar set, $(\mathfrak{S}_B, d_B)$, associated with another Banach or Jordan-Banach algebra B, extends to a surjective real-linear isometry from A onto B. In case of a positive answer to this question, the problem of discussing whether in such a case the algebras A and B are in fact isomorphic or Jordan isomorphic is the subsequent question. The main results presented here will cover the cases in which the sets $(\mathfrak{S}_A, d_A)$ and $(\mathfrak{S}_B, d_B)$ are in one of the following situations:

(✓) Subsets of the set of invertible elements in a unital complex Banach algebra or in a unital complex Jordan-Banach algebra with the metric induced by the norm. Specially in the cases of unital C*- and JB*-algebras.

(✓) The sets of positive invertible elements in unital C*- or JB*-algebras with respect to the metric induced by the norm and with respect to the Thompson's metric.

(✓) Subsets of the set of unitary elements in unital C*- and JB*-algebras.

Contents

2020 *Mathematics Subject Classification.* Primary 47B49, 46B20; Secondary 46A22, 46H70, 46L70.

Key words and phrases. Metric invariants, invertible elements, unitaries, C*-algebra, JB*-algebra.

Author partially supported by grant PID2021-122126NB-C31 funded by MCIN/AEI/ 10.13039/501100011033, Junta de Andalucía grant FQM375, MOST (Ministry of Science and Technology of China) grant G2023125007L, and by the IMAG–María de Maeztu grant CEX2020-001105-M/AEI/10.13039/501100011033.

1. Introduction

This paper, originally thought as an expository essay on the results establishing metric invariants on associative Banach algebras and Jordan-Banach algebras, gathers in a unified perspective the main conclusions obtained during the last fifteen years in this line. Nevertheless, what do we mean by a metric invariant? The idea behind a metric invariant is to find "natural" subsets, $\mathfrak{S}_A$, associated with each complex Banach or Jordan-Banach algebra A, sets which when equipped with a certain metric, d_A, enjoy the property that each surjective isometry from $(\mathfrak{S}_A, d_A)$ to a similar set $(\mathfrak{S}_B, d_B)$, associated with another Banach or Jordan-Banach algebra B, extends to a surjective real-linear isometry from A onto B. In case of a positive answer to this question, the problem of discussing whether in such a case the algebras A and B are in fact isomorphic or Jordan isomorphic is the subsequent question. We shall see that examples of metric invariants include many natural subsets, like certain subsets of the set of invertible elements in a unital complex Banach algebra or in a unital complex Jordan-Banach algebra with the metric given by the norm, the set of positive invertible elements in a unital C*- or JB*-algebra with the distance given by the norm or with respect to the Thompson's metric, and certain subsets of the set of unitary elements in unital C*- or JB*-algebras with the distance given by the norm. These are typical problems on preservers.

During the elaboration of this paper we struggled for completeness and the final version contains most of the results with complete proofs and references to prior works and basic background. The manuscript actually contains some new results, many new or modified and simplified arguments, and a detailed revision of the similarities and differences between the different conclusions for Banach and Jordan-Banach algebras. All Banach algebras and spaces in this note will be over the complex field.

Why metric invariants? A natural candidate as metric invariant is the set, A^{-1}, of all invertible elements in a Banach algebra A with the metric given by the norm. However, according to a well known result which dates back to the seventies, there are examples of commutative complex Banach algebras A and B whose groups of invertible elements are topologically isomorphic as groups but A and B are not isomorphic as Banach algebras nor isometric as Banach spaces (cf. Example 3.1). It should be noted that the existence of a surjective linear isometry between the Banach spaces underlying to Banach algebras does not necessarily imply that both algebras are isomorphic (see Example 3.3 and Anti-theorem 4.6 for an example in the setting of JB*-algebras).

Section 3 is devoted to the results studying the sets of invertible elements as a metric invariant. In [**57**] Hatori established the first result in this line showing the following: Let A and B be unital Banach algebras. We additionally assume that A is semisimple and commutative. Let $\mathfrak{A}$ and $\mathfrak{B}$ be open subgroups of A^{-1} and B^{-1}, respectively. Let $\Delta : \mathfrak{A} \to \mathfrak{B}$ be a surjective isometry for the distances given by the norms of A and B, respectively. Then B is semisimple and commutative, and the mapping $\Delta(\mathbf{1}_A)^{-1}\Delta$ admits an extension to an isometric real-linear algebra isomorphism from A onto B. In particular, A^{-1} and B^{-1} are isometrically isomorphic as metric groups (see Theorem 3.2). The hypotheses concerning commutativity and semisimplicity cannot be relaxed in the just quoted result (see Example 3.4). If we relax the hypotheses on A, another result by Hatori shows that if $\mathfrak{A}$ and $\mathfrak{B}$ are open subgroups of A^{-1} and B^{-1}, respectively, and $\Delta : \mathfrak{A} \to \mathfrak{B}$ is a surjective isometry for the distances given by the norms of A and B, respectively, then there exist a surjective real-linear isometry $T : A \to B$, and an element u_0 in the Jacobson radical of B which satisfy that $\Delta(a) = T(a) + u_0$, for every a in $\mathfrak{A}$ (cf. Theorem 3.5). If $\Delta : \mathfrak{A} \to \mathfrak{B}$ is a surjective isometric group isomorphism, then there exists a real-linear isometric isomorphism $T : A \to B$ satisfying $\Delta(a) = T(a)$, for every a in $\mathfrak{A}$. In particular, A^{-1} and B^{-1} are isometrically isomorphic as metrizable groups (see Corollary 3.6).

Among the consequences of the above results we conclude in Corollary 3.7 that the following statements are equivalent for any pair of unital Banach algebras A and B:

(*a*) A and B are isometrically isomorphic as real Banach algebras.
(*b*) There exist two open subgroups $\mathfrak{A}$ and $\mathfrak{B}$ of A^{-1} and B^{-1}, respectively, which are isometrically isomorphic as groups.
(*c*) A^{-1} and B^{-1} are isometrically isomorphic as metric groups.

We shall obtain the previous results from the corresponding versions in the wider setting of Jordan-Banach algebras. This will be a characteristic of this note because we try to emphasize the links and divergences between both settings. To facilitate the accessibility for novice readers, section 2 is devoted to present the basic notions, definitions and facts in the theory of Jordan-Banach algebras with a complete list of references. We shall pay especial attention to the McCrimmon radical of Jordan-Banach algebras and its different characterizations (see Theorem 2.8 and Proposition 2.9), the connected components of the invertible elements in a unital Jordan-Banach algebra (see Theorem 2.12), and the characterization of the principal component of the set of all invertible elements in a Jordan-Banach algebra M as the least quadratic subset of M^{-1} containing $\exp(M)$ (see Lemma 2.15).

Continuing with the results about subsets of invertible elements in a unital complex Jordan-Banach algebra M, the main conclusion known until now is treated in Theorem 3.8 and asserts the following: Let M and N be unital Jordan-Banach algebras. Suppose that $\mathfrak{M} \subseteq M^{-1}$ and $\mathfrak{N} \subseteq N^{-1}$ are clopen quadratic subsets which are closed for powers and inverses. Let $\Delta : \mathfrak{M} \to \mathfrak{N}$ be a surjective isometry. Then there exist a surjective real-linear isometry $T_0 : M \to N$ and an element u_0 in the McCrimmon radical of N such that $\Delta(a) = T_0(a) + u_0$, for all $a \in \mathfrak{M}$. The Jordan product of two invertible elements is not, in general, an invertible element. This is the reason for considering quadratic subsets. We recall that a subset $\mathfrak{M}$ of M^{-1} is called a *quadratic subset* if it is non-empty and $U_a(b) \in \mathfrak{M}$ for all $a, b \in \mathfrak{M}$ (see Definition 2.14). The main results concerning Jordan-Banach algebras in this section are inspired by those in [**104**].

Among the new results in section 3 we establish that if A is a semisimple and commutative unital Banach algebra, N is a unital Jordan-Banach algebra, $\mathfrak{A}$ and $\mathfrak{N}$ are clopen quadratic subsets closed for powers and inverses of A^{-1} and N^{-1}, respectively, and $\Delta : \mathfrak{A} \to \mathfrak{N}$ is a surjective isometry for the distances given by the norms of A and N, then N is semisimple and associative, and the mapping $\Delta(\mathbf{1}_A)^{-1}\Delta$ admits an extension to an isometric real-linear algebra isomorphism from A onto N. In particular, A^{-1} and N^{-1} are isometrically isomorphic as metric groups (see Theorem 3.9). In the proof we employ a version of the Gleason-Kahane-Żelazko theorem for Jordan-Banach algebras from [**42**] and a new subtle variant for linear maps valued in the exponential spectrum (see Lemma 3.14).

Another new result derived from Theorem 3.9 reads as follows: Let A be a unital and commutative Banach algebra and let N be a semisimple unital Jordan-Banach algebra. Suppose that $\mathfrak{A} \subseteq A^{-1}$ and $\mathfrak{N} \subseteq N^{-1}$ are clopen quadratic subsets which are closed for powers and inverses. Let $\Delta : \mathfrak{A} \to \mathfrak{N}$ be a surjective isometry. Then A is semisimple and N is associative, and the mapping $\Delta(\mathbf{1}_{A})^{-1}\Delta$ admits an extension to an isometric real-linear algebra isomorphism from A onto N. In particular, A^{-1} and N^{-1} are isometrically isomorphic as metric groups (see Corollary 3.15).

In the setting of Jordan-Banach algebras preservers of Jordan products of invertible elements do not make any sense, the natural hypothesis is the preservation of expressions of the form $U_a(b)$. In Proposition 3.16 it is established that if M and N are unital Jordan-Banach algebras, $\mathfrak{M} \subseteq M^{-1}$ and $\mathfrak{N} \subseteq N^{-1}$ are clopen quadratic subsets which are closed for powers and inverses, and $\Delta : \mathfrak{M} \to \mathfrak{N}$ is a surjective isometry satisfying

$$\Delta(U_a(b)) = U_{\Delta(a)}(\Delta(b)), \text{ for all } a, b \in \mathfrak{M},$$

then setting $u = \Delta(\mathbf{1}_{M})$, there exists a real-linear isometric Jordan isomorphism J from M onto the u-isotope $N_{(u)}$ such that $\Delta(a) = J(a)$ for all $a \in \mathfrak{M}$.

The final conclusion in the setting of Jordan-Banach algebras is surveyed in Corollary 3.17 and shows that for any pair M, N of unital complex Jordan-Banach algebras the following statements are equivalent:

(a) There exists an isometric real-linear Jordan isomorphism from M onto N.
(b) There exist two clopen quadratic subsets $\mathfrak{M} \subseteq M^{-1}$ and $\mathfrak{N} \subseteq N^{-1}$ which are closed for powers and inverses and a surjective isometry $\Delta : \mathfrak{M} \to \mathfrak{N}$ satisfying $\Delta(\mathbf{1}_{M}) = \mathbf{1}_{N}$ and $\Delta(U_a(b)) = U_{\Delta(a)}(\Delta(b))$, for all $a, b \in \mathfrak{M}$.
(c) There exists a surjective isometry $\Delta : M^{-1} \to N^{-1}$ satisfying $\Delta(\mathbf{1}_{M}) = \mathbf{1}_{N}$ and $\Delta(U_a(b)) = U_{\Delta(a)}(\Delta(b))$, for all $a, b \in M^{-1}$.

We have already commented the existence of non-isomorphic Banach algebras which are isometrically isomorphic as Banach spaces (see Example 3.4). A celebrated result of Kadison proves that this pathology cannot happen if both algebras are C*-algebras (see Theorem 4.1).

We devote the first part of section 4 to survey the results on surjective isometries between subsets of invertible elements in two unital C*-algebras obtained by Hatori and Watanabe in [**64**]. It is worth recalling that every C*-algebra is semisimple.

Let A and B be unital C*-algebras, and let $\mathfrak{A}$ and $\mathfrak{B}$ be open subgroups of A^{-1} and B^{-1}, respectively. Under these hypotheses, a bijection $\Delta : \mathfrak{A} \to \mathfrak{B}$ is an isometry if and only if $\Delta(\mathbf{1}_{A})$ is unitary in B and there exist a projection p in the centre of B, and a complex-linear Jordan *-isomorphism $\tilde{J}$ from A onto B such that

$$\Delta(a) = \Delta(\mathbf{1}_{A}) p \tilde{J}(a) + \Delta(\mathbf{1}_{A})(\mathbf{1}_{N} - p)\tilde{J}(a)^*, \text{ for all } a \in \mathfrak{A}.$$

Therefore Δ admits an extension to a surjective real-linear isometry from A onto B (cf. Theorem 4.2).

JB*-triple theory with Kaup's Banach–Stone theorem (Theorem 4.3) and the theory of isotopes for unital JB*-algebras are subsequently employed to establish the following result: Let M and N be unital JB*-algebras. Suppose that $\mathfrak{M} \subseteq M^{-1}$ and $\mathfrak{N} \subseteq N^{-1}$ are clopen quadratic subsets which are closed for powers and inverses. Let $\Delta : \mathfrak{M} \to \mathfrak{N}$ be a surjective isometry. Then $\Delta(\mathbf{1}_M) = u$ is a unitary element in N and there exists a projection p in the centre of M and a complex-linear Jordan *-isomorphism J from M onto the u^*-homotope $N_{(u^*)} = (N, \circ_u, *_u)$ such that

$$\Delta(a) = J(p \circ a) + J((\mathbf{1}_M - p) \circ a^*), \text{ for all } a \in \mathfrak{M}.$$

Assuming that there exists a unitary ω_0 in N such that the identity $U_{\omega_0}(\Delta(\mathbf{1})) = \mathbf{1}$ holds, then there exist a central projection $p \in M$ and a complex-linear Jordan *-isomorphism Φ from M onto N such that

$$\Delta(a) = U_{w_0^*}\left(\Phi(p \circ a) + \Phi((\mathbf{1}_M - p) \circ a^*)\right),$$

for all $a \in \mathfrak{M}$ (see Theorem 4.8).

The final affirmation can be better understood after a couple of comments. It is well known that the group of surjective linear isometries in a unital C*-algebra A acts transitively on the set of unitaries of A. However, this fact is, in general, false in the setting of JB*-algebras, actually there exists a unital JC*-algebra M containing two unitaries u_1 and u_2 satisfying $T(u_1) \neq u_2$ for all $T \in Iso(M)$. Furthermore, u_1 and u_2 induce two structures of unital JB*-algebra $M_{(u_1^*)} = (M, \circ_{u_1}, *_{u_1})$ and $M_{(u_2^*)} = (M, \circ_{u_2}, *_{u_2})$ which are not Jordan *-isomorphic but have the same underlying Banach space (cf. Anti-theorem 4.6). In a very informal language, the same complex Banach space M can be equipped with many non-Jordan *-isomorphic structures of JB*-algebra, although by Kaup's Banach–Stone theorem (Theorem 4.3) there is only one triple product making M a JB*-triple.

The idea of considering the set of all positive invertible elements in a unital C*-algebra as a metric invariant was considered by Hatori and Molnár in [**63**, page 167]. If we equip this set with the usual metric given by the norm, the conclusion can be achieved via a key theorem due to Mankiewicz (see Theorem 3.18). Here we extend the result to the setting of unital JB*-algebras by showing that each surjective isometry between sets of positive invertible elements in two unital JB*-algebras M and N admits an extension to a Jordan *-isomorphism $J : M \to N$ (see Proposition 4.9).

The first part of section 5 is devoted to present an outstanding geometric tool due to Hatori, Hirasawa, Miura and Molnár in [**61**], which provides a fix point theorem type for maps between abstract spaces preserving a certain mild relation d between elements, which is a far-reaching generalization of the usual notion of distance (cf. Theorem 5.2). The arguments are related to certain ideas in Väisälä's simplified proof of the celebrated Mazur–Ulam theorem, and the statement of this theorem is so general that permits many different applications of this exceptional geometric tool. Here we employ it in the study of surjective isometries between sets of positive invertible elements in two unital JB*-algebras when equipped with the Thompson's metric (subsection 5.1), and in the study of surjective isometries between sets of unitary elements in two unital JB*-algebras with respect to the metric given by the norm (section 6).

Thompson's metric on the interior of the positive cone A^+ (i.e. the set of all Archimedean order units) in an order unit space A equipped with an Archimedean order unit is defined as follows: For $a, b \in \operatorname{int}(A^+)$, we set

$$M(a/b) := \inf\{\beta > 0 \colon a \leq \beta b\} (< \infty),$$

and the Thompson's metric is defined by

$$d_T(a,b) = \log\left(\max\{M(a/b), M(b/a)\}\right).$$

One of the most fruitful sources of examples in this paper is coming from the cones of positive elements in a C*-algebra A, or more generally, in a JB*-algebra M. The interior of M^+ is non-empty if and only if M is unital and in such a case $\operatorname{int}(M^+) = M^{++}$ (see Proposition 5.10). Andruchow, Corach, and Stojanoff gave in [**4**] an explicit formula to compute Thompson's metric on the set positive invertible elements in a unital C*-algebra A via differential geometric methods. An alternative explanation for $B(H)^{++}$ was given by Molnár in [**93**]. The concrete expression affirms that for every $a, b \in A^{++}$ the Thompson's metric satisfies

$$d_T(a,b) = \left\|\log\left(a^{-1/2} b a^{-1/2}\right)\right\| \quad \text{(cf. Proposition 5.11).}$$

By applying the celebrated Shirshov–Cohn theorem (see Theorem 5.14) we shall see that an appropriate Jordan version of the formula proved by Andruchow, Corach, and Stojanoff remains valid for unital JB*-algebras. The Jordan version of this formula, which was originally established by Lemmens, Roelands, and Wortel in [**82**], asserts that for every unital JB*-algebra M, the Thompson's metric on M^{++} satisfies

$$d_T(a,b) = \left\|\log\left(U_{a^{-1/2}}(b)\right)\right\|, \text{ for every } a, b \in M^{++}.$$

Furthermore,

$$d_T\left(U_b(a^{-1}), a\right) = 2 d_T(b,a), \text{ and}$$
$$d_T(U_c(a^{-1}), U_c(b^{-1})) = d_T(a^{-1}, b^{-1}) = d_T(a,b),$$

for every $a, b, c \in M^{++}$ (see Proposition 5.15).

In subsection 5.1 we revisit the results exploring the set of positive invertible elements with the Thompson's metric as a metric invariant for unital C*-algebras and JB*-algebras. A change in the metric produces a important difference in the arguments with respect to those employed in subsection 4.1, where this set was equipped with the metric given by the norm.

In the case of C*-algebras, a result due to Hatori and Molnár [**63**] proves that if A and B are unital C*-algebras, then a mapping $\Delta : (A^{++}, d_T) \to (B^{++}, d_T)$ is a surjective isometry if and only if there exist a central projection p in B and a Jordan *-isomorphism $J : A \to B$ such that Δ is of the form

$$\Delta(a) = \Delta(\mathbf{1}_A)^{1/2}\left(pJ(a) + (1-p)J(a^{-1})\right)\Delta(\mathbf{1}_A)^{1/2}, \quad (a \in A^{++})$$

(see Theorem 5.19). A forerunner to this result in the case that $A = B = B(H)$ was previously established by Molnár (see [**93**, Theorem 1]).

The reader has probably realized that, since the question makes perfectly sense in the case of unital JB*-algebras, it also seems interesting to determine if the set of positive invertible elements in a unital JB*-algebra is a metric invariant when equipped with the Thompson's metric. This problem was studied, and solved, by Lemmens, Roelands, and Wortel via techniques based on differential geometry and geodesics. Here, we present an adaptation of the ideas by Hatori and Molnár in [**63**] to conclude in Theorem 5.20 that if M and N are unital JB*-algebras, then a mapping $\Delta : (M^{++}, d_T) \to (N^{++}, d_T)$ is a surjective isometry if and only if there exist a central projection p in N and a Jordan *-isomorphism $J : M \to N$ such that Δ is of the form

$$\Delta(a) = U_{\Delta(\mathbf{1}_M)^{1/2}} \left(p \circ J(a) + (\mathbf{1}_N - p) \circ J(a^{-1})\right), \quad a \in M^{++}.$$

In the final section 6 we study the set $\mathcal{U}(A)$, of all unitary elements in a unital C*-algebra A, or more generally in a unital JB*-algebra, as a metric invariant when equipped with the metric given by the norm. In strict chronological order, Hatori and Molnár made the first contribution in the case that A coincides with $B(H)$ in 2012 (see [**62**]). Concretely, if H is a complex Hilbert space and $\Delta : \mathcal{U}(B(H)) \to \mathcal{U}(B(H))$ is a surjective isometry with respect to the metric given by the operator norm, then there exists a surjective real-linear isometry (equivalently, a triple automorphism) T on $B(H)$ such that $\Delta(u) = T(u)$ for all $u \in \mathcal{U}(B(H))$ (cf. Theorem 6.1). In this special case we can further conclude that there exist unitaries $v, w \in \mathcal{U}(B(H))$ such that Δ is of one of the following forms:

(*a*) $\Delta(u) = vuw^*$, for all $u \in \mathcal{U}(B(H))$,
(*b*) $\Delta(u) = vu^*w^*$, for all $u \in \mathcal{U}(B(H))$,
(*c*) $\Delta(u) = vu^tw^*$, for all $u \in \mathcal{U}(B(H))$,
(*d*) $\Delta(u) = v\overline{u}w^*$, for all $u \in \mathcal{U}(B(H))$,

where the maps $a \mapsto \overline{a}, a^t$ denote an abstract conjugation and transposition on $B(H)$, respectively.

In a general unital C*-algebra A the set $\mathcal{U}(A)$ is a multiplicative group (for the product inherited from A), which need not be connected, and the extension of a surjective isometry between the unitary groups of two unital C*-algebras is not always possible (cf. [**58**]). However, in a subsequent paper, Hatori and Molnár established that for each surjective isometry $\Delta : \mathcal{U}(A) \to \mathcal{U}(B)$, where A and B are unital C*-algebras and we consider the metrics given by the C*-norms, we have $\Delta(e^{iA_{sa}}) = \Delta(\mathbf{1}_A)\, e^{iB_{sa}}$, and there exist a central projection $p \in B$ and a Jordan *-isomorphism $J : A \to B$ such that

$$\Delta(e^{ix}) = \Delta(\mathbf{1}_A)(pJ(e^{ix}) + (\mathbf{1}_B - p)J(e^{ix})^*) \quad (x \in A_{sa}). \tag{1.1}$$

Under the additional assumptions that A or B is a von Neumann algebra, then both algebras are von Neumann algebras and Δ admits an extension to a real-linear isometry from A onto B. More precisely, there exists a central projection $p \in B$ and a Jordan *-isomorphism $J : A \to B$ such that the conclusion in (1.1) holds when e^{ix} is replaced with an arbitrary unitary u in A (cf. Theorem 6.2).

We can therefore arrive to the striking conclusion that two unital C*-algebras are isomorphic as Jordan *-algebras if and only if their unitary groups are isometric as metric spaces (see Corollary 6.3).

We shall also study in this note the set of unitary elements in a unital JB*-algebra M as a metric invariant with respect to the metric given by the norm. Subsections 6.1, 6.2 and 6.3 contain the results in this direction obtained by Cueto, Enami, Hirota, Miura and the author of this article in [**31**].

As in the case of invertible elements, the Jordan product of two unitaries in a unital JB*-algebra M need not be, in general, a unitary. The key property in the setting of unital JB*-algebras assures that given $u, v \in \mathcal{U}(M)$, the element $U_u(v)$ lies in $\mathcal{U}(M)$ (cf. Lemma 2.1). So, the key hypotheses in this setting are given in terms of expressions of the form $U_u(v)$ and quadratic subsets. As in the case of C*-algebras, the set $\mathcal{U}(M)$ is not, in general, connected. The connected component of $\mathcal{U}(M)$ containing the unit element is called the *principal connected component* or simply the *principal component*, and it will be denoted by $\mathcal{U}^0(M)$.

The principal component of the unitary elements in a unital C*-algebra A is a well studied object, whose description can be found in books (cf. [**74**, Exercises 4.6.6 and 4.6.7], [**110**, Exercise 2, page 56], and [**58**, Lemma 3.2]). It is known that

$$\begin{aligned}\mathcal{U}^0(A) &= \{e^{ih_1}\cdot\ldots\cdot e^{ih_n} : n\in\mathbb{N},\ h_1,\ldots,h_n\in A_{sa}\}\\ &= \{e^{ih_1}\cdot\ldots\cdot e^{ih_n}\mathbf{1}_A e^{ih_n}\cdot\ldots\cdot e^{ih_1} : n\in\mathbb{N},\ h_1,\ldots,h_n\in A_{sa}\}\end{aligned}$$

is open, closed, and path connected in the (relative) norm topology on $\mathcal{U}(A)$. Furthermore, by the continuity of the modulus mapping $x\mapsto |x| := (x^*x)^{\frac{1}{2}}$, we have

$$A_{\mathbf{1}}^{-1}\cap\mathcal{U}(A) = \mathcal{U}^0(A),$$

where $A_{\mathbf{1}}^{-1}$ stands for the principal component of the invertible elements in A (see Theorem 6.11).

An algebraic characterization of the principal component of the set of unitaries in a unital JB*-algebra has not been available until the recent reference [**31**]. Here the result is presented in Theorem 6.14 and Proposition 6.17, where it is shown that for each unital JB*-algebra M with principal component of $\mathcal{U}(M)$ denoted by $\mathcal{U}^0(M)$, the following statements are equivalent for every unitary u in M:

(*a*) $u\in M_{\mathbf{1}}^{-1}\cap\mathcal{U}(M)$.
(*b*) There exists a continuous path $\Gamma : [0,1]\to\mathcal{U}(M)$ with $\Gamma(0) = \mathbf{1}_M$ and $\Gamma(1) = u$.
(*c*) $u\in\mathcal{U}^0(M)$.
(*d*) $u = U_{e^{ih_n}}\cdots U_{e^{ih_1}}(\mathbf{1}_M)$, for some $n\in\mathbb{N}$, $h_1,\ldots,h_n\in M_{sa}$.
(*e*) There exists $w\in\mathcal{U}^0(M)$ such that $\|u-w\|<2$.

Consequently,

$$\begin{aligned}\mathcal{U}^0(M) &= M_{\mathbf{1}}^{-1}\cap\mathcal{U}(M)\\ &= \{U_{e^{ih_n}}\cdots U_{e^{ih_1}}(\mathbf{1}_M)\colon n\in\mathbb{N},\ h_j\in M_{sa}\ \forall\, 1\le j\le n\}\\ &= \{u\in\mathcal{U}(M) : \text{ there exists } w\in\mathcal{U}^0(M) \text{ with } \|u-w\|<2\}\end{aligned}\tag{1.2}$$

is analytically arc-wise connected. Moreover, $\mathcal{U}^0(M)$ is a self-adjoint quadratic subset of $\mathcal{U}(M)$. $U_{\mathcal{U}^0(M)}$ is precisely the smallest quadratic subset of $\mathcal{U}(M)$ containing the set $e^{iM_{sa}}$.

After a series of technical results, we present a characterization of all surjective isometries between the principal components of the unitary sets in two unital JB*-algebras in Theorem 6.23. Suppose $\Delta : \mathcal{U}^0(M)\to\mathcal{U}^0(N)$ is a surjective isometry between the principal components of two unital JB*-algebras. Then there exist self-adjoint elements $k_1,\ldots,k_n$ in N, a central projection p in N, and a Jordan *-isomorphism $\Phi : M\to N$ such that

$$\Delta(u) = p\circ U_{e^{ik_n}}\ldots U_{e^{ik_1}}\Phi(u) + (\mathbf{1}_N - p)\circ\left(U_{e^{-ik_n}}\ldots U_{e^{-ik_1}}\Phi(u)\right)^*,$$

for all $u\in\mathcal{U}^0(M)$. Moreover, the elements $k_1,\ldots,k_n\in N_{sa}$ can be replaced by any finite collection of self-adjoint elements $\tilde{k}_1,\ldots,\tilde{k}_m$ such that $\Delta(\mathbf{1}_M) = U_{e^{i\tilde{k}_m}}\cdots U_{e^{i\tilde{k}_1}}(\mathbf{1}_N)$. Under these hypotheses M and N are Jordan *-isomorphic, and there exists a surjective real-linear isometry (i.e., a real-linear triple isomorphism) from M onto N whose restriction to $\mathcal{U}^0(M)$ is Δ.

If under the previous hypotheses, we also assume that there exists $v \in \mathcal{U}(N)$ such that $U_v(\Delta(\mathbf{1}_M)) = \mathbf{1}_N$. Then there exist a central projection p in N and a Jordan *-isomorphism $\Phi : M \to N$ such that

$$\Delta(w) = p \circ U_{v^*}\Phi(w) + (\mathbf{1}_N - p) \circ (U_v\Phi(w))^*,$$

for all $w \in \mathcal{U}^0(M)$ (cf. Corollary 6.26).

In Theorem 6.30 we obtain an strengthened version of a result from the recent reference [**31**], showing that if M and N are two unital JB*-algebras such that M or N is a JBW*-algebra, and $\Delta : \mathcal{U}(M) \to \mathcal{U}(N)$ is a surjective isometry, then M and N are JBW*-algebras and there exist a self-adjoint element k_1 in N, a central projection $q \in N$, and a Jordan *-isomorphism $\Phi : M \to N$ such that

$$\Delta(u) = q \circ U_{e^{ik_1}}(\Phi(u)) + (\mathbf{1}_N - q) \circ U_{e^{ik_1}}(\Phi(u)^*),$$

for all $u \in \mathcal{U}(M)$. If Δ is unital we can take $k_1 = 0$. Consequently, Δ admits a (unique) extension to a surjective real-linear isometry from M onto N. A version of this result assuming that M and N are von Neumann algebras was previously obtained by Hatori and Molnár in [**63**].

The consequences of the results in subsection 6.1 are summarized in Corollary 6.33. To present the conclusion, let M and N be unital JB*-algebras. Consider the following statements

(1) M and N are isometrically isomorphic as (complex) Banach spaces.
(2) M and N are isometrically isomorphic as real Banach spaces.
(3) $\mathcal{U}(M)$ and $\mathcal{U}(N)$ are isometrically isomorphic as metric spaces.

(a) M and N are Jordan *-isomorphic.
(b) There exists a surjective isometry $\Delta : \mathcal{U}(M) \to \mathcal{U}(N)$ satisfying $\Delta(\mathbf{1}_M) \in \mathcal{U}^0(N)$.
(c) There exists a surjective isometry $\Delta : \mathcal{U}^0(M) \to \mathcal{U}^0(N)$.

Then (1) $\Leftrightarrow$ (2) $\Leftrightarrow$ (3) $\Leftarrow$ (a) $\Leftrightarrow$ (b) $\Leftrightarrow$ (c). The implication (3) $\Rightarrow$ (a) holds when M and N are unital C*-algebras (cf. Corollary 6.35). We shall see in Corollary 6.36 that the implication (3) $\Rightarrow$ (a) is, in general, false for unital JB*-algebras.

Example 6.10 shows that a surjective isometry between the sets of unitary elements in two unital JB*-algebras is not always extendable to a surjective real-linear isometry between the algebras. A complete characterization of those surjective isometries between sets of unitary elements in two unital JB*-algebras admitting an extension to a surjective real-linear isometry is obtained in Corollary 6.38.

The concluding result in this article gives a sufficient condition to guarantee that a surjective isometry between the unitary elements in two unital JB*-algebras M and N is extendable to a surjective real-linear isometry. Indeed, let $\Delta : \mathcal{U}(M) \to \mathcal{U}(N)$ be a surjective isometry and let $\{\mathcal{U}^0(M)\} \cup \{\mathcal{U}^j(M) : j \in \Lambda\}$ be the collection of all different connected components of $\mathcal{U}(M)$. Assume that for each $j \in \Lambda$ there exists $u_j \in \mathcal{U}(M)$ with $u_j^2 \in \mathcal{U}^j(M)$. Suppose additionally that

$$\Delta\{u, v, u\} = \Delta U_u(v^*) = U_{\Delta(u)}(\Delta(v)^*) = \{\Delta(u), \Delta(v), \Delta(u)\}, \tag{1.3}$$

for all $u, v \in \mathcal{U}(M)$. Then Δ admits an extension to a surjective (real-linear) isometry from M onto N (see Proposition 6.40).

2. A taste of radicals in Banach and Jordan-Banach algebras

We devote this section to recall the basic notions and results for Jordan-Banach algebras and the McCrimmon radical. Our aim is to present the main results needed in later sections following a parallelism between associative Banach and Jordan-Banach algebras. Perhaps the notion of Jordan-Banach algebra has already reached to a wide enough audience after almost a century of developments, and it does not need a detailed presentation. Anyway, in order to make this paper accessible to newcomers (or reader who are less familiarized with the theory of Jordan algebras) we shall equip the reader with full information and bibliographic sources. For a full basic-background on Jordan-Banach algebras and any result or notation not explicitly presented here, the reader is referred to [**2, 7, 54, 65, 70, 90–92**] and the most updated monograph [**22**].

By quoting a phrase from Jacobson's book [**70**], "The Jordan theory had its birth in an attempt by Jordan and subsequently by Jordan, von Neumann and Wigner to formulate the foundation of quantum mechanics". These renowned mathematicians gave a complete determination of the finite-dimensional Jordan algebras over the reals which are formally real [**72**].

A *Jordan algebra* over $\mathbb{K} = \mathbb{R}$ or $\mathbb{C}$, is a linear space M over $\mathbb{K}$ equipped with a bilinear product $M \times M \to M$, $(a, b) \mapsto a \circ b$ $(a, b \in M)$ satisfying the following two axioms

(1) (commutativity) $a \circ b = b \circ a$ $(a, b \in M)$.
(2) (*Jordan identity*) $(a \circ b) \circ a^2 = a \circ (b \circ a^2)$ $(a, b \in M)$.

By replacing the usual associativity of the product by the weaker assumption in the Jordan identity we get a wider class of algebras. A Jordan algebra M is called *unital* if there exists an element $\mathbf{1} = \mathbf{1}_M \in M$ (called the unit of M) satisfying $\mathbf{1}_M \circ a = a$ for all $a \in M$. A subalgebra N of M is said to be a *strongly associative subalgebra* if

$$(b \circ x) \circ a - b \circ (x \circ a) = 0, \text{ for all } a, b \in N \text{ and } x \in M.$$

A *normed Jordan algebra* is a Jordan algebra M equipped with a norm, $\|.\|$, satisfying $\|a \circ b\| \leq \|a\| \, \|b\|$ $(a, b \in M)$. A *Jordan-Banach algebra* is a normed Jordan algebra M whose norm is complete. Unless otherwise stated, all Banach spaces and algebras in this paper are over the complex field. If M is a unital Jordan-Banach algebra with unit $\mathbf{1}_M$, we shall also require that $\|\mathbf{1}_M\| = 1$. When M lacks of a unit, it can be always embedded in a unital Jordan-Banach algebra.

In most of the results in this paper, we shall employ the following notation for the powers of an element a in a Jordan algebra M:

$$a^1 = a; \quad a^{n+1} = a \circ a^n, \quad n \geq 1.$$

If M is unital, we set $a^0 = \mathbf{1}_M$. The Jordan identity assures that each Jordan algebra M is *power associative*, that is, each subalgebra generated by a single element is associative, equivalently, $a^m \circ a^n = a^{m+n}$, for all $a \in M$, $m, n \in \mathbb{N}$ (see [**54**, Lemma 2.4.5]).

From a purely algebraic point of view, elements a, b in a Jordan algebra M are said to *operator commute* if

$$(a \circ c) \circ b = a \circ (c \circ b),$$

for all $c \in M$ (cf. [**54**, 4.2.4]). The *centre* of M is, by definition, the set of all elements of M which operator commute with any other element in M, and will be denoted by $Z(M)$. Elements in the centre of M are called *central.*

By a *Banach algebra* we mean an associative Banach algebra. There is a natural way to regard every real or complex Banach algebra as a real or complex Jordan-Banach algebra by just considering the natural Jordan product defined by $a \circ b := \frac{1}{2}(ab + ba)$. Norm closed Jordan subalgebras of associative Banach algebras are called *special.* Jordan algebras which are not special will be called exceptional. Examples of the latter can be found in [**70**, Chapter IX]. So, the class of Jordan algebras is strictly wider than the mere collection of all special Jordan algebras (more concrete examples will appear later).

Along this paper, we shall assume that our Jordan-Banach algebras are unital. A (real-linear) Jordan homomorphism between Jordan-Banach algebras is a (real-linear) linear map preserving Jordan products (equivalently, squares of elements).

Invertibility in Jordan-Banach algebras requires some extra attention. We recall that an element a in a unital Jordan-Banach algebra M is said to be *invertible* if there exists $b \in M$ satisfying $a \circ b = \mathbf{1}_M$ and $a^2 \circ b = a$. In such a case, the element b is unique, it is called the inverse of a, and will be denoted by a^{-1} (cf. [**54**, 3.2.9] and [**22**, Definition 4.1.2]). The symbol M^{-1} will stand for the set of all invertible elements in M.

In an associative Banach algebra A, each element a gives rise to the left and right multiplication operators defined by $L_a(x) = ax$ and $R_a(x) = xa$ $(x \in A)$. When A is unital, the invertibility of a is equivalent to the invertibility of L_a or R_a. In a Jordan-Banach algebra M left and right multiplication operators are essentially indistinguishable. The (Jordan) multiplication operator by an element $a \in M$ (denoted by M_a) is defined by $M_a(x) = a \circ x$ $(x \in M)$, while the U_a operator is given by

$$U_a(x) = (2L_a^2 - L_{a^2})(x) = 2(a \circ x) \circ a - a^2 \circ x \quad (x \in M).$$

Furthermore, given $a, b \in M$, we shall write $U_{a,b} : M \to M$ for the linear mapping on M defined by

$$U_{a,b}(x) = (a \circ x) \circ b + (b \circ x) \circ a - (a \circ b) \circ x, \quad (x \in M).$$

The "fundamental formula" in Jordan algebras asserts that

$$U_a U_b U_a = U_{U_a(b)}, \tag{2.1}$$

for all a, b in M (cf. [**54**, 2.4.18] or [**90**] or [**22**, Proposition 3.4.15]). We shall also employ the following version of the fundamental formula, which can be easily obtained via induction,

$$U_{U_{a_1} U_{a_2} \cdots U_{a_n}(b)} = U_{a_1} U_{a_2} \ldots U_{a_n} U_b U_{a_n} \ldots U_{a_2} U_{a_1}, \tag{2.2}$$

for all $n \in \mathbb{N}$, $a_1, \ldots, a_n, b$ in M.

We gather in the next lemma some starting properties of invertible elements in unital Jordan-Banach algebras. From now on we shall write $L(X)$ for the space of all linear maps on a linear space X.

LEMMA 2.1 ([**22**, Theorem 4.1.3, Proposition 4.1.6, and Theorem 4.1.7]). *Let M be a unital Jordan-Banach algebra. Then the following statements hold:*

(a) *An element $a \in M$ is invertible if and only if U_a is a bijective mapping (i.e. invertible in $L(M)$), and in such a case $U_a^{-1} = U_{a^{-1}}$.*
(b) *The set M^{-1} of all invertible elements in M is open, and consequently M^{-1} is locally connected and all its connected components are open.*
(c) *An element of the form $U_x(y)$ is invertible if and only if both x and y belong to M^{-1}, and such a case, $(U_x(y))^{-1} = U_x^{-1}(y^{-1}) = U_{x^{-1}}(y^{-1})$.*
(d) *If $a \in M^{-1}$, then a^n and $(a^{-1})^m$ operator commute for all $n, m \in \mathbb{N}$. Furthermore, by defining $a^{-n} := (a^{-1})^n$ for all natural n and $a^0 := \mathbf{1}_M$, then $a^k \circ a^l = a^{k+l}$ for all integers k, l.*
(e) *If A is a unital complex Banach algebra regarded as a Jordan-Banach algebra with respect to the natural Jordan product, and element $a \in A$ is invertible in the Jordan sense if and only if it is invertible in the usual sense, and in such a case the inverse is the same in both senses.*
(f) *The mapping $M^{-1} \to M^{-1}$, $a \mapsto a^{-1}$ is a homeomorphism.*

The reader should be warned that the set invertible elements in a Jordan-Banach algebra is not, in general, stable under Jordan products.

As in the setting of Banach algebras, the spectrum of an element a of a complex Jordan-Banach algebra M ($Sp(a)$, $Sp(M, a)$ or J-$\sigma(a)$) is defined as the set of all $\lambda \in \mathbb{C}$ for which $a - \lambda \mathbf{1}_M$ is not invertible. An appropriate Gelfand theorem for Jordan-Banach algebras assures that $Sp(a)$ is a non-empty compact subset of the complex plane (see, for example, [**22**, Theorem 4.1.17], where even a Jordan version of the celebrated Gelfand-Beurling formula is stated). The multi-valued mapping $a \mapsto Sp(a)$ is upper semi-continuous on M. We shall write $\rho(a)$ for the *spectral radius* of a, that is, $\rho(a) = \max\{|\lambda| : \lambda \in Sp(a)\}$. It should be noted that if A is a complex Banach algebra, the spectrum of each element in A coincides with its spectrum when A is regarded as a unital complex Jordan-Banach algebra.

From now on, given a Banach space X, the open (respectively, closed) ball of radius ρ and centre $a \in X$ will be denoted by $B_X(a, \rho)$ or by $B(a, \rho)$ (respectively, $\overline{B}_X(a, \rho)$).

Continuous and holomorphic functional calculus for elements in a complex Jordan-Banach algebra M are relatively simple and natural to understand. Indeed, the closed subalgebra, M_a, of a unital Jordan-Banach algebra M generated by $\mathbf{1}_M$ and an element $a \in M$ is an associative and commutative Banach algebra (cf. [**8**, Theorem 2.3] and [**22**, Theorems 4.1.88 and 4.1.93]). Therefore, the standard holomorphic functional calculus in complex Banach algebras can be perfectly understood in the setting of complex Jordan-Banach algebras. We gather in the next result some of the basic properties of this holomorphic functional calculus.

THEOREM 2.2 (Holomorphic functional calculus, [**8**, Theorem 2.3] and [**22**, Theorems 4.1.88 and 4.1.93]). *Let a be an element in a unital complex Jordan-Banach algebra M. Suppose Ω is an open neighbourhood of $Sp(a)$, and let h_0 denote the function defined by the inclusion mapping $\Omega \hookrightarrow \mathbb{C}$. Then there is a unique continuous unit-preserving algebra homomorphism $f \mapsto f(a)$ from the complex algebra $\mathcal{H}(\Omega)$, of all complex-valued holomorphic functions on Ω, into M taking h_0 to a and satisfying the following properties:*

(*i*) $f(a) = \frac{1}{2\pi i}\int_\Gamma f(\lambda)(\lambda \mathbf{1}_M - a)^{-1} d\lambda$, *for any positively orientated curve included in* Ω *and surrounding* $Sp(a)$.

(*ii*) $f(a)$ *is contained in the smallest closed strongly associative subalgebra of* M *containing* $\mathbf{1}_M$ *and* a.

(*iii*) *For each* $f \in \mathcal{H}(\Omega)$ *we have* $\mathrm{Sp}(M, f(a)) = f(\mathrm{Sp}(M,a))$ (*Spectral mapping theorem*).

(*iv*) *The set*

$$M_\Omega := \{x \in M : \mathrm{Sp}(M,x) \subseteq \Omega\}$$

is a non-empty open subset of M, *and for each* $f \in \mathcal{H}(\Omega)$ *the mapping* $\widetilde{f} : x \mapsto f(x)$ *from* M_Ω *to* M *is holomorphic (i.e. there exists a bounded linear operator* $T : M \to M$ *such that* $\lim_{h\to 0} \frac{\|f(x+h) - f(x) - T(h)\|}{\|h\|} = 0$ *for any* $x \in M_\Omega$).

(*v*) *Let* Ω_1 *be another open subset in* $\mathbb{C}$, $f \in \mathcal{H}(\Omega)$, $g \in \mathcal{H}(\Omega_1)$ *such that* $f(\Omega) \subseteq \Omega_1$. *Then* $(g \circ f)(a) = g(f(a))$.

2.1. Jacobson and McCrimmon Radicals.

We begin this subsection by recalling the notion of Jacobson radical for Banach algebras (see [**118**, §12]). Given a Banach algebra A and a complex vector space X, a *representation* of A in the space X is a homomorphism $\pi : A \to L(X)$, $x \mapsto \pi(x)$. A subspace X_0 of X is called *invariant* for the representation π if $\pi(x)(X_0) \subseteq X_0$ for all $x \in A$. A representation is called irreducible if the only invariant subspaces are $\{0\}$ and X. A two-sided ideal $I \subseteq A$ is said to be a *left-* (respectively, *right-* or *modular*) *modular ideal* when there exists an element $e \in A$ such that $x - xe \in I$ (respectively, $x - ex \in I$ or $x - xe, x - ex \in I$) for any $x \in A$. Finally, I is called *primitive* if it is of the form $I = \{x \in A : xA \subseteq J\}$, where J is a maximal left-modular ideal in A. A celebrated result by Rickart shows that a two-sided ideal is primitive if and only if it is the kernel of an irreducible representation. Moreover, every maximal two-sided modular ideal is primitive. If A is commutative, primitive ideals are precisely the maximal modular ideals (see [**108**, Chapter II]).

Assuming that A is non-unital, an element x in A is called quasi-regular if there exists an element $y \in A$ (called the quasi-inverse of x) such that $x + y - xy = x + y - yx = 0$, i.e., $\mathbf{1}_A - x$ is invertible in the unitization of A with inverse $\mathbf{1}_A - y$.

Definition 2.3 ([**118**, §12], [**108**, Chapter II] or [**100**, §4.3]). The Jacobson radical of any Banach algebra A, $\mathrm{rad}(A)$, is defined as the intersection of all primitive ideals of A. If $A \neq \mathrm{rad}(A)$, the Jacobson radical coincides with the intersection of all maximal left-modular ideals, or, equivalently, with the intersection of all right-modular ideals. Furthermore

$$\begin{aligned}\mathrm{rad}(A) &= \{x \in A : \forall \lambda \in \mathbb{C} \text{ the element } \lambda x + yx \text{ is quasi-regular } \forall y \in A\}\\ &= \{x \in A : \forall \lambda \in \mathbb{C} \text{ the element } \lambda x + xy \text{ is quasi-regular } \forall y \in A\}.\end{aligned}$$

We say that A is semisimple if $\mathrm{rad}(A) = \{0\}$.

An element a in a Banach algebra A is called *quasi-nilpotent* if its spectral radius, $\rho(a)$, is zero. It is known that the Jacobson radical of A is contained in the set of all quasi-nilpotent elements of A, however the Jacobson radical is, in general, smaller. More analytic characterizations of the Jacobson radical in terms of the spectral radius were obtained by Słodkowski, Wojtyński, and Zemzánek, and subsequently extended by Aupetit.

THEOREM 2.4 ([**100**, Theorem 2.3.3] and [**7**, Theorem 1]). *For each complex Banach algebra A we have*

$$\begin{aligned}\operatorname{rad}(A) &= \{x \in A : \rho(a+x) = \rho(a),\ \forall a \in A\} = \{x \in A : \rho(ax) = 0,\ \forall a \in A\}\\ &= \{x \in A : \sigma(a+x) = \sigma(a),\ \forall a \in A\}\\ &= \{x \in A :\ \textit{for each } a \in A \textit{ the set } \{\rho(a+\lambda x) : \lambda \in \mathbb{C}\} \textit{ is bounded}\}.\end{aligned}$$

We know, by an observation due to Hatori [**59**], that the second characterization of the radical presented in Theorem 2.4 admits a weaker version in case that A is unital. We shall present here an alternative proof to that given by Hatori. Let us write $A_{\mathbf{1}}^{-1}$ for the principal connected component of A^{-1}.

LEMMA 2.5 ([**59**, Lemma 3.1]). *Let A be a unital (associative) complex Banach algebra. Then*

$$\operatorname{rad}(A) = \left\{x \in A : \rho(ax) = 0,\ \forall a \in A^{-1}\right\} = \left\{x \in A : \rho(ax) = 0,\ \forall a \in A_{\mathbf{1}}^{-1}\right\}.$$

PROOF. The containing $\subseteq$ is clearly true in both cases. So, we only need to show that the set on the right hand side is inside the Jacobson radical. Suppose that $\rho(ax) = 0$, for all $a \in A_{\mathbf{1}}^{-1}$. Fix an arbitrary $b \in A$ and $\mu \in \mathbb{C}$ with $|\mu| > \rho(b)$. Clearly, $\mathbf{1}_A - \mu^{-1}b$ is invertible, and since $\lim_{|\mu|\to\infty}(\mathbf{1}_A - \mu^{-1}b)^{-1} = \mathbf{1}_A$, we deduce that $(\mathbf{1}_A - \mu^{-1}b)^{-1} \in A_{\mathbf{1}}^{-1}$, which implies that $\beta(\mathbf{1}_A - \mu^{-1}b)^{-1} \in A_{\mathbf{1}}^{-1}$ for all $\beta \in \mathbb{C}\backslash\{0\}$. It follows from the assumptions that $\rho\left(\beta(\mathbf{1}_A - \mu^{-1}b)^{-1}x\right) = 0$, for all $\beta \in \mathbb{C}\backslash\{0\}$.

By applying the previous conclusion we get

$$\begin{aligned}\sigma\left((\mathbf{1}_A - \mu^{-1}b)^{-1}(\mathbf{1}_A - \mu^{-1}b - \beta x)\right) &= \sigma(\mathbf{1}_A - \beta(\mathbf{1}_A - \mu^{-1}b)^{-1}x)\\ &= 1 - \sigma(\beta(\mathbf{1}_A - \mu^{-1}b)^{-1}x) = \{1\},\end{aligned}$$

and thus $(\mathbf{1}_A - \mu^{-1}b - \beta x)$ must be invertible for all $\beta \in \mathbb{C}\backslash\{0\}$. However, it is clear that we also have an invertible element for $\beta = 0$. This is equivalent to say that $-\mu\mathbf{1}_A + b + \lambda x = -\mu((\mathbf{1}_A - \mu^{-1}b - \lambda\mu^{-1}x)) \in A^{-1}$, equivalently, $\mu \notin \sigma(b+\lambda x)$, for all $\lambda \in \mathbb{C}$. We have shown that

$$\mathbb{C}\backslash B_{\mathbb{C}}(0, \rho(b)+\varepsilon) \subseteq \mathbb{C}\backslash\sigma(b+\lambda x),$$

for all $\lambda \in \mathbb{C}$ and $\varepsilon > 0$, and consequently, $\rho(b+\lambda x) \leq \rho(b)$, for all $\lambda \in \mathbb{C}$. The arbitrariness of b and the last characterization in Theorem 2.4 prove that $x \in \operatorname{rad}(A)$. □

We turn now to Jordan-Banach algebras. An element a in a unital Jordan-Banach algebra M is *quasi-invertible* with *quasi-inverse* $b \in M$ if $\mathbf{1}_M - a$ is invertible with inverse $\mathbf{1}_M - b$. A subset of M is called quasi-invertible if all its elements are quasi-invertible in M.

An *outer ideal* of a Jordan-Banach algebra M is a subspace J such that $U_M(J) \subseteq J$; an *inner ideal* or a *strict inner ideal* is a subspace I satisfying $U_I(M) \subseteq I$ and $I^2 \subseteq I$ (if M is unital the second condition clearly follows from the first one). An ideal J is a subspace that is both an outer and a strict inner ideal (i.e. $J^2 \subset J$). It is known that if J is an ideal or an inner ideal, then the quasi-inverse of any quasi-invertible element in J also belongs to J (see [**91**, page 672]).

The existence of an appropriate notion of radical in the setting of Jordan-Banach algebras was proved by McCrimmon [**91**].

THEOREM 2.6 ([**91**, Theorem 1], [**65**, Theorem 1.1], [**22**, Proposition 4.4.11 and Definition 4.4.12]). *Let M be a Jordan-Banach algebra. Then there exists a unique maximum quasi-invertible ideal $\mathcal{R}ad(M)$ of M, called the* McCrimmon radical *of M, which contains all quasi-invertible ideals. Actually the McCrimmon radical of a unital Jordan-Banach algebra M coincides with the intersection of all maximal inner ideals in M.*

A Jordan-Banach algebra M is called *Jordan-semisimple*, *J-semisimple*, or simply *semisimple* if $\mathcal{R}ad(M) = \{0\}$.

REMARK 2.7 ([**22**, Theorem 3.6.21]). Let A be a Banach algebra, and let M denote the Jordan-Banach algebra obtained when A is equipped with its natural Jordan product. Then the McCrimmon radical of M coincides with the usual Jacobson radical of A.

Our next result, originally established by Aupetit in [**7**], is a Jordan version of the characterizations of the radical in terms of the spectral radius presented in Theorem 2.4.

THEOREM 2.8 ([**7**, Theorem 2, Corollaries 1 and 2]). *Let x be an element in a Jordan-Banach algebra M. Then the following statements are equivalent:*

(*a*) *x is in the McCrimmon radical of M.*
(*b*) *For every $a \in M$ the set $\{\rho(a + \lambda x) : \lambda \in \mathbb{C}\}$ is bounded.*
(*c*) *$\rho(U_a(x)) = 0$, for every a in M.*
(*d*) *There exists $C \geq 0$ such that $\rho(a) \leq C\|a - x\|$, for all a in a neighbourhood of x.*

Our next goal is a Jordan version of Lemma 2.5, which improves the characterization established by Aupetit in the previous theorem. We also include a slightly different proof from that in the original source. According to the usual notation, we shall write $M_{\mathbf{1}}^{-1}$ for the principal connected component of the invertible elements in a unital Jordan-Banach algebra M.

PROPOSITION 2.9 ([**104**, Proposition 2.3]). *Let M be a unital complex Jordan-Banach algebra. Then*

$$\begin{aligned}\mathcal{R}ad(M) &= \left\{x \in M : \rho(U_a(x)) = 0,\ \forall a \in M^{-1}\right\}\\ &= \left\{x \in M : \rho(U_a(x)) = 0,\ \forall a \in M_{\mathbf{1}}^{-1}\right\}.\end{aligned}$$

PROOF. As in the associative case, it suffices to prove that the last set is contained in the McCrimmon radical of M. Assume that $x \in M$ satisfies $\rho(U_a(x)) = 0$, for all $a \in M_{\mathbf{1}}^{-1}$.

Fix an arbitrary $b \in M$ and $\mu \in \mathbb{C}$ with $|\mu| > \rho(b)$. In such a case $b - \mu \mathbf{1}_{M} \in M^{-1}$, equivalently, $-\mu^{-1}b + \mathbf{1}_{M} \in M^{-1}$.

Let us make some observations on the spectrum of the elements of the form $-\mu^{-1}b + \mathbf{1}_{M} \in M_{\mathbf{1}}^{-1}$. Clearly

$$\mathrm{Sp}\left(-\mu^{-1}b + \mathbf{1}_{M}\right) = 1 - \mu^{-1}\mathrm{Sp}(b) \subseteq 1 - B_{\mathbb{C}}(0,1) \subseteq \{z \in \mathbb{C} : \Re\mathrm{e}(z) > 0\} = \Omega.$$

We are in a position to apply the holomorphic functional calculus (cf. Theorem 2.2), via the principal branch of the square root $f:\Omega\to\mathbb{C}$, $f(\lambda)=\frac{1}{\sqrt{\lambda}}$, to define the element $\left(\mathbf{1}_M-\mu^{-1}b\right)^{-\frac{1}{2}}:=f\left(\mathbf{1}_M-\mu^{-1}b\right)$ ($|\mu|>\rho(b)$). By the continuity of the holomorphic functional calculus, $\lim_{|\mu|\to\infty}\left(\mathbf{1}_M-\mu^{-1}b\right)^{-\frac{1}{2}}=\mathbf{1}_M$, and consequently $\beta\left(\mathbf{1}_M-\mu^{-1}b\right)^{-\frac{1}{2}}\in M^{-1}_{\mathbf{1}_M}$, for all $\beta\in\mathbb{C}\backslash\{0\}$.

By the assumptions $\rho\left(U_{\beta\left(\mathbf{1}_M-\mu^{-1}b\right)^{-\frac{1}{2}}}(x)\right)=0$, for all non-zero $\beta\in\mathbb{C}$, $|\mu|>\rho(b)$). Since $\mathrm{Sp}\left(U_{\beta\left(\mathbf{1}_M-\mu^{-1}b\right)^{-\frac{1}{2}}}(x)\right)=\{0\}$, the identity

$$U_{\left(\mathbf{1}_M-\mu^{-1}b\right)^{-\frac{1}{2}}}\left(\mathbf{1}_M-\mu^{-1}b-\beta^2x\right)=\mathbf{1}_M-U_{\beta\left(\mathbf{1}_M-\mu^{-1}b\right)^{-\frac{1}{2}}}(x),$$

implies that $U_{\left(\mathbf{1}_M-\mu^{-1}b\right)^{-\frac{1}{2}}}\left(\mathbf{1}_M-\mu^{-1}b-\beta^2x\right)$ is an invertible element for all $\beta\in\mathbb{C}\backslash\{0\}$. But the same conclusion trivially holds for $\beta=0$. We therefore deduce that the element $U_{\left(\mathbf{1}_M-\mu^{-1}b\right)^{-\frac{1}{2}}}\left(\mathbf{1}_M-\mu^{-1}b-\alpha x\right)$ is invertible for all $\alpha\in\mathbb{C}$. The fundamental identity (2.1) shows that

$$U_{\left(\mathbf{1}_M-\mu^{-1}b\right)^{-\frac{1}{2}}}\left(\mathbf{1}_M-\mu^{-1}b-\alpha x\right)=U_{\left(\mathbf{1}_M-\mu^{-1}b\right)^{-\frac{1}{2}}}U_{\mathbf{1}_M-\mu^{-1}b-\alpha x}U_{\left(\mathbf{1}_M-\mu^{-1}b\right)^{-\frac{1}{2}}},$$

and since $\left(\mathbf{1}_M-\mu^{-1}b\right)^{-\frac{1}{2}}$ is clearly invertible, the element $\mathbf{1}_M-\mu^{-1}b-\alpha x$, equivalently $-\mu\mathbf{1}_M+b+\alpha x$, must be invertible for all $\alpha\in\mathbb{C}$ (cf. Lemma 2.1(*c*)). This shows that $\mu\notin\mathrm{Sp}(b+\alpha x)$ ($\alpha\in\mathbb{C}$, $|\mu|>\rho(b)$), and hence $\mathbb{C}\backslash\overline{B}_{\mathbb{C}}(0,\rho(b))\subseteq\mathbb{C}\backslash\mathrm{Sp}(b+\alpha x)$, for all $\alpha\in\mathbb{C}$. We have therefore shown that $\rho(b+\alpha x)\leq\rho(b)$ for all $\alpha\in\mathbb{C}$. The arbitrariness of b and Theorem 2.8 imply that x lies in the McCrimmon radical of M. □

2.2. On the connected components of the set of invertible elements.

The aim of this subsection is to revisit the topological properties of the set of invertible elements in a unital Jordan-Banach algebra and its connected components. We have already made use of the principal connected component in the latest characterizations of the Jacobson and McCrimmon radicals in unital Banach and Jordan-Banach algebras.

By employing holomorphic functional calculus, or by analytic series, for each element a in a unital complex Jordan-Banach algebra M, we can define the exponential of a, $\exp(a)=\displaystyle\sum_{n=0}^{\infty}\frac{a^n}{n!}$. Clearly, the mapping $a\mapsto\exp(a)$ is analytic. If an associative Banach algebra A is regarded as a Jordan-Banach algebra with its natural Jordan product, the exponential of each element $a\in A$ coincides with the usual notion in the setting of associative Banach algebras.

Since the Jordan-Banach subalgebra M_a of M generated by a and $\mathbf{1}$ is a commutative unital Banach algebra with respect to the inherited Jordan product [**70**, 1.1] and contains $\exp(a)$, we deduce that $\exp(sa)\ \exp(ta)=\exp((s+t)a)$ for all $s,t\in\mathbb{C}$. We also have $U_a(c)=a^2\circ c$ for $a,c\in M_a$, which yields

$$U_{\exp(ta)}(\exp(-ta))=\exp(ta),\ \text{ and } U_{\exp(ta)}(\exp(-ta)^2)=\mathbf{1}.$$

Invertibility is actually a local notion, i.e., a is invertible in M if and only if it is invertible in M_a. Moreover, $\exp(a) \in M^{-1}$ with inverse $\exp(-a)$ for all $a \in M$ (cf. [**70**, 1.6.1]).

We gather in the next result a list of topological and algebraic properties of the connected components of the invertible elements in a unital complex Banach algebra.

PROPOSITION 2.10 ([**15**, Propositions 8.6 and 8.7] or [**118**, §7]). *Let A be a complex unital Banach algebra, and let $A_{\mathbf{1}}^{-1}$ denote the principal connected component of A^{-1}. Then the following statements hold:*

(*a*) *$A_{\mathbf{1}}^{-1}$ is the least subgroup of A^{-1} containing* $\exp(A)$.
(*b*) *$A_{\mathbf{1}}^{-1}$ is a normal subgroup (i.e. $aba^{-1} \in A_{\mathbf{1}}^{-1}$, for all $a \in A^{-1}$, $b \in A_{\mathbf{1}}^{-1}$) and is a maximal connected subgroup of A^{-1}.*
(*c*) *If A is commutative, $A_{\mathbf{1}}^{-1}$ is* $\exp(A)$.

REMARK 2.11. Suppose A is a unital complex Banach algebra. Let $\mathfrak{A}$ be an open subgroup of A^{-1}. It is well known that $\mathfrak{A}$ must be also closed, and consequently $\mathfrak{A}$ contains the principal component of A^{-1}. To see this it suffices to show that $A^{-1}\backslash\mathfrak{A}$ is open. Indeed, for each $b \in A^{-1}\backslash\mathfrak{A}$, the set $\mathfrak{A}b$ is an open neighbourhood of b, whose intersection with $\mathfrak{A}$ must be empty, since if $a \in \mathfrak{A} \cap \mathfrak{A}b$, then $a = cb$ with $c \in \mathfrak{A}$ and, by applying that $\mathfrak{A}$ is a subgroup, we have $b = c^{-1}(cb) \in \mathfrak{A}$, which is impossible.

Furthermore, since A^{-1} is open, and hence locally connected and its connected components are open, and $\mathfrak{A}$ is clopen a classic result in topology (see, for example, [**96**, Theorem 5]) asserts that $\mathfrak{A}$ must coincide with the union of some connected components of A^{-1}.

If $\mathcal{C}$ is another connected component of A^{-1} and we pick $b \in \mathcal{C}$, it is easy to see that $\mathcal{C} = bA_{\mathbf{1}}^{-1} = A_{\mathbf{1}}^{-1}b$ (just apply that the left and right multiplication operators by b are homeomorphisms mapping 1 to b). The component $\mathcal{C}$ need not be a subgroup, however, if we replace the product of A with the one defined by $x \cdot_{b^{-1}} y = xb^{-1}y$, we get another another associative complete normed algebra $A_{b^{-1}}$, called the b^{-1}*-homotope* of A, with unit b. The main connected component of $A_{b^{-1}}$ is precisely $\mathcal{C}$, which is therefore a subgroup of $A_{b^{-1}}$.

The topological and algebraic properties of the principal connected component, $M_{\mathbf{1}}^{-1}$, of the invertible elements in a unital complex Jordan-Banach algebra M have not been completely clarified until quite recently. For example, Aupetit asked in [**8**, page 484] whether $M_{\mathbf{1}}^{-1}$ is connected by analytic arcs, as it is in Banach algebras. An algebraic characterization was finally found by Loos in 1996.

THEOREM 2.12 ([**86**]). *For each unital complex Jordan-Banach algebra M we have*

$$M_{\mathbf{1}}^{-1} = \left\{U_{\exp(a_1)} \cdots U_{\exp(a_n)}(\mathbf{1}_M) : a_i \in M,\ n \geq 1\right\}.$$

Consequently, $M_{\mathbf{1}}^{-1}$ is open and closed in M^{-1}. Moreover each connected component of M^{-1} is analytically arc-wise connected.

It should be noted that in Loos's proof of the previous result the norm is not required to satisfy $\|x \circ y\| \leq \|x\|\ \|y\|$ and $\|\mathbf{1}\| = 1$.

The arguments in Loos' paper rely on the fact that passing to an appropriate isotope we can avoid the use of left or right multiplication by an invertible element like in an associative Banach algebra, operation which has no direct Jordan analogue. Suppose c is an invertible element in a unital Jordan-Banach algebra M then the vector space M becomes a complete normed Jordan algebra $M_{(c)}$ with unit element c^{-1} and Jordan product $a \circ_c b := U_{a,b}(c)$, for all $a, b \in M$. This algebra is called the *c-isotope* of M. It is known that the quadratic operator in the u-isotope satisfies

$$U_a^{(c)} = U_a U_c, \text{ for all } a, b \in M. \tag{2.3}$$

REMARK 2.13. Let c be an invertible element in a unital complex Jordan-Banach algebra M, and let $M_{(c)}$ stand for the c-isotope of M. Then $M^{-1} = M_{(c)}^{-1}$. If $a \in M^{-1}$, having in mind that the corresponding $U_a^{(c)}$ operator in the c-isotope is given by $U_a^{(c)} = U_a U_c$, where the two operators on the right-hand-side are invertible, we deduce that $a \in M_{(c)}^{-1}$. The reciprocal inclusion follows from the identity $U_a^{(c)} U_c^{-1} = U_a$. Moreover, $\left(U_a^{(c)}\right)^{-1} = U_c^{-1} U_{a^{-1}}$, and by the fundamental identity, we have

$$U^{(c)}_{U_{c^{-1}}(a^{-1})} = U_{U_{c^{-1}}(a^{-1})} U_c = U_{c^{-1}} U_{a^{-1}} U_{c^{-1}} U_c = U_{c^{-1}} U_{a^{-1}},$$

which shows that $U_{c^{-1}}(a^{-1})$ is the inverse of a in $M_{(c)}$.

The transfer and repercussion between the results for Banach algebras and Jordan-Banach algebras is not always fluent, especially from the Jordan setting to the associative case. A similar conclusion to that in Loos' theorem was independently discovered by Hatori for the principal component of the unitary group in a unital C*-algebra [**63**, Lemma 3.2].

The Jordan product of two invertible elements in a unital Jordan-Banach algebra M is not, in general, invertible. The best we can conclude is that $U_a(b) \in M^{-1}$ for all $a, b \in M^{-1}$ (cf. Lemma 2.1). This motivates us to introduce the following concept.

DEFINITION 2.14. A subset $\mathfrak{M}$ of M^{-1} will be called a *quadratic subset* if it is non-empty and $U_a(b) \in \mathfrak{M}$ for all $a, b \in \mathfrak{M}$.

We can now present a recent algebraic characterization of the principal component of the invertible elements in a unital Jordan-Banach algebra.

LEMMA 2.15 ([**104**, Lemma 2.5]). *Let M be a unital complex Jordan-Banach algebra. Then the principal component of M^{-1} is precisely the least quadratic subset of M^{-1} containing* $\exp(M)$. *Consequently, for each natural number n and each $a \in M_{\mathbf{1}}^{-1}$ the element a^n lies in $M_{\mathbf{1}}^{-1}$.*

PROOF. Obviously, $M_{\mathbf{1}}^{-1}$ contains $\exp(M)$. To prove that $M_{\mathbf{1}}^{-1}$ is a quadratic subset of M^{-1} Loos' Theorem 2.12 plays a central role, since we can restrict to elements of the form $a = U_{\exp(a_1)} \cdots U_{\exp(a_n)}(\mathbf{1})$ and $b = U_{\exp(b_1)} \cdots U_{\exp(b_m)}(\mathbf{1})$ in $M_{\mathbf{1}}^{-1}$. By the fundamental formula (2.2) we have

$$\begin{aligned} U_a(b) &= U_{U_{\exp(a_1)} \cdots U_{\exp(a_n)}(\mathbf{1})}(U_{\exp(b_1)} \cdots U_{\exp(b_m)}(\mathbf{1})) \\ &= U_{\exp(a_1)} \cdots U_{\exp(a_n)} U_{\mathbf{1}} U_{\exp(a_n)} \cdots U_{\exp(a_1)} U_{\exp(b_1)} \cdots U_{\exp(b_m)}(\mathbf{1}) \in M_{\mathbf{1}}^{-1}. \end{aligned}$$

Given $a \in M_{\mathbf{1}}^{-1}$, it follows from the above that $a^2 = U_a(\mathbf{1}) \in M_{\mathbf{1}}^{-1}$, $a^3 = U_a(a) \in M_{\mathbf{1}}^{-1}$, and by an induction argument $a^n = U_a(a^{n-2}) \in M_{\mathbf{1}}^{-1}$ for all $n \geq 3$.

We finally assume that $\mathfrak{M}$ is a *quadratic subset* of M^{-1} containing $\exp(M)$. Since, clearly $\mathbf{1} \in \mathfrak{M}$, we have $U_{\exp(a)}(\mathbf{1}) \in \mathfrak{M}$. Suppose that $U_{\exp(a_n)} \cdots U_{\exp(a_1)}(\mathbf{1}) \in \mathfrak{M}$ for $n \geq 2$, $a_i \in M$. Then, by the assumption on $\mathfrak{M}$ and the induction hypothesis, we have

$$U_{\exp(a_{n+1})} U_{\exp(a_n)} \cdots U_{\exp(a_1)}(\mathbf{1}) \in \mathfrak{M}.$$

A new application of Loos' Theorem 2.12 implies that $M_{\mathbf{1}}^{-1} \subseteq \mathfrak{M}$. □

REMARK 2.16. Let M be a unital complex Jordan-Banach algebra. Since the inverse mapping $a \mapsto a^{-1}$ is a homeomorphism on M^{-1} (cf. Lemma 2.1(f)), and maps $\mathbf{1}_M$ to itself, it fixes the principal component of M^{-1}, that is, $a^{-1} \in M_{\mathbf{1}}^{-1}$ for all $a \in M_{\mathbf{1}}^{-1}$. Therefore, $a^n \in M_{\mathbf{1}}^{-1}$ for all $a \in M_{\mathbf{1}}^{-1}$ and every integer number n.

PROBLEM 2.17. For a Jordan-Banach algebra M, it would be interesting to know if every open quadratic subset $\mathfrak{M}$ of M^{-1} containing the unit must be also closed, and hence $M_{\mathbf{1}}^{-1} \subseteq \mathfrak{M}$.

The set M^{-1} is a clopen quadratic subset of itself closed for powers and inverses.

We shall conclude this section by showing that clopen quadratic subsets of the invertible elements in a unital Jordan-Banach algebra which are closed for powers and inverses are also stable under translations by elements in the McCrimmon radical.

LEMMA 2.18. *Let M be a unital Jordan-Banach algebra, and let $\mathfrak{M} \subseteq M^{-1}$ be a clopen quadratic subset of M^{-1} which is closed for powers and inverses. Then for each element u_0 in the McCrimmon radical of M the identity $\mathfrak{M} - u_0 = \mathfrak{M}$ holds.*

PROOF. Clearly, $\mathfrak{M}$ contains the unit of M because $U_{a^{-1}}(a^2) = \mathbf{1}_M \in \mathfrak{M}$ for all $a \in \mathfrak{M}$. Since $\mathfrak{M}$ is clopen and the connected components of M^{-1} are open, we deduce that $\mathfrak{M}$ coincides with the union of some of the connected components of M^{-1} and $M_{\mathbf{1}}^{-1} \subseteq \mathfrak{M}$.

Fix an arbitrary $b \in \mathfrak{M}$. A result by McCrimmon assures that, the McCrimmon radical of the b-homotope algebra $M_{(b)}$ satisfies

$$\mathcal{R}ad(M_{(b)}) = \{z \in M : U_b(z) \in \mathcal{R}ad(M)\} \text{ (see [\textbf{92}, Proposition 3 in page 110]).}$$

Consequently, $U_{b^{-1}}(u_0) = U_b^{-1}(u_0) \in \mathcal{R}ad(M_{(b)})$. It follows that

$$\rho_{M_{(b)}}(U_x^{(b)}(U_{b^{-1}}(u_0))) = 0, \text{ for all } x \in M_{(b)} \text{ (cf. Theorem 2.8).}$$

Therefore $\rho_{M_{(b)}}(tU_{b^{-1}}^{(b)}(U_{b^{-1}}(u_0))) = 0$, which implies that

$$tU_{b^{-1}}^{(b)}(U_{b^{-1}}(u_0)) + b^{-1} \in M_{(b)}^{-1},$$

for all $t \in \mathbb{R}_0^+$ (because b^{-1} is the unit of the homotope $M_{(b)}$). Moreover, since $U_{b^{-1}}^{(b)}$ is the identity mapping because b^{-1} is the unit of the homotope $M_{(b)}$, the previous conclusion implies that $tU_{b^{-1}}(u_0) + b^{-1} \in (M_{(b)})^{-1} = M^{-1}$, for all $t \in \mathbb{R}_0^+$. This proves that $U_{b^{-1}}(u_0) + b^{-1}$ is in the connected component of M^{-1} containing b^{-1}, and thus the element

$$u_0 + b = U_b(U_{b^{-1}}(u_0) + b^{-1})$$

is in the connected component of M^{-1} containing b, and hence $u_0 + b \in \mathfrak{M}$. The arbitrariness of b and the fact that $-b = U_{ib}(b^{-1}) \in \mathfrak{M}$ give the desired conclusion. □

3. Sets of invertible elements as a metric invariant

In the seeking of a certain subset of unital Banach algebras which can serve as an invariant to identify or distinguish them, the first natural candidate is the set of invertible elements. If A is a (associative) unital Banach algebra, the set A^{-1} of all invertible elements in A is a multiplicative group which contains the unit element $\mathbf{1}_A$. In the next example we illustrate the well known fact in spectral theory assuring that there exist non-isomorphic unital commutative Banach algebras whose groups of invertible elements are topologically homeomorphic as topological groups.

EXAMPLE 3.1. [**118**, Remark I.7.8] The commutative Banach algebras $A = C([-1,-\frac{1}{2}]\cup[\frac{1}{2},1])$ and $B = C([0,1]\cup\{2\})$ are not isomorphic since $I = \{b \in B : b|_{[0,1]} \equiv 0\}$ is a minimal ideal in B, and A lacks of minimal ideals. Actually, the Banach spaces A and B are not isometric. The mapping $a \mapsto b$ with

$$b(t) := \begin{cases} a(t-1), & \text{if } t \in [0,\frac{1}{2}];\\ a\left(-\frac{1}{2}\right)/a\left(\frac{1}{2}\right) a(t), & \text{if } t \in [\frac{1}{2},1];\\ a\left(\frac{1}{2}\right)/a\left(-\frac{1}{2}\right), & \text{if } t = \frac{1}{2}, \end{cases}$$

is a group homeomorphism from A^{-1} onto B^{-1}.

Roughly speaking, the topological properties of the group of invertible elements in a unital Banach is not an invariant to distinguish the whole algebra.

Hatori introduced in [**57**] the problem of determining whether the set of invertible elements regarded as a metric space with the induced distance is an invariant. The first successful answer appears in the setting of unital semisimple commutative Banach algebras.

THEOREM 3.2 ([**57**]). *Let A and B be unital complex Banach algebras. We additionally assume that A is semisimple and commutative. Let $\mathfrak{A}$ and $\mathfrak{B}$ be open subgroups of A^{-1} and B^{-1}, respectively. Let $\Delta : \mathfrak{A} \to \mathfrak{B}$ be a surjective isometry for the distances given by the norms of A and B, respectively. Then B is semisimple and commutative, and the mapping $\Delta(\mathbf{1}_A)^{-1}\Delta$ admits an extension to an isometric real-linear algebra isomorphism from A onto B. In particular, A^{-1} and B^{-1} are isometrically isomorphic as metric groups.*

There are certain aspects of the previous result that should be commented more in detail. We begin with a handicap, the Banach space structure of a complex Banach algebra is not, in general, an invariant by itself, that is, there exist (unital commutative and semisimple) Banach algebras which are isometrically isomorphic as Banach spaces but not isomorphic as Banach algebras.

EXAMPLE 3.3 ([**57**, Example 3.2]). The existence of a surjective linear isometry between two Banach algebras does not imply that these algebras are isomorphic as Banach algebras, even if we assume that they are unital, semisimple, and commutative. For example, the Wiener algebra, $A(\mathbb{T})$, is usually defined as the Banach space of absolutely convergent Fourier series, that is,

$$A(\mathbb{T}) := \left\{ f \in C(\mathbb{T}) : \|f\| := \sum_{k=-\infty}^{\infty} |\widehat{f}(n)| < \infty \right\},$$

equipped with the pointwise product, where $\mathbb{T}$ stands for the unit sphere of $\mathbb{C}$ and $\widehat{f}(n)$ denotes the nth Fourier coefficient of f. It is known from Gelfand theory that $A(\mathbb{T})$ is isometrically isomorphic, via the isomorphism given by the Fourier transform, to the the discrete group Banach algebra $\ell_1(\mathbb{Z})$ with convolution product (cf. [**5**, page 30]). Furthermore, the spectrum of $A(\mathbb{T})$ identifies with $\mathbb{T}$.

On the other hand, we can consider the subalgebra

$$\ell_1(\mathbb{Z}^+) := \left\{ f \in A(\mathbb{T}) : \widehat{f}(n) = 0, \text{ for all } n < 0 \right\},$$

whose spectrum is the closed unit disc $\overline{\mathbb{D}}$ (cf. [**95**]). Since $\overline{\mathbb{D}}$ is not homeomorphic to $\mathbb{T}$, $A(\mathbb{T})$ and $\ell_1(\mathbb{Z}^+)$ are not isomorphic as complex Banach algebras, neither are as Jordan-Banach algebras.

The mapping $T_W : A(\mathbb{T}) \to \ell_1(\mathbb{Z}^+)$ defined by

$$(T_W(f))(e^{it}) = \sum_{n=0}^{\infty} \widehat{f}(n) e^{2nit} + \sum_{n=1}^{\infty} \widehat{f}(-n) e^{(2n-1)it}$$

is a surjective linear isometry.

Another example, also due to Hatori, shows that a surjective unital isometry between two unital Banach algebras need not be, in general, multiplicative.

EXAMPLE 3.4. ([**57**, Example 3.4] and [**59**, Example 5.3]) Let B denote the unitization of the algebra of all 3×3 upper triangular matrices with the usual matrix product, that is,

$$B = \left\{ \begin{pmatrix} \alpha & a & b \\ 0 & \alpha & c \\ 0 & 0 & \alpha \end{pmatrix} : \alpha, a, b, c \in \mathbb{C} \right\}.$$

Let A denote the algebra obtained from the same linear space of B equipped with the trivial product given by $(\alpha\mathbf{1} + y)(\beta\mathbf{1} + x) = \alpha\beta\mathbf{1} + \alpha x + \beta y$ for all $\alpha, \beta \in \mathbb{C}$ and every pair of upper triangular matrices x, y. Both algebras are equipped with the usual operator norm. Clearly A and B are unital but not semisimple with A being commutative. Thus, the identity mapping $Id : x \mapsto x$ is a unital surjective linear isometry from A to B. It is easy to check that Id maps invertible elements to invertible elements, where $A^{-1} = B^{-1} = \left\{ \begin{pmatrix} \alpha & a & b \\ 0 & \alpha & c \\ 0 & 0 & \alpha \end{pmatrix} : \alpha, a, b, c \in \mathbb{C}, \alpha \neq 0 \right\}$. Actually, A^{-1} and B^{-1} cannot be isomorphic as groups because one is commutative and the other one is not.

We shall see in the subsequent Anti-theorem 4.6 the existence of two unital JB*-algebras which are not Jordan *-isomorphic but share the same underlying Banach space.

The weakest hypotheses that one could consider for Theorem 3.2 in the associative setting are those relaxing the assumptions that A is commutative and semisimple.

THEOREM 3.5 ([**59**]). *Let A and B be unital Banach algebras. Assume that $\mathfrak{A}$ and $\mathfrak{B}$ are open subgroups of A^{-1} and B^{-1}, respectively. Let $\Delta : \mathfrak{A} \to \mathfrak{B}$ be a surjective isometry for the distances given by the norms of A and B, respectively. Then there exist a surjective real-linear isometry $T : A \to B$, and an element u_0 in* *the Jacobson radical of B which satisfy that $\Delta(a) = T(a) + u_0$, for every a in $\mathfrak{A}$. Furthermore, the element u_0 can be obtained as $\lim\limits_{\mathfrak{A}\ni a\to 0} \Delta(a)$.*

Instead of revisiting the original arguments leading to the proofs of Theorems 3.2 and 3.5, we shall obtain them as a consequence of a more general conclusion in the wider setting of Jordan-Banach algebras. First, we shall obtain some corollaries.

It should be observed now that in Theorem 3.5 the hypotheses on the mapping Δ do not require that it preserves the group structures of $\mathfrak{A}$ and $\mathfrak{B}$. If we assume this extra hypothesis the conclusion is stronger.

COROLLARY 3.6 ([**59**]). *Let A and B be unital Banach algebras. Assume that $\mathfrak{A}$ and $\mathfrak{B}$ are open subgroups of A^{-1} and B^{-1}, respectively. Let $\Delta : \mathfrak{A} \to \mathfrak{B}$ be a surjective isometric group isomorphism. Then there exist a real-linear isometric isomorphism $T : A \to B$ satisfying $\Delta(a) = T(a)$, for every a in $\mathfrak{A}$. In particular, A^{-1} and B^{-1} are isometrically isomorphic as metric groups.*

PROOF. By Theorem 3.5 we can find a surjective real-linear isometry $T : A \to B$ and $u_0 = \lim\limits_{\mathfrak{A}\ni a\to 0} \Delta(a) \in \operatorname{rad}(B)$ satisfying $\Delta(a) = T(a) + u_0$ for all $a \in \mathfrak{A}$. By the surjectivity of Δ we can find $a_0 \in \mathfrak{A}$ such that $\Delta(a_0) = 2\,\mathbf{1}_B$. Since $\mathfrak{A}$ is a subgroup and Δ is multiplicative, we have

$$u_0 = \lim_{\mathfrak{A}\ni a\to 0} \Delta(a) = \lim_{\mathfrak{A}\ni a\to 0} \Delta(a_0 a) = \lim_{\mathfrak{A}\ni a\to 0} \Delta(a_0)\Delta(a) = \lim_{\mathfrak{A}\ni a\to 0} 2\,\mathbf{1}_B\, \Delta(a) = 2u_0,$$

which shows that $u_0 = 0$, and hence Δ agrees with T on $\mathfrak{A}$. In particular, $T(\mathbf{1}_A) = \mathbf{1}_B$. Finally, fix $a, b \in A$, for $\delta > 0$ small enough, the elements $\mathbf{1}_A + \delta a$ and $\mathbf{1}_A + \delta b$ lie in $\mathfrak{A}$, and hence

$$\begin{aligned}(\mathbf{1}_B + \delta T(a))(\mathbf{1}_B + \delta T(b)) &= T(\mathbf{1}_A + \delta a)T(\mathbf{1}_A + \delta b) = \Delta(\mathbf{1}_A + \delta a)\Delta(\mathbf{1}_A + \delta b)\\ &= \Delta\left((\mathbf{1}_A + \delta a)(\mathbf{1}_A + \delta b)\right) = T\left((\mathbf{1}_A + \delta a)(\mathbf{1}_A + \delta b)\right)\\ &= \mathbf{1}_B + \delta T(a) + \delta T(b) + \delta^2 T(ab),\end{aligned}$$

implying that $T(a)T(b) = T(ab)$ for all $a, b \in A$, by the arbitrariness of a and b. Therefore, T is a real-linear algebra isometric algebra isomorphism, whose restriction to A^{-1} gives an isometric group isomorphism onto B^{-1}. □

The next corollary is clear.

COROLLARY 3.7 ([**59**]). *Let A and B be unital Banach algebras. The following statements are equivalent:*

(a) *A and B are isometrically isomorphic as real Banach algebras.*
(b) *There exist two open subgroups $\mathfrak{A}$ and $\mathfrak{B}$ of A^{-1} and B^{-1}, respectively, which are isometrically isomorphic as groups.*
(c) *A^{-1} and B^{-1} are isometrically isomorphic as metric groups.*

The last main result in this section is an extension of Theorem 3.5 to the setting of Jordan-Banach algebras. The reader who is familiar with the theory of Jordan-Banach algebras, or has previously taken a look to section 2, will easily appreciate the similarities between the corresponding statements and conclusions up to the necessary adaptations required in the Jordan setting. We result reads as follows (see section 2 for any non-clear notion).

THEOREM 3.8 ([**104**]). *Let M and N be unital Jordan-Banach algebras. Suppose that $\mathfrak{M} \subseteq M^{-1}$ and $\mathfrak{N} \subseteq N^{-1}$ are clopen quadratic subsets which are closed for powers and inverses. Let $\Delta : \mathfrak{M} \to \mathfrak{N}$ be a surjective isometry. Then the following statements hold:*

(a) *For each w_0 in the McCrimmon radical of N the mapping $a \mapsto \Delta(a) - w_0$ is a surjective isometry from $\mathfrak{M}$ to $\mathfrak{N}$.*
(b) *The norm limit $\lim_{\mathfrak{M} \ni a \to 0} \Delta(a)$ exists and defines an element u_0 in the McCrimmon radical of N.*
(c) *There exists a surjective real-linear isometry $T_0 : M \to N$ such that $\Delta(a) = T_0(a) + u_0$, for all $a \in \mathfrak{M}$.*

The proof will be obtained in a series of results in the next subsection 3.1. Before presenting it, we shall see how Theorems 3.2 and 3.5 can be obtained from the previous Theorem 3.8. Theorem 3.5 is easier to obtain since each open subgroup of the invertible elements of a unital Banach algebra A is a clopen, quadratic subset which is closed for powers and inverses (cf. Remark 2.11) and the McCrimmon radical of A regarded as a Jordan-Banach algebra is precisely the Jacobson radical (see Remark 2.7).

The proof of Theorem 3.2 is a bit more laborious. We shall actually prove a bit more general statement. Let us remark that the hypotheses are fully justified in view of Example 3.3.

THEOREM 3.9. *Let A be a semisimple and commutative unital Banach algebra, and let N be a unital Jordan-Banach algebra. Let $\mathfrak{A}$ and $\mathfrak{N}$ be clopen quadratic subsets closed for powers and inverses of A^{-1} and N^{-1}, respectively. Let $\Delta : \mathfrak{A} \to \mathfrak{N}$ be a surjective isometry for the distances given by the norms of A and N, respectively. Then N is semisimple and associative, and the mapping $\Delta(\mathbf{1}_A)^{-1}\Delta$ admits an extension to an isometric real-linear algebra isomorphism from A onto N. In particular, A^{-1} and N^{-1} are isometrically isomorphic as metric groups.*

For the proof we shall need some technical tools. Let us begin with the celebrated Gleason–Kahane–Żelazko theorem.

THEOREM 3.10 ([**100**, Theorem 2.4.13]). *Let A be a complex unital Banach algebra, and let $\varphi : A \to \mathbb{C}$ be a non-zero linear functional. Then the following statements are equivalent:*

(a) *φ is spectral-valued, that is, $\varphi(a) \in \sigma(a)$ for all $a \in A$.*
(b) *φ is unital and maps invertible elements in A to non-zero complex numbers.*
(c) *φ is multiplicative.*

An appropriate version of the Gleason–Kahane–Żelazko theorem for complex Jordan-Banach algebras has been recently established in [**42**] via local theory.

THEOREM 3.11 ([**42**, Theorem 3.2]). *Let M be a complex Jordan-Banach algebra, and let $\varphi : M \to \mathbb{C}$ be a non-zero linear functional. Then the following statements are equivalent:*

(a) *$\varphi(a) \in \mathrm{Sp}(a)$ for all $a \in M$.*
(b) *φ is unital or admits a unital extension to the unitization of M and maps invertible elements in M to non-zero complex numbers.*
(c) *φ is (Jordan-)multiplicative.*

As noted by Hatori in [**57**], the arguments in most of available proofs of the Gleason–Kahane–Żelazko theorem are also valid for linear maps valued in the exponential spectrum [**56**]. We consider this exponential spectrum in the setting of Jordan-Banach algebras.

DEFINITION 3.12. Let a be an element in a unital complex Jordan-Banach algebra M. The exponential spectrum of a is defined as

$$\mathrm{Sp}_{exp}(a) := \{\lambda \in \mathbb{C} : a - \lambda \mathbf{1}_M \notin M_{\mathbf{1}_M}^{-1}\}.$$

Clearly, $\mathrm{Sp}(a) \subseteq \mathrm{Sp}_{exp}(a)$ is a non-empty compact subset of $\mathbb{C}$ bounded by $\|a\|$.

LEMMA 3.13. *Let $\varphi : M \to \mathbb{C}$ be a non-zero linear functional, where M is a complex unital Jordan-Banach algebra. Suppose that $\varphi(a) \in \mathrm{Sp}_{exp}(a)$ for all $a \in M$. Then φ is (Jordan-)multiplicative.*

PROOF. We observe that for each $a \in M$ and $\lambda \in \mathbb{C}$ the element $e^{\lambda a}$ lies in $M_{\mathbf{1}}^{-1}$, and hence $\varphi(a) \in \mathrm{Sp}_{exp}(a) \in \mathbb{C}\backslash\{0\}$. We also know that $\varphi(\mathbf{1}_M) \subseteq \mathrm{Sp}_{exp}(\mathbf{1}_M) = \{1\}$. Therefore, the mapping $\varphi : \mathbb{C} \to \mathbb{C}$, $f(\lambda) = \varphi(e^{\lambda a})$ never vanishes and $f(0) = 1$. We can now follow the standard reasoning. Having in mind that φ is linear and continuous we deduce that $f(\lambda) = \sum\limits_{n=0}^{\infty} \frac{\varphi(a^n)}{n!}\lambda^n$ for all $\lambda \in \mathbb{C}$. Clearly, f is an entire function that never vanishes, and hence it admits an entire "logarithm", that is, there exists an entire function $g : \mathbb{C} \to \mathbb{C}$ satisfying $f(\lambda) = e^{g(\lambda)}$ $(\lambda \in \mathbb{C})$. The inequality

$$|f(\lambda)| \leq \sum_{n=0}^{\infty} \frac{|\varphi(a^n)|}{n!}|\lambda|^n \leq \|\varphi\|\ e^{|\lambda|\ \|a\|},\ (\lambda \in \mathbb{C})$$

implies that f has exponential type bounded by one. It follows from Hadamard's Factorization theorem (see [**14**, Theorem 2.7.1]) that $g(\lambda) = \alpha\lambda + \beta$ for suitable complex numbers α and β. By applying that $f(0) = 1$, we get $\beta = 0$. Consequently,

$$\sum_{n=0}^{\infty} \frac{\varphi(a^n)}{n!}\lambda^n = f(\lambda) = e^{\alpha\lambda} = \sum_{n=0}^{\infty} \frac{\alpha^n}{n!}\lambda^n, \text{ for all } \lambda \in \mathbb{C},$$

which proves that $\varphi(a^n) = \alpha^n$ for all natural n, and thus $\varphi(a) = \alpha$ and $\varphi(a^2) = \alpha^2 = \varphi(a)^2$. The final conclusion follows from the identity $2a \circ b = (a+b)^2 - a^2 - b^2$ and the linearity of φ. □

The next result was established by Hatori in the case of Banach algebras. The idea comes from a result by Kowalski and Słodkowski (cf. [**79**, Lemma 2.1]).

LEMMA 3.14. *Let $\varphi : M \to \mathbb{C}$ be a non-zero real-linear functional, where M is a complex unital Jordan-Banach algebra. Suppose that $\varphi(a) \in \mathrm{Sp}_{exp}(a)$ for all $a \in M$. Then φ is complex linear and (Jordan-)multiplicative.*

PROOF. The desired conclusion can be obtained by fixing an element a in M and considering the Jordan-Banach subalgebra M_a generated by a and $\mathbf{1}_M$, which is a commutative unital complex Banach algebra. Since $(M_a)_{\mathbf{1}}^{-1} \subseteq M_{\mathbf{1}}^{-1}$, and hence

$$\mathrm{Sp}_{exp}(M, b) \subseteq \mathrm{Sp}_{exp}(M_a, b),$$

for every $b \in M_a$, we can apply [**57**, Lemma 2.2] to the restricted mapping $\varphi|_{M_a} : M_a \to \mathbb{C}$. For the sake of self-containment, we revisit all details.

As in the proof of [**79**, Lemma 2.1] and [**57**, Lemma 2.2] (see also [**85**, Lemma 3.3]), the functionals

$$\varphi_1(x) := \Re\mathrm{e}\varphi(x) + i\Im\mathrm{m}(-i\varphi(ix)) = \Re\mathrm{e}\varphi(x) - i\Re\mathrm{e}\varphi(ix)$$
$$\varphi_2(x) := \Re\mathrm{e}(-i\varphi(ix)) + i\Im\mathrm{m}\varphi(x) = \Im\mathrm{m}\varphi(ix) + i\Im\mathrm{m}\varphi(x)$$

are continuous, complex linear, $\Re\mathrm{e}(\varphi(x)) = \Re\mathrm{e}(\varphi_1(x))$ and $\Im\mathrm{m}(\varphi(x)) = \Im\mathrm{m}(\varphi_2(x))$ for all $x \in M$, and $\varphi_j(x) \in \mathrm{Sp}_{exp}(x)$, for every $x \in M$, $j = 1, 2$. By Lemma 3.13, φ_1 and φ_2 are Jordan multiplicative functionals on M. The desired conclusion will follow as soon as we prove that $\varphi_1 = \varphi_2$. If that is not the case, there exists $a \in M$ with $\varphi_1(a) = 1$ and $\varphi_2(a) = 0$. The element $b = \exp(\frac{\pi i a}{2}) - \mathbf{1}_M \in M$ satisfies that $b + \mathbf{1}_M = \exp(\frac{\pi i a}{2}) \in M_{\mathbf{1}}^{-1}$, and thus $-1 \notin \mathrm{Sp}_{exp}(b)$. On the other hand by applying that φ_1 and φ_2 are continuous homomorphisms we deduce that $\varphi_1(b) = i - 1$, $\varphi_2(b) = 0$, and $\varphi(b) = -1$. The hypotheses on φ imply that $-1 \in \mathrm{Sp}_{exp}(b)$, which is impossible. □

We can now present the proof of Theorem 3.9 which follows the lines in [**57**, Theorem 3.3]. The proof is a bit technical, but perhaps the reader will appreciate a complete access to all details.

Proof of Theorem 3.9. Let $\Delta : \mathfrak{A} \to \mathfrak{N}$ be a surjective isometry satisfying the hypotheses of the theorem. We observe that under these assumptions $\mathfrak{A} \supseteq A_{\mathbf{1}}^{-1}$ and $\mathfrak{N} \supseteq N_{\mathbf{1}}^{-1}$. By Theorem 3.8 applied to Δ^{-1}, there exists a real-linear surjective isometry $T : N \to A$ such that $\Delta^{-1} = T|_{\mathfrak{N}}$. We need to prove that T is a Jordan homomorphism. Let Ω_A denote the character space of A, and let $G_A : A \to C(\Omega_A)$ be the Gelfand transform, which is an isomorphism of Banach algebras.

In a first step we shall prove that

$$|G_A(T(\lambda\mathbf{1}_N))(t)| = 1, \text{ for all } \lambda \in \mathbb{T}, t \in \Omega_A. \tag{3.1}$$

Clearly, $|G_A(T(\lambda\mathbf{1}_N))(t)| \leq \|G_A(T(\lambda\mathbf{1}_N))\| \leq \|T(\lambda\mathbf{1}_N)\| = \|\lambda\mathbf{1}_N\| = 1$. If there exists $t \in \Omega_A$ with $|G_A(T(\lambda\mathbf{1}_N))(t)| < 1$, for $\alpha = G_A(T(\lambda\mathbf{1}_N))(t)$ we have

$$1 > |\alpha| = \|T(\lambda\mathbf{1}_N) - (T(\lambda\mathbf{1}_N) - \alpha\mathbf{1}_A)\| = \|\lambda\mathbf{1}_N - T^{-1}(T(\lambda\mathbf{1}_N) - \alpha\mathbf{1}_A)\|,$$

and consequently $T^{-1}(T(\lambda\mathbf{1}_N) - \alpha\mathbf{1}_A) \in N_{\mathbf{1}}^{-1} \subseteq \mathfrak{N}$. Since $T(\mathfrak{N}) = \Delta^{-1}(\mathfrak{N}) = \mathfrak{A} \subseteq A^{-1}$, we deduce that $T(\lambda\mathbf{1}_N) - \alpha\mathbf{1}_A \in \mathfrak{A} \subseteq A^{-1}$, which is impossible since

$$G_A\left(T(\lambda\mathbf{1}_N) - \alpha\mathbf{1}_A\right)(t) = G_A\left(T(\lambda\mathbf{1}_N)\right)(t) - G_A\left(T(\lambda\mathbf{1}_N)\right)(t) = 0.$$

This finishes the proof of (3.1).

Define now a new real-linear bijection $S : N \to A$ given by

$$S(b) := T(\mathbf{1}_N)^{-1}T(b) = \left(\Delta^{-1}(\mathbf{1}_N)\right)^{-1} T(b).$$

Clearly $S(\mathbf{1}_N) = \mathbf{1}_A$ and $\|b\| = \|T(b)\| \leq \|T(\mathbf{1}_N)\| \, \|T(\mathbf{1}_N)^{-1}T(b)\| = \|S(b)\|$. Furthermore

$$S(\mathfrak{N}) = T(\mathbf{1}_N)^{-1}T(\mathfrak{N}) = T(\mathbf{1}_N)^{-1}\mathfrak{A} \subseteq A^{-1}. \tag{3.2}$$

Next we show that

$$G_A(S(i\,\mathbf{1}_N))(t) \in \{\pm i\}, \text{ for all } t \in \Omega_A. \tag{3.3}$$

This will be simpler from (3.1), since by this result we get

$$\begin{aligned}|1 \pm G_A(S(i\mathbf{1}_N))(t)| &= |G_A(T(\mathbf{1}_N))(t) \pm G_A(T(i\mathbf{1}_N))(t)| \\ &\leq \|T(\mathbf{1}_N \pm i\mathbf{1}_N)\| = \|(1+i)\mathbf{1}_N\| = \sqrt{2}.\end{aligned}$$

Having in mind that $G_A(S(i\,\mathbf{1}_{_N}))(t) \in \mathbb{T}$, we arrive to the desired conclusion in (3.3).

Therefore, the set Ω_A decomposes as the disjoint union of the (possibly empty) clopen sets

$$\Omega_A^+ = \{t \in \Omega_A : G_A(S(i\,\mathbf{1}_{_N}))(t) = i\} \text{ and } \Omega_A^- = \{t \in \Omega_A : G_A(S(i\,\mathbf{1}_{_N}))(t) = -i\}.$$

We shall next prove the following statement:

(3.4) For each $t \in \Omega_A^+$ the mapping $\varphi_t := \delta_t \circ G_A \circ S : N \to \mathbb{C}$ is a real-linear functional with $\varphi_t(b) \in \mathrm{Sp}_{exp}(b)$ for each element $b \in N$, and hence it is a complex linear Jordan homomorphism by Lemma 3.14.

To get the desired conclusion, let $\mu = \varphi_t(b)$ for some $b \in N$. Clearly, $G_A(S(b)-\mu\mathbf{1}_{_A})(t) = 0$. If $\mu \notin \mathrm{Sp}_{exp}(N, b)$ the element $b-\mu\mathbf{1}_{_N} \in N_{\mathbf{1}}^{-1} \subseteq \mathfrak{N}$. Having in mind (3.2), the previous properties imply that $G_A(S(b - \mu\mathbf{1}_{_N})) \in G_A(S(\mathfrak{N}) = G_A(T(\mathbf{1}_{_N})^{-1}\mathfrak{A}) \subseteq G_A(A^{-1}) \subseteq C(\Omega)^{-1}$. However, since $t \in \Omega_A^+$ we get

$$\begin{aligned} G_A(S(b - \mu\mathbf{1}_{_N}))(t) &= G_A(S(b - \Re\mathrm{e}(\mu)\mathbf{1}_{_N} - \Im\mathrm{m}(\mu)i\mathbf{1}_{_N}))(t) \\ &= (G_A(S(b)) - \Re\mathrm{e}(\mu)G_A(S(\mathbf{1}_{_N})) - \Im\mathrm{m}(\mu)iG_A(S(\mathbf{1}_{_N})))(t) \\ &= (G_A(S(b)) - \mu G_A(S(\mathbf{1}_{_N})))(t) = G_A(S(b) - \mu\mathbf{1}_{_A})(t), \end{aligned}$$

contradicting that $G_A(S(b) - \mu\mathbf{1}_{_A})(t) = 0$.

Similar arguments to those given in the proof of (3.4) lead to

(3.5) For each $t \in \Omega_A^-$ the mapping $\varphi_t := \overline{\delta_t \circ G_A \circ S} : N \to \mathbb{C}$ is a real-linear functional with $\varphi_t(b) \in \mathrm{Sp}_{exp}(b)$ for each element $b \in N$, and hence it is a complex linear Jordan homomorphism by Lemma 3.14.

Now, by combining (3.4) and (3.5) and the fact that G_A is a Jordan isomorphism, we deduce that S is a Jordan isomorphism. This implies that N is a semisimple associative unital complex Jordan-Banach algebra (i.e. a semisimple commutative unital Banach algebra), and S is an expansive isomorphism of Banach algebras.

Finally, having in mind that N is a unital commutative semsimple complex Banach algebra, the arguments in the first part also assure that the mapping $R = \Delta(\mathbf{1}_{_A})^{-1}T^{-1} = \left(T^{-1}(\mathbf{1}_{_A})\right)^{-1}T^{-1} : A \to N$ is an isomorphism of Banach algebras and expansive (that is, $\|R(a)\| \geq \|a\|$ for all $a \in A$). It is not hard to check that $R = S^{-1}$. Namely,

$$\begin{aligned} R(S(b)) &= R\left(\left(\Delta^{-1}(\mathbf{1}_{_N})\right)^{-1}T(b)\right) = R\left(\left(\Delta^{-1}(\mathbf{1}_{_N})\right)^{-1}\right)R(T(b)) \\ &= R\left(\Delta^{-1}(\mathbf{1}_{_N})\right)^{-1}R(T(b)) \\ &= \left[\Delta(\mathbf{1}_{_A})^{-1}T^{-1}T(\mathbf{1}_{_N})\right]^{-1}\Delta(\mathbf{1}_{_A})^{-1}T^{-1}T(b) = b, \end{aligned}$$

for all $b \in N$, and similarly $S(R(a)) = a$, for all $a \in A$. All together gives $\|b\| \leq \|S(b)\| \leq \|R(S(b))\| = \|b\|$, for all b in N, which proves that S is an isometry. $\square$

The next result is essentially due to Hatori (see [**59**, Corollary 5.2]), here we state it in a wider setting.

COROLLARY 3.15. *Let A be a unital and commutative Banach algebra and let N be a semisimple unital Jordan-Banach algebra. Suppose that $\mathfrak{A} \subseteq A^{-1}$ and $\mathfrak{N} \subseteq N^{-1}$ are clopen quadratic subsets which are closed for powers and inverses. Let $\Delta : \mathfrak{A} \to \mathfrak{N}$ be a surjective isometry. Then A is semisimple and N is associative, and the mapping $\Delta(\mathbf{1}_A)^{-1}\Delta$ admits an extension to an isometric real-linear algebra isomorphism from A onto N. In particular, A^{-1} and N^{-1} are isometrically isomorphic as metric groups.*

PROOF. The desired conclusion will follow from Theorem 3.9 as soon as we prove that A is semisimple. Suppose, on the contrary that there exists $u_0 \in \mathcal{R}ad(A)\backslash\{0\}$.

Since N is semisimple its McCrimmon radical reduces to $\{0\}$. Then an application of Theorem 3.8 gives the existence of a surjective real-linear isometry $T : A \to N$ such that $\Delta = T|_{\mathfrak{A}}$. By Lemma 2.18, and the hypotheses, the mapping $a \mapsto \Delta(a - u_0)$ is a surjective isometry from $\mathfrak{A}$ onto $\mathfrak{N}$. A new application of Theorem 3.8 guarantees the existence of a surjective real-linear isometry $T_{u_0} : A \to N$ such that $\Delta(a - u_0) = T_{u_0}(a)$ for all $a \in \mathfrak{A}$. Having in mind that $\mathbf{1}_A \in A_{\mathbf{1}}^{-1} \subseteq \mathfrak{A}$, we have

$$T_{u_0}\left(\frac{1}{n}\mathbf{1}_A\right) = \Delta\left(\frac{1}{n}\mathbf{1}_A - u_0\right) = T\left(\frac{1}{n}\mathbf{1}_A - u_0\right), \text{ for all natural } n.$$

The continuity of the surjective real-linear isometries T and T_{u_0} assures that $T(-u_0) = 0$, and thus $u_0 = 0$, which is impossible. □

The set of invertible elements in a unital complex Jordan-Banach algebra is not stable under Jordan products. So, there is no hope to expect a result in the line of Corollary 3.6, instead of that we can consider preservers of expressions of the form $U_a(b)$.

PROPOSITION 3.16 ([**104**, Proposition 3.9]). *Let M and N be unital Jordan-Banach algebras. Suppose that $\mathfrak{M} \subseteq M^{-1}$ and $\mathfrak{N} \subseteq N^{-1}$ are clopen quadratic subsets which are closed for powers and inverses. Let $\Delta : \mathfrak{M} \to \mathfrak{N}$ be a surjective isometry and set $u = \Delta(\mathbf{1}_M)$. We shall also assume that Δ satisfies the following property:*

$$\Delta(U_a(b)) = U_{\Delta(a)}(\Delta(b)), \text{ for all } a, b \in \mathfrak{M}. \tag{3.6}$$

Then there exists a real-linear isometric Jordan isomorphism J from M onto the u-isotope $N_{(u)}$ such that $\Delta(a) = J(a)$ for all $a \in \mathfrak{M}$.

PROOF. Recall that $N_{(u)}^{-1} = N^{-1}$ (cf. Remark 2.13). Then $\mathfrak{N} \subseteq N_{(u)}^{-1}$. Let us prove that $\mathfrak{N}$ satisfies the same hypotheses as a subset of $N_{(u)}^{-1}$ in the u-homotope. Indeed, let us take $a, b \in \mathfrak{N}$. It is known that $U_{u^{-1}}(a^{-1})$ is the inverse of a in the u-homotope N_u (see Remark 2.13) and belongs to $\mathfrak{N}$ by the hypotheses. Furthermore, by definition $U_a^{(u)}(b) = U_a U_u(b) \in \mathfrak{N}$, because, by the hypotheses, $U_u(b) \in \mathfrak{N}$. Furthermore $a \circ_u a = U_a(u) \in \mathfrak{N}$ by assumptions, and a simple induction argument shows that $\mathfrak{N}$ is closed for powers in $N_{(u)}$.

We shall next show that Δ also satisfies property (3.6) for the Jordan product in the u-homotope. Namely, by the hypotheses,

$$\begin{aligned} U_{\Delta(a)}^{(u)}(\Delta(b)) &= U_{\Delta(a)}U_{\Delta(1)}(\Delta(b)) = U_{\Delta(a)}\Delta\left(U_1(b)\right) \\ &= U_{\Delta(a)}\Delta\left(b\right) = \Delta\left(U_a(b)\right), \end{aligned} \tag{3.7}$$

for all $a, b \in \mathfrak{M}$, which proves the claim. Therefore $\Delta : \mathfrak{M} \subseteq M^{-1} \to \mathfrak{N} \subseteq (N_{(u)})^{-1}$ is a unital mapping satisfying the same hypotheses in our statement. By applying Theorem 3.8(c) to the latter mapping, we deduce the existence of a surjective real-linear isometry $T_0 : M \to N$ and an element u_0 in the McCrimmon radical of $N_{(u)}$ such that $\Delta(a) = T_0(a) + u_0$ for all $a \in \mathfrak{M}$.

Pick $a \in \mathfrak{M}$ with $\Delta(a) = 2u$. Theorem 3.8(b) and (3.6) imply that for each norm-null sequence $(a_n)_n$ in $\mathfrak{M}$ we have

$$u_0 = \lim_n \Delta\left(U_a(a_n)\right) = \lim_n U^{(u)}_{\Delta(a)}(\Delta(a_n)) = \lim_n U^{(u)}_{2u}(\Delta(a_n)) = \lim_n 2\Delta(a_n) = 2u_0,$$

which assures that $u_0 = 0$. Therefore $\Delta(a) = T_0(a)$ for all $a \in \mathfrak{M}$.

It is now time to explore the properties of the real-linear isometry $T_0 : M \to N_u$. Fix an arbitrary $a \in M$. Since for $r \in \mathbb{R}$ with $|r|$ large enough we have $a + r\,\mathbf{1}_M \in M_{\mathbf{1}}^{-1} \subseteq \mathfrak{M}$, we deduce from (3.6) and the conclusions in the previous paragraphs that

$$\begin{aligned} U^{(u)}_{T_0(a+r\mathbf{1}_M)}(T_0(a+r\mathbf{1}_M)) &= U^{(u)}_{\Delta(a+r\mathbf{1}_M)}(\Delta(a+r\mathbf{1}_M)) = \Delta\left(U_{a+r\mathbf{1}_M}(a+r\mathbf{1}_M)\right) \\ &= T_0\left(U_{a+r\mathbf{1}_M}(a+r\mathbf{1}_M)\right). \end{aligned} \tag{3.8}$$

By expanding the extreme terms in the previous identity we get

$$\begin{aligned} U^{(u)}_{T_0(a+r\mathbf{1}_M)}(T_0(a+r\mathbf{1})) &= U^{(u)}_{T_0(a)+ru}(T_0(a)+ru) = U^{(u)}_{T_0(a)}(T_0(a)) + rU^{(u)}_{T_0(a)}(u) \\ &\quad + r^2T_0(a) + r^3u + 2rU^{(u)}_{T_0(a),u}(T_0(a)) + 2r^2U^{(u)}_{T_0(a),u}(u) \\ &= U^{(u)}_{T_0(a)}(T_0(a)) + 3rT_0(a)\circ_u T_0(a) + 3r^2T_0(a) + r^3u \end{aligned}$$

and

$$U_{a+r\mathbf{1}_M}(a+r\mathbf{1}_M) = U_a(a) + 3ra^2 + 3r^2a + r^3\mathbf{1}_M,$$

which combined with (3.8) lead to

$$\begin{aligned} &U^{(u)}_{T_0(a)}(T_0(a)) + 3rT_0(a)\circ_u T_0(a) + 3r^2T_0(a) + r^3u \\ &= T_0(U_a(a)) + 3rT_0(a\circ a) + 3r^2T_0(a) + r^3u, \end{aligned}$$

for all $r \in \mathbb{R}$ with $|r|$ big enough. Replacing r with $-r$ and subtracting we get $T_0(a)\circ_u T_0(a) = T_0(a\circ a)$ for all $a \in M$. It is standard to check that T_0 is a Jordan isomorphism from M onto $N_{(u)}$. □

The last result in this section is a clear consequence of the previous proposition.

Corollary 3.17. *Let M and N be unital complex Jordan-Banach algebras. The following statements are equivalent:*

(a) *There exists an isometric real-linear Jordan isomorphism from M onto N.*
(b) *There exist two clopen quadratic subsets $\mathfrak{M} \subseteq M^{-1}$ and $\mathfrak{N} \subseteq N^{-1}$ which are closed for powers and inverses and a surjective isometry $\Delta : \mathfrak{M} \to \mathfrak{N}$ satisfying $\Delta(\mathbf{1}_M) = \mathbf{1}_N$ and $\Delta(U_a(b)) = U_{\Delta(a)}(\Delta(b))$, for all $a, b \in \mathfrak{M}$.*
(c) *There exists a surjective isometry $\Delta : M^{-1} \to N^{-1}$ satisfying $\Delta(\mathbf{1}_M) = \mathbf{1}_N$ and $\Delta(U_a(b)) = U_{\Delta(a)}(\Delta(b))$, for all $a, b \in M^{-1}$.*

3.1. Proof of Theorem 3.8.

The proof presented in this section relies, among other tools, on a powerful result established by Mankiewicz in [**87**] on the affine behaviour of certain surjective isometries between convex sets. All the arguments have been borrowed from [**104**]. We recall that a mapping F from a convex subset $\mathcal{C}$ of a Banach space X to another Banach space Y is said to be *real affine*, or simply *affine*, if for each $t \in [0,1]$, and $x, y \in \mathcal{C}$ we have

$$F(tx + (1-t)y) = tF(x) + (1-t)F(y).$$

A *convex body* in a normed space is a closed convex set with non-empty interior.

The mentioned result by Mankiewicz, which is generalization of the celebrated Mazur–Ulam theorem, is the following:

THEOREM 3.18 ([**87**, Theorem 5]). *Every surjective isometry between convex bodies in two arbitrary normed spaces can be uniquely extended to an affine function between the spaces.*

It is worth noting that the above theorem by Mankiewicz motivated the introduction of the so-called strong Mankiewicz property in [**94**], a property which plays a fundamental role in the positive solution to Tingley's problem on the extension of surjective isometries from the unit sphere of a unital C*-algebra (respectively, a JBW*-triple, a compact C*-algebra and unital JB*-algebra) onto the unit sphere of another Banach space [**94**] (respectively, [**12, 75, 105, 107**]). We shall see further applications in a subsequent section.

In the rest of this section M, N, $\mathfrak{M}$, $\mathfrak{N}$, and Δ will have the meaning in the hypotheses of Theorem 3.8. Our arguments start with an observation on the local affine behaviour of Δ.

LEMMA 3.19 ([**104**]). *Let $\Delta : \mathfrak{M} \to \mathfrak{N}$ be a surjective isometry. Then Δ is a local affine mapping, concretely, for each $a \in \mathfrak{M}$ there exists a positive δ_a, depending on a, such that $B(a, \delta_a) \subseteq \mathfrak{M}$, $B(\Delta(a), \delta_a) \subseteq \mathfrak{N}$, $\Delta(B(a, \delta_a)) = B(\Delta(a), \delta_a)$ and there exists a surjective affine isometry $F_{a,\delta_a} : M \to N$ such that $\Delta|_{B(a,\delta_a)} = F_{a,\delta_a}|_{B(a,\delta_a)}$ (and hence $\Delta|_{B(a,\delta_a)}$ is an affine mapping). Furthermore, suppose that $\gamma : [0,1] \to \mathfrak{M}$ is a continuous path. Then there exists a surjective affine isometry $F : M \to N$ and an open neighbourhood U of $\gamma([0,1])$ such that $U \subseteq \mathfrak{M}$ and $F|_U = \Delta$.*

PROOF. Since Δ is a surjective isometry, for each $\delta > 0$, $a \in \mathfrak{M}$, we have

$$\Delta\left(B(a, \delta) \cap \mathfrak{M}\right) = B(\Delta(a), \delta) \cap \mathfrak{N}.$$

Since $\mathfrak{M}$ and $\mathfrak{N}$ are open we can find $\delta_a > 0$ satisfying $B(a, \delta_a) \subset \mathfrak{M}$, $B(\Delta(a), \delta_a) \subset \mathfrak{N}$ and $\Delta(B(a, \delta_a)) \subseteq B(\Delta(a), \delta_a)$. Clearly, the restricted mapping

$$\Delta|_{B(a,\delta_a)} : B(a, \delta_a) \to B(\Delta(a), \delta_a)$$

is a surjective isometry. The existence of a surjective affine isometry $F_{a,\delta_a} : M \to N$ extending the mapping $\Delta|_{B(a,\delta_a)}$ (and thus proving that the latter is an affine mapping) follows from Mankiewicz theorem (Theorem 3.18).

In order to prove the final statement we employ an argument based on compactness. Suppose $\gamma : [0,1] \to \mathfrak{M}$ is continuous path. By the conclusion in the first part, for each $t \in [0,1]$, there exist $\delta_t > 0$ and a surjective affine isometry $F_t : M \to N$

satisfying $B(\gamma(t),\delta_t) \subseteq \mathfrak{M}$, $B(\Delta(\gamma(t)),\delta_t) \subseteq \mathfrak{N}$, $\Delta(B(\gamma(t),\delta_t)) = B(\Delta(\gamma(t)),\delta_t)$, and $\Delta|_{B(\gamma(t),\delta_t)} = F_t|_{B(\gamma(t),\delta_t)}$.

By applying that $\gamma([0,1])$ is compact we deduce the existence of $t_0 = 0 < t_1 < \ldots < t_{n-1} < t_n = 1$, such that $\gamma([0,1]) \subseteq \bigcup_{k=0}^{n} B(\gamma(t_k),\delta_{t_k})$. The existence of the surjective affine isometry F will be clear if we show that the corresponding maps F_{t_j} satisfy $F_{t_k} = F_{t_l}$ for all $k,l \in \{0,\ldots,n\}$. A very basic identity principle assures that two surjective affine isometries between normed spaces are the same if and only if they coincide on an open subset. So, if the open set $B(\gamma(t_k),\delta_{t_k}) \cap B(\gamma(t_l),\delta_{t_l})$ is non-empty we clearly have $F_{t_k} = F_{t_l}$, otherwise, by the connectedness of $\gamma([0,1])$, we can find a finite collection $t_k = t_{j_1},\ldots,t_{j_m} = t_l \in \{t_0 = 0, t_1,\ldots,t_{n-1},t_n\}$ such that

$$B(\gamma(t_{j_k}),\delta_{t_{j_k}}) \cap B(\gamma(t_{j_{k+1}}),\delta_{t_{j_{k+1}}}) \neq \emptyset,$$

for all $k \in \{1,\ldots,m-1\}$. It then follows that $F_{t_k} = F_{t_{j_1}} = F_{t_{j_2}} = \ldots = F_{t_{j_m}} = F_{t_l}$, which gives the desired statement. To conclude the proof it suffices to take $F = F_{t_0}$ and the open neighbourhood of $\gamma([0,1])$ given by $U = \bigcup_{k=0}^{n} B(\gamma(t_k),\delta_{t_k})$. □

The next tool, probably known in the folklore of affine maps, was designed ad hoc to play a specific role in the proof of Theorem 3.8. Here we present a more general version with a simplified proof.

LEMMA 3.20. *Let X and Y be real normed spaces with dim$(Y) \geq 2$. Suppose $G : Y \to X$ is a continuous mapping satisfying:*

(*a*) $G(0) = 0$;

(*b*) *The restriction of G to each segment not containing zero is an affine map.*

Then G is real-linear.

PROOF. It suffices to show that G is affine on each segment containing 0. If $0 \in [x,y]$, the vectors x,y are linearly dependent. We may assume that $x \neq 0$. Take $z \in Y$ such that $\{x,z\}$ is a linearly independent set. Clearly $0 \notin [x,y] + \varepsilon z = [x+\varepsilon z, y+\varepsilon z]$ for all $\varepsilon > 0$. By hypothesis (*b*), the restriction of G to $[x+\varepsilon z, y+\varepsilon z]$ is an affine map, meaning that

$$G(t(x+\varepsilon z) + (1-t)(y+\varepsilon z)) = tG((x+\varepsilon z)) + (1-t)G(y+\varepsilon z)$$

for all $t \in [0,1]$. Letting $\varepsilon \to 0$ we deduce from the continuity of G that

$$G(t(x) + (1-t)(y)) = tG(x) + (1-t)G(y),$$

for all $t \in [0,1]$. □

Let us observe that the conclusion of the above lemma is not true when $\dim(Y) = 1$, the absolute value on $\mathbb{R}$ is a counterexample.

Before entering into the details of the proof of Theorem 3.8 we recall the definition and some basic facts on numerical ranges. By a *numerical range space* we mean a (real or complex) Banach space X with a distinguished element u in its unit sphere. The *state space* of the numerical range space (X,u) is the defined as the set

$$D(X) = D(X,u) = \{\phi \in X^* : \|\phi\| = \phi(u) = 1\},$$

and the *numerical range* of an element $x \in X$ is the non-empty compact and convex set given by

$$V(X,x) = V(x) = \{\phi(x) : \phi \in D(X,u)\}.$$

Finally, the *numerical radius* of $x \in X$ is the number given by

$$v(x) = \max\{|\lambda| : \lambda \in V(x)\},$$

while the *numerical index* of X is defined by

$$\begin{aligned} n(X,u) = n(X) &= \inf\{v(x) : x \in X,\ \|x\| = 1\} \\ &= \max\{\alpha \geq 0 : \alpha\|x\| \leq v(x) \text{ for all } x \in X\}. \end{aligned}$$

It may occur that $n(X,u) = 0$. Those elements u in the unit sphere of X satisfying $n(X,u) > 0$ are called *geometrically unitaries*. The Bohnenblust–Karlin theorem is a milestone result in the theory of complex Banach algebras.

THEOREM 3.21 (Bohnenblust–Karlin theorem ([**100**, Theorem 2.6.4] and [**22**, Proposition 2.1.11])). *Let A be a unital Banach algebra with unit $\mathbf{1}_A$ (and $\|\mathbf{1}_A\| = 1$). Then the numerical radius of the numerical range space $(A, \mathbf{1}_A)$ is a norm on A which is equivalent to the original norm of A. Moreover, $n(A, \mathbf{1}_A) \geq \frac{1}{e}$ and thus $v(a) \leq \|a\| \leq e\, v(a)$ for all $a \in A$. Furthermore, the same conclusion holds when A is replaced with a unital (non-necessarily associative) normed complex algebra M with $\|\mathbf{1}_A\| = 1$.*

We proceed next to present the promised proof of Theorem 3.8.

PROOF OF THEOREM 3.8. (a) Let w_0 be an element in the McCrimmon radical of N. Lemma 2.18 proves that $\mathfrak{N} - w_0 = \mathfrak{N}$, and hence $a \mapsto \Delta(a) - w_0$ is a surjective isometry from $\mathfrak{M}$ onto $\mathfrak{N}$.

(b) Observe first that 0 always lies in the norm adherence of $\mathfrak{M}$. Take any sequence $(a_n)_n \subseteq \mathfrak{M}$ converging to 0 in norm. Since Δ is an isometry, the sequence $(\Delta(a_n))_n$ is Cauchy in N, and hence there exists $u_0^a \in N$, depending on $(a_n)_n$, such that $(\Delta(a_n))_n \to u_0^a$. Actually the limit u_0^a does not depend on the chosen sequence. Namely, if $(b_n)_n \subseteq \mathfrak{M}$ with $(\Delta(b_n))_n \to u_0^b$, by the assumptions on Δ, we have

$$\begin{aligned} \|u_0^a - u_0^b\| &\leq \|u_0^a - \Delta(a_n)\| + \|\Delta(b_n) - \Delta(a_n)\| + \|\Delta(b_n) - u_0^b\| \\ &= \|u_0^a - \Delta(a_n)\| + \|b_n - a_n\| + \|\Delta(b_n) - u_0^b\|, \end{aligned}$$

which implies that $u_0^a = u_0^b$ because the sequences on the right hand side of the above inequality tend to zero. We have therefore proven the existence of the limit $\lim\limits_{\mathfrak{M} \ni a \to 0} \Delta(a) = u_0 \in N$.

We shall finally show that u_0 lies in the McCrimmon radical of N. The key result is the characterization given in Proposition 2.9. By the just quoted result, it suffices to show that $Sp(U_b(u_0)) = \{0\}$ for every $b \in N_{\mathbf{1}}^{-1} \subseteq \mathfrak{N}$ (observe that under our assumptions $\mathfrak{M}$ must coincide with the union of several connected components of M^{-1} including $M_{\mathbf{1}}^{-1}$ among them).

Arguing by contradiction, we assume the existence of $b \in N_{\mathbf{1}}^{-1} \subseteq \mathfrak{N}$ and a non-zero $\lambda \in Sp(U_b(u_0))$. By applying that $\mathfrak{N}$ is a quadratic subset of N^{-1} with $\mathbf{1}_N \in \mathfrak{N}$, we have $\alpha^2 c = U_{\alpha \mathbf{1}_N}(c) \in \mathfrak{N}$ for all $\alpha \in \mathbb{C}\backslash\{0\}$, $c \in \mathfrak{N}$, that is, $(\mathbb{C}\backslash\{0\})\,\mathfrak{N} = \mathfrak{N}$. Now, having in mind that $\mathfrak{N}$ is closed for powers and inverses, and the previous comment, we get $-\lambda b^{-2} \in \mathfrak{N} = \Delta(\mathfrak{M})$. Therefore, there exists (a unique) $c_\lambda \in \mathfrak{M}$

such that $\Delta(c_\lambda) = -\lambda b^{-2}$. Actually, by Lemma 2.15 and Remark 2.16 we know that $-\lambda b^{-2} \in N_{\mathbf{1}}^{-1}$.

By the arguments above, for each $0 < s < \frac{1}{2}$, the segment $[sc_\lambda, (1-s)c_\lambda]$ lies in $\mathfrak{M}$. By Lemma 3.19 we can find an open neighbourhood U of $[sc_\lambda, (1-s)c_\lambda]$ with $U \subseteq \mathfrak{M}$ and a surjective affine isometry $F : M \to N$ such that $\Delta|_U = F|_U$. The properties of these maps assure that

$$
\begin{aligned}
\Delta\left(\frac{c_\lambda}{2}\right) &= F\left(\frac{1}{2}sc_\lambda + \frac{1}{2}(1-s)c_\lambda\right) = \frac{F((1-s)c_\lambda) + F(sc_\lambda)}{2} \\
&= \frac{\Delta((1-s)c_\lambda) + \Delta(sc_\lambda)}{2}.
\end{aligned} \tag{3.9}
$$

Having in mind that $\lim\limits_{\mathfrak{M}\ni a\to 0} \Delta(a) = u_0$ and the continuity of Δ, we allow $s \to 0$ in (3.9) to deduce that

$$
\Delta\left(\frac{c_\lambda}{2}\right) = \frac{\Delta(c_\lambda) + u_0}{2} = \frac{-\lambda b^{-2} + u_0}{2},
$$

and consequently,

$$
-\lambda \mathbf{1}_N + U_b(u_0) = -\lambda U_b(b^{-2}) + U_b(u_0) = U_b\left(-\lambda b^{-2} + u_0\right) = 2U_b\left(\Delta\left(\frac{c_\lambda}{2}\right)\right). \tag{3.10}
$$

Since b and $\Delta\left(\frac{c_\lambda}{2}\right)$ belong to $\mathfrak{N}$, and the latter is a quadratic subset of N^{-1}, the element $2U_b\left(\Delta\left(\frac{c_\lambda}{2}\right)\right)$ lies in $U_{\mathfrak{N}}(\mathfrak{N}) \subseteq N^{-1}$. The identity in (3.10) implies that $\lambda \notin Sp(U_b(u_0))$, which contradicts our initial assumption.

(c) Let $u_0 = \lim\limits_{\mathfrak{M}\ni a\to 0} \Delta(a) \in \mathcal{R}ad(N)$ be the element given by (b). The conclusion in (a) implies that the mapping $\Delta_0 : \mathfrak{M} \to \mathfrak{N}$, $\Delta_0(a) = \Delta(a) - u_0$ is a surjective isometry. Furthermore, $\exists \lim\limits_{\mathfrak{M}\ni a\to 0} \Delta_0(a) = 0$.

The set $\mathfrak{M}$ is not a linear subspace. However, as we have seen in the proof of (b), $(\mathbb{C}\backslash\{0\})\,\mathfrak{M} = \mathfrak{M}$. In particular, for each $a \in \mathfrak{M}$, $(\mathbb{C}\backslash\{0\})\,a \subseteq \mathfrak{M}$ and hence the mapping $\lambda \mapsto G_a(\lambda) = \Delta_0(\lambda a)$ is well defined on $\mathbb{C}\backslash\{0\}$. We extend the definition to the whole complex plane by setting $G_a(0) = 0$. We claim that

$$
\text{for each } a \in \mathfrak{M} \text{ the mapping } G_a : \mathbb{C} \to N \text{ is real-linear.} \tag{3.11}
$$

G_a is continuous since $\lim\limits_{\mathfrak{M}\ni a\to 0} \Delta_0(a) = 0$ and Δ_0 is an isometry. Moreover, if $[\lambda, \mu]$ is a segment in $\mathbb{C}$ not containing zero, Lemma 3.19 applied to Δ_0 on $[\lambda a, \mu a]$, implies that $\Delta_0|_{[\lambda a, \mu a]}$ is an affine mapping (note that $[\lambda a, \mu a]$ is even convex). Therefore, G_a is affine on each segment not containing 0. Lemma 3.20 now gives the desired statement.

We have actually proved a better property for Δ_0. Suppose $a, b \in \mathfrak{M}$ with $[a, b] \subset \mathfrak{M}$. A new application of Lemma 3.19 implies that $\Delta_0|_{[a,b]}$ is affine. By combining this property with (3.11) we obtain

$$
\frac{1}{2}\Delta_0(a+b) = \Delta_0\left(\frac{a+b}{2}\right) = \frac{\Delta_0(a) + \Delta_0(b)}{2}, \ \forall a, b \in M \text{ with } [a, b] \subset \mathfrak{M}. \tag{3.12}
$$

For the final step we consider the set

$$
\Omega = \bigcup_{\alpha>0} B(\alpha \mathbf{1}_M, \alpha) = \{x \in M : \|x - \alpha \mathbf{1}_M\| < \alpha \text{ for some } \alpha > 0\} \subseteq M_{\mathbf{1}}^{-1} \subset \mathfrak{M}.
$$

Clearly Ω is open since it is the union of a family of open balls. Given $x, y \in \Omega$ there exist positive α, β such that $\|x - \alpha\mathbf{1}\| < \alpha$ and $\|y - \beta\mathbf{1}\| < \beta$, and consequently, for each $0 < t < 1$ we have

$$\begin{aligned}\|tx + (1-t)y - (\alpha t - \beta(1-t))\mathbf{1}_M\| &\leq t\,\|x - \alpha\mathbf{1}_M\| + (1-t)\,\|y - \beta\mathbf{1}_M\| \\ &< t\alpha t + (1-t)\beta,\end{aligned}$$

which implies that Ω is convex. The set Ω also enjoys the following property: $\mathbb{R}^+\Omega = \Omega$ (just observe that for $t > 0$, we have $\|x - \alpha\mathbf{1}_M\| < \alpha \Rightarrow \|tx - t\alpha\mathbf{1}_M\| < t\alpha$).

The next step consists in showing that $\Delta_0(\Omega)$ also is convex. Given $a, b \in \Omega$ and $0 < t < 1$, the elements ta and $(1-t)b$ belong to Ω, and hence $[ta, (1-t)b] \subset \Omega$ by convexity. Applying (3.11) we arrive to

$$\begin{aligned}t\Delta_0(a) + (1-t)\Delta_0(b) &= tG_a(1) + (1-t)G_b(1) = G_a(t) + G_b(1-t) \\ &= \Delta_0(ta) + \Delta_0((1-t)b).\end{aligned}$$

Having in mind that $[ta, (1-t)b] \subset \Omega \subseteq \mathfrak{M}$, the conclusion in (3.12) leads to $\Delta_0(ta) + \Delta_0((1-t)b) = \Delta_0(ta + (1-t)b)$, and thus $t\Delta_0(a) + (1-t)\Delta_0(b) = \Delta_0(ta + (1-t)b)$. This guarantees that $\Delta_0(\Omega)$ is convex.

The mapping $\Delta_0|_\Omega : \Omega \to \Delta_0(\Omega)$ is a surjective isometry between two open convex sets. Similar arguments to those in the proof of (b), show that $\Delta_0|_\Omega$ admits an extension to a surjective isometry from the closure of Ω onto the closure of $\Delta_0(\omega)$. Mankiewicz theorem (see Theorem 3.18) applied to the extension of Δ_0 to the corresponding norm closures, implies the existence of a surjective affine isometry $T_0 : M \to N$ such that $\Delta_0|_\Omega = T_0|_\Omega$. As in (b), $0 = \lim_{\Omega \ni a \to 0} \Delta_0(a) = \lim_{\Omega \ni a \to 0} T_0(a) = T_0(0)$. Therefore, T_0 is a surjective real-linear isometry.

The desired statement in (c) will be proved if we show that $T_0|_{\mathfrak{M}} = \Delta_0$. Pick an arbitrary $a \in \mathfrak{M}$. Since $\| \pm a + 2\|a\|\mathbf{1}_M - 2\|a\|\mathbf{1}_M\| = \|a\| < 2\|a\|$, we deduce that $\pm a + 2\|a\|\mathbf{1}_M \in \Omega$, and thus $T_0(\pm a + 2\|a\|\mathbf{1}_M) = \Delta_0(\pm a + 2\|a\|\mathbf{1}_M)$.

Now, having in mind that T_0 is a surjective real-linear isometry, Δ_0 is an isometry and $\Delta_0|_\Omega = T_0|_\Omega$, we arrive to

$$\begin{aligned}2\|a\| &= \|a + 2\|a\|\mathbf{1}_M - a\| = \|\Delta_0(a + 2\|a\|\mathbf{1}_M) - \Delta_0(a)\| \\ &= \|a + 2\|a\|\mathbf{1}_M - T_0^{-1}\Delta_0(a)\|,\end{aligned} \tag{3.13}$$

and if we also apply (3.11) we get

$$\begin{aligned}2\|a\| &= \| - a + 2\|a\|\mathbf{1}_M + a\| = \|\Delta_0(-a + 2\|a\|\mathbf{1}_M) - \Delta_0(-a)\| \\ &= \| - a + 2\|a\|\mathbf{1}_M - T_0^{-1}\Delta_0(-a)\| = \| - a + 2\|a\|\mathbf{1}_M + T_0^{-1}\Delta_0(a)\|.\end{aligned} \tag{3.14}$$

Let us take a state $\phi \in D(M, \mathbf{1}_M) = \{\phi \in M^* : \|\phi\| = \phi(\mathbf{1}_M) = 1\}$. Now, combining (3.13) and (3.14) we deduce that

$$\begin{aligned}|2\|a\| \pm \phi(a - T_0^{-1}\Delta_0(a))| &= |\phi(2\|a\|\mathbf{1}_M \pm (a - T_0^{-1}\Delta_0(a)))| \\ &\leq \|2\|a\|\mathbf{1}_M \pm (a - T_0^{-1}\Delta_0(a))\| = 2\,\|a\|,\end{aligned}$$

which proves that $\phi(a - T_0^{-1}\Delta_0(a)) = 0$. The arbitrariness of ϕ proves that the numerical radius of $a - T_0^{-1}\Delta_0(a)$ is zero, that is, $v(a - T_0^{-1}\Delta_0(a)) = 0$. The Bohnenblust–Karlin theorem (see Theorem 3.21) implies that $a = T_0^{-1}\Delta_0(a)$, and hence $\Delta(a) = T_0(a) + u_0$ for all $a \in \mathfrak{M}$, as desired. □

4. The geometric goodness of C*- and JB*-algebras

C*-algebras form a very well known and deeply studied model in mathematics, particularly in functional analysis and physics, whose origins go back to their use in quantum mechanics too. A C*-algebra is a complex Banach algebra A equipped with an algebra involution $a \mapsto a^*$ satisfying the Gelfand–Naimark axiom $\|aa^*\| = \|a\|^2$ for all $a \in A$. The Gelfand–Naimark theorem (1943) affirms that each C*-algebra admits a concrete representation as a norm closed self-adjoint subalgebra of the space $B(H)$ of all bounded linear operators on a complex Hilbert space H. These objects have been intensively studied since the forties of the last century, a wide collection of books and structure results have been completely devoted to study them from many different points of view. For the sake of brevity we shall not employ too much time on introducing examples nor basic results, which seem to be widely known. We refer to the monographs [**15, 74, 100, 102, 108, 110**] as reference sources on C*-algebras.

It was soon realised, since the early stages of the theory of C*-algebras, that these objects enjoy very special geometric properties. We can trace back the good geometric properties to an earlier result. The Banach–Stone theorem asserts that given two compact Hausdorff spaces K_1 and K_2 and a surjective linear isometry $T : C(K_1) \to C(K_2)$, then there exists a homeomorphism $\varphi : K_2 \to K_1$ and a unimodular function $u \in C(K_2)$ satisfying

$$(Tf)(s) = u(s)f(\varphi(s)) \text{ for all } s \in K_2,\ f \in C(K_1).$$

The composition operator $f \mapsto f\circ\varphi$ is a *-isomorphism from $C(K_1)$ onto $C(K_2)$. An appropriate version of the Banach–Stone theorem works for non-unital commutative C*-algebras (i.e. $C_0(L)$ spaces where L is a locally compact Hausdorff space, see [**46**, Corollary 2.3.12]). In particular, the counterexample in Example 3.3 can never occur when both algebras are commutative C*-algebras. This is just the starting point of a series of deep geometric results. The first step was the following result proved by Kadison in the unital case, and by Paterson and Sinclair in the non-unital setting. We first recall that the *multiplier algebra*, $M(A)$, of a C*-algebra A is the idealizer of A in its bidual A^{**}, that is, the largest C*-subalgebra of the von Neumann algebra A^{**} containing A as an ideal). In other words $M(A) = \{x \in A^{**} : xA, Ax \subseteq A\}$ [**102**, §3.12].

THEOREM 4.1 ([**73**, Theorem 7] and [**101**, Theorem 1]). *Let A and B be C*-algebras. Then for each surjective linear isometry $T : A \to B$ there exists a unitary element u in the multiplier algebra of B inside B^{**}, and a Jordan $*$-isomorphism $\Phi : A \to B$ such that*

$$T(x) = u\ \Phi(x) \quad (x \in A).$$

The above result shows that two C*-algebras are Jordan isomorphic if and only if their underlying Banach spaces are isometrically isomorphic, so there is no place for counterexamples like Example 3.3 in the setting of C*-algebras.

Theorem 4.1 also implies that every surjective linear isometry between C*-algebras preserves triple products of the form

$$\{x, y, z\} = 2^{-1}(xy^*z + zy^*x). \tag{4.1}$$

Another consequence of the just-quoted theorem proves that every unital surjective isometry between unital C*-algebras is a Jordan *-isomorphism.

Hatori and Watanabe explored in [**64**] an appropriate version of Theorem 3.5 in the case that A and B are unital C*-algebras and they found that a stronger conclusion is valid under these extra hypotheses. It should be recalled that every C*-algebra is semisimple (cf. [**100**, Propositions 2.3.17 and 2.3.18]). So, the mapping Δ in Theorem 3.5 extends to a surjective real-linear isometry T from A into B. We shall see that T admits a particular form.

THEOREM 4.2 ([**64**, Theorem 2.2]). *Let A and B be unital C*-algebras, and let $\mathfrak{A}$ and $\mathfrak{B}$ be open subgroups of A^{-1} and B^{-1}, respectively. Suppose Δ is a bijection from $\mathfrak{A}$ onto $\mathfrak{B}$. Then Δ is an isometry if and only if $\Delta(\mathbf{1}_A)$ is unitary in B and there exist a projection p in the centre of B, and a complex-linear Jordan *-isomorphism $\tilde{J}$ from A onto B such that*

$$\Delta(a) = \Delta(\mathbf{1}_A)p\tilde{J}(a) + \Delta(\mathbf{1}_A)(\mathbf{1}_N - p)\tilde{J}(a)^*, \text{ for all } a \in \mathfrak{A}.$$

Furthermore the operator $\Delta(\mathbf{1}_A)p\tilde{J}(\cdot) + \Delta(\mathbf{1}_M)(\mathbf{1}_N - p)\tilde{J}(\cdot)^*$ defines a surjective real-linear isometry from A onto B.

As in the previous section, we shall obtain Theorem 4.2 above from its version for JB*-algebras that will be presented in Theorem 4.8.

JB*-algebras are the Jordan alter ego of C*-algebras. The notion was by proposed by Kaplansky during his celebrated final lecture to the 1976 St. Andrews Colloquium of the Edinburgh Mathematical Society. Concretely, a *JB*-algebra* is a complex Jordan-Banach algebra M equipped with an algebra involution * satisfying the following version of the Gelfand–Naimark axiom:

$$\|\{a,a,a\}\| = \|a\|^3, \text{ for all } a \in M, \tag{4.2}$$

(where $\{a,a,a\} = U_a(a^*) = 2(a \circ a^*) \circ a - a^2 \circ a^*$). A basic result in the theory of JB*-algebras asserts that the involution of every JB*-algebra is an isometry (cf. [**116**, Lemma 4] or [**22**, Proposition 3.3.13]).

Among the starring examples of JB*-algebras we find all C*-algebra when they are equipped with the same norm and involution and the natural Jordan product. In such a case $U_a(a^*) = aa^*a$, and hence $\|aa^*a\| = \|U_a(a^*)\| = \|a\|^3$, which is an equivalent reformulation of the Gelfand–Naimark axiom. JB*-subalgebras of C*-algebras are called *JC*-algebras.* We should note the existence of exceptional JB*-algebras which cannot be embedded as Jordan *-subalgebras of a C*-algebra (see, for example, [**54**, Corollary 2.8.5]).

JB*-algebras are indissolubly linked with JB-algebras. A *JB-algebra* is a real Jordan algebra $\mathfrak{J}$ equipped with a complete norm satisfying

$$\|a^2\| = \|a\|^2, \text{ and } \|a^2\| \leq \|a^2 + b^2\| \text{ for all } a, b \in \mathfrak{J}. \tag{4.3}$$

Wright proved in [**113**] that every JB-algebra $\mathfrak{J}$ corresponds uniquely to the self-adjoint part $M_{sa} = \{x \in M : x^* = x\}$ of a JB*-algebra M, answering in this way a conjecture by Kaplansky.

A JBW*-algebra (respectively, a JBW-algebra) is a JB*-algebra (respectively, a JB-algebra) which is also a dual Banach space. Every JBW*-algebra (respectively, each JBW-algebra) contains a unit element (see [**54**, §4] or [**2**]). It is worth to note that JBW-algebras are precisely the self-adjoint parts of JBW*-algebras (see [**40**, Theorems 3.2 and 3.4] or [**88**, Corollary 2.12]). In particular every von Neumann algebra is a JBW*-algebra.

Concerning surjective linear isometries, Wright and Youngson established that every unital and surjective linear isometry between two unital JB*-algebras is a Jordan $*$-isomorphism [**115**, Theorem 6]. A result by Isidro and Rodríguez Palacios proves that every surjective linear isometry T between two JB-algebras $\mathfrak{J}_1$ and $\mathfrak{J}_2$ is of the form $T(x) = s \circ \Phi(x)$, where s is a central symmetry in the multiplier algebra of $\mathfrak{J}_2$ and $\Phi : \mathfrak{J}_1 \to \mathfrak{J}_2$ is a surjective Jordan isomorphism (cf. [**69**, Theorem 1.9]). We recall that an element s in a unital JB-algebra is called a *symmetry* if we can write it in the form $s = p - q$ with $pq = 0$ and $p + q = \mathbf{1}$ (equivalently, $s^2 = \mathbf{1}$). We also recall that the *(Jordan) multiplier algebra* of a JB*-algebra (respectively, a JB-algebra) M is defined as

$$M(M) := \{x \in M^{**} : x \circ M \subseteq M\}.$$

According to the result by Isidro and Rodríguez Palacios, a bijective linear map Φ between two JB-algebras is an isometry if and only if Φ is a triple-isomorphism with respect to the triple product

$$\{x, y, z\} = (x \circ y) \circ z + (z \circ y) \circ x - (x \circ z) \circ y. \tag{4.4}$$

There is no better way to understand surjective isometries between JB*-algebras than when they are regarded inside the strictly wider class of JB*-triples. The complex Banach spaces in this class are characterized by holomorphic property. Namely, that their open unit balls are bounded symmetric domains [**76**], a property already observed by Harris for the open unit ball of each C*-algebra [**55**]. The contribution of Kaup gives an axiomatic description of these complex Banach spaces. A JB*-triple is a complex Banach space E equipped with a continuous triple product $\{.,.,.\} : E \times E \times E \to E$, $(a, b, c) \mapsto \{a, b, c\}$, which is bilinear and symmetric in (a, c) and conjugate linear in b, and satisfies the following axioms for all $a, b, x, y \in E$:

(a) (Jordan identity)

$$L(a, b)L(x, y) = L(x, y)L(a, b) + L(L(a, b)x, y) - L(x, L(b, a)y),$$

where $L(a, b) : E \to E$ is the operator defined by $L(a, b)x = \{a, b, x\}$.

(b) $L(a, a)$ is a hermitian operator with non-negative spectrum.

(c) $\|\{a, a, a\}\| = \|a\|^3$.

The triple product of every JB*-triple is a continuous mapping with norm 1, that is,

$$\|\{a, b, c\}\| \leq \|a\|\|b\|\|c\|, \quad \text{for all } a, b, c \text{ (cf. [49, Corollary 3])}. \tag{4.5}$$

The class of JB*-triples is rich of interesting examples, complex Hilbert spaces, spaces of operators of the form $B(H, K)$, C*-algebras and JB*-algebras are among them. The triple products for C*-algebras and JB*-algebras are given by

$$\{x, y, z\} = \frac{1}{2}(xy^*z + zy^*x), \tag{4.6}$$

and

$$\{x, y, z\} = (x \circ y^*) \circ z + (z \circ y^*) \circ x - (x \circ z) \circ y^*, \tag{4.7}$$

respectively (cf. [**55**] and [**17**, Theorem 3.3]). Note that we have already employed the triple product given in (4.6) after the Kadison–Paterson-Sinclair theorem in (4.1). Additional examples can be found by recalling that each element in the widely studied class of Hilbert C*-modules is a JB*-triple (cf. [**68, 80**]).

For each element a in a JB*-triple E, we shall write $Q(a)$ for the conjugate linear operator on E defined by $Q(a)(x) = \{a, x, a\}$ $(x \in E)$. Another important connection with JB*-algebras and U operators affirms that if M is a JB*-algebra regarded as a JB*-triple with the triple product given in (4.7), for each $a \in M$, we have $Q(a)(x) = U_a(x^*)$ $(x \in M)$.

Let E and F denote two JB*-triples. A linear mapping $T : E \to F$ is a *triple homomorphism* if it preserves triple products. A triple homomorphism which is also a linear bijection is called a *triple isomorphism.* If the corresponding maps are only real-linear we call them *real-linear triple homomorphisms* and *real-linear triple isomorphisms*, respectively.

There is an undoubted attractiveness in the geometric properties of JB*-triples expressed in the following Kaup's Banach–Stone theorem.

THEOREM 4.3 ([**76**, Proposition 5.5] or [**23**, Theorem 5.6.57]). *Let E and F be JB*-triples. Then a linear bijection $T : E \to F$ is an isometry if and only if it is a triple isomorphism.*

Alternative proofs and additional information can be found in [**37**], [**44**, Theorem 2.2], [**13**, Corollary 3.4], and [**45**]. Kaup's Banach–Stone theorem reaffirms the deep connection between the triple product and the geometric properties of the underlying Banach space.

In order to understand the implications of Kaup's Banach–Stone Theorem, we need to recall the notion of tripotent and some variants like unitary and complete tripotents. An element e in a JB*-triple E is said to be a *tripotent* if $\{e, e, e\} = e$. In such a case, the tripotent e produces a *Peirce decomposition* of E,

$$E = E_2(e) \oplus E_1(e) \oplus E_0(e),$$

in terms of the eigenspaces of the mapping $L(e, e)$, that is,

$$E_j(e) = \{x \in E : \{e, e, x\} = \frac{j}{2}x\} \quad (j \in \{0, 1, 2\}).$$

Triple products among elements in different Peirce subspaces follow the following patterns known under the name of *Peirce arithmetic* or *Peirce rules.*

$$\{E_i(e), E_j(e), E_k(e)\} \subseteq E_{i-j+k}(e) \ \text{ if } i - j + k \in \{0, 1, 2\},$$
$$\{E_i(e), E_j(e), E_k(e)\} = \{0\} \text{ if } i - j + k \notin \{0, 1, 2\},$$

and $\{E_2(e), E_0(e), E\} = \{E_0(e), E_2(e), E\} = 0$. Consequently, each Peirce subspace $E_j(e)$ is a JB*-subtriple of E.

The natural projection of E onto $E_k(e)$ is called the *Peirce k-projection*, and is denoted by $P_k(e)$. Among their properties, we highlight that Peirce projections are contractive (cf. [**48**, Corollary 1.2]), and can be algebraically recovered by the following identities: $P_2(e) = Q(e)^2$, $P_1(e) = 2(L(e, e) - Q(e)^2)$, and $P_0(e) = \mathrm{Id}_E - 2L(e, e) + Q(e)^2$.

In case that a C*-algebra A is regarded as a JB*-triple with respect to the triple product in (4.6), the identity $e = \{e, e, e\} = ee^*e$ is an equivalent reformulation of the fact that e is a partial isometry. Therefore partial isometries and tripotents in A are the same elements. In this particular case, Peirce projections are given by the left and right projections of e, that is, $P_2(e)(a) = ee^*ae^*e$, $P_1(e)(a) = (\mathbf{1} - ee^*)ae^*e + ee^*a(\mathbf{1} - e^*e)$, and $P_0(e)(a) = (\mathbf{1} - ee^*)a(\mathbf{1} - e^*e)$ $(a \in A)$.

For our purposes here, we note that for each tripotent e in a JB*-triple E, the Peirce 2-subspace $E_2(e)$ is a unital JB*-algebra with unit e, Jordan product $x \circ_e y := \{x, e, y\}$ and involution $x^{*_e} := \{e, x, e\}$, respectively (see [**22**, Theorem 4.1.55]).

Following standard notation, a tripotent e in in a JB*-triple E is called *unitary* (respectively, *complete* or *maximal*) if $E_2(e) = E$. (respectively, $E_0(e) = \{0\}$). There exists complete tripotents which are not unitaries, for example, a non-surjective isometry in $B(H)$. For each unitary u in E, the space E is a unital JB*-algebra with unit u, Jordan product $\circ_u$, and involution $*_u$.

We could have produce a certain polysemy for the word unitary since the term is already in use in unital C*-algebras and unital JB*-algebras. We recall that an element u in a unital C*-algebra A is called *unitary* if $uu^* = u^*u = \mathbf{1}_A$, equivalently, u is invertible with inverse u^*. The latter reformulation literally makes sense in unital JB*-algebras. Since the class of JB*-triples includes all unital JB*-algebras, we might have a possible conflict of meanings. Fortunately, the concerns are completely clarified by a result of Braun, Kaup, and Upmeier showing that unitary elements in a unital JB*-algebra M are precisely the unitary tripotents in M when the latter is regarded as a JB*-triple *(*cf. [**17**, Proposition 4.3] or [**22**, Theorem 4.2.24, Definition 4.2.25 and Fact 4.2.26]*)*. The set of all unitary elements in a unital C*-algebra A or in a unital JB*-algebra M will be denoted by $\mathcal{U}(A)$ and $\mathcal{U}(M)$, respectively. Note that given a unitary u in a unital JB*-algebra M, the unital JB*-algebra $(M, \circ_u, *_u)$ is precisely the u^*-isotope $M_{(u^*)}$ of M defined in page 268. The next lemma also is a straight consequence of [**17**, Proposition 4.3].

LEMMA 4.4 ([**17**, Proposition 4.3]). *Let M be a unital JB*-algebra and let u be a unitary element in M. The unitary elements of the JB*-algebras M and $M_{(u^*)} = (M, \circ_u, *_u)$ are the same, and they are precisely the unitary tripotents of M when the latter is regarded as a JB*-triple.*

A *JBW*-triple* is a JB*-triple which is also a dual Banach space. Examples of JBW*-triples include von Neumann algebras and JBW*-algebras. More examples can be obtained by having in mind that the second dual of a JB*-triple E is a JBW*-triple under a triple product extending the product of E [**39**]. Among the additional properties of JBW*-triples we note that every JBW*-triple admits a unique isometric predual, and its triple product is separately weak* continuous [**10**]. The celebrated Dixmier-Ng theorem characterizes when a normed space is in fact a dual Banach space. Concretely, a normed space X is a dual Banach space if and only if there exists a Hausdorff locally convex topology τ on X so that the closed unit ball of X is τ-compact. It follows that if two JB*-triples are isometrically isomorphic as Banach spaces (equivalently, JB*-triple isomorphic by Theorem 4.3), then one is a JBW*-triple if and only if the other one is.

Let us go back to surjective linear isometries between JB*-algebras. Suppose $T : M \to N$ is a surjective linear isometry between two JB*-algebras. Clearly, $T^{**} : M^{**} \to N^{**}$ is a surjective linear isometry between two JBW*-algebras (which are always unital, cf. [**2**, Proposition 2.7]). Kaup's Banach–Stone theorem (Theorem 4.3) assures that T^{**} is a triple isomorphism and hence $T^{**}(\mathbf{1}_{M^{**}}) = u$ must be a unitary in N^{**} and $T^{**} : M^{**} \to (N^{**}, \circ_u, *_u)$ is a unital Jordan *-isomorphism (cf. [**115**]). The restriction $T^{**}|_M : M \to (N, \circ_u, *_u)$ is a Jordan

*-isomorphism. We have obtained a Jordan *-isomorphism at the cost of replacing the Jordan product and involution in the JB*-algebra N.

Since Theorem 3.8 only assures the existence of a real-linear surjective isometry extending the mapping Δ, we shall also require an additional result. Henceforth, we shall say that a JB*-algebra M is the (orthogonal) direct sum of two JB*-subalgebras M_1 and M_2 (denoted by $M = M_1 \oplus^{\ell_\infty} M_2$) if $M = M_1 \oplus M_2$ and $ab = 0$ for all $a \in M_1$, $b \in M_2$.

A more precise notion of orthogonality between general elements in a JB*-triple has been widely study in the setting of preservers (cf. [**19, 20, 97–99**]). Elements a, b in a JB*-triple E are called *orthogonal* (written $a \perp b$) if $L(a, b) = 0$. It is known that $a \perp b \Leftrightarrow \{a, a, b\} = 0 \Leftrightarrow \{b, b, a\} = 0 \Leftrightarrow b \perp a$ (see [**18**, Lemma 1]).

THEOREM 4.5 ([**36**, Corollary 3.2] or by [**45**, Corollary 3.4]). *Every surjective real-linear isometry T between JB*-algebras M and N is a real-linear triple homomorphism. Moreover, there exist two JB*-subalgebras M_1 and M_2 of M such that M is the (orthogonal) direct sum $M = M_1 \oplus^{\ell_\infty} M_2$, $T|_{M_1} : M_1 \to N$ is a complex-linear triple homomorphism and $T|_{M_2} : M_2 \to N$ is a conjugate-linear triple homomorphism.*

For each Banach space X we shall write $Iso(X)$ for the group of all surjective linear isometries on X. It is well known that for each unital C*-algebra A the group $Iso(A)$ acts transitively on $\mathcal{U}(A)$. Namely, if $u, v \in \mathcal{U}(A)$, the mapping $x \mapsto vu^*x$ is a surjective linear isometry on A mapping u to v. At this point we must remark one of the main divergences in C*-algebra theory and JB*-algebra theory. When A is replaced with a unital JB*-algebra M, this conclusion is no longer true.

ANTI-THEOREM 4.6 ([**17**, Example 5.7]) and [**22**, Antitheorem 3.4.34]). There exists a unital JC*-algebra M containing two unitaries u_1 and u_2 satisfying $T(u_1) \neq u_2$ for all $T \in Iso(M)$. Furthermore, the unital JC*-algebras $M_{(u_1^*)} = (M, \circ_{u_1}, *_{u_1})$ and $M_{(u_2^*)} = (M, \circ_{u_2}, *_{u_2})$ are linearly isometric but not Jordan *-isomorphic.

We have seen after Kadison–Paterson–Sinclair theorem that there is no place for counterexamples like Example 3.3 in the setting of C*-algebras. The previous anti-theorem shows that these pathologies are possible for JB*-algebras.

Anti-theorem 4.6 also proves that the same complex Banach space M may admit two different structures of JB*-algebras which are not Jordan *-isomorphic. However, by Kaup's Banach–Stone theorem (Theorem 4.3) there is only one structure of JB*-triple.

COROLLARY 4.7. *Let u be a unitary in a unital JB*-algebra M, then there is a unique triple product $\{.,.,.\}$ on M making it a JB*-triple, and it satisfies*

$$\begin{aligned}\{x, y, z\} &= (x \circ y^*) \circ z + (z \circ y^*) \circ x - (x \circ z) \circ y^* \\ &= (x \circ_u y^{*_u}) \circ_u z + (z \circ_u y^{*_u}) \circ_u x - (x \circ_u z) \circ_u y^{*_u},\end{aligned}$$

for all $x, y, z \in M$.

We are now in a position to determine the concrete form of a surjective isometry between two clopen quadratic subsets of the invertible elements in two unital JB*-algebras which are closed for powers and inverses.

THEOREM 4.8 ([**104**, Theorem 4.1]). *Let M and N be unital JB*-algebras. Suppose that $\mathfrak{M} \subseteq M^{-1}$ and $\mathfrak{N} \subseteq N^{-1}$ are clopen quadratic subsets which are* *closed for powers and inverses. Let* $\Delta : \mathfrak{M} \to \mathfrak{N}$ *be a surjective isometry. Then* $\Delta(\mathbf{1}_M) = u$ *is a unitary element in* N *and there exists a projection* p *in the centre of* M *and a complex-linear Jordan* **-isomorphism* J *from* M *onto the* u^**-homotope* $N_{(u^*)} = (N, \circ_u, *_u)$ *such that*

$$\Delta(a) = J(p \circ a) + J((\mathbf{1}_M - p) \circ a^*), \text{ for all } a \in \mathfrak{M}.$$

If we additionally assume that there exists a unitary ω_0 *in* N *such that* $U_{\omega_0}(\Delta(\mathbf{1})) = \mathbf{1}$*, then there exist a central projection* $p \in M$ *and a complex-linear Jordan* **-isomorphism* Φ *from* M *onto* N *satisfying*

$$\Delta(a) = U_{w_0^*}\left(\Phi(p \circ a) + \Phi((\mathbf{1}_M - p) \circ a^*)\right),$$

for all $a \in \mathfrak{M}$.

PROOF. We recall that every JB*-algebra M is (Jordan-)semisimple, that is $\mathcal{R}ad(M) = \{0\}$ (cf. [**22**, Lemma 4.4.28(iii)]). Thus, by Theorem 3.8 there exists a surjective real-linear isometry $T_0 : M \to N$ such that $\Delta(a) = T_0(a)$ for all $a \in \mathfrak{M}$. Theorem 4.5 implies that T_0 is a real-linear triple isomorphism, that is, T_0 preserves triple products of the form $\{a, b, c\} = (a \circ b^*) \circ c + (c \circ b^*) \circ a - (a \circ c) \circ b^*$ $(a, b, c \in M)$. Actually, there exist two JB*-subalgebras M_1 and M_2 of M such that M is the (orthogonal) direct sum $M = M_1 \oplus^{\ell_\infty} M_2$, $T_0|_{M_1} : M_1 \to N$ is a complex-linear triple homomorphism and $T_0|_{M_2} : M_2 \to N$ is a conjugate-linear triple homomorphism.

It is now easy to check that the element $u = T_0(\mathbf{1}_M)$ is a unitary in N (observe that $\{u, u, u\} = T_0(\{\mathbf{1}_M, \mathbf{1}_M, \mathbf{1}_M\}) = T(\mathbf{1}_M) = u$, and, by the surjectivity of T_0, for each z in N we can find $x \in M$ with $T_0(x) = z$, and hence $\{u, u, z\} = T_0(\{\mathbf{1}_M, \mathbf{1}_M, x\}) = T_0(x) = z$).

As we have seen above the u^*-isotope $N_{(u^*)} = (N, \circ_u, *_u)$ is a unital JB*-algebra with unit u. Since

$$T_0(x \circ y) = T_0(\{x, \mathbf{1}_M, y\}) = \{T_0(x), T_0(\mathbf{1}_M), T_0(y)\} = T_0(x) \circ_{u^*} T_0(y)$$

and

$$T_0(x^*) = T_0(\{\mathbf{1}_M, x, \mathbf{1}_M\}) = \{T_0(\mathbf{1}_M), T_0(x), T_0(\mathbf{1}_M)\} = T_0(x)^{*_u}, \ (x, y \in M),$$

we deduce that $T_0 : M \to N_{u^*}$ is an isometric real-linear unital Jordan *-isomorphism.

Let p denote the unit of M_1. Clearly, p is a central projection in M and $\mathbf{1}_M - p$ is precisely the unit of M_2. The arguments in the previous paragraphs show that $T_0|_{M_1} : M_1 \to N_u$ is a complex-linear Jordan *-monomorphism and $T_0|_{M_2} : M_2 \to N_u$ is a conjugate-linear Jordan *-monomorphism. Furthermore, bearing in mind that M_1 and M_2 are orthogonal in M and that $T_0 : M \to N_u$ is a real-linear Jordan *-isomorphism, we deduce that $N_1 = T_0(M_1)$ and $N_2 = T_0(M_2)$ are orthogonal JB*-subalgebras of $N_{(u^*)}$ with $N_{(u^*)} = N_1 \oplus^{\ell_\infty} N_2$. Moreover, the mapping $J : M \to N_{(u^*)}$,

$$J(x) = T_0(p \circ x) + T_0((\mathbf{1}_M - p) \circ x)^{*_u} = T_0(p \circ x) + T_0((\mathbf{1}_M - p) \circ x^*) \ \ (x \in M)$$

is a complex-linear Jordan *-isomorphism (just apply that $x \mapsto p \circ x$ and $x \mapsto (\mathbf{1}_M - p) \circ x$ are the natural projections of M onto M_1 and M_2, respectively), and the identity

$$\Delta(a) = J(p \circ a) + J((\mathbf{1}_M - p) \circ a^*),$$

holds for all $a \in \mathfrak{M}$.

To prove the final statement, we assume the existence of a unitary $w_0 \in N$ such that $U_{w_0}(\Delta(\mathbf{1}_M)) = \mathbf{1}_N$. Under this assumption, the mapping $U_{w_0} : N \to N$ is a surjective complex-linear isometry and a triple isomorphism sending u to $\mathbf{1}_N$ (cf. [**22**, Theorem 4.2.28]), therefore $U_{w_0} : N_{(u^*)} \to N$ is a complex-linear Jordan *-isomorphism. Let $p \in M$ and $J : M \to N_{(u^*)}$ be the central projection and the complex-linear Jordan *-isomorphism given by our first conclusions. Clearly, $\Phi = U_{w_0} \circ J : M \to N$ is a Jordan *-isomorphism and the equality

$$\Delta(a) = U_{w_0^{-1}}(\Phi(p \circ a) + \Phi((\mathbf{1}_M - p) \circ a^*)) = U_{w_0^*}(\Phi(p \circ a) + \Phi((\mathbf{1}_M - p) \circ a^*)),$$

holds for all $a \in \mathfrak{M}$. □

We shall next obtain Theorem 4.2 as a consequence of the just proved theorem.

Proof of Theorem 4.2. Let $\Delta : \mathfrak{A} \to \mathfrak{B}$ be a surjective isometry satisfying the hypothesis of the theorem. We observe that $\mathfrak{A}$ and $\mathfrak{B}$ clearly satisfy the hypothesis of Theorem 4.8 (cf. page 273). By Theorem 4.8, $\Delta(\mathbf{1}_A) = u$ is a unitary element in B and there exists a projection p in the centre of A and a complex-linear Jordan *-isomorphism J from A onto the u^*-homotope $B_{(u^*)} = (B, \circ_u, *_u)$ such that

$$\Delta(a) = J(p \circ a) + J((\mathbf{1}_A - p) \circ a^*), \text{ for all } a \in \mathfrak{A}.$$

The mapping $L_{u^*} : B \to B$, $L_{u^*}(x) = u^* x$ is a surjective isometry mapping u to $\mathbf{1}_A$, and hence it is a Jordan *-isomorphism from $B_{(u^*)} = (B, \circ_u, *_u)$ onto B. Therefore the mapping $\tilde{J} = L_{u^*} \circ J$ is a Jordan *-isomorphism from A onto B with $J = L_u \circ \tilde{J} = L_{\Delta(\mathbf{1}_A)} \circ \tilde{J}$, and since p is central,

$$\Delta(a) = \Delta(\mathbf{1}_A)\left(\tilde{J}(pa) + \tilde{J}((\mathbf{1}_A - p)a)^*\right),$$

for all $a \in \mathfrak{A}$. □

4.1. The cone of positive elements as a metric invariant.

An element a in a JB*-algebra M is called positive if $a^* = a$ and the spectrum of a is contained in $\mathbb{R}_0^+$. The set M^+ of all positive elements in M is a closed convex cone [**54**, Lemma 3.3.7]. If M is a C*-algebra, M^+ is the usual cone of positive elements. If M^+ is regarded inside M_{sa} and M is unital, we can also conclude that M^+ has non-empty interior, and hence it is a convex body. In this case, we shall write M^{++} for the set of all positive and invertible elements in M.

The idea of considering the cone of positive elements or the set of positive invertible elements in a unital C*-algebra as metric invariants with respect to the original norm was considered by Hatori and Molnár in [**63**, page 167], the same ideas work for unital JB*-algebras.

Proposition 4.9. *Let M and N be two unital JB*-algebras, and let $\Delta : M^{++} \to N^{++}$ be a surjective isometry with respect to the distances induced by the JB*-norms. Then there exits a Jordan *-isomorphism $J : M \to N$ whose restriction to M^{++} is Δ.*

Proof. We can extend Δ to a surjective isometry $\tilde{\Delta}$ from M^+ onto N^+. Since M^+ and N^+ are closed convex bodies of M_{sa} and N_{sa} respectively, we can apply Mankiewicz theorem (see Theorem 3.18) to deduce that $\tilde{\Delta}$ extends to a surjective affine isometry $F : M_{sa} \to N_{sa}$. Since 0 is the unique extreme point of M^+ and N^+, we have $\tilde{\Delta}(0) = F(0) = 0$, and thus F is a linear isometry. Now applying

[**69**, Theorem 1.9], and having in mind that N_{sa} is unital, it follows that there exists a central symmetry $s \in N$ and a surjective Jordan isomorphism $J : M_{sa} \to N_{sa}$ such that $T(x) = s \circ J(x)$, for all $x \in M_{sa}$. Since $s = F(\mathbf{1}_M) = \Delta(\mathbf{1}_M) \in N^+$, we obtain $s = \mathbf{1}_M$ and $F(x) = J(x)$. □

As we already commented, Mankiewicz theorem (Theorem 3.18) motivated the introduction of the so-called Mankiewicz property by Mori and Ozawa, who established the following result.

THEOREM 4.10 ([**94**, Theorem 2]). *Let X be a Banach space such that the closed convex hull of the set $\partial_e(B_X)$, of all extreme points of the closed unit ball, B_X, of X has non-empty interior in X. Then every convex body $C \subseteq X$ has the strong Mankiewicz property, that is, every surjective isometry from C onto an arbitrary convex subset L in a normed space Y is affine.*

Mori and Ozawa applied this result to show that every convex body in a unital C*-algebra, or in a (unital) real von Neumann algebra satisfies the strong Mankiewicz property [**94**, Corollary 2]. The same occurs for convex bodies in JBW*-triples [**12**, Corollary 2.2], convex bodies in $C(\Omega, \mathcal{H})$, where Ω is a compact Hausdorff space, and $\mathcal{H}$ is a real Hilbert space with $\dim(\mathcal{H}) \geq 2$ [**32**], and convex bodies in unital JB*-algebras and real JBW*-algebras [**107**, pages 15 and 20].

It is worth recalling that a *real JB*-algebra* is a closed *-invariant real subalgebra of a (complex) JB*-algebra. Unital real JB*-algebras were introduced by K. Alvermann in [**3**]. Besides the just quoted reference an axiomatic intrinsic definition of unital real JB*-algebras can be found in [**103**]. Additional examples of real JB*-algebras include all JB-algebras, which are precisely those real JB*-algebras whose involution is the identity mapping, and all real C*-algebras (i.e., closed *-invariant real subalgebras of C*-algebras) equipped with their natural Jordan product [**53**, **84**]. A real JBW*-algebra is a real JB*-algebra which is also a dual Banach space. It follows from the results commented above that every JBW-algebra $\mathfrak{J}$ is a real JBW*-algebra, every convex body in $\mathfrak{J}$ has the strong Mankiewicz property, and in particular $\mathfrak{J}^+$ enjoys this property. We have just discovered the following corollary.

COROLLARY 4.11. *Let M be a JBW*-algebra and let L be a convex subset in a normed space Y. Suppose $\Delta : M^+ \to L$ is a surjective isometry. Then Δ extends to a surjective isometry from M_{sa} onto a complete subspace of Y.*

PROOF. Since M^+ satisfies the strong Mankiewicz property as a convex body inside M_{sa}, the mapping Δ is affine. It is part of the folklore of affine maps that in such a case Δ can be uniquely extended to an affine isometry from M_{sa} onto a complete subspace of Y. □

5. A geometric tool: preservers of inverted Jordan triple products

In [**111**] Väisälä gave an astonishing-simplified proof of the celebrated Mazur–Ulam theorem based on the idea that a surjective isometry between normed real-linear spaces preserves algebraic midpoints, i.e. $T\left(\frac{x+y}{2}\right) = \frac{T(x)+T(y)}{2}$ (cf. the forerunner due to Vogt in [**112**]). By replacing preservation of midpoints with preservation of inverted Jordan triple products of the form $yx^{-1}y$, Hatori, Hirasawa, Miura and Molnár established in [**61**] a very abstract and general result concerning a local property for maps between abstract spaces preserving a certain mild relation

d between elements, which is a far-reaching generalization of the usual notion of distance.

The simplicity of the statement and the hypotheses are striking. Let us begin by recalling some definitions. A *metric* on a set X is a function $d: X \times X \to \mathbb{R}$ satisfying the following axioms:

(1) (Semi-positivity) $d(x,y) \geq 0$;
(2) (Identity of indiscernibles) $x = y \Leftrightarrow d(x,y) = 0$;
(3) (Symmetry) $d(x,y) = d(y,x)$;
(4) (Triangle inequality) $d(x,z) \leq d(x,y) + d(y,z)$,

for all $x, y, z \in X$. There is a wide range of generalizations of the notion of metric (see, for instance, [**38**, §1.1]). For example, a *distance* or a *dissimilarity* on X is a function $d: X \times X \to \mathbb{R}$ which is semi-positive, symmetric and that instead of the identity of indiscernibles, it satisfies the following weaker version

(2′) (Reflexivity) $d(x,x) = 0$ for all $x \in X$.

A *semi-metric* is a mapping d which is semi-positive, symmetric, reflexive and satisfies the triangular inequality. The mildest set of axioms leads us to the notion of *quasi-distance* or *parametric* which is a mapping $d: X \times X \to \mathbb{R}$ which is merely semi-positive and reflexive. Here, we shall work with an even weaker notion, a *non-reflexive quasi-distance* or a *non-reflexive parametric* (*nr-quasi-distance* or an *nr-parametric* in short) on X is a semi-positive mapping $d: X \times X \to \mathbb{R}$. Here, the term non-reflexive actually means non-necessarily reflexive.

Along this section, X will be a non-empty set equipped with a nr-quasi-distance d_X. The pair (X, d_X) will be called a nr-quasi-distance space. If (Y, d_Y) is another nr-quasi-distance space, a mapping $f: X \to Y$ is said to be *d-preserving* if

$$d_X(x,y) = d_Y(f(x), f(y)), \text{ for every } (x,y) \in X \times X.$$

Clearly every metric space is a nr-quasi-distance space. Furthermore, a mapping between metric spaces with respect to the d maps given by the distances is d-preserving precisely when it is an isometry. If no confusion arises, we shall omit the subindex in d_X.

Our first theorem goes back to a result by Vogt [**112**, Theorem 1.2] under stronger hypotheses which were later simplified by Hatori, Hirasawa, Miura and Molnár in [**61**, Lemma 2.3], from where we have borrowed the proof.

THEOREM 5.1 ([**61, 112**]). *Let (X,d) be a nr-quasi-distance space. Suppose that there exists a point c in X, a bijective d-preserving mapping $\varphi: X \to X$, and a constant $K > 1$ satisfying:*

(a) $d(\varphi(x), x) \geq K d(x,c)$, for every x in X;
(b) the function $d(\cdot, c): X \to \mathbb{R}_0^+$, $x \mapsto d(x,c)$ is bounded.

Then we have $d(f(c), c) = 0$ for every bijective d-preserving map $f: X \to X$.

PROOF. It follows from hypothesis (b) that the set

$$\{d(f(c),c) : f: X \to X \text{ bijective and } d\text{-preserving}\}$$

is bounded. Let λ denote the supremum of this set. For each bijective d-preserving mapping $f: X \to X$, the map $\tilde{f} = f^{-1} \circ \varphi \circ f$ is a d-preserving bijection. Therefore, by (a), we have

$$\lambda \geq d(\tilde{f}(x), c) = d(\varphi f(x), f(c)) \geq K d(f(x), c).$$

The arbitrariness of f shows that $\lambda \geq K\lambda$, which leads to $\lambda = 0$. □

If we assume that in Theorem 5.1 above the mapping d satisfies the additional hypothesis $d(x,y)=0 \Rightarrow x=y$, like, for example, when d is a metric, the previous theorem proves that c is a fixed point for every d-preserving bijection $f : X \to X$. In [**112**, Theorem 1.2], X is a bounded metric space for which there exists a surjective isometry $\varphi : X \to X$ and $K > 1$ satisfying hypothesis (a) in Theorem 5.1 for certain point $c \in X$, then the conclusion is that every surjective isometry f on M fixes c (note that hypothesis (b) clearly follows from the boundedness of X).

It should be also added that a weaker form of Vogt's result [**112**, Theorem 1.2], asserting that if a bounded metric space (X,d) is symmetric with respect to some $c\in X$– i.e., there exists a surjective isometry g on X such that $d(g(x),x) = 2d(x,c)$ for all $x \in X$– then c must be a fixed point of every surjective isometry on X, was employed by Mankiewicz in his celebrated extension of the Mazur–Ulam theorem on the extension of surjective isometries between arbitrary convex subsets with non-empty interior in real normed spaces to affine surjective isometries between the spaces (see [**87**]).

We are now in a position to revisit the main result of this section which is due to Hatori, Hirasawa, Miura and Molnár [**61**].

THEOREM 5.2 ([**61**, Theorem 2.4]). *Let (X, d_X) and (Y, d_Y) be two nr-quasi-distance spaces. Let a and c be two arbitrary points in X. Suppose we can find a period-2 (i.e. $\varphi^2 = \varphi\circ\varphi = Id_X$), d-preserving mapping $\varphi : X \to X$ admitting c as a fixed point. Let L denote the set of all points in X such that the nr-quasi-distance from a and from $\varphi(a)$ to them is equal to $d_X(a,c)$, that is,*

$$L = \{x\in X : d_X(a,x) = d_X(\varphi(a),x) = d_X(a,c)\}.$$

We shall additionally assume that the set $\{d_X(x,c) : x\in L\}$ is bounded, and there exists a constant $K>1$ such that $d_X(\varphi(x),x) \geq K d_X(x,c)$ for every $x \in L$. Then, if T is a bijective d-preserving map from X onto Y, and ψ is a bijective d-preserving map from Y onto itself such that one of the next statements holds

$$\psi(T(a)) = T(\varphi(a)), \text{ and } \psi(T(\varphi(a))) = T(a), \tag{5.1}$$

$$\psi(T(a)) = T(a), \text{ and } \psi(T(\varphi(a))) = T(\varphi(a)), \tag{5.2}$$

we have $d_Y(\psi(T(c)), T(c)) = 0$.

PROOF. The hypotheses $\varphi(c) = c$ and φ being d-preserving imply that $c\in L\neq \emptyset$. By applying that φ is a period-2, d-preserving mapping, it is easy to check that $\varphi(L)\subseteq L$, and thus $\varphi(L) = L$ because $\varphi^2 = Id_X$.

The alter ego of L in Y is the set

$$L_Y = \{y\in Y : d_Y(T(a),y) = d_Y(T(\varphi(a)),y) = d_X(a,c)\}.$$

We claim that $T(L) = L_Y$. Indeed, for each $x\in L$ the assumptions on T give

$$d_Y(T(a),T(x)) = d_X(a,x) = d_X(a,c), \text{ and}$$
$$d_Y(T(\varphi(a)),T(x)) = d_X(\varphi(a),x) = d_X(a,c),$$

proving that $T(x) \in L_Y$ and hence $T(L)\subseteq L_Y$. For the reciprocal, we observe that $\tilde\varphi := T\circ\varphi\circ T^{-1} : Y\to Y$ is a period-2, d-preserving mapping admitting $T(c)$ as a fixed point, $T^{-1} : Y \to X$ is a d-preserving bijection for which L_Y plays the same role of L for T. So, it follows from the previous arguments that $T^{-1}(L_Y)\subseteq L$, and hence $T(L) = L_Y$.

The stability of L_Y under the mapping ψ follows from the hypotheses in (5.1) or (5.2). We shall only prove the case in which (5.1) holds, the other case is similar. In such a case, for each $y \in L_Y$ we have

$$d_Y(T(a), \psi(y)) = d_Y(\psi(T(\varphi(a))), \psi(y)) = d_Y(T(\varphi(a)), y) = d_X(a, c), \text{ and}$$
$$d_Y(T(\varphi(a)), \psi(y)) = d_Y(\psi(T(a)), \psi(y)) = d_Y(T(a), y) = d_X(a, c).$$

The last two identities show that $\psi(L_Y) \subseteq L_Y$. On the other hand, given $y \in L_Y$, by applying that ψ^{-1} also is d-preserving we get

$$d_X(a, c) = d_Y(T(a), y) = d_Y(\psi^{-1}T(a), \psi^{-1}(y)) = d_Y(T(\varphi(a)), \psi^{-1}(y)), \text{ and}$$
$$d_X(a, c) = d_Y(T(\varphi(a)), y) = d_Y(\psi^{-1}T(\varphi(a)), \psi^{-1}(y)) = d_Y(T(a), \psi^{-1}(y)),$$

proving that $\psi^{-1}(y) \in L_Y$, and thus $\psi(L_Y) = L_Y$.

Finally, the mapping $\tilde{\psi} = T^{-1} \circ \psi \circ T$ is a d-preserving bijection on X with $\tilde{\psi}(L) = L$, and $\varphi|_L : L \to L$ is a d-preserving bijection. Since $d_X(\varphi(x), x) \geq K d_X(x, c)$ for every $x \in L$ (with $K > 1$) and the set $\{d_X(x, c) : x \in L\}$ is bounded, Theorem 5.1, applied to $\tilde{\psi}|_L$ and $\varphi|_L$, assures that $0 = d_X(\tilde{\psi}(c), c) = d_Y(\psi(T(c)), T(c))$, as desired. □

REMARK 5.3. Note that if d_X is a metric, the triangle inequality guarantees that the set $\{d_X(x, c) : x \in L\}$ in Theorem 5.2 is bounded.

The refinement in the previous theorem is to consider the set L to guarantee the hypothesis concerning boundedness in Theorem 5.1.

As the reader might guess, Vogh and Väisälä arguments in [**111, 112**] leading to an optimal proof of the Mazur–Ulam theorem, can be now rediscovered from the previous results. Namely, suppose $F : X \to Y$ is a surjective isometry between real normed spaces. We consider the metrics given by the norms on X and Y, respectively. For each $c \in X$, let $\varphi_c : X \to X$ denote the mapping given by the reflection of X in c, that is, $\varphi_c(x) := 2c - x$ ($x \in X$). Fix $a, b \in X$ and denote by φ_c and $\tilde{\varphi}_{\frac{F(a)+F(b)}{2}}$ the reflections of X and Y in $c = \frac{a+b}{2}$ and $\frac{F(a)+F(b)}{2}$, respectively. We consider the mapping $f = \varphi_c \circ F^{-1} \circ \tilde{\varphi}_{\frac{F(a)+F(b)}{2}} \circ F : X \to X$ which is a d-preserving bijection. Clearly, φ_c is a period-2, d-preserving bijection with $\varphi_c(c) = c$, $\varphi_c(a) = b$, and $\varphi_c(b) = a$. Furthermore,

$$d_X(\varphi_c(x), x) = \|2c - x - x\| = 2\|c - x\|, \text{ for all } x \in X.$$

In principle, the function $d_X(\cdot, c)$ is not bounded on X. To avoid this obstacle we restrict our maps to a suitable subset $L = \{x \in X : d_X(a, x) = d_X(\varphi_c(a), x) = d_X(a, c)\}$. As we saw in the previous theorem, or with simpler arguments, $\varphi_c(L) = L$. By the triangular inequality $\sup\{d_X(x, c) : x \in L\} < \infty$.

The mapping $\psi = F \circ f \circ F^{-1} = F \circ \varphi_c \circ F^{-1} \circ \tilde{\varphi}_{\frac{F(a)+F(b)}{2}}$ is a d-preserving bijection on Y with

$$\psi(F(a)) = F \circ \varphi_c \circ F^{-1} \circ \tilde{\varphi}_{\frac{F(a)+F(b)}{2}}(F(a)) = F(a),$$

and

$$\psi(F(\varphi_c(a))) = F \circ \varphi_c \circ F^{-1} \circ \tilde{\varphi}_{\frac{F(a)+F(b)}{2}}(F(\varphi_c(a))) = F(b) = F(\varphi_c(a)).$$

Theorem 5.2 implies that $F(c) = \psi(F(c)) = F(f(c)) = F \circ \varphi_c \circ F^{-1} \circ \tilde{\varphi}_{\frac{F(a)+F(b)}{2}} \circ F(c)$, or equivalently, $\tilde{\varphi}_{\frac{F(a)+F(b)}{2}} \circ F(c) = F(c)$. Therefore $F(c)$ is the unique fixed point for $\tilde{\varphi}_{\frac{F(a)+F(b)}{2}}$, that is, $\frac{F(a)+F(b)}{2} = F(\frac{a+b}{2})$. A standard argument shows that F is affine.

After a first application to rediscover the classical Mazur–Ulam theorem, we should not forget that Theorem 5.2 is a device thought to prove the preservation of inverted Jordan triple products of the form $yx^{-1}y$. We shall see different applications in various contexts. We shall adopt the notation from [**61**].

Along this section, we shall assume that $\mathcal{G}$ is a group. Unless another assumptions are made, we shall work with a nr-quasi-distance space (X, d) satisfying the mere assumptions that X is a subset of $\mathcal{G}$ and

$$yx^{-1}y \in X \text{ for all } x, y \in X. \tag{5.3}$$

The hypotheses do not necessarily imply that X is a subgroup of $\mathcal{G}$. We continue by recalling certain extra algebraic-geometric properties.

DEFINITION 5.4. Let us fix a, b in X. We shall say that (X, d) satisfies property $B(a, b)$ if the following three statements hold:

$(B.1)$ $d(bx^{-1}b, by^{-1}b) = d(x, y)$, for all $x, y \in X$.

$(B.2)$ $\sup\{d(x, b) : x \in L_{a,b}\} < \infty$, where we set

$$L_{a,b} = \{x \in X : d(a, x) = d(ba^{-1}b, x) = d(a, b)\}.$$

$(B.3)$ There exists a constant $K > 1$ satisfying

$$d(bx^{-1}b, x) \geq Kd(x, b),$$

for all $x \in L_{a,b}$.

In case that d satisfies the triangular inequality (for example, when d is a semi-metric or a metric), condition $(B.2)$ above is trivially true by the triangular inequality. Following the notation of the pioneering paper [**61**], by an *algebraic mid-point* of a couple of elements $x, y \in \mathcal{G}$ we mean any element $c \in \mathcal{G}$ satisfying $y = cx^{-1}c$. Let us note that b is an algebraic mid-point of a and $ba^{-1}b$, and in case that X satisfies property $B(a, b)$ the element b lies in the set $L_{a,b}$ by definition.

DEFINITION 5.5. Let us fix $a, b \in X$. We shall say that (X, d) satisfies property $C_1(a, b)$ if the following statements hold:

$(C.1)$ $ax^{-1}b, bx^{-1}a \in X$, for every $x \in X$.

$(C.2)$ $d(ax^{-1}b, ay^{-1}b) = d(x, y)$, for all $x, y \in X$.

DEFINITION 5.6. Let us fix $a, b \in X$. We shall say that (X, d) has property $C_2(a, b)$ if there exists $c \in X$ such that $ca^{-1}c = b$ and $d(cx^{-1}c, cy^{-1}c) = d(x, y)$ for all $x, y \in X$.

Let us remark that properties $B(a, b)$, $C_1(a, b)$ and $C_2(a, b)$ above depend on the points a, b.

An element $x \in X$ is called *2-divisible* if there exists $y \in X$ such that $y^2 = x$. X is called *2-divisible* if every element in X is 2-divisible. We observe that X contains the unit of $\mathcal{G}$ whenever it is 2-divisible. Namely, if we take $y \in X$, by assumptions there is $x \in X$ with $x^2 = y$ and hence $e = xy^{-1}x \in X$. Moreover, let us take $a, b \in X$, by the property on X, there exist $d, f \in X$ with $d^2 = a$ and $f^2 = db^{-1}d$, and hence setting $c = df^{-1}d$ we have $c \in X$ and $ca^{-1}c = df^{-1}da^{-1}df^{-1}d = d(f^{-1})^2d = b$.

Finally, X is a called *2-torsion free* if it contains the unit of $\mathcal{G}$ and the condition $x^2 = 1$ with $x \in X$ implies $x = 1$.

DEFINITION 5.7. Let us additionally assume that X is a subgroup of $\mathcal{G}$. We say that d is inverse and translation invariant if $d(x^{-1}, y^{-1}) = d(ax, ay) = d(x, y)$ for every triplet x, y, a in X.

If X is actually a subgroup of $\mathcal{G}$, then d is inverse and translation invariant if and only if (X, d) satisfies $C_1(a, b)$ for every $a, b \in X$. If X is a subgroup of $\mathcal{G}$, 2-divisible, and d is inverse and translation invariant, then (X, d) satisfies $C_2(a, b)$ for every $a, b \in X$.

The assumptions until now are so flexible that may spaces can serve as examples. The first one is a basic example. Suppose X is an additive subgroup of a normed space $(Z, \|.\|)$ with the metric given by the norm, then X satisfies properties $B(a, b)$ and $C_1(a, b)$ for every a, b in X, moreover X satisfies $C_2(a, b)$ in the case that X is 2-divisible.

The next three examples have been borrowed from [**61**].

EXAMPLE 5.8. Let X denote the n-dimensional torus $\mathbb{T}^n$ with the usual pointwise multiplication and metric defined by $d(x, y) = \max\{|x_j - y_j| : 1 \leq j \leq n\}$ for $x = (x_1, \ldots, x_n)$, $y = (y_1, \ldots, y_n)$ in $\mathbb{T}^n$. It is shown in [**61**, Example 3.7] that for any $a, b \in \mathbb{T}^n$ with $d(a, b) < \sqrt{2}$, (X, d) satisfies $B(a, b)$ with $K = \sqrt{2}$. (X, d) does not satisfy property $B(a, -a)$ (the third condition fails). It is also shown that $X = \mathbb{T}^n$ is 2-divisible and d is inverse and translation invariant, and hence (X, d) satisfies $C_1(a, b)$ and $C_2(a, b)$ for all $a, b \in X$.

EXAMPLE 5.9. [**61**, Example 3.8] Let K be a compact Hausdorff space and E any real-linear subspace of the space $C(K)$ of all complex-valued continuous functions on K. Define

$$d(f, g) = \max\left\{\|f/g - 1\|_\infty, \|g/f - 1\|_\infty\right\}$$

for all $f, g \in \exp(E)$, where $\|\cdot\|_\infty$ stands for the supremum norm $C(K)$. Then $\exp(E)$ is a (multiplicative) group and $(\exp(E), d)$ is a nr-quasi-distance space. However d is not, in general, a metric, it is semi-positive, symmetric and satisfies the identity of indiscernibles but it fails, in general, the triangle inequality. It is known that $\exp(E)$ is 2-divisible, and d is inverse and translation invariant. The nr-quasi-distance space $(\exp(E), d)$ satisfies properties $C_1(a, b)$ and $C_2(a, b)$ for every a, b in $\exp(E)$.

The next examples are based on metrics and distances which are not directly related with the norm of a Banach space. In our approach to the Hilbert projective metric and to the Thompson's metric we shall consider them in the setting of order unit spaces as in [**61, 81–83, 109**]. We refer to [**1, 6, 54, 82**] and the survey [**11**, §1] for the basic theory any undefined terms and details used on ordered Banach spaces. By a *a partially ordered space* we mean a real vector space A with a *proper cone* A^+ (i.e., A^+ is convex, $\mathbb{R}^+ A^+ \subseteq A^+$, and $A^+ \cap (-A^+) = \{0\}$). The *partial ordering* "$\leq$" on A is defined by $a \leq b$ if $b - a \in A^+$. Given $a, b \in A$ the *order interval* $[a, b]$ is the set of all $x \in A$ such that $a \leq x \leq b$. An element $u \in A^+$ is called an *order unit*, if it satisfies $A = \cup\{[-\lambda\, u, \lambda\, u] : \lambda \geq 0\}$. For each order unit u in A^+, the expression

$$\|x\|_u = \inf\{\lambda > 0 : -\lambda u \leq x \leq \lambda u\} \quad (x \in A),$$

defines a seminorm on A. The order unit u is called *Archimedean* if $na \leq u$ for all $n \in \mathbb{N}$, then $-a \in A^+$. If u is an Archimedean order unit, the seminorm $\|\cdot\|_u$ is actually a norm, A^+ is a closed convex cone, and the closed unit ball with respect to this norm is the order interval $[-u, u]$ (cf. [**6**, Theorem 2.2.5]). It is further known that u is a norm interior point of A^+ [**6**, Proposition 2.2.1]. Actually, by the just quoted results, the set $\mathrm{int}(A^+)$ of norm interior points of A^+ coincides with the set of *order units* in A (this was also rediscovered in [**11**, §1.4] and [**82**, Lemma 2.1]).

In the case that we have a partial order space A which is also a Banach space with respect to a given norm $\|\cdot\|$ (usually called an *ordered Banach space*), for each Archimedean order unit u in A we find a new norm $\|\cdot\|_u$. It is natural to ask what is the relation of these two norms.

For our goals in this note a typical example of an ordered Banach space can be given by the self-adjoint part, A_{sa}, of a C*-algebra A with the natural C*-norm, or the self-adjoint part, $\mathfrak{A}_{sa}$, of a JB*-algebra $\mathfrak{A}$ with the natural JB*-norm, with the cone of positive elements. We recall that the cone of positive elements in a JB*-algebra is a proper closed convex cone (see [**110**, Theorem I.6.1] and [**54**, §3.3]).

It is shown in [**52**, Example 2.15] that the cone of positive elements in a C*-algebra A admits an interior point with respect to the topology generated by the norm if and only if A is unital, and in such a case, the interior of A^+ with respect to the norm topology is the set of all positive invertible elements in A.

A first natural question now is to determine when the self-adjoint part of a C*-algebra (respectively, a JB*-algebra) admits an Archimedean order unit u. Let u be an Archimedean order unit in a C*-algebra A, and let $\|\cdot\|_u$ denote the corresponding order norm on A_{sa} generated by u. Let us take $a \in A_{sa}$. It is well known that a writes in the form $a = a^+ - a^-$ with $a^+, a^- \geq 0$, $a^+a^- = 0$ and $\|a\| = \|a^+\| \vee \|a^-\|$. There is no loss of generality in assuming that $\|a\| = \|a^+\|$. If $\delta > 0$ satisfies $-\delta u \leq a \leq \delta u$, by applying that the operator $x \mapsto a^+ x a^+$ is positive, we get

$$0 \leq \left(a^+\right)^3 = U_{a^+}(a^+) \leq \delta U_{a^+}(u),$$

and hence

$$\|a\|^3 = \|a^+\|^3 = \|\left(a^+\right)^3\| \leq \delta \|a^+\|^2 \|u\|, \text{ and } \|a\| = \|a^+\| \leq \|u\| \delta.$$

Taking infimum in δ we arrive to

$$\|a\| \leq \|u\| \|a\|_u, \text{ for all } a \in A_{sa}. \tag{5.4}$$

Similar arguments to those given above, but applying that for each hermitian element h in a JB*-algebra the operator U_h is positive [**54**, Proposition 3.3.6], prove that (5.4) holds for every self-adjoint element in a JB*-algebra $\mathfrak{A}$. However, the inequality in (5.4) nor its version for JB*-algebras allow us to apply the above commented [**52**, Example 2.15] because it does not imply that an interior point of A^* for the norm topology is an interior point for the norm $\|\cdot\|_u$.

PROPOSITION 5.10. *Let M be a JB*-algebra (in particular a C*-algebra). Then M_{sa} admits an Archimedean order unit u if and only if it is unital. In such a case u is a positive invertible element in M and $\|\cdot\|_u$ is equivalent to the JB*-norm on M_{sa}. Furthermore, the set of all Archimedean order units in M is precisely the set*

M^{++} of all positive invertible elements in M, which coincides with the interior of M^{+} with respect to the JB-norm or with respect to $\|.\|_u$.*

PROOF. In the first statement the "if" implication is clear. Suppose u is an Archimedean order unit in M_{sa}. Let N denote the JB*-subalgebra of M generated by u. It is known that M_{sa} is isometrically isomorphic as JB-algebras to $C_0(Sp(u)\backslash\{0\},\mathbb{R})$ and under this identification u corresponds to the inclusion of $Sp(u)$ into $\mathbb{C}$, where $Sp(u)$ denotes the spectrum of u (cf. [**54**, Theorem 3.2.4]). Since u must be also an order unit in N_{sa}, it follows that there exits a positive δ such that $u^{\frac{1}{2}} \leq \delta u$, which implies that 0 is isolated in $Sp(u)$, and hence $N_{sa} \cong C(Sp(u),\mathbb{R})$ is an associative unital JB-algebra and u is positive and invertible in N. Let p denote the unit of N. Let us show that p is the unit in M. Given $a \in M_{sa}$, the element $a - U_p(a)$ lies in M_{sa}, and by [**54**, Lemma 2.4.21]

$$U_p(a - U_p(a)) = U_p(a) - U_p^2(a) = U_p(a) - U_{p^2}(a) = 0.$$

Having in mind that u is an order unit in M_{sa} and $u \leq \|u\|p$, we deduce that p also is an order unit in M_{sa}, and so $\beta p \leq a - U_p(a) \leq \beta p$ for some $\beta > 0$. An standard argument in spectral theory shows that $U_p(a - U_p(a)) = a - U_p(a)$, and thus $a = U_p(a)$ for all $a \in M_{sa}$. Therefore p is the unit in M. This concludes the proof of the first statement.

For the second statement, suppose that u is an Archimedean order unit in M and the latter is unital with unit $\mathbf{1}_M$. In such a case $\mathbf{1}_M \leq \delta u$ for some $\delta > 0$, which clearly implies that u is invertible, positive, and lies in the interior of M^+ with respect to the JB*-norm. Let us finally observe that for each $a \in M_{sa}$ we have

$$\frac{1}{\|u^{-1}\|}\ \{\beta > 0 : -\beta \mathbf{1}_M \leq a \leq \beta \mathbf{1}_M\} \subseteq \{\delta > 0 : -\delta u \leq a \leq \delta u\},$$

and consequently $\|a\|_u \leq \frac{1}{\|u^{-1}\|}\|a\|$, which combined with (5.4) concludes the proof. □

The set of invertible elements in a unital C*-algebra A is a multiplicative group. However, the cone of positive elements in A is not, in general, stable under products. The same difficulty occurs to the set of invertible elements and the set of positive elements in a unital JB*-algebra M. However, as we have commented before, the sets M^{-1} of all invertible elements, M^+ of positive elements, and the set M^{++} of all positive and invertible elements in M enjoy another stability property

$$U_a(b) \in M^\sigma, \text{ for all } a, b \in M^\sigma \text{ and } \sigma \in \{+, -1, ++\}. \tag{5.5}$$

If we only consider invertible elements, the JB*-algebra M can be replaced with any unital complex Jordan-Banach algebra. Furthermore, the conclusion is also true when M^σ is replaced by the set $\mathcal{U}(M)$ of all unitary elements in M. The stability exhibited in (5.5) reaffirms the usefulness of the inverted Jordan triple product.

Let us fix a C*-algebra A. We equip the sets A^{-1}, A^{++}, and $\mathcal{U}(A)$, of invertible, positive and invertible and unitary elements in A, respectively, with the metric given by the norm, and we consider the multiplicative structure of A. Clearly, by (5.5), these three sets satisfy the condition in (5.3). It is easy to check that A^{-1} and A^{++} fail the properties $B(a,b)$, $C_1(a,b)$ and $C_2(a,b)$ in definitions 5.4, 5.5 and 5.6 for most pairs a, b in these sets. As we shall see later, unitaries exhibit a better behaviour. To avoid the difficulties mathematicians have considered alternative metrics on A^{++}.

It seems an appropriate moment to introduce the Hilbert projective metric and the Thompson's metric on the interior of the positive cone A^+ (i.e. the set of all Archimedean order units) in an order unit space equipped with an Archimedean order unit. For $a, b \in \text{int}(A^+)$, by applying the properties of order units, we set

$$M(a/b) := \inf\{\beta > 0\colon a \leq \beta b\}(< \infty).$$

On $\text{int}(A^+)$ Hilbert's projective metric and Thompson's metric are given by

$$(5.6)\quad d_H(a,b) = \log M(a/b)M(b/a), \text{ and } d_T(a,b) = \log\left(\max\{M(a/b), M(b/a)\}\right),$$

respectively. The mapping d_T satisfies all properties required for a metric on $\text{int}(A^+)$, while Hilbert's projective metric is a semi-metric (i.e., it is semi-positive, reflexive, symmetric, and satisfies the triangular inequality), however it has a certain handicap in what respect the identity of indiscernibles, since $d_H(\lambda a, \mu b) = d_H(a,b)$ for all $\lambda, \mu > 0$ and $a, b \in \text{int}(A^+)$ and $d_H(a,b) = 0$ with $a, b \in \text{int}(A^+)$ if and only if $a = \lambda b$ for some $\lambda > 0$ (see [**81**, Proposition 2.1.1 and comments in page 30]).

One of the most fruitful source of examples in this paper is coming from the cones of positive elements in a C*-algebra A, or more generally, in a JB*-algebra M. The interior of M^+ is non-empty if and only if M is unital and in such a case $\text{int}\,(M^+) = M^{++}$ (see Proposition 5.10). Do we know a spectral formula for the Thompson's metric?

We borrowed some paragraphs from a paper by Andruchow, Corach, and Stojanoff [**4**] on the differential geometry of the cone A^{++} of positive invertible elements in a unital C*-algebra A (see [**4, 26–30**]). A^{++} is an open convex subset of A_{sa}, and hence it can be considered as an open submanifold of A_{sa}, and the tangent space $(TA^+)_a$ at a point $a \in A^{++}$ will be identified with A_{sa}. There is a natural action of A^{-1} on A^{++} given by $(x,a) \mapsto xax^*$ $(x \in A^{-1}, a \in A^{++})$. This is a transitive action, that is, given $a, c \in A^{++}$, $x = c^{1/2}a^{-1/2}$ verifies $xax^* = c$. For every $a \in A^{++}$, the map $\tau_a : A^{-1} \to A^{++}$, $\tau_a(x) = xax^*$ is a principal fibre bundle with a natural connection. The unique geodesic γ such that $\gamma(0) = a$ and $\dot\gamma(0) = x \in A_{sa}$ is

$$\gamma(t) = e^{(t/2)xa^{-1}} = a^{1/2}e^{ta^{-1/2}xa^{-1/2}}a^{1/2},$$

and for every $a, b \in A^{++}$ there is a unique geodesic $\gamma_{a,b}$ such that $\gamma_{a,b}(0) = a$ and $\gamma_{a,b}(1) = b$, which is explicitly given by the expression

$$\gamma_{a,b}(t) = a^{1/2}(a^{-1/2}ba^{-1/2})^t a^{1/2},$$

corresponding to the vector $x = a^{1/2}\log\left(a^{-1/2}ba^{-1/2}\right)a^{1/2}$.

Although A^{++} is not a Riemannian manifold, there is a natural Finsler structure on it. The *geodesic or rectifiable distance* on A^{++} defined by

$$d_g(a,b) = \inf\{\text{ length } \gamma : \gamma \text{ is a smooth curve in } A^{++} \text{ joining } a \text{ and } b\},$$

satisfies $d_g(a,b) = \text{length}\gamma_{a,b} = \left\|\log\left(a^{-1/2}ba^{-1/2}\right)\right\|$. If a and b commute we have $d_g(a,b) = \left\|\log\left(ba^{-1}\right)\right\|$.

Andruchow, Corach, and Stojanoff proved in [**4**] that the geodesic distance obtained by differential geometric methods coincides with the Thompson's metric. An alternative explanation for $B(H)^{++}$ was given by Molnár in [**93**].

PROPOSITION 5.11 ([**4**, Proposition in page 1035]). *Let A be a unital C*-algebra. For every $a, b \in A^{++}$ the Thompson's metric satisfies*

$$d_T(a,b) = \left\| \log \left(a^{-1/2} b a^{-1/2} \right) \right\|. \tag{5.7}$$

EXAMPLE 5.12. [**61**, Example 3.6] Let d_T denote the Thompson's metric on the set A^{++} of all positive invertible elements in a unital C*-algebra A. The condition $B(a,b)$ holds for every pair $a, b \in A^{++}$. Given $a, b, c \in A^{++}$, since the mapping $z \mapsto bzb$ ($z \in A$) is positive, we can easily see that $a \leq tc$ for some $t > 0$ if and only if $bab \leq tbcb$, if and only if, $t^{-1}c^{-1} \leq a^{-1}$ [**102**, Proposition 1.3.6]. Thus the equality $d_T(ba^{-1}b, bc^{-1}b) = d_T(a^{-1}, c^{-1}) = d_T(a, c)$ follows directly from the definition of the Thompson's metric in (5.6). For the next property we take an argument from [**93**, Proof of Theorem 1]. Since $(ba^{-1}b)^{-1}a = (b^{-1}a)^2$, we are dealing with positive elements, by the basic properties of the spectrum we have

$$\begin{aligned} \sigma\left((ba^{-1}b)^{-1/2} a (ba^{-1}b)^{-1/2} \right) &= \sigma\left((ba^{-1}b)^{-1} a \right) = \sigma\left((b^{-1}a)^2 \right) \\ &= \sigma\left(b^{-1}a \right)^2 = \sigma\left(b^{-1/2} a b^{-1/2} \right)^2, \end{aligned}$$

and hence

$$\begin{aligned} \sigma \left(\log \left((ba^{-1}b)^{-1/2} a (ba^{-1}b)^{-1/2} \right) \right) &= \log \sigma \left((ba^{-1}b)^{-1/2} a (ba^{-1}b)^{-1/2} \right) \\ = 2 \log \sigma \left(b^{-1/2} a b^{-1/2} \right) &= 2\sigma \left(\log \left(b^{-1/2} a b^{-1/2} \right) \right). \end{aligned}$$

Taking supremum in the above sides of the above identity, by (5.7), we get

$$\begin{aligned} d_T\left(ba^{-1}b, a \right) &= \left\| \log \left((ba^{-1}b)^{-1/2} a (ba^{-1}b)^{-1/2} \right) \right\| \\ &= 2 \left\| \log \left(b^{-1/2} a b^{-1/2} \right) \right\| = 2 d_T\left(b, a \right). \end{aligned} \tag{5.8}$$

Since d_T is a metric, property $B(a,c)$ holds for every $a, c \in A^{++}$. Moreover, since A^{++} is 2-divisible, the conclusions in the first part of this example assure that property $C_2(a,c)$ also holds for every $a, c \in A^{++}$.

REMARK 5.13. The Thompson's metric does not change when computed among positive invertible elements in a unital JB*-subalgebra. That is, if M is a unital JB*-algebra and N is a JB*-subalgebra containing the unit of M, then for all $a, b \in N^{++}$ we have $d_T^N(a,b) = d_T^M(a,b)$. This is clear from the definition in (5.6).

We deal next with the case of JB*-algebras. There is a central result in the theory of JB*-algebras which has not been considered in this note yet. The so-called Shirshov–Cohn theorem gives a detailed information about the structure of the JB*-subalgebra generated by two self-adjoint elements.

THEOREM 5.14 (Shirshov–Cohn theorem ([**54**, Theorem 7.2.5], [**113**, Proposition 2.1])). *Let M be a unital JB*-algebra. Then the unital JB*-subalgebra $M_{a,b,\mathbf{1}}$ generated by two self-adjoint elements a, b and the unit in M is a JC*-algebra, that is, there exists a unital C*-algebra A such that $M_{a,b,\mathbf{1}}$ is a JB*-subalgebra of A.*

The first consequence of this result is a Jordan version of the formula proved by Andruchow, Corach, and Stojanoff in Proposition 5.11. The result was established by Lemmens, Roelands, and Wortel in [**82**], here we present a slightly different proof.

PROPOSITION 5.15 ([**82**, Proposition 2.4]). *Let M be a unital JB*-algebra. Then the Thompson's metric on M^{++} satisfies*

$$d_T(a,b) = \|\log(U_{a^{-1/2}}(b))\|, \text{ for every } a,b \in M^{++}. \tag{5.9}$$

Furthermore,

$$\begin{aligned} d_T\left(U_b(a^{-1}), a\right) &= 2 d_T(b,a), \text{ and} \\ d_T(U_c(a^{-1}), U_c(b^{-1})) &= d_T(a^{-1}, b^{-1}) = d_T(a,b), \end{aligned} \tag{5.10}$$

for every $a,b,c \in M^{++}$. Therefore (M^{**}, d_T) satisfies the Jordan analogues of properties $B(a,b)$ and $C_2(a,b)$ for all $a,b \in M^{++}$.

Let us observe, before dealing with the proof, that a direct proof of the fact that the formula on the right-hand-side of (5.9) is a distance on M^{++} does not seem an easy task.

PROOF. Let $N_{a,b,\mathbf{1}}$ denote the JB*-subalgebra of M generated by a,b and $\mathbf{1}$. Clearly, $a^{-1}, b^{-1} \in N_{a,b,\mathbf{1}}$. By the Shirshov-Cohn theorem (Theorem 5.14) there exists a unital C*-algebra A such that $M_{a,b,\mathbf{1}}$ is a JB*-subalgebra of A. By Proposition 5.11 we have

$$d_T^M(a,b) = d_T^{N_{a,b,\mathbf{1}}}(a,b) = d_T^A(a,b) = \|\log(U_{a^{-1/2}}(b))\|_A = \|\log(U_{a^{-1/2}}(b))\|_M .$$

Finally, by (5.8),

$$d_T^M\left(U_b(a^{-1}), a\right) = d_T^{N_{a,b,\mathbf{1}}}\left(U_b(a^{-1}), a\right) = d_T^A\left(ba^{-1}b, a\right) = 2d_T^A(b,a) = 2d_T^M(b,a).$$

The rest is clear by the properties of the cone of positive elements in M via the same arguments given in Example 5.12 (observe that M^{++} is 2-divisible). $\square$

Besides Theorem 5.2, the reference [**61**] contains other appropriate tools to determine preservers of inverted Jordan triple products.

THEOREM 5.16 ([**61**, Corollary 3.9]). *Let (X, d_X) and (Y, d_Y) be two non-empty nr-quasi-distance spaces with the additional property that $d_Y(y_1, y_2) = 0 \Rightarrow y_1 = y_2$. We also assume that X and Y are subsets of two groups $\mathcal{G}_1$ and $\mathcal{G}_2$, respectively, satisfying (5.3), that is, $yx^{-1}y \in X$ for all $x,y \in X$ and similarly for Y. Let $\Delta : X \to Y$ be a bijective d-preserving map, and fix $a,b \in X$. Suppose that the following hypotheses hold:*

(a) *X satisfies property $B(a,b)$.*
(b) *Y satisfies property $C_1(\Delta(a), \Delta(ba^{-1}b))$.*

Then we have $\Delta(ba^{-1}b) = \Delta(b)\Delta(a)^{-1}\Delta(b)$.

THEOREM 5.17 ([**61**, Corollary 3.10]). *Let (X, d_X) and (Y, d_Y) be two non-empty nr-quasi-distance spaces, and suppose additionally that $d_Y(y_1, y_2) = 0 \Rightarrow y_1 = y_2$. We also assume that X and Y are subsets of two groups $\mathcal{G}_1$ and $\mathcal{G}_2$, respectively, satisfying (5.3), that is, $yx^{-1}y \in X$ for all $x,y \in X$ and similarly for Y. Let $\Delta : X \to Y$ be a bijective d-preserving map, and fix $a,b \in X$. Suppose that the following hypotheses hold:*

(a) *X satisfies property $B(a,b)$.*
(b) *Y satisfies $C_2(\Delta(a), \Delta(ba^{-1}b))$.*
(c) *Y is 2-divisible and 2-torsion free.*

Then we have $\Delta(ba^{-1}b) = \Delta(b)\Delta(a)^{-1}\Delta(b)$.

The previous theorem was employed by Hatori and Molnár in [**62**, Theorem 9] to prove that every surjective isometry $\Delta : (A^{++}, d_T^A) \to (B^{++}, d_T^B)$, where A and B are unital C*-algebras, preserves inverted Jordan triple products. In this case A^{++} can be regarded inside A^{-1}, and the latter is a multiplicative group. However, for a unital JB*-algebra, the set M^{++} cannot be easily embedded inside a group, and Theorem 5.17 is not useful (there are subtle difficulties for properties $B(a, b)$, $C_1(a, b)$ and $C_2(a, b)$ in the Jordan setting). Despite of this handicap, we establish next a Jordan version of [**62**, Theorem 9].

THEOREM 5.18. *Let $\Delta : (M^{++}, d_T^M) \to (N^{++}, d_T^N)$ be a surjective isometry, where M and N are two unital JB*-algebras. Then Δ preserves inverted Jordan triple products, that is,*

$$\Delta(U_b(a^{-1})) = U_{\Delta(b)}(\Delta(a)^{-1}), \text{ for all } a, b \in M^{++}.$$

PROOF. Consider the mapping $\varphi : M^{++} \to M^{++}$ defined by $\varphi(x) := U_b(x^{-1})$. We claim that φ satisfies the hypotheses in Theorem 5.2. Namely,

$$\varphi^2(x) = \varphi\left(U_b(x^{-1})\right) = U_b\left((U_b(x^{-1}))^{-1}\right) = U_b(U_{b^{-1}}(x)) = x, \ (x \in X),$$

and hence φ is a period-2 bijection. We deduce that φ is d_T^M-preserving by (5.10) in Proposition 5.15. It is also clear that $\varphi(b) = b$.

Pick $c \in N^{++}$ with $U_c(\Delta(a)^{-1}) = \Delta\left(U_b(a^{-1})\right)$. Clearly, $d_T^N(U_c(x^{-1}), U_c(y^{-1})) = d_T^N(x, y)$ for all $x, y \in N^{++}$ (cf. Proposition 5.15). Define a mapping $\psi : N^{++} \to N^{++}$ given by $\psi(y) := U_c(y^{-1})$. As before, ψ is period-2 and d-preserving.

It is not difficult to check that

$$\psi(\Delta(a)) = U_c(\Delta(a)^{-1}) = \Delta\left(U_b(a^{-1})\right) = \Delta(\varphi(a))$$

and

$$\psi(\Delta(\varphi(a))) = U_c\left(\Delta\left(U_b(a^{-1})\right)^{-1}\right) = U_c\left(\left(U_c(\Delta(a)^{-1})\right)^{-1}\right) = \Delta(a).$$

The boundedness of the set $\{d_T^M(x, c) : x \in L\}$ in Theorem 5.2 follows from the fact that d_T^M is a metric (cf. Remark 5.3). Having in mind (5.10) we have $d_T^M(\varphi(x), x) = d_T^M\left(U_b(x^{-1}), x\right) = 2d_T^M(b, x)$ for all $x \in M^{++}$.

We are in a position to apply Theorem 5.2 to deduce that

$$d_T^N(\psi(\Delta(b)), \Delta(b)) = 0,$$

and hence $U_c(\Delta(b)^{-1}) = \Delta(b)$, because d_T^N is a metric. In such a case, by the fundamental identity we arrive to

$$\begin{aligned}\mathbf{1}_N &= U_{\Delta(b)^{-1/2}}(\Delta(b)) = U_{\Delta(b)^{-1/2}}U_c(\Delta(b)^{-1}) = U_{\Delta(b)^{-1/2}}U_cU_{\Delta(b)^{-1/2}}(\mathbf{1}_N)\\ &= U_{U_{\Delta(b)^{-1/2}}(c)}(\mathbf{1}_N) = \left(U_{\Delta(b)^{-1/2}}(c)\right)^2,\end{aligned}$$

which implies that $U_{\Delta(b)^{-1/2}}(c) = \mathbf{1}_N$, since $U_{\Delta(b)^{-1/2}}(c) \in N^{++}$. However,

$$U_{\Delta(b)^{-1/2}}(c) = \mathbf{1}_N = U_{\Delta(b)^{-1/2}}(\Delta(b)),$$

which combined with the invertibility of $U_{\Delta(b)^{-1/2}}$ implies that $c = \Delta(b)$. Finally, it follows from the definition of c that

$$\Delta(U_b(a^{-1})) = U_{\Delta(b)}(\Delta(a)^{-1}).$$

□

5.1. Positive invertible elements with the Thompson's metric.

We have already seen in subsection 4.1 that, thanks to Mankiewicz theorem (Theorem 3.18), for each unital JB*-algebra M, the metric space M^{++} equipped with the distance given by the norm is an invariant to identify unital JB*-algebras (cf. Theorem 4.9). In this subsection we shall consider the metric space (M^{++}, d_T).

The first result in this line was established by Molnár in [**93**, Theorem 1], who proved that if H is a complex Hilbert space with $\dim(H) \geq 3$, and $\Delta : (B(H)^{++}, d_T) \to (B(H)^{++}, d_T)$ is a bijective isometry, then Δ is either of the form $\Delta(a) = bab^*$, or of the form $\Delta(a) = bab^*$, $(a \in B(H)^{++})$, where b is an invertible bounded linear or conjugate-linear operator on H. The result was latter extended to general unital C*-algebras by Hatori and Molnár [**63**].

THEOREM 5.19 ([**63**, Theorem 9]). *Let A and B be unital C*-algebras. Then a mapping $\Delta : (A^{++}, d_T) \to (B^{++}, d_T)$ is a surjective isometry if and only if there exist a central projection p in B and a Jordan *-isomorphism $J : A \to B$ such that Δ is of the form*

$$\Delta(a) = \Delta(\mathbf{1}_A)^{1/2} \left(pJ(a) + (1-p)J(a^{-1})\right) \Delta(\mathbf{1}_A)^{1/2}, \quad a \in A^{++}.$$

A similar problem was studied by Lemmens, Roelands, and Wortel in the wider setting of unital JB*-algebras.

THEOREM 5.20 ([**82**, Theorem 3.2]). *Let M and N be unital JB*-algebras. Then a mapping $\Delta : (M^{++}, d_T) \to (N^{++}, d_T)$ is a surjective isometry if and only if there exist a central projection p in N and a Jordan *-isomorphism $J : M \to N$ such that Δ is of the form*

$$\Delta(a) = U_{\Delta(\mathbf{1}_M)^{1/2}} \left(p \circ J(a) + (\mathbf{1}_N - p) \circ J(a^{-1})\right), \quad a \in M^{++}. \tag{5.11}$$

The proof of the above result given by Lemmens, Roelands, and Wortel in [**82**] is based on differential geometry and geodesics. Here we present an adaptation of the ideas by Hatori and Molnár in [**63**].

PROOF. Let us begin by showing that expressions of the form in (5.11) preserve Thompson's metric. It is clear from the definition in (5.6) that every Jordan *-isomorphism $J : M \to N$ preserves Thompson's metric. The second identity in (5.10) in Proposition 5.15 assures that maps of the form $a \mapsto a^{-1}$ and $a \mapsto U_c(a)$ (with c fixed in M^{++}) preserve Thompson's metric. All these statements together prove the "if" implication.

Let us now deal with the "only if" implication. By replacing Δ with the mapping $\Delta_0(x) = U_{\Delta(\mathbf{1}_M)^{-1/2}}(\Delta(x))$, $(x \in M^{++})$, we can assume that Δ is unital. Theorem 5.18 assures that Δ preserves inverted Jordan triple products, that is,

$$\Delta(U_a(b^{-1})) = U_{\Delta(a)}(\Delta(b)^{-1}), \text{ for all } a, b \in M^{++}. \tag{5.12}$$

Combining this property with the fact that Δ is unital we arrive to $\Delta(a^2) = \Delta(a)^2$ for all $a \in M^{++}$, and by induction and a standard argument

$$\Delta(a^n) = \Delta(a)^n, \text{ and } \Delta(a^{\frac{1}{n}}) = \Delta(a)^{\frac{1}{n}}, \text{ for all } a \in M^{++},\ n \in \mathbb{N}. \tag{5.13}$$

Now, as in the proof of [**63**, Theorem 9] and [**82**, Theorem 3.2] we consider the next mapping $S : M_{sa} \to N_{sa}$, $S(x) = \log \Delta(e^x)$. Clearly, $S(0) = \log \Delta(\mathbf{1}_M) = 0$. We claim that S is a surjective isometry.

To see the claim, let us fix x, y in M_{sa}. By the Lie–Trotter formula established in [**42**, Corollary 2.2] (which is valid in the wider setting of unital complex Jordan-Banach algebras), we have

$$\lim_{n\to\infty}\left(U_{e^{-\frac{x}{2n}}}\left(e^{\frac{y}{n}}\right)\right)^n = e^{y-x}, \text{ and hence } \lim_{n\to\infty}\frac{\log\left(U_{e^{-\frac{x}{2n}}}\left(e^{\frac{y}{n}}\right)\right)}{n^{-1}} = y-x.$$

It then follows that

$$\lim_{n\to\infty}\frac{d_T^M\left(e^{\frac{x}{n}},e^{\frac{y}{n}}\right)}{n^{-1}} = \lim_{n\to\infty}\frac{\left\|\log\left(U_{e^{-\frac{x}{2n}}}\left(e^{\frac{y}{n}}\right)\right)\right\|}{n^{-1}} = \|x-y\|, \tag{5.14}$$

which asserts that we can asymptotically approximate the usual distance between self-adjoint elements by Thompson's distances of some exponentials.

By applying (5.13) we deduce that $e^{\frac{S(x)}{n}} = e^{\frac{\log\Delta(e^x)}{n}} = \Delta(e^{\frac{x}{n}})$, for all natural n. Having in mind that Δ is an isometry for the Thompson metrics we also have

$$d_T^N\left(e^{\frac{S(x)}{n}},e^{\frac{S(y)}{n}}\right) = d_T^N\left(\Delta(e^{\frac{x}{n}}),\Delta(e^{\frac{x}{n}})\right) = d_T^M\left(e^{\frac{x}{n}},e^{\frac{x}{n}}\right),$$

identity which combined with the conclusion in (5.14) leads to $\|S(x)-S(y)\| = \|x-y\|$, witnessing that S is an isometry. The basic properties of the holomorphic functional calculus show that the exponential is a bijection from M_{sa} onto M^{++}, and hence $S : M_{sa}\to N_{sa}$ is a surjective isometry with $S(0) = 0$. The Mazur–Ulam theorem now affirms that S is a surjective real-linear isometry between two unital JB-algebras. The result by Isidro and Rodríguez Palacios in [**69**, Theorem 1.9] proves the existence of a central symmetry s in N_{sa} and surjective Jordan isomorphism $J : M_{sa}\to N_{sa}$ such that $S(x) = s\circ J(x)$, for all $x\in M_{sa}$. It is known that $s = p-(\mathbf{1}_N - p)$, where p is a central projection in N. Therefore, for each $x\in M_{sa}$, since elements of the form $p\circ c$ and $(\mathbf{1}_N-p)\circ d$ with $c,d\in N_{sa}$ operator commute and have zero Jordan product, we have

$$\begin{aligned}\Delta(e^x) = e^{S(x)} &= e^{(p-(\mathbf{1}-p))\circ J(x)} = e^{p\circ J(x)}\circ e^{-(\mathbf{1}-p)\circ J(x)}\\ &= (\mathbf{1}_N - p + p\circ e^{J(x)})\circ(p+(\mathbf{1}_N-p)\circ e^{-J(x)})\\ &= (\mathbf{1}_N - p + p\circ J(e^x))\circ(p+(\mathbf{1}_N-p)\circ J(e^{-x}))\\ &= p\circ J(e^x)+(\mathbf{1}_N-p)\circ J(e^{-x}),\end{aligned}$$

for all $x\in M_{sa}$, where we have used that J is a Jordan isomorphism. This proves that $\Delta(a) = p\circ J(a)+(\mathbf{1}_N-p)\circ J(a^{-1})$ for all $a\in M^{++}$. □

It is clear that Theorem 5.19 can be rediscovered as a consequence of the result for JB*-algebras in Theorem 5.20.

6. Unitary elements as a metric invariant

After having explored the sets of invertible elements and positive invertible elements as metric invariants (with respect to the usual norm and with respect to the Thompson metric) in unital C*- and JB*-algebras in sections 4 and 5, our goal in this section is to explore whether the subset of unitary elements can be also employed to identify these algebras.

The first precedent in the literature is a result by Hatori and Molnár for $B(H)$. We first recall the notion of abstract transposition on $B(H)$. Let H be a complex Hilbert space. Suppose we fix a conjugation (i.e. conjugate linear isometry of period-2 on H). For each $a \in B(H)$ we define $\overline{a}, a^t \in B(H)$ by $\overline{a}(\xi) := \overline{a(\overline{\xi})}$, and $a^t := \overline{a^*}$ ($\xi \in H$). We can define an abstract transposition and a conjugation on $B(H)$ defined by $a \mapsto a^t$ and $a \mapsto \overline{a}$, respectively. The transposition is complex linear, satisfies $a^{tt} = a$ and $(ab)^t = b^t a^t$, while the conjugation is conjugate-linear and satisfies $\overline{\overline{a}} = a$ and $\overline{ab} = \overline{a}\overline{b}$, for all $a, b \in B(H)$.

THEOREM 6.1 ([**62**, Theorem 8]). *Let H be a complex Hilbert space. Suppose that $\Delta : \mathcal{U}(B(H)) \to \mathcal{U}(B(H))$ is a surjective isometry with respect to the metric given by the operator norm. Then there exists a surjective real-linear isometry (equivalently, a triple automorphism) T on $B(H)$ such that $\Delta(u) = T(u)$ for all $u \in \mathcal{U}(B(H))$. Furthermore, there exist unitaries $v, w \in \mathcal{U}(B(H))$ such that Δ is of one of the following forms:*

(*a*) $\Delta(u) = vuw^*$, *for all* $u \in \mathcal{U}(B(H))$,
(*b*) $\Delta(u) = vu^*w^*$, *for all* $u \in \mathcal{U}(B(H))$,
(*c*) $\Delta(u) = vu^tw^*$, *for all* $u \in \mathcal{U}(B(H))$,
(*d*) $\Delta(u) = v\overline{u}w^*$, *for all* $u \in \mathcal{U}(B(H))$.

It is worth to note that the concrete description of the mapping Δ in (a) to (d) above is a consequence of the first conclusion and a classic result affirming that every surjective real-linear isometry (equivalently, triple isomorphism cf. Theorem 4.5) T on $B(H)$ is of the one of the forms given in (a) to (d) for all $u \in B(H)$ in this case (see [**47**, page 162]).

The same authors of the previous theorem also determined the precise form of all surjective isometries between the unitary groups of two von Neumann algebras.

THEOREM 6.2 ([**63**, Theorem 1]). *Let A and B be unital C*-algebras. Assume that $\Delta : \mathcal{U}(A) \to \mathcal{U}(B)$ is a surjective isometry between the corresponding groups of unitary elements (with respect to the C*-norms). Then we have $\Delta(e^{iA_{sa}}) = \Delta(\mathbf{1}_A)\, e^{iB_{sa}}$ and there exist a central projection $p \in B$ and a Jordan *-isomorphism $J : A \to B$ such that*

$$\Delta(e^{ix}) = \Delta(\mathbf{1}_A)(pJ(e^{ix}) + (\mathbf{1}_B - p)J(e^{ix})^*),\ x \in A_{sa}. \tag{6.1}$$

*Furthermore, if A or B is a von Neumann algebra, then both algebras are von Neumann algebras and Δ admits an extension to a real-linear isometry from A onto B, more precisely, there exists a central projection $p \in B$ and a Jordan *-isomorphism $J : A \to B$ such that the conclusion in* (6.1) *holds when e^{ix} is replaced with an arbitrary unitary u in A.*

COROLLARY 6.3 ([**63**, Corollary 2]). *Two unital C*-algebras are isomorphic as Jordan *-algebras if and only if their unitary groups are isometric as metric spaces.*

As in previous sections, we shall include in this note versions of these results in the wider setting of unital JB*-algebras with a complete proof (see subsection 6.2). Before working with JB*-algebras, we shall turn back our view to the geometric tools in section 5 with the aim of establishing certain local properties of surjective isometries between subgroups of surjective linear isometries which will be useful in subsequent results.

The powerful geometric tool developed in Theorem 5.2 and the subsequent combinations with properties $B(a, b)$, $C_1(a, b)$ and $C_2(a, b)$ in Theorems 5.16 and 5.17 played a fundamental role in the study of surjective isometries between sets positive invertible elements in unital C*-algebras [**63**, **93**] and JB*-algebras [**82**]. These tools will be also very important in the results presented in this subsection.

Keeping the notation in previous sections, for each Banach space X, the symbol $\mathrm{Iso}(X)$ will stand for the surjective linear isometries on X, which is clearly a multiplicative group inside $B(X)$. The group $\mathrm{Iso}(X)$ serves as a test bench for properties $B(a,b)$ and $C_1(a,b)$. It will be checked in the proof of the next result, whose proof has been slightly simplified.

THEOREM 6.4. [**62**, Theorem 6] *Let X and Y be complex Banach spaces and let $\mathcal{G}$ and $\tilde{\mathcal{G}}$ be a subgroup of Iso(X) and Iso(Y), respectively, equipped with the metrics coming from the operator norm. Then $\mathcal{G}$ (respectively, $\tilde{\mathcal{G}}$) always satisfies property $C_1(U,V)$ for all $U,V\in\mathcal{G}$ (respectively, for all $U,V\in\tilde{\mathcal{G}}$). Suppose we take $U,V\in\mathcal{G}$ with $d(U,V)<1/2$. Then $\mathcal{G}$ satisfies property $B(U,V)$. Furthermore, every surjective isometry $\Delta:\mathcal{G}\to\tilde{\mathcal{G}}$ satisfies*

$$\Delta(VU^{-1}V)=\Delta(V)\Delta(U)^{-1}\Delta(V)$$

for all $U,V\in\mathcal{G}$ with $d(U,V)<1/2$.

PROOF. The three statements are intrinsically related. Since $\mathcal{G}$ is a subgroup of $\mathrm{Iso}(X)$ and the operator norm satisfies

$$\|UR^{-1}V-US^{-1}V\|=\|R^{-1}-S^{-1}\|=\|R-S\|$$

for all $R,S,U,V\in\mathrm{Iso}(X)$, it follows that $\mathcal{G}$ satisfies property $C_1(U,V)$ for all $U,V\in\mathcal{G}$. The final conclusion will follow from Theorem 5.16 as soon as we prove that $\mathcal{G}$ satisfies property $B(U,V)$ for all $U,V\in\mathcal{G}$ with $\|U-V\|<\frac{1}{2}$.

For $(B.1)$ we observe that

$$\|VW^{-1}V-VR^{-1}V\|=\|W^{-1}-R^{-1}\|=\|I-WR^{-1}\|=\|R-W\|,$$

for all $U,V,R,W\in\mathrm{Iso}(X)$. Property $(B.2)$ trivially holds because the distance given by the operator norm is a metric. The non-trivial property is $(B.3)$. For this purpose take $W\in L_{U,V}$ (that is, $\|U-W\|=\|VU^{-1}V-W\|=\|U-V\|$). The constant $K=2-2\|U-V\|>1$ by the hypothesis.

We shall first prove that $V+W$ is invertible. Namely

$$\|W-V\|\leq\|W-U\|+\|U-V\|=2\|U-V\|<\frac{1}{2},\tag{6.2}$$

and consequently, $\frac{1}{2}>\left\|\frac{V+W}{2}-V\right\|=\left\|\frac{V+W}{2}V^{-1}-I\right\|$, which gives the desired statement.

We claim next that

$$\|(W+V)^{-1}\|\leq\frac{1}{K}.\tag{6.3}$$

Indeed,

$$\begin{aligned}1&=\|(W+V)(W+V)^{-1}\|=\|(2W-(W-V))(W+V)^{-1}\|\\&\geq\|2W(W+V)^{-1}\|-\|(W-V)(W+V)^{-1}\|\\&\geq2\|(W+V)^{-1}\|-\|W-V\|\|(W+V)^{-1}\|\\&\geq(\text{by }(6.2))\geq2\|(W+V)^{-1}\|-2\|U-V\|\|(W+V)^{-1}\|\\&=(2-2\|U-V\|)\|(W+V)^{-1}\|=K\|(W-V)^{-1}\|.\end{aligned}$$

Now, we employ (6.3) and the identity $\left(W(W+V)^{-1}\right)^{-1} = I + VW^{-1}$ to deduce that

$$\left\|\left(I+VW^{-1}\right)^{-1}\right\| = \|W(W+V)^{-1}\| = \|(W+V)^{-1}\| \leq \frac{1}{K}. \tag{6.4}$$

Finally, by (6.4),

$$\begin{aligned} K\|W-V\| &= K\|VW^{-1}-I\| = K\|(VW^{-1}-I)(VW^{-1}-I)(VW^{-1}-I)^{-1}\| \\ &\leq K\|(VW^{-1}-I)^{-1}\|\|(VW^{-1}-I)(VW^{-1}-I)\| \\ &\leq \|(VW^{-1}-I)(VW^{-1}-I)\| = \|VW^{-1}VW^{-1}-I\| \\ &= \|VW^{-1}V-W\|, \end{aligned}$$

which concludes the proof of property $B(U,V)$, and hence the arguments in this proof. □

REMARK 6.5. Theorem 6.4 above implies the following result in the setting of unital C*-algebras: Let A be a unital C*-algebra, and let $\mathcal{U}(A)$ be the group of unitary elements in A. Since, by the Gelfand–Naimark theorem A can be regarded as a C*-subalgebra of some $B(H)$ and the unit of A coincides with the identity on H. Since $\mathcal{U}(A)$ can be regarded as a subgroup of Iso(H), Theorem 6.4 implies that $\mathcal{U}(A)$ satisfies $C_1(a,b)$ for all $a,b\in\mathcal{U}(A)$, and $B(a,b)$ for all $a,b\in\mathcal{U}(A)$ with $\|a-b\|<\frac{1}{2}$. Furthermore, if B is another unital C*-algebra and $\Delta:\mathcal{U}(A)\to\mathcal{U}(B)$ is a surjective isometry, we have $\Delta(ab^{-1}a) = \Delta(a)\Delta(b)^{-1}\Delta(a)$ for all $a,b\in\mathcal{U}(A)$ with $\|a-b\|<\frac{1}{2}$.

REMARK 6.6. As observed by Hatori and Molnár in [**62**, page 2130], the hypothesis $d(U,V)<\frac{1}{2}$ is absolutely necessary for the final conclusions in Theorem 6.4. Consider, for example, a compact Hausdorff space Ω with at least two points and the Banach space $C(\Omega)$. Let $\mathcal{G} = \text{Aut}(C(\Omega))$ denote the subgroup of Iso$(C(\Omega))$ of all composition operators on $C(\Omega)$ of the form $T_\varphi(f) = f\circ\varphi$ $(f\in C(\Omega))$, where ϕ runs in the set of all self-homeomorphisms on Ω (i.e., the group of all *-automorphisms on $C(\Omega)$). By applying Urysohn's lemma it can be checked that $\|T_\varphi - T_\phi\| = 0$ if $\phi=\varphi$ and $\|T_\varphi - T_\phi\| = 2$, otherwise. Therefore, every bijective transformation of $\mathcal{G}$ onto itself is a surjective isometry.

It is worth commenting at this point that finding an extension of every surjective isometry between the unitary groups of two unital C*-algebras is not always possible (that is, the final conclusion in Theorem 6.2 is not, in general valid when von Neumann algebras are replaced with mere unital C*-algebras). A counter example can be found even in the setting of commutative unital C*-algebras. A complete characterization of those surjective isometries between the unitary groups of two commutative unital C*-algebras was established by Hatori and Molnár in [**63**, Corollary 8]. We shall provide concrete methods to produce counterexamples in the next subsection.

6.1. Unitaries in unital JB*-algebras and the principal component.

We recall that an element u in a unital JB*-algebra M is called unitary if u is invertible with $u^* = u^{-1}$, which is equivalent to say to say that u is a unitary tripotent in M when the latter is regarded as JB*-triple (cf. [**17**, Proposition 4.3] or the comments in page 288).

The Jordan product of two unitaries need not be, in general, a unitary. However, given $u, v \in \mathcal{U}(M)$, the element $U_u(v)$ lies in $\mathcal{U}(M)$ (cf. Lemma 2.1). As in the case of C*-algebras, the set $\mathcal{U}(M)$ is not, in general, connected. The connected component of $\mathcal{U}(M)$ containing the unit element is called the *principal connected component* or simply the *principal component* and will be denoted by $\mathcal{U}^0(M)$.

Along this note we shall employ on several occasions that for each $u \in \mathcal{U}(M)$ the mapping U_u is a surjective linear isometry (cf. [**22**, Theorem 4.2.28(*vii*)]).

As in previous sections, we shall present the results in the setting of JB*-algebras in an analogy with the corresponding versions for C*-algebras, to allow the reader to taste similarities and divergences.

Let u be a unitary element in a unital C*-algebra A. A well known result in C*-algebra theory shows that $\|\mathbf{1}_A - u\| < 2$ is a sufficient condition to guarantee that u admits a "logarithm", that is, $u = e^{ih}$ for some $h \in A_s$ (see [**74**, Exercise 4.6.6]). The value 2 in the inequality is optimal; actually the conclusion is not true without this assumption (see the discussion preceding Proposition 4.4.10 in [**74**]). We present next an analogue of this result in the setting of JB*-algebras. Along the next technical results, the Shirshov-Cohn theorem will play a central role.

In the next lemma we shall only need the results affirming that a JB*-subalgebra generated by a single self-adjoint element and the unit element is isometrically isomorphic to a commutative unital C*-algebra [**54**, 3.2.4. The spectral theorem]. Let M be a unital JB*-algebra, and let u be a unitary in M. To avoid confusions, when we compute the exponential of an element a in the u-isotope JB*-algebra $M_{(u^*)}$ we shall denote $exp_{(u)}(a)$ or simply $\exp(a) = e^a$ if no confusion arises.

LEMMA 6.7. *Let M be a unital JB*-algebra M. Let u and v be unitaries in M satisfying $\|u - v\| = \eta < 2$. Then the following statements hold:*

(*a*) *There exists a self-adjoint element h in the u-isotope JB*-algebra $M(u) = M_{(u^*)} = (M, \circ_u, *_u)$ such that $v = \exp_{(u)}(ih)$, where the exponential is computed in the JB*-algebra $M(u)$. In particular u and v lie in the same connected component of $\mathcal{U}(M)$.*

(*b*) *There exists a unitary w in M satisfying $U_w(u^*) = v$.*

(*c*) *The equality $v^2 = u$ is equivalent to $U_{v^*}(u) = \mathbf{1}_M$.*

(*d*) *If $U_v(u^*) = \{v, u, v\} = u$ we have $w = u$.*

Moreover, if $\|u - v\| = \eta = \left|1 - e^{it_0}\right| = \sqrt{2}\sqrt{1 - \cos(t_0)}$ for some $t_0 \in (-\pi, \pi)$, we can further assume that $\|w - u\|, \|w - v\| \leq \sqrt{2}\sqrt{1 - \cos(\frac{t_0}{2})}$.

PROOF. (*a*) We consider the unital JB*-algebra $M(u) = M_{(u^*)} = (M, \circ_u, *_u)$. Let $\mathcal{C}$ denote the JB*-subalgebra of $M(u)$ generated by v and its unit –i.e. u–. Let us observe that the product and involution on $\mathcal{C}$, which will be denoted by "$\cdot$" and $*_u$, respectively, are precisely $\circ_u|_{\mathcal{C}\times\mathcal{C}}$ and $*_u|_{\mathcal{C}}$, respectively. Since v also is unitary in $M(u)$ (cf. Lemma 4.4), the JB*-subalgebra $\mathcal{C}$ must be isometrically Jordan *-isomorphic to a unital and commutative C*-algebra, that is, to some $C(\Omega)$ for an appropriate compact Hausdorff space Ω, and under this identification, u corresponds to the unit (cf. [**54**, 3.2.4. The spectral theorem] or [**22**, Proposition 3.4.2 and Theorem 4.1.3(*v*)]).

The hypothesis $\|u - v\| < 2$ is of pure geometric nature, and so it holds in $\mathcal{C}$. Since v is a unitary in $\mathcal{C}$ and u is the unit element of the C*-algebra $\mathcal{C} \equiv C(\Omega)$, we can find a self-adjoint element $h \in \mathcal{C}_{sa}$ such that $v = \exp_{(u)}(ih)$, where the exponential is computed in $\mathcal{C}$ or equivalently in $M(u) = M_{(u^*)}$ (cf. [**74**, Exercise 4.6.6]).

To see the final statement, we simply observe that the curve $\gamma(t) = \exp_{(u)}(ith)$ ($t \in [0,1]$), where the exponential is computed in $M(u)$, takes values in $\mathcal{U}(M)$ with $\gamma(0) = u$ and $\gamma(1) = v$.

(b) We keep the notation employed in the proof of (a). We observe that $w = \exp_{(u)}(i\frac{h}{2})$ is a unitary element in $\mathcal{C}$ satisfying $w \cdot u^{*_u} \cdot w = w \cdot u \cdot w = v$ (let us observe that the involution of $\mathcal{C}$ is precisely the restriction of $*_u$ to $\mathcal{C}$). Let $U_a^{(u)}$ denote the U-operator on the unital JB*-algebra $M(u) = M_{(u^*)}$ associated with an element a. Since $\mathcal{C}$ is a unital JB*-subalgebra of $M(u)$, we compute the corresponding operations in $\mathcal{C}$ to deduce, thanks to the uniqueness of the triple product, that

$$v = w \cdot u^{*_u} \cdot w = U_w^{(u)}(u^{*_u}) = \{w, u, w\} = U_w(u^*),$$

and clearly w is a unitary in M because it is a unitary in $M(u)$ (cf. Lemma 4.4).

(c) If $v^2 = u$, it is easy to see that $U_{v^*}(u) = U_{v^*}(v^2) = U_{v^*}U_v(\mathbf{1}_M) = \mathbf{1}_M$. Reciprocally, if $U_{v^*}(u) = \mathbf{1}_M$, we get

$$v^2 = U_v(\mathbf{1}_M) = U_v U_{v^*}(u) = u.$$

(d) We consider in this case the JB*-subalgebra $\mathcal{D}$ of $M(v) = M_{(v^*)} = (M, \circ_v, *_v)$ generated by u and the unit (v in this case). As before, we identify $\mathcal{D}$ with an appropriate $C(\Omega)$ in such a way that u is a unitary in this commutative unital C*-algebra with $u^{*_v} = U_v(u^*) = u$ (i.e. u self-adjoint in $\mathcal{D}$). The hypothesis $\|u - v\| < 2$ implies that $v = u$.

To see the final conclusion we employ, once again, the identification of $\mathcal{C}$ with $C(\Omega)$ in which u corresponds to the unit and v and w with $\exp_{(u)}(ih)$ and $\exp_{(u)}(i\frac{h}{2})$, respectively. □

The next observation is a straightforward consequence of Lemma 6.7(a). For the proof we simply observe that for any unitary u in a unital JB*-algebra M, the set $\mathbb{T}u$ lies in the same connected component of $\mathcal{U}(M)$ that u.

COROLLARY 6.8. *Let u and v be unitaries in a unital JB*-algebra. Then if u and v lie in two different connected components of $\mathcal{U}(M)$ we have $\|u \pm v\| = 2$.*

Thanks to the previous corollary, we can now exhibit examples of surjective isometries between sets of unitaries in unital JB*-algebras which cannot be extended to surjective real-linear isometries between the corresponding algebras.

REMARK 6.9. [**31**] Let u be a unitary element in a unital JB*-algebra M. We cannot always guarantee that $u = e^{ih}$ for some $h \in M_{sa}$ (cf. [**74**, Exercise 4.6.9]). Suppose that u satisfies the weaker condition that $u = v^2$ for some unitary $v \in M$ –this automatically holds if $u = e^{ih}$. Let $\mathcal{U}(M)^c$ denote the connected component of $\mathcal{U}(M)$ containing u. Since the mapping $U_v : M \to M(u)$ is a unital surjective linear isometry –and hence a Jordan *-isomorphism between these two algebras (see Theorem 4.3)– $U_v(\mathcal{U}^0(M))$ is a connected component of $\mathcal{U}(M) = \mathcal{U}(M(u))$ (see Lemma 4.4) and contains $u = U_v(\mathbf{1})$. It then follows $\mathcal{U}^c(M) = U_v(\mathcal{U}^0(M))$.

The existence of a unitary square root does not necessarily hold for any unitary element, even in the setting of commutative C*-algebras. Let $\mathbb{T}$ denote the unit sphere of the complex plane. The principal component of $\mathcal{U}(C(\mathbb{T}))$ is precisely the subgroup $\exp(iC_{\mathbb{R}}(\mathbb{T}))$, which is the set of all functions $u : \mathbb{T} \to \mathbb{T}$ which are deformable or homotopic to the unit element (cf. [**74**, Exercise 4.6.7]). The quotient group $\mathcal{U}(C(\mathbb{T}))/\mathcal{U}^0(C(\mathbb{T})) = \mathcal{U}(C(\mathbb{T}))/\exp(iC_{\mathbb{R}}(\mathbb{T}))$ –known as the Bruschlinsky group– identifies with the ring $\mathbb{Z}$ of integers (see [**67**, §II.3] or [**110**, Exercise I.11.3]). It is known that the maps $u_1, u_2 : \mathbb{T} \to \mathbb{T}$, $u_1(\lambda) = \lambda$ and $u_2(\lambda) = \lambda^2$ are not in the principal component of $\mathcal{U}(C(\mathbb{T}))$, they are actually in two different connected components (cf. [**67**, §II, Lemma 3.2]). Therefore, $u_2 \notin \exp(iC_{\mathbb{R}}(\mathbb{T}))$ but $u_2 = u_1^2$ admits a unitary square root.

We can now present an example of a surjective isometry between the unitary sets of two unital JB*-algebras which cannot be extended to a surjective real-linear isometry.

EXAMPLE 6.10. Let M be a unital JB*-algebra whose set of unitaries is not connected. Let $\mathcal{U}^c(M)$ be a connected component of $\mathcal{U}(M)$ such that $\mathcal{U}^0(M) \neq \mathcal{U}^c(M)$. Let us take an element $u \in \mathcal{U}^c(M)$ and any non-zero $h \in \left(M_{(u)}\right)_{sa}$. Set $w = \exp_{(u)}(ih), v = \exp_{(u)}(i\frac{h}{2}) \in \mathcal{U}(M) = \mathcal{U}(M_{(u)})$. Clearly, w, v lie in the connected component of $\mathcal{U}(M)$ containing u, that is, $u, v, w \in \mathcal{U}^c(M)$, and since $U_v(u) = \{v, u, v\} = v \circ_u v = \exp_{(u)}(i\frac{h}{2}) \circ_u \exp_{(u)}(i\frac{h}{2}) = \exp_{(u)}(ih)$, the surjective linear isometry U_v maps $\mathcal{U}^c(M)$ onto itself. Clearly, $U_v \neq Id_M$. Define a surjective mapping $\Delta : \mathcal{U}(M) \to \mathcal{U}(M)$ by $\Delta(a) = U_v(a)$ if $a \in \mathcal{U}^c(M)$ and $\Delta(a) = a$ otherwise. Δ is an isometry thanks to Corollary 6.8. If there exists an extension of Δ to a surjective real-linear isometry $T : M \to M$, the conditions $\Delta(\mathbf{1}_M) = \mathbf{1}_N$, and $\Delta(\lambda a) = \lambda a$ for all $a \in \mathcal{U}(M)\backslash\mathcal{U}^c(M)$ and Theorem 4.5 assure that T must be a Jordan*-homomorphism, Furthermore, since $T|_{\mathcal{U}^0(M)} = Id_M|_{\mathcal{U}^0(M)}$ and $\exp(iM_{sa}) \subseteq \mathcal{U}^0(M)$, the mapping T must be the identity, which contradicts that $\Delta|_{\mathcal{U}^c(M)} = U_v|_{\mathcal{U}^c(M)}$. See [**63**, Corollary 8]) and [**58**, Corollary 5.1] for additional examples.

We have described the principal component of the invertible elements in a unital Jordan-Banach algebra in subsection 2.2. Now it is the turn of the principal component of the unitary elements in a unital JB*-algebra. It is perhaps worthwhile to revisit the structure in the case of unital C*-algebras. The next theorem gathers some classical results and a more recent reinterpretation due to Hatori.

THEOREM 6.11 ([**74**, Exercises 4.6.6 and 4.6.7], [**110**, Exercise 2, page 56], and [**58**, Lemma 3.2]). *Let A be a unital C*-algebra, and let $\mathcal{U}^0(A)$ denote the principal component of the set $\mathcal{U}(A)$ of all unitary elements in A. Then*

$$\begin{aligned}\mathcal{U}^0(A) &= \{e^{ih_1} \cdot \ldots \cdot e^{ih_n} : n \in \mathbb{N},\ h_1, \ldots, h_n \in A_{sa}\} \\ &= \{e^{ih_1} \cdot \ldots \cdot e^{ih_n} \mathbf{1}_A e^{ih_n} \cdot \ldots \cdot e^{ih_1} : n \in \mathbb{N},\ h_1, \ldots, h_n \in A_{sa}\}\end{aligned} \tag{6.5}$$

is open, closed, and path connected in the (relative) norm topology on $\mathcal{U}(A)$. Furthermore, by the continuity of the modulus mapping $x \mapsto |x| := (x^*x)^{\frac{1}{2}}$, we have

$$A_{\mathbf{1}}^{-1} \cap \mathcal{U}(A) = \mathcal{U}^0(A),$$

where $A_{\mathbf{1}}^{-1}$ stands for the principal component of the invertible elements in A.

The final conclusion in the previous theorem is usually obtained via "polar decompositions", this tool is not, in general, available in the setting of JB*-algebras.

Our next goal is to present a recent description of the principal component of the set of unitaries in a unital JB*-algebra. To this end, let us fix a unital JB*-algebra

M. Let us consider a unitary element u in M, and the u-isotope $M(u) = M_{(u^*)} = (M, \circ_u, *_u)$ as defined in previous sections. We know that $\mathcal{U}(M) = \mathcal{U}(M(u))$ (cf. Lemma 4.4). Since the connected components of a topological space are disjoint, we can affirm that

$$\mathcal{U}^0(M(u)) = \mathcal{U}^0(M), \text{ for every } u \in \mathcal{U}^0(M). \tag{6.6}$$

Our arguments below require the employment of another tool coming from JB*-triple theory. The subtriple generated by a single element admits an useful concrete description. This is not natural for single generated C*-subalgebras, which cannot be identified, via Gelfand theory, with a commutative C*-algebra. The JB*-subtriple of a JB*-triple E generated by a single element $a \in E$ (denoted by E_a) coincides with the norm closure of the linear span of the odd powers of a defined as $a^{[1]} = a$, $a^{[3]} = \{a, a, a\}$, and $a^{[2n+1]} := \{a, a, a^{[2n-1]}\}$ $(n \in \mathbb{N})$. The Gelfand theory for commutative JB*-triples shows that E_a is isometrically JB*-triple isomorphic to some $C_0(\Omega_a)$ for a unique compact Hausdorff space Ω_a contained in the set $[0, \|a\|]$, such that 0 cannot be an isolated point in Ω_a, where, $C_0(\Omega_a)$ stands for the Banach space of all complex-valued continuous functions on Ω_a vanishing at 0 in case that $0 \in \Omega_a$, and all continuous functions on Ω_a otherwise. We can further assume that under the corresponding isomorphism a identifies with the inclusion function of Ω_a in $\mathbb{C}$ (cf. [**77**, Lemma 3.2, Corollary 3.4 and Proposition 3.5], [**76**, Corollary 1.15], see also [**22**, Theorem 4.2.9]). The set Ω_a is called *the triple spectrum* of a (in E), and it does not change when computed with respect to any JB*-subtriple $F \subseteq E$ containing a [**77**, Proposition 3.5]. The reader should be warned that in some reference, like [**22**, Theorem 4.2.9], the triple spectrum of an element a is defined as the locally compact set $\Omega \backslash \{0\}$. Here we follow the notation in [**77**], where, in a more natural terminology, the triple spectrum is a compact set.

We can naturally define the *continuous triple functional calculus* at the element a. Let us consider, via Gelfand theory, the triple isomorphism $\Psi : E_a \to C_0(\Omega_a)$. For each continuous function $f \in C_0(\Omega_a)$, we set $f_t(a) := \Psi^{-1}(f)$, and we call it the *continuous triple functional calculus* of f at a. If $p(\lambda) = \alpha_1\lambda + \alpha_3\lambda^3 + \ldots + \alpha_{2n-1}\lambda^{2n-1}$ is an odd polynomial with complex coefficients, it is easy to check that $p_t(a) = \alpha_1 a + \alpha_3 a^{[3]} + \ldots + \alpha_{2n-1}a^{[2n-1]}$. However, for more general functions in $C_0(\Omega_a)$ the continuous triple functional calculus is not so obvious. For example if a is a non-self-adjoint element in a C*-algebra A, the element $a^{[2]} = h_t(a)$ with $h(t) = t^2$ does not coincide with a^2 in A.

Another example: the function $g(t) = \sqrt[3]{t}$ produces $g_t(a) = a^{[\frac{1}{3}]}$. Let us note that $g_t(a) = a^{[\frac{1}{3}]}$ is the unique *cubic root* of a in E_a, i.e. the unique element $a^{[\frac{1}{3}]} \in E_a$ satisfying

$$\{a^{[\frac{1}{3}]}, a^{[\frac{1}{3}]}, a^{[\frac{1}{3}]}\} = a. \tag{6.7}$$

The sequence $(a^{[\frac{1}{3^n}]})_n$ can be recursively defined by $a^{[\frac{1}{3^{n+1}}]} = \left(a^{[\frac{1}{3^n}]}\right)^{[\frac{1}{3}]}$, $n \in \mathbb{N}$. It is easy to see that the sequence $(a^{[\frac{1}{3^n}]})_n$ need not be, in general, norm convergent in E (consider, for example, the element $a(t) = t$ in $E = C([0, 1])$). We shall need another topology to assure the convergence of the latter sequence.

If we strength the assumptions on E, for example if a is an element in a JBW*-triple W (in particular when E is regarded inside E^{**}), the sequence $(a^{[\frac{1}{3^n}]})_n$ converges in the weak*-topology of W to a (unique) tripotent denoted by $r(a)$. The tripotent $r(a)$ is called the *range tripotent* of a in W. This range tripotent $r(a)$ can be characterized as the smallest tripotent $e \in W$ satisfying that a is positive in the JBW*-algebra $W_2(e)$ (compare [**41**, Lemma 3.3]). When E is just a JB*-triple, we shall denote by $r(a)$ the range tripotent of a in E^{**}.

The range tripotents of von Neumann regular elements in a JB*-triple E always provide elements in E. We recall that an element a in a JB*-triple E is called *von Neumann regular* if $a \in Q(a)(E)$; if $a \in Q(a)^2(E)$ we say that a is *strongly von Neumann regular* (cf. [**21, 43, 71, 77**]). For each von Neumann regular element a in a JB*-triple E there might exist many different elements c in E satisfying $Q(a)(c) = a$. According to [**43**, Theorem 1], [**77**, Lemma 4.1, Lemma 3.2 and comments after its proof], [**21**, Theorem 2.3 and Corollary 2.4], the following five statements are equivalent reformulations:

(*a*) a is von Neumann regular.
(*b*) There exists $b \in E$ such that $[Q(a), Q(b)] = Q(a)Q(b) - Q(b)Q(a) = 0$, $Q(a)(b) = a$, and $Q(b)(a) = b$.
(*c*) 0 does not belong to the triple spectrum, Ω_a, of a.
(*d*) $Q(a)$ has norm-closed range.
(*e*) The range tripotent $r(a)$ of a lies in E and a is positive and invertible in the JB*-algebra $E_2(r(a))$.

The element b appearing in statement (*b*) above is unique. It is called the *generalized inverse* of a in E and is denoted by $a^\dagger$. The symbol $\mathrm{Reg}_{vN}(E)$ will stand for the set of all von Neumann regular elements in E. Another interesting property assures that

$$L(a, a^\dagger) = L(a^\dagger, a) = L(r(a), r(a)), \text{ and } Q(a)Q(a^\dagger) = Q(a^\dagger)Q(a) = P_2(r(a))$$

(see, for example, [**71**, page 589] or [**21**, comments in page 192]).

As a illustrative example, we mention that in a unital JB*-algebra M, the set $\mathrm{Reg}_{vN}(M)$ contains all invertible elements in M, and for each $z \in M^{-1}$, its range tripotent, $r(z)$, lies in $\mathcal{U}(M)$:

$$z \in M^{-1} \Rightarrow r(z) \in \mathcal{U}(M) \quad \text{(cf. [71, Lemma 2.2 and Remark 2.3])}. \tag{6.8}$$

A brief argument reads as follows:

$$\begin{aligned} Q(z)((z^{-1})^*) &= \{z, (z^{-1})^*, z\} = U_z(z^{-1}) = z; \\ Q((z^{-1})^*)(z) &= \{(z^{-1})^*, z, (z^{-1})^*\} = U_{(z^{-1})^*}(z^*) = (z^{-1})^*; \\ Q(z)Q((z^{-1})^*)(x) &= U_z\left(U_{(z^{-1})^*}(x^*)\right)^* = U_z U_{z^{-1}}(x) = U_z U_z^{-1}(x) = x; \\ Q((z^{-1})^*)Q(z)(x) &= U_{(z^{-1})^*}\left(U_z(x^*)\right)^* = U_{(z^*)^{-1}} U_{z^*}(x) = U_{z^*}^{-1} U_{z^*}(x) = x, \end{aligned}$$

for all $x \in M$, which proves that z is von Neumann regular with $z^\dagger = (z^{-1})^* = (z^*)^{-1}$, and since $P_2(r(z)) = Q(z)Q(z^\dagger) = Id$, the tripotent $r(z)$ is a unitary in M.

In principle the continuous triple functional calculus seems to be limited to odd powers of the involved elements. Fix an element a in a JB*-triple E and a triple isomorphism $\Psi : E_a \to C_0(\Omega_a)$ satisfying $\Psi(a)(t) = t$ $(t \in \Omega_a)$. The function $g(t) = t^2$ lies in $C_0(\Omega_a)$, and hence $g_t(a) = a^{[2]}$ is an element in E_a. The reader should be warned that $a^{[2]}$ is only a square for a "local binary product" defined only on E_a, depending on the element a.

If a is von Neumann regular (i.e. $\Omega_a \subset \mathbb{R}^+$), the function $h(t) = \frac{1}{t}$ lies in $C_0(\Omega_a)$ and $a^{[-1]} = h_t(a) \in E_a$. It is easy to see that

$$\{a, a, a^{[-1]}\} = a, \{a, r(a), a^{[-1]}\} = r(a), \text{ and } \{r(a), a^{[-1]}, r(a)\} = a^{[-1]}.$$

By definition $a^{[2]}$ belongs to the subtriple of E generated by a. Since the range tripotent of a satisfies that $a \in E_2(r(a))$, it follows that $E_a \subseteq E_2(r(a))$ because the latter is a JB*-subtriple of E. The elements $a, a^{[-1]}$, $a^{[2]}$ and $a^\dagger$ belong to the unital JB*-algebra $E_2(r(a))$. By Kaup's Banach–Stone theorem (Theorem 4.3) the restriction of the triple product of E to $E_2(r(a))$ coincides with

$$\{x, y, z\} = (x \circ_{r(a)} y^{*_{r(a)}}) \circ_{r(a)} z + (z \circ_{r(a)} y^{*_{r(a)}}) \circ_{r(a)} x - (x \circ_{r(a)} z) \circ_{r(a)} y^{*_{r(a)}},$$

for all $x, y, z \in E_2(r(a))$. We consequently have

$$\begin{aligned} r(a) &= \{a^\dagger, a, r(a)\} = L(a^\dagger, a)(r(a)) = a^\dagger \circ_{r(a)} a, \\ a &= L(a^\dagger, a)(a) = \{a^\dagger, a, a\} \\ &= (a^\dagger \circ_{r(a)} a^{*_{r(a)}}) \circ_{r(a)} a + (a \circ_{r(a)} a^{*_{r(a)}}) \circ_{r(a)} a^\dagger - (a^\dagger \circ_{r(a)} a) \circ_{r(a)} a^{*_{r(a)}} \\ &= (a \circ_{r(a)} a) \circ_{r(a)} a^\dagger, \end{aligned}$$

which proves that $a^\dagger = a^{[-1]}$.

Among the consequences of the basic properties of the holomorphic functional calculus for complex Jordan-Banach algebras reviewed in Theorem 2.2, we see that for each holomorphic function f on an open set $\Omega \subseteq \mathbb{C}$, the mapping $a \mapsto f(a)$ is a continuous application from the set of all elements in a complex Jordan-Banach M whose spectrum in inside Ω into M. We insert now a result assuring a similar property for the continuous triple functional calculus (cf. the arguments in the proof of [**110**, Proposition I.4.10]).

PROPOSITION 6.12 ([**31**, Proposition 2.1]). *Let $\Omega \subset \mathbb{R}_0^+$ be a compact set, and let E be a JB*-triple. Suppose $\mathcal{E}_\Omega$ denotes the set of all elements $a \in E$ whose triple spectrum is inside Ω, that is*

$$\mathcal{E}_\Omega = \{a \in E : \Omega_a \subseteq \Omega\}.$$

If $f : \Omega \to \mathbb{C}$ is a continuous function, then the mapping defined by the continuous triple functional calculus:

$$\mathcal{E}_\Omega \longrightarrow E, \; a \mapsto f_t(a)$$

is (norm) continuous. In particular, the mapping $a \in E \mapsto a^{[2]} \in E$ is continuous.

PROOF. Let $\iota : \Omega \to \mathbb{C}$ denote the natural inclusion mapping. By the sharp refinement of Stone-Weierstrass theorem in [**76**] and [**22**, Lemma 4.2.8] the set

$$\{p(\iota^2)\iota : p \text{ a polynomial with complex coefficients}\}$$

is dense in $C_0(\Omega)$. Therefore, given $\varepsilon > 0$ there exists a polynomial q with complex coefficients such that for $q(\lambda) = p(\lambda^2)\lambda$ we have $\sup_{t\in\Omega} |q(t) - f(t)| < \varepsilon$. The norm continuity of q follows from the norm continuity of the triple product of E (cf. (4.5)). We can therefore find $\delta > 0$ such that $\|q_t(a) - q_t(b)\| < \varepsilon$ for all $a, b \in E$ with $\|a - b\| < \delta$ and $\|a\|, \|b\| \leq \max \Omega$. Taking $a, b \in \mathcal{E}_\Omega$ with $\|a - b\| < \delta$ we have

$$\|f_t(a) - f_t(b)\| \leq \|f_t(a) - q_t(a)\| + \|q_t(a) - q_t(b)\| + \|q_t(b) - f_t(b)\| < 3\varepsilon.$$

The rest is standard. □

Despite the lacking of polar decompositions in the setting of JB*-algebras, the previous result assures the continuity of the mapping which sends invertible elements in a unital JB*-algebra to their corresponding range tripotents.

COROLLARY 6.13 ([**31**, Corollary 2.2]). *Let M be a unital JB*-algebra. Then the mapping*

$$M^{-1} \longrightarrow \mathcal{U}(M), \ a \mapsto r(a) \text{ is continuous.}$$

PROOF. By considering the JB*-subtriple of M generated by an element a, and the properties of the continuous triple functional calculus, we see that $r(a) = Q(a^\dagger)(a^{[2]})$.

The set M^{-1} is open, and the mapping $a \mapsto a^{-1}$ is continuous on M^{-1} (cf. Lemma 2.1). Therefore, the mapping $a \in M^{-1} \mapsto a^\dagger = (a^{-1})^* \in M^{-1}$ is continuous. Consequently, by the continuity of the triple product (see (4.5)), the mapping

$$a \in M^{-1} \mapsto r(a) = Q(a^\dagger)(a^{[2]})$$

is continuous. Let us finally note that $r(a) \in \mathcal{U}(M)$ for all $a \in M^{-1}$ (cf. (6.8)). □

The pursued algebraic characterization of the principal component of the set of unitaries in a unital JB*-algebra can be now established.

THEOREM 6.14 ([**31**, Theorem 2.3.]). *Let M be a unital JB*-algebra, let $\mathcal{U}^0(M)$ denote the principal component of the set of unitaries in M and let u be an element in $\mathcal{U}(M)$. Then the following statements are equivalent:*

(*a*) $u \in M_{\mathbf{1}}^{-1} \cap \mathcal{U}(M)$.
(*b*) *There exists a continuous path* $\Gamma : [0,1] \to \mathcal{U}(M)$ *with* $\Gamma(0) = \mathbf{1}_M$ *and* $\Gamma(1) = u$.
(*c*) $u \in \mathcal{U}^0(M)$.
(*d*) $u = U_{e^{ih_n}} \cdots U_{e^{ih_1}}(\mathbf{1}_M)$, *for some* $n \in \mathbb{N}$, $h_1, \ldots, h_n \in M_{sa}$.
(*e*) *There exists* $w \in \mathcal{U}^0(M)$ *such that* $\|u - w\| < 2$.

Consequently,

$$\begin{aligned} \mathcal{U}^0(M) &= M_{\mathbf{1}}^{-1} \cap \mathcal{U}(M) \\ &= \{U_{e^{ih_n}} \cdots U_{e^{ih_1}}(\mathbf{1}_M) \colon n \in \mathbb{N},\ h_j \in M_{sa}\ \forall\, 1 \le j \le n\} \\ &= \big\{u \in \mathcal{U}(M) : \text{ there exists } w \in \mathcal{U}^0(M) \text{ with } \|u - w\| < 2\big\} \end{aligned} \tag{6.9}$$

is analytically arc-wise connected.

PROOF. (*a*) ⇒ (*b*) Suppose $u \in M_{\mathbf{1}}^{-1} \cap \mathcal{U}(M)$. Since by Theorem 2.12, the set $M_{\mathbf{1}}^{-1}$ is analytically arc-wise connected, there exists a continuous path $\gamma : [0,1] \to M_{\mathbf{1}}^{-1}$ satisfying $\gamma(0) = \mathbf{1}_M$ and $\gamma(1) = u$. Corollary 6.13 assures that the mapping $\Gamma : [0,1] \to \mathcal{U}(M)$, $\Gamma(t) := r(\gamma(t))$ is continuous with $\Gamma(0) = \mathbf{1}_M$ and $\Gamma(1) = u$.

The implications (*b*) ⇒ (*c*) and (*c*) ⇒ (*e*) are clear.

(*c*) ⇒ (*a*) It follows from (*a*) ⇒ (*b*) ⇒ (*c*) that $M_{\mathbf{1}}^{-1} \cap \mathcal{U}(M) \subseteq \mathcal{U}^0(M)$. On the other hand, $M_{\mathbf{1}}^{-1}$ is a clopen subset of M^{-1} (cf. Theorem 2.12), and thus $M_{\mathbf{1}}^{-1} \cap \mathcal{U}(M)$ is a clopen subset of $\mathcal{U}(M)$ which contains the unit and is contained in $\mathcal{U}^0(M)$. Having in mind that $\mathcal{U}^0(M)$ is a connected set, we deduce that $M_{\mathbf{1}}^{-1} \cap \mathcal{U}(M) = \mathcal{U}^0(M)$.

The arguments up to this point assure that statements (*a*), (*b*) and (*c*) are equivalent, and moreover, $\mathcal{U}^0(M) = M_{\mathbf{1}}^{-1} \cap \mathcal{U}(M)$ is analytically arc-wise connected.

$(b) \Rightarrow (d)$ Suppose there exists a continuous path $\Gamma : [0,1] \to \mathcal{U}(M)$ with $\Gamma(0) = \mathbf{1}_M$ and $\Gamma(1) = u$. Let us find, by continuity and compactness, $0 = t_0 < t_1 < \ldots < t_n < 1 = t_{n+1}$ such that $\|\Gamma(t_i) - \Gamma(t_{i+1})\| < 2$ for all $0 \leq i \leq n$. Set $u_i = \Gamma(t_i)$ for all $0 \leq i \leq n+1$. By applying Lemma 6.7(a) to the element u_1 –which satisfies $\|u_1 - \mathbf{1}_M\| < 2$– we deduce the existence of $h_1 \in M_{sa}$ such that $U_{e^{ih_1}}(\mathbf{1}_M) = u_1$. Since $2 > \|u_1 - u_2\| = \|\mathbf{1}_M - U_{e^{-ih_1}}(u_2)\|$, a new application of Lemma 6.7(a) shows the existence of $h_2 \in M_{sa}$ satisfying $U_{e^{-ih_1}}(u_2) = U_{e^{ih_2}}(\mathbf{1}_M)$, or equivalently, $u_2 = U_{e^{ih_1}} U_{e^{ih_2}}(\mathbf{1}_M)$. Suppose, by an induction argument on n, that we have found $h_1, \ldots, h_n \in M_{sa}$ with $u_n = U_{e^{ih_1}} \cdots U_{e^{ih_n}}(\mathbf{1}_M)$. As before, the condition

$$2 > \|u_{n+1} - u_n\| = \|U_{e^{-ih_n}} \cdots U_{e^{-ih_1}}(u_{n+1}) - \mathbf{1}_M\|,$$

implies, via Lemma 6.7(a), the existence of $h_{n+1} \in M_{sa}$ such that

$$u = u_{n+1} = U_{e^{ih_1}} U_{e^{ih_2}} \cdots U_{e^{ih_{n+1}}}(\mathbf{1}_M),$$

which gives (d).

The implication $(d) \Rightarrow (c)$ is also easy because given $u = U_{e^{ih_n}} \cdots U_{e^{ih_1}}(\mathbf{1}_M)$, with $n \in \mathbb{N}$, $h_1, \ldots, h_n \in M_{sa}$, the mapping $\Gamma : [0,1] \to \mathcal{U}(M) \subseteq M^{-1}$, $\Gamma(t) = U_{e^{ith_n}} \cdots U_{e^{ith_1}}(\mathbf{1}_M)$ is an analytic curve with $\Gamma(0) = \mathbf{1}_M$ and $\Gamma(1) = u$. This also proves the second equality in (6.9). Therefore, (d) also is equivalent to the first three statements.

Suppose finally that statement (e) holds, that is, there exists $w \in \mathcal{U}^0(M)$ satisfying $\|w - u\| < 2$. By the first two equalities in (6.9), $w = U_{e^{ih_n}} \cdots U_{e^{ih_1}}(\mathbf{1}_M)$ for some $h_1, \ldots, h_n$ in M_{sa}. A similar argument to that in the proof of $(b) \Rightarrow (d)$ implies that $u = U_{e^{ih_{n+1}}} U_{e^{ih_n}} \cdots U_{e^{ih_1}}(\mathbf{1}_M) \in \mathcal{U}^0(M)$. □

In a unital C*-algebra A all connected components of $\mathcal{U}(A)$ are determined by the principal component. Namely, for each $u \in \mathcal{U}(A)$, the left multiplication operator $L_u : A \to A$ is a surjective linear isometry mapping the unit to u. Consequently, $L_u(\mathcal{U}^0(A)) = u\mathcal{U}^0(A)$ is precisely the connected component of $\mathcal{U}(A)$ containing u. However, the left multiplication operator admits no direct Jordan analogue. The good behaviour exhibited by unital C*-algebras fails in the setting of unital JB*-algebras, not only due to the lacking of a left (or right) multiplication operator, but for the possibility that two different connected components of the unitary set of a unital JB*-algebra are not isometric as metric spaces (see Remark 6.36 below).

REMARK 6.15. Let w be a unitary element in a unital JB*-algebra M. Suppose $\mathcal{U}^c(M)$ is a connected component of $\mathcal{U}(M)$, $w \in \mathcal{U}^c(M)$ and M_w^{-1} is the connected component of M^{-1} containing w. Let $M(w)$ denote the w-isotope admitting w as unit element. Since $\mathcal{U}(M) = \mathcal{U}(M(w))$ and $M^{-1} = (M(w))^{-1}$ as sets (cf. Lemma 2.13), $\mathcal{U}^c(M)$ and M_w^{-1} are the principal components of $\mathcal{U}(M(w))$ and $(M(w))^{-1}$, respectively. Therefore the previous Theorem 6.14 applied to the w-isotope $M(w)$ gives

(6.10)
$$\begin{aligned}
\mathcal{U}^c(M) &= M_w^{-1} \cap \mathcal{U}(M) \\
&= \left\{ U^{(w)}_{\exp_{(w)}(ih_n)} \cdots U^{(w)}_{\exp_{(w)}(ih_1)}(w) \colon n \in \mathbb{N}, h_j \in (M(w))_{sa}, \forall\, 1 \leq j \leq n \right\} \\
&= \{u \in \mathcal{U}(M) : \text{ there exists } v \in \mathcal{U}^c(M) \text{ with } \|u - v\| < 2\}
\end{aligned}$$

is analytically arc-wise connected. We can actually take any $\widetilde{w} \in \mathcal{U}^c(M)$ and the same conclusion holds for the connected component $\mathcal{U}^c(M)$ and the $\widetilde{w}$-isotope $M(\widetilde{w})$.

The existence of a square root for a unitary u in a unital JB*-algebra M facilitates the arguments to understand the connected component of $\mathcal{U}(M)$ containing u.

REMARK 6.16. Let u be a unitary element in a unital JB*-algebra M. Suppose that u satisfies that $u = v^2$ for some unitary $v \in M$. Let $\mathcal{U}(M)^{(c)}$ denote the connected component of $\mathcal{U}(M)$ containing u. Since the mapping $U_v : M \to M(u) = M_{(u^*)}$ is a unital surjective linear isometry –and hence a Jordan *-isomorphism between these two algebras– $U_v(\mathcal{U}^0(M))$ is a connected component of $\mathcal{U}(M) = \mathcal{U}(M(u))$ and contains u. It then follows from Theorem 6.14 that

$$\mathcal{U}^c(M) = U_v(\mathcal{U}^0(M)) = \{U_v U_{e^{ih_n}} \cdots U_{e^{ih_1}}(\mathbf{1}_M) \colon n \in \mathbb{N},\ h_j \in M_{sa}\ \forall\, 1 \le j \le n\}.$$

Previous studies have searched for sufficient conditions to assure that a unitary element u in a JB*-algebra M admits a square root (see, for example, [**17**, Problem 5.1 and Lemma 5.2], [**60**], [**24**] and [**89**].

A consequence of the previous Theorem 6.14 allows us to describe the principal component of $\mathcal{U}(M)$ in terms of quadratic subsets.

PROPOSITION 6.17 ([**31**, Proposition 2.6]). *Let M be a unital JB*-algebra. Then $\mathcal{U}^0(M)$ is a self-adjoint quadratic subset of $\mathcal{U}(M)$, that is, for all $u, w \in \mathcal{U}^0(M)$ the elements w^* and $U_w(u)$ belong to $\mathcal{U}^0(M)$. Furthermore, $\mathcal{U}^0(M)$ is the smallest quadratic subset of $\mathcal{U}(M)$ containing the set $e^{iM_{sa}}$. In particular,*

$$U_{\mathcal{U}^0(M)}\left(\mathcal{U}^0(M)\right) = \mathcal{U}^0(M) = \left\{U_{e^{ih_n}} \cdots U_{e^{ih_1}}(e^{ih_0}) \colon n \in \mathbb{N},\ h_j \in M_{sa}\ \forall\, 0 \le j \le n\right\}.$$

PROOF. Since $u, w \in \mathcal{U}^0(M)$, Theorem 6.14 guarantees the existence of hermitian elements $h_m, \ldots, h_1, k_n, \ldots, k_1 \in M_{sa}$, with $n, m \in \mathbb{N}$, such that $w = U_{e^{ih_m}} \cdots U_{e^{ih_1}}(\mathbf{1}_M)$ and $u = U_{e^{ik_n}} \cdots U_{e^{ik_1}}(\mathbf{1}_M)$. By applying that the involution is a conjugate linear Jordan *-isomorphism we get

$$w^* = (U_{e^{ih_m}} \cdots U_{e^{ih_1}}(\mathbf{1}_M))^* = U_{e^{-ih_m}} \cdots U_{e^{-ih_1}}(\mathbf{1}_M),$$

which, by virtue of Theorem 6.14, proves that $w^* \in \mathcal{U}^0(M)$.

On the other hand, by the fundamental identity (2.2) we arrive to

$$\begin{aligned} U_w(u) &= U_w\left(U_{e^{ik_n}} \cdots U_{e^{ik_1}}(\mathbf{1}_M)\right) = U_{U_{e^{ih_m}} \cdots U_{e^{ih_1}}(\mathbf{1}_M)}\left(U_{e^{ik_n}} \cdots U_{e^{ik_1}}(\mathbf{1}_M)\right) \\ &= U_{e^{ih_m}} \cdots U_{e^{ih_1}} U_{\mathbf{1}_M} U_{e^{ih_1}} \cdots U_{e^{ih_m}}\left(U_{e^{ik_n}} \cdots U_{e^{ik_1}}(\mathbf{1}_M)\right) \\ &= U_{e^{ih_m}} \cdots U_{e^{ih_1}} U_{e^{ih_1}} \cdots U_{e^{ih_m}} U_{e^{ik_n}} \cdots U_{e^{ik_1}}(\mathbf{1}_M) \in \mathcal{U}^0(M), \end{aligned}$$

by Theorem 6.14. We have therefore proved that $\mathcal{U}^0(M)$ is a self-adjoint quadratic subset of M.

Suppose now that $\mathcal{V}$ is a quadratic subset of $\mathcal{U}(M)$ containing the set $e^{iM_{sa}}$. Clearly for each $h_1, \ldots, h_n$ in M_{sa} an induction argument shows that

$$U_{e^{ih_n}} \ldots U_{e^{ih_1}}(\mathbf{1}_M) \in \mathcal{V},$$

and thus, by Theorem 6.14, $\mathcal{U}^0(M) \subseteq \mathcal{V}$. The rest is clear. □

The next remark leads to a natural open problem.

REMARK 6.18. Let M be a unital JB*-algebra and let u be an element in $\mathcal{U}(M)\backslash\mathcal{U}^0(M)$. suppose $\mathcal{U}(M)^{(u)}$ denotes the connected component of $\mathcal{U}(M)$ containing u. The set

$$\mathcal{E}_u := \left\{ U_{e^{ih_1}} \dots U_{e^{ih_n}}(u) : \begin{array}{c} n \in \mathbb{N},\ h_j \in M_{sa} \\ \text{for all } 1 \leq j \leq n \end{array} \right\}$$

is arc-wise connected and contains u. Thus, $\mathcal{E}_u \subseteq \mathcal{U}(M)^{(u)}$. We do not know if the reverse inclusion holds, in general. If $M = A$ is a unital C*-algebra both inclusions hold, a fact which is essentially due to the existence of left and right multiplication operators. Namely, since $\mathcal{U}^0(A) = \{e^{ih_1} \cdot \ldots \cdot e^{ih_n} : n \in \mathbb{N},\ h_1, \ldots, h_n \in A_{sa}\}$ (cf. (6.5)), and the right multiplication operator $R_u : A \to A$, $x \mapsto xu$ is a surjective linear isometry mapping the unit to u, we deduce that

$$\{e^{ih_1} \cdot \ldots \cdot e^{ih_n} u : n \in \mathbb{N},\ h_1, \ldots, h_n \in A_{sa}\} = R_u(\mathcal{U}^0(A)) = \mathcal{U}(A)^{(u)}.$$

Similarly, the products of the form $we^{ih_n} \cdot \ldots \cdot e^{ih_1}$ lie in $\mathcal{U}(A)^{(u)}$ for all $w \in \mathcal{U}(A)^{(u)}$, $h_1, \ldots, h_n \in A_{sa}$. So, given $h_1, \ldots, h_n \in A_{sa}$, the mapping $R_{e^{ih_1} \cdot \ldots \cdot e^{ih_n}}$ is a surjective linear isometry on A mapping $\mathcal{U}(A)^{(u)}$ onto itself, and hence

$$\begin{aligned} \mathcal{U}(A)^{(u)} &= \{e^{ih_1} \cdot \ldots \cdot e^{ih_n} u e^{ih_n} \cdot \ldots \cdot e^{ih_1} : n \in \mathbb{N},\ h_1, \ldots, h_n \in A_{sa}\} \\ &= \left\{ U_{e^{ih_1}} \dots U_{e^{ih_n}}(u) : \begin{array}{c} n \in \mathbb{N},\ h_j \in A_{sa} \\ \text{for all } 1 \leq j \leq n \end{array} \right\}. \end{aligned}$$

Fortunately, the inclusion $\mathcal{E}_u \subseteq \mathcal{U}^c(M)$ is enough to prove the following rule concerning the U-operators

$$U_{\mathcal{U}^0(M)}\left(\mathcal{U}(M)^{(u)}\right) = \mathcal{U}(M)^{(u)} \tag{6.11}$$

for any connected component $\mathcal{U}(M)^{(u)} \subseteq \mathcal{U}(M)$. Indeed, by Theorem 6.14, every $w \in \mathcal{U}^0(M)$ is of the form $w = U_{e^{ih_n}} \cdots U_{e^{ih_1}}(\mathbf{1}_M)$ with $n \in \mathbb{N}$, $h_j \in M_{sa}$. Therefore, given $\tilde{u} \in \mathcal{U}(M)^{(u)}$, it follows from (2.2) that

$$\begin{aligned} U_w(\tilde{u}) &= U_{U_{e^{ih_n}} \cdots U_{e^{ih_1}}(\mathbf{1}_M)}(\tilde{u}) \\ &= U_{e^{ih_n}} \dots U_{e^{ih_1}} U_{e^{ih_1}} \dots U_{e^{ih_n}}(\tilde{u}) \in \mathcal{E}_{\tilde{u}} \subseteq \mathcal{U}(M)^{(\tilde{u})} = \mathcal{U}(M)^{(u)}. \end{aligned}$$

We finally observe that the equality $\displaystyle\bigcup_{w \in \mathcal{U}(M)^{(u)}} \mathcal{E}_w = \mathcal{U}(M)^{(u)}$ follows trivially from the above arguments.

The subsection about properties of the set of unitary elements in a unital JB*-algebras is not complete without a short discussion on one-parameter unitary quadratic maps.

In the setting of operators, a *one-parameter group* of bounded linear operators on a Banach space Z is a mapping $\mathbb{R} \to B(Z)$, $t \mapsto E(t)$ satisfying

$$E(0) = I, \text{ and } E(t + s) = E(s)E(t), \text{ for all } s, t \in \mathbb{R}.$$

There are two notions of continuity. In the first one, we say that a one-parameter group $\{E(t) : t \in \mathbb{R}\}$ is uniformly continuous at the origin if $\lim_{t \to 0} \|E(t) - I\| = 0$. This is the strongest assumption, and it is further known that this is equivalent to the existence of a bounded linear operator $R \in B(Z)$ such that $E(t) = e^{tR}$ for all $t \in \mathbb{R}$, where the exponential is computed in the Banach algebra $B(Z)$ (cf., for example, [**16**, Proposition 3.1.1]). The one-parameter group $\{E(t) : t \in \mathbb{R}\}$ on Z is called *strongly continuous* if for each z in Z the mapping $t \mapsto E(t)(z)$ is (norm) continuous ([**25**, Definition 5.3, Chapter X]). A *one-parameter unitary group* on a complex Hilbert space H is a one-parameter group on H such that $E(t)$ is a unitary element for each $t \in \mathbb{R}$.

The celebrated Stone's one-parameter theorem affirms that for each strongly continuous one-parameter unitary group $\{E(t) : t \in \mathbb{R}\}$ on a complex Hilbert space H there exists a self-adjoint operator $h \in B(H)$ such that $E(t) = e^{ith}$, for every $t \in \mathbb{R}$ ([**25**, 5.6, Chapter X]).

Let us recall that the existence of exceptional JB*-algebras which cannot be represented inside $B(H)$ makes hopeless a direct application of Stone's one-parameter theorem. Uniformly continuous one-parameter groups of surjective isometries (i.e. triple isomorphisms compare Kaup's Banach–Stone theorem 4.3), Jordan *-isomorphisms and orthogonality preserving operators on JB*-algebras have been recently recently described in terms of triple derivations in [**50**, **51**].

It should be recalled that a *triple derivation* on a JB*-triple E is a linear mapping $\delta : E \to E$ satisfying a ternary version of Leibniz' rule

$$\delta\{a, b, c\} = \{\delta(a), b, c\} + \{a, \delta(b), c\} + \{a, b, \delta(c)\}, \quad (a, b, c \in E).$$

Triple derivations on JB*-triples are automatically continuous (cf. [**9**, Corollary 2.2] or [**66**, **106**]). Another interesting property affirms that if $\delta : M \to M$ is a triple derivation on a unital JB*-algebra, we have $\delta(\mathbf{1}_M)^* = -\delta(\mathbf{1}_M)$ (cf. [**66**, Proof of Lemma 1]).

The natural limitations of the set of unitary elements in a unital JB*-algebra (e.g. the Jordan product of two unitary elements is not, in general, a unitary) produces what seems, a priori, an obstacle for a literal transcription of Stone's one-parameter theorem to the setting of unital JB*-algebras. The next appropriate Jordan version of Stone's one-parameter theorem has been taken from [**35**].

THEOREM 6.19 ([**35**, Theorem 3.1]). *Let M be a unital JB*-algebra. Suppose that $\{u(t) : t \in \mathbb{R}\}$ is a family in $\mathcal{U}(M)$ such that $u(0) = \mathbf{1}_M$ and the correspondence $t \mapsto u(t)$ is (norm) continuous. Then the following statements are equivalent:*

(*a*) *There exists $h \in M_{sa}$ such that $u(t) = e^{ith}$ for all $t \in \mathbb{R}$.*
(*b*) *$u(s + t) = u(s) \circ u(t)$, for all $t, s \in \mathbb{R}$.*
(*c*) *$U_{u(t)}(u(s)) = u(2t + s)$, for all $t, s \in \mathbb{R}$.*

PROOF. Since the JB*-subalgebra generated by the element h is a commutative (and associative) Banach algebra, and the exponential is taken in this subalgebra, the implication $(a) \Rightarrow (b)$ is clear since $u(s + t) = e^{i(s+t)h} = e^{ish}e^{ith} = u(s) \circ u(t)$. The implications $(b) \Rightarrow (c)$ is also clear since, by (b) we have

$$\begin{aligned} U_{u(t)}(u(s)) &= 2(u(t) \circ u(s)) \circ u(t) - u(t)^2 \circ u(s) \\ &= 2u(t + s) \circ u(t) - u(2t) \circ u(s) = u(2t + s). \end{aligned}$$

$(c) \Rightarrow (a)$. In order to apply the results on one-parameter group of surjective linear isometries in [**50**, **51**], we consider the family of surjective linear isometries on M defined by $\{U_{u(t)} : t \in \mathbb{R}\}$. Clearly, the mapping $t \mapsto U_{u(t)}$ is (norm) continuous with $U_{u(0)} = Id_M$ by hypotheses. The fundamental identity (2.1) asserts that

$$U_{u(t)}U_{u(s)}U_{u(t)} = U_{U_{u(t)}(u(s))} = U_{u(2t+s)},$$

for all $s, t \in \mathbb{R}$. It then follows that

$$U_{u(t)}^2 = U_{u(t)}U_{u(0)}U_{u(t)} = U_{u(2t)},\ U_{u(t)}^3 = U_{u(t)}U_{u(t)}U_{u(t)} = U_{u(3t)},$$

and if we assume that $U_{u(kt)} = U_{u(t)}^k$ for all $k < n$ and $n \geq 3$ we get

$$U_{u(nt)} = U_{u(2t+(n-2)t)} = U_{u(t)}U_{u((n-2)t)}U_{u(t)} = U_{u(t)}U_{u(t)}^{n-2}U_{u(t)} = U_{u(t)}^n.$$

An induction arguments proves that

$$U_{u(nt)} = U_{u(t)}^n \text{ for all } t \in \mathbb{R},\ n \in \mathbb{N}.$$

Furthermore, for each real t, $U_{u(t)}U_{u(-t)}U_{u(t)} = U_{u(t)}$ and $U_{u(t)} \in \text{Iso}(M)$, and hence $U_{u(-t)} = U_{u(t)}^{-1}$ for all $t \in \mathbb{R}$. It follows that, for a negative integer n and each real t, we have

$$U_{u(nt)} = U_{u((-n)(-t))} = U_{u(-t)}^{-n} = U_{u(t)}^n,$$

an identity which then holds for all $n \in \mathbb{Z}$.

We shall prove that the family $\{U_{u(t)} : t \in \mathbb{R}\}$ is a one-parameter group of surjective linear isometries on M. To this end, let us fix two rational numbers $\frac{n}{m}, \frac{n'}{m'}$ with $m, m' \in \mathbb{N}$, $n, n' \in \mathbb{Z}$. The properties proved above imply that

$$U_{u\left(\frac{n}{m}t+\frac{n'}{m'}t\right)} = U_{u\left(\frac{nm'+n'm}{mm'}t\right)} = U_{u\left(\frac{1}{mm'}t\right)}^{nm'+n'm} = U_{u\left(\frac{1}{mm'}t\right)}^{nm'}U_{u\left(\frac{1}{mm'}t\right)}^{n'm} = U_{u\left(\frac{n}{m}t\right)}U_{u\left(\frac{n'}{m'}t\right)},$$

and thus it follows by continuity that

$$U_{u(t+s)} = U_{u(t)}U_{u(s)}, \text{ for all } t, s \in \mathbb{R},$$

that is, the family $\{U_{u(t)} : t \in \mathbb{R}\}$ is a uniformly continuous one-parameter group of surjective isometries on M. By [**51**, Lemma 3.2] there exists a triple derivation $\delta : M \to M$ satisfying

$$U_{u(t)} = e^{t\delta} = \sum_{n=0}^{\infty} \frac{t^n}{n!}\delta^n \text{ for all } t \in \mathbb{R}, \tag{6.12}$$

where the exponential is computed in the Banach algebra $B(M)$.

We continue with the next observation:

$$e^{\frac{t}{2}\delta}(\mathbf{1}_M) = U_{u\left(\frac{t}{2}\right)}(\mathbf{1}_M) = u\left(\frac{t}{2}\right)^2 = U_{u\left(\frac{t}{2}\right)}(u(0)) = u(t), \tag{6.13}$$

for all $t \in \mathbb{R}$, and $u(t)^2 = U_{u(t)}(\mathbf{1}_M) = U_{u(t)}(u(0)) = u(2t)$. It follows that

$$\left(\sum_{k=0}^{\infty} \frac{t^n}{2^n n!}\delta^n(\mathbf{1}_M)\right)^2 = \left(e^{\frac{t}{2}\delta}(\mathbf{1}_M)\right)^2 = (u(t))^2 = u(2t) = e^{t\delta}(\mathbf{1}_M) = \sum_{k=0}^{\infty} \frac{t^n}{n!}\delta^n(\mathbf{1}_M). \tag{6.14}$$

By comparing both extremes in the previous identity we arrive to

$$\frac{1}{2^2}\delta(\mathbf{1}_M)^2 + 2\frac{1}{2^3}\delta^2(\mathbf{1}_M) = \frac{1}{2}\delta^2(\mathbf{1}_M),$$

which assures that $\delta(\mathbf{1}_M)^2 = \delta^2(\mathbf{1}_M)$. If we combine this identity with that in (6.14) it follows that

$$\frac{1}{2^2 3!}\delta^3(\mathbf{1}_M) + \frac{1}{2^3}\delta(\mathbf{1}_M)^3 = 2\frac{1}{2^3 3!}\delta^3(\mathbf{1}_M) + 2\frac{1}{2^3}\frac{1}{2}\delta^2(\mathbf{1}_M)\delta(\mathbf{1}_M) = \frac{1}{3!}\delta^3(\mathbf{1}_M),$$

and hence $\delta(\mathbf{1}_M)^3 = \delta^3(\mathbf{1}_M)$. An easy induction argument leads to

$$\delta(\mathbf{1}_M)^n = \delta^n(\mathbf{1}_M), \text{ for all natural } n. \tag{6.15}$$

It follows from (6.12), (6.13), and (6.15) that

$$u(t) = e^{\frac{t}{2}\delta}(\mathbf{1}_M) = \sum_{n=0}^{\infty} \frac{t^n}{2^n n!}\delta^n(\mathbf{1}_M) = \sum_{n=0}^{\infty} \frac{t^n}{2^n n!}\delta(\mathbf{1}_M)^n = e^{t\frac{\delta(\mathbf{1}_M)}{2}},$$

for all $t \in \mathbb{R}$. As we commented above $\delta(\mathbf{1}_M)^* = -\delta(\mathbf{1}_M)$ in M (cf. [**66**, Proof of Lemma 1]), and hence $u(t) = e^{ith}$ with $h = \frac{\delta(\mathbf{1}_M)}{2i} \in M_{sa}$. □

6.2. Surjective isometries between connected components of unitary sets.

In the previous subsection 6.1 we have studied the properties of the unitary set of a unital JB*-algebra and its principal connected component. We have seen that the existence of multiple connected components can be applied to find surjective isometries between unitary sets which cannot be extended to surjective real-linear isometries between the algebras (cf. Example 6.10). In this subsection we survey the results studying the extensibility of a surjective isometry between two connected components of the unitary sets of two unital JB*-algebras.

We would like to apply the geometric tools in section 5. For this purpose we need to check appropriate Jordan versions of properties $B(a,b)$, $C_1(a,b)$ and $C_2(a,b)$. We should observe that the existence of exceptional JB*-algebras which cannot be embedded inside $B(H)$ makes it impossible to apply Theorem 6.4 as we did in Remark 6.5.

Our first result gathers ideas from [**35**] and [**31**].

THEOREM 6.20 ([**35**, Theorem 2.9] and [**31**, Theorem 3.1]). *Let M and N be two unital JB*-algebras, and let $\Delta : \mathcal{U}^0(M) \to \mathcal{U}^0(N)$ be a surjective isometry. Suppose $u, v \in \mathcal{U}^0(M)$ with $\|u - v\| < \frac{1}{2}$. Then the following statements hold:*

(1) *For all $x, y \in \mathcal{U}^0(M)$ we have*

$$\left\|U_v(x^{-1}) - U_v(y^{-1})\right\| = \|U_v(x^*) - U_v(y^*)\| = \|x^* - y^*\| = \|x - y\|.$$

(2) *The constant $K = 2 - 2\|u - v\| > 1$ satisfies that*

$$\|U_v(w^*) - w\| = \left\|U_v(w^{-1}) - w\right\| \geq K\|w - v\|,$$

for all w in the set

$$L^0_{u,v} = \{w \in \mathcal{U}^0(M) : \|u - w\| = \|U_v(u^{-1}) - w\| = \|U_v(u^*) - w\| = \|u - v\|\}.$$

(3) *$\mathcal{U}^0(N)$ satisfies a Jordan version of $C_2(\Delta(u), \Delta(U_u(v^*)))$, that is, there exists w in $\mathcal{U}^0(N)$ satisfying*

$$U_w(\Delta(u)^*) = \Delta\left(U_v(u^*)\right), \ \|w - \Delta(v)\| < 2,$$

and

$$\|U_w(x^*) - U_w(y^*)\| = \|x - y\|, \quad \forall x, y \in \mathcal{U}^0(N).$$

(4) *The equalities $\Delta(U_v(u^*)) = \Delta(U_v(u^{-1})) = U_{\Delta(v)}(\Delta(u)^{-1}) = U_{\Delta(v)}(\Delta(u)^*)$ hold.*

PROOF. (1) The desired conclusion follows from the fact that U_v and the involution are isometries on M (cf. [**22**, Theorem 4.2.28(*vii*)]).

(2) Let us consider the u-isotope JB*-algebra $M(u) = (M, \circ_u, *_u)$ of M. The U-operator on $M(u)$ will be denoted by U^u. We fix an element $w \in L_{u,v}$. Since $\|u - w\| = \|u - v\| < \frac{1}{2}$, we deduce from Lemma 6.7 the existence of two self-adjoint

elements $h_1, h_2 \in M(u)$ such that $v = e^{ih_1}$ and $w = e^{ih_2}$, where the exponentials are computed in $M(u)$. Let $\mathcal{B}$ denote the JB*-subalgebra of $M(u)$ generated by u, h_1, h_2 –u is the unit here–. The Shirshov-Cohn theorem (see Theorem 5.14) assures the existence of a complex Hilbert space H such that $\mathcal{B}$ embeds as a JB*-subalgebra of $B(H)$ and both share the same unit u (the product of $B(H)$ will be denoted by mere juxtaposition and the involution by $\sharp$). Obviously $u, v, w \in \mathcal{U}(\mathcal{B}) \subseteq \mathcal{U}(B(H)) \subseteq \mathrm{Iso}(H)$ with $\|u-w\|_{\mathcal{B}} = \|u-w\|_M = \|u-v\|_M = \|u-v\|_{\mathcal{B}}$. Let us compute

$$\begin{aligned} U_v^u(u) &= 2(v \circ_u u) \circ_u v - (v \circ_u v) \circ_u u \\ &= 2v \circ_u v - v \circ_u v = v \circ_u v = \{v,u,v\} = U_v(u^*), \end{aligned}$$

and

$$\begin{aligned} \|vu^{-1}v - w\|_{B(H)} &= \|vu^{-1}v - w\|_{\mathcal{B}} = \|U_v^u(u^\sharp) - w\|_{\mathcal{B}} = \|U_v^u(u^{*_u}) - w\|_{\mathcal{B}} \\ &= \|U_v^u(u) - w\|_{\mathcal{B}} = \|U_v(u^*) - w\|_{\mathcal{B}} = \|U_v(u^*) - w\|_M. \end{aligned}$$

Remark 6.5 proves that for $K = 2 - 2\|u - v\| > 1$ we have

$$\|U_v^u(w^{*_u}) - w\|_{\mathcal{B}} = \left\|U_v^u(w^{-1}) - w\right\|_{\mathcal{B}} = \|vw^{-1}v - w\|_{B(H)} \geq K\|w-v\|_{\mathcal{B}} = K\|w-v\|.$$

On the other hand, by the uniqueness of the triple product (see Theorem 4.3) we have

$$U_v^u(w^{*_u}) = \{v,w,v\}_{\mathcal{B}} = \{v,w,v\}_{M(u)} = \{v,w,v\}_M = U_v(w^*).$$

All together gives

$$\begin{aligned} \left\|U_v(w^{-1}) - w\right\| &= \|U_v(w^*) - w\|_M = \|U_v(w^*) - w\|_{\mathcal{B}} \\ &= \|U_v^u(w^{*_u}) - w\|_{\mathcal{B}} \geq K\|w - v\|, \end{aligned}$$

which completes the proof of (2).

(3) Let us consider the u-isotope $M(u)$, and observe that $v \in \mathcal{U}(M(u))$ (cf. Lemma 4.4). Let $\mathcal{B}$ denote the JB*-subalgebra of $M(u)$ generated by v and the unit of $M(u)$–u in this case–. The Shirshov-Cohn theorem (see Theorem 5.14) assures that $\mathcal{B}$ is a JB*-subalgebra of $B(H)$, for some complex Hilbert space H, and we can further assume that u is the unit in $B(H)$ (cf. [**54**, Theorem 7.2.5] and [**113**, Corollary 2.2]). We shall use juxtaposition for the product in $B(H)$, and the symbol $\sharp$ for its involution.

It is worth noting that when M and $\mathcal{B}$ are regarded as JB*-triples, the triple product in $\mathcal{B}$ is precisely $\{\cdot,\cdot,\cdot\}_M|_{\mathcal{B}\times\mathcal{B}\times\mathcal{B}}$. At the same time, the triple product in $\mathcal{B}$ can be expressed in terms of the associative product of $B(H)$. By Proposition 6.17 the element $U_v(u^*)$ lies in $\mathcal{U}^0(M)$, and hence we can apply Δ at this element and compute the next distances:

$$\begin{aligned} \|\Delta(u) - \Delta(U_v(u^*))\|_N &= \|u - U_v(u^*)\|_M = \|u - \{v,u,v\}_M\|_M = \|u - \{v,u,v\}_{\mathcal{B}}\|_{\mathcal{B}} \\ &= \|u - \{v,u,v\}_{B(H)}\|_{B(H)} = \|u - vu^\sharp v\|_{B(H)} \\ &= \|u - vv\|_{B(H)} = \|v^\sharp - v\|_{B(H)} \leq \|v^\sharp - u\|_{B(H)} + \|u - v\|_{B(H)} \\ &= 2\|u-v\|_{B(H)} = 2\|u-v\|_{\mathcal{B}} = 2\|u-v\|_M < 1. \end{aligned}$$

We are in a position to apply Lemma 6.7(b) to deduce the existence of a unitary $w \in \mathcal{U}(N)$ satisfying

$$U_w(\Delta(u)^*) = \Delta\left(U_v(u^*)\right).$$

However, by hypotheses $\|\Delta(u) - \Delta(v)\| = \|u - v\| < \frac{1}{2}$, and consequently, by the final statement in Lemma 6.7 we may assume that

$$\|w - \Delta(v)\| \leq \|w - \Delta(u)\| + \|\Delta(u) - \Delta(v)\| \leq \|\Delta(u) - \Delta(U_v(u^*))\| + \frac{1}{2} < 1 + \frac{1}{2} < 2.$$

Since $\Delta(v) \in \mathcal{U}^0(N)$, the characterization of the principal component given in Theorem 6.14 implies that $w \in \mathcal{U}^0(N)$.

Now, it follows from the fact that the mapping U_w is an isometry that

$$\|U_w(x^*) - U_w(y^*)\| = \|U_w(x^{-1}) - U_w(y^{-1})\| = \|x^{-1} - y^{-1}\| = \|x^* - y^*\| = \|x - y\|,$$

for all $x, y \in \mathcal{U}^0(N) \subseteq \mathcal{U}(N)$. We can hence conclude that the Jordan version of property $C_2(\Delta(u), \Delta(U_u(v^*)))$ holds for $\mathcal{U}^0(N)$.

(4) Let $w \in \mathcal{U}^0(N)$ be the element given by (3). We consider the mappings defined by

$$\varphi(x) := U_v(x^{-1}) = U_v(x^*), \quad x \in \mathcal{U}^0(M),$$

and

$$\psi(y) := U_w(y^{-1}) = U_w(y^*), \quad y \in \mathcal{U}^0(N).$$

Proposition 6.17 assures that $\varphi\left(\mathcal{U}^0(M)\right) = \mathcal{U}^0(M)$ and $\psi\left(\mathcal{U}^0(N)\right) = \mathcal{U}^0(N)$.

We aim to apply the main geometric tool presented in Theorem 5.2 to φ, ψ and Δ at $a = u$ and $c = v$. For this purpose let us observe that, since v and w are unitaries lying in the principal components of $\mathcal{U}(M)$ and $\mathcal{U}(N)$, respectively, φ and ψ are distance preserving bijections because the U-maps the involutions are surjective isometries.

It is easy to check from the definitions that

$$\varphi \circ \varphi(x) = U_v((U_v(x^{-1}))^{-1}) = U_v((U_v(x^*))^*) = U_v(U_{v^*}(x)) = U_v(U_{v^{-1}}(x)) = x,$$

for every $x \in \mathcal{U}^0(M)$. That is, $\varphi \circ \varphi$ is the identity mapping on $\mathcal{U}^0(M)$. By applying that $U_w(\Delta(u)^*) = \Delta\left(U_v(u^*)\right)$ we have

$$\psi(\Delta(u)) = U_w(\Delta(u)^*) = \Delta\left(U_v(u^*)\right) = \Delta\left(\varphi(u)\right), \text{ and}$$
$$\psi\left(\Delta\left(\varphi(u)\right)\right) = U_w\left(\left(\Delta\left(U_v(u^*)\right)\right)^*\right) = \left(U_{w^*}\left(\Delta\left(U_v(u^*)\right)\right)\right)^* = \Delta(u)^{**} = \Delta(u).$$

Therefore the mappings φ, ψ, Δ, $a = u$ and $c = v$ satisfy all hypotheses in Theorem 5.2, and hence the mentioned result implies that $U_w(\Delta(v)^*) = \Delta(v)$. We also know from (3) that $\|w - \Delta(v)\| < 2$, and hence Lemma 6.7(*d*) gives $w = \Delta(v)$. Having in mind that $U_w(\Delta(u)^*) = \Delta\left(U_v(u^*)\right)$ we get $\Delta(U_v(u^*)) = U_{\Delta(v)}(\Delta(u)^*)$, as desired. □

Before presenting the result describing surjective isometries between the principal components of the unitaries in two unital JB*-algebras, we state a couple of technical tools.

The statement of the next result has been borrowed from [**31**, Lemma 3.2] while the proof is coming from [**35**, Lemma 3.3].

LEMMA 6.21 ([**35**, Lemma 3.3]). *Let M and N be two unital JB*-algebras. Let $\{u_k : 0 \leq k \leq 2^n\}$ be a subset of $\mathcal{U}^0(M)$ (with $n \in \mathbb{N}$) and let $\Phi : \mathcal{U}^0(M) \to \mathcal{U}^0(N)$ be a mapping such that*

$$U_{u_{k+1}}(u_k^*) = u_{k+2}, \text{ and } \Phi(U_{u_{k+1}}(u_k^*)) = U_{\Phi(u_{k+1})}(\Phi(u_k)^*),$$

for all $0 \leq k \leq 2^n - 2$. *Then* $U_{u_{2^n-1}}(u_0^*) = u_{2^n}$ *and*

$$\Phi\left(U_{u_{2^n-1}}(u_0^*)\right) = U_{\Phi(u_{2^n-1})}\left(\Phi(u_0)^*\right).$$

PROOF. We shall argue by induction on n. The statement is clear for $n = 1$. Suppose that our statement is true for every family with $2^n + 1$ elements satisfying the conditions above. Let $\{w_k : 0 \leq k \leq 2^{n+1}\}$ be a subset of $\mathcal{U}^0(M)$ with $2^{n+1}+1$ elements such that $U_{w_{k+1}}(w_k^*) = w_{k+2}$ and

$$\Phi(U_{w_{k+1}}(w_k^*)) = U_{\Phi(w_{k+1})}(\Phi(w_k)^*),$$

for all $0 \leq k \leq 2^{n+1}-2$. We consider the family $\{u_k = w_{2k} : 0 \leq K \leq 2^n\} \subseteq \mathcal{U}^0(M)$. We shall show that the latter family satisfies the desired properties in the hypotheses of the lemma, and hence we can apply the induction hypothesis to it.

Fix $0 \leq k \leq 2^n - 2$, by an application of the fundamental identity (2.1) we get

$$\begin{aligned} u_{k+2} = w_{2k+4} = U_{w_{2k+3}}(w_{2k+2}^*) &= U_{U_{w_{2k+2}}(w_{2k+1}^*)}(w_{2k+2}^*) \\ &= U_{w_{2k+2}} U_{w_{2k+1}^*} U_{w_{2k+2}}(w_{2k+2}^*) = U_{w_{2k+2}} U_{w_{2k+1}^*}(w_{2k+2}) \\ &= U_{w_{2k+2}} U_{w_{2k+1}^*} U_{w_{2k+1}}(w_{2k}^*) = U_{w_{2k+2}}(w_{2k}^*) = U_{u_{k+1}}(u_k^*). \end{aligned}$$

Actually, the previous identities and the induction hypothesis combined with the fundamental identity (2.1) lead to

$$\begin{aligned} \Phi(U_{u_{k+1}}(u_k^*)) = \Phi(U_{w_{2k+2}}(w_{2k}^*)) &= \Phi(U_{w_{2k+3}}(w_{2k+2}^*)) = U_{\Phi(w_{2k+3})}(\Phi(w_{2k+2})^*) \\ &= U_{\Phi(U_{w_{2k+2}}(w_{2k+1}^*))}(\Phi(w_{2k+2})^*) = U_{U_{\Phi(w_{2k+2})}(\Phi(w_{2k+1})^*)}(\Phi(w_{2k+2})^*) \\ &= U_{\Phi(w_{2k+2})} U_{\Phi(w_{2k+1})^*} U_{\Phi(w_{2k+2})}(\Phi(w_{2k+2})^*) \\ &= U_{\Phi(w_{2k+2})} U_{\Phi(w_{2k+1})^*}(\Phi(w_{2k+2})) \\ &= U_{\Phi(w_{2k+2})} U_{\Phi(w_{2k+1})^*}(\Phi(U_{w_{2k+1}}(w_{2k}^*))) \\ &= U_{\Phi(w_{2k+2})} U_{\Phi(w_{2k+1})^*} U_{\Phi(w_{2k+1})}(\Phi(w_{2k})^*) \\ &= U_{\Phi(w_{2k+2})}(\Phi(w_{2k})^*) = U_{\Phi(u_{k+1})}(\Phi(u_k)^*). \end{aligned}$$

We have proved that the family formed by the u_k's satisfies the properties in the statement of the lemma.

The induction hypothesis applied to the family $\{u_k : 0 \leq K \leq 2^n\} \subseteq \mathcal{U}^0(M)$ gives

$$U_{w_{2^n}}(w_0^*) = U_{u_{2^n-1}}(u_0^*) = u_{2^n} = w_{2^{n+1}},$$

and

$$\Phi\left(U_{w_{2^n}}(w_0^*)\right) = \Phi\left(U_{u_{2^n-1}}(u_0^*)\right) = U_{\Phi(u_{2^n-1})}\Phi(u_0)^* = U_{\Phi(w_{2^n})}\Phi(w_0)^*,$$

which concludes the proof. □

As we have seen in Proposition 6.17, the principal component of the set of unitary elements in a unital JB*-algebra M is the smallest quadratic subset of $\mathcal{U}(M)$ containing $e^{iM_{sa}}$. So, in order to understand the precise form of a surjective isometry between principal components the behaviour on finite compositions of U-maps of elements in $e^{iM_{sa}}$ seems to be crucial. We recall that a mapping between unital JB*-algebras is called *unital* if it maps the unit to the unit.

PROPOSITION 6.22 ([**31**, Proposition 3.3]). *Let M and N be unital JB*-algebras. Let the symbol $Iso_1(\mathcal{U}^0(M),\mathcal{U}^0(N))$ denote the set of all surjective unital isometries from $\mathcal{U}^0(M)$ onto $\mathcal{U}^0(N)$. Then the identity*

$$\Delta_0(U_{e^{ih_n}}\cdots U_{e^{ih_1}}(e^{ih_0})) = U_{\Delta_0(e^{ih_n})}\cdots U_{\Delta_0(e^{ih_1})}(\Delta(e^{ih_0})),$$

holds for all $\Delta_0 \in \mathrm{Iso}_1(\mathcal{U}^0(M),\mathcal{U}^0(N))$, all natural n, and all $h_n,\ldots,h_1,h_0 \in M_{sa}$.

Furthermore, there exists a mapping

$$k : M_{sa} \times \mathrm{Iso}_1(\mathcal{U}^0(M),\mathcal{U}^0(N)) \to N_{sa}$$

satisfying the following properties:

(a) *If we fix $\Delta_0 \in \mathrm{Iso}_1(\mathcal{U}^0(M),\mathcal{U}^0(N))$, the mapping $k(\cdot,\Delta_0) : M_{sa} \to N_{sa}$ is a surjective linear isometry.*
(b) *$k(\cdot,\Delta_0)^{-1} = k(\cdot,\Delta_0^{-1})$, for every $\Delta_0 \in \mathrm{Iso}_1(\mathcal{U}^0(M),\mathcal{U}^0(N))$.*
(c) *$\Delta_0(e^{ith}) = e^{itk(h,\Delta_0)}$, for all $\Delta_0 \in \mathrm{Iso}_1(\mathcal{U}^0(M),\mathcal{U}^0(N))$, $h \in M_{sa}$ and $t \in \mathbb{R}$.*

PROOF. We begin the proof by establishing a series of properties and claims. Fix an arbitrary $\Delta_0 \in \mathrm{Iso}_1(\mathcal{U}^0(M),\mathcal{U}^0(N))$ and $h \in M_{sa}$, and consider the continuous mapping $\mathbb{R} \to \mathcal{U}^0(M)$, $t \mapsto v_t = v_t^h := e^{ith}$. In a first step we shall show that

$$\Delta_0(U_{v_t}(v_s)) = U_{\Delta_0(v_t)}(\Delta_0(v_s^*)^*) \tag{6.16}$$

for every $s,t \in \mathbb{R}$. In order to prove (6.16), choose a positive integer m such that

$$e^{\frac{\|i(s+t)h\|_M}{2^m}} - 1 < \frac{1}{2}, \text{ and hence } \left\| e^{\frac{i(s+t)h}{2^m}} - \mathbf{1}_M \right\|_M \le e^{\frac{\|i(s+t)h\|_M}{2^m}} - 1 < \frac{1}{2}. \tag{6.17}$$

Given $0 \le l \le 2^{m+1}$, we set

$$u_l := v_s^* \circ e^{\frac{il(s+t)h}{2^m}} = e^{-ish} \circ e^{\frac{il(s+t)h}{2^m}} = e^{\left(-is+\frac{il(s+t)}{2^m}\right)h},$$

and consider the family $\{u_l : 0 \le l \le 2^{m+1}\} \subseteq \mathcal{U}^0(M)$.

By applying the properties of the holomorphic functional calculus, the reader can check that the following identities hold:

$$u_0 = v_s^*,\ u_{2^m} = v_t,\ u_{2^{m+1}} = U_{v_t}(v_s),\ U_{u_{l+1}}(u_l{}^*) = U_{u_{l+1}}(u_l^{-1}) = u_{l+2}, \tag{6.18}$$

for any $0 \le l \le 2^{m+1} - 2$, and

$$\begin{aligned}
\|u_{l+1} - u_l\| &= \left\| v_s{}^* \circ e^{\frac{i(l+1)(s+t)h}{2^m}} - v_s{}^* \circ e^{\frac{il(s+t)h}{2^m}} \right\| \\
&\le \|v_s{}^*\| \left\| e^{\frac{il(s+t)h}{2^m}} \circ e^{\frac{i(s+t)h}{2^m}} - e^{\frac{il(s+t)h}{2^m}} \right\| \\
&\le \left\| e^{\frac{il(s+t)h}{2^m}} \right\| \left\| e^{\frac{i(s+t)h}{2^m}} - \mathbf{1}_M \right\| \\
&\le e^{\frac{\|i(s+t)h\|}{2^m}} - 1 < \frac{1}{2}, \text{ for all } 0 \le l \le 2^{m+1} - 1.
\end{aligned}$$

Now, Theorem 6.20(3), applied to Δ_0 and the unitaries u_{l+1} and u_l, gives

$$\Delta_0(U_{u_{l+1}}(u_l{}^*)) = U_{\Delta_0(u_{l+1})}(\Delta_0(u_l)^*), \text{ for every } l \in \{0,\ldots,2^{m+1}-1\}. \tag{6.19}$$

It follows from (6.18) and (6.19) that we are in a position to apply Lemma 6.21 with $\Phi = \Delta_0$ and $n = m+1$ to deduce that

$$\begin{aligned}\Delta_0(U_{v_t}(v_s)) &= \Delta_0(U_{v_t}(({v_s}^*)^*)) = \Delta_0(U_{u_{2^m}}({u_0}^*))\\ &= U_{\Delta_0(u_{2^m})}(\Delta_0(u_0)^*) = U_{\Delta_0(v_t)}(\Delta_0({v_s}^*)^*),\end{aligned}$$

which concludes the proof of (6.16).

By taking $t = 1$, $s = 0$ and $h = h_1 \in M_{sa}$ in (6.16) we obtain

$$\Delta_0(e^{i2h_1}) = \Delta_0(U_{e^{ih_1}}(\mathbf{1}_M)) = U_{\Delta_0(e^{ih_1})}(\mathbf{1}_N) = \Delta_0(e^{ih_1})^2. \tag{6.20}$$

On the other hand, by replacing $t = 0$ in (6.16) we get

$$\Delta_0(e^{ish}) = \Delta_0(e^{-ish})^*, \quad \forall h \in M_{sa},\ \forall s \in \mathbb{R}. \tag{6.21}$$

The most important consequence derived from (6.16) (and (6.21)) assures that, for each $h \in M_{sa}$, the mapping

$$\Gamma : \mathbb{R} \to \mathcal{U}^0(N),\ t \mapsto \Gamma(t) = \Delta_0(e^{ith})$$

is a uniformly continuous one-parameter family of unitaries satisfying

$$\begin{aligned}U_{\Gamma(t)}(\Gamma(s)) &= U_{\Delta_0(e^{ith})}(\Delta_0(e^{ish})) = U_{\Delta_0(e^{ith})}(\Delta_0(e^{-ish})^*)\\ &= \Delta_0(U_{e^{ith}}(e^{ish})) = \Delta_0(e^{i(2t+s)h}) = \Gamma(2t+s),\end{aligned}$$

and hence Γ satisfies the hypotheses in the Jordan version of Stone's one-parameter theorem presented in Theorem 6.19. Therefore there exists a unique $k(h, \Delta_0) \in N_{sa}$, depending on h and Δ_0, satisfying

$$\Delta_0(e^{ith}) = e^{itk(h,\Delta_0)}, \text{ for all } t \in \mathbb{R}. \tag{6.22}$$

We observe that the mapping $k(\cdot, \Delta_0) : M_{sa} \to N_{sa}$ is well defined, and moreover, by (6.22), $\mathbf{1}_N = \Delta_0(e^{it0}) = e^{itk(0,\Delta_0)}$, for all $t \in \mathbb{R}$, which proves that $k(0, \Delta_0) = 0$.

We claim that

$$k(\cdot, \Delta_0) : M_{sa} \to N_{sa} \text{ is a surjective linear isometry.} \tag{6.23}$$

Namely, the previous conclusion applied to $\Delta_0^{-1} : \mathcal{U}^0(N) \to \mathcal{U}^0(M)$, show that for each $k_0 \in N_{sa}$ there exists a well defined mapping $h(\cdot, \Delta_0^{-1}) : N_{sa} \to M_{sa}$ satisfying

$$\Delta_0^{-1}(e^{itk_0}) = e^{ith(k_0,\Delta_0^{-1})}, \text{ for all } t \in \mathbb{R}. \tag{6.24}$$

The identities (6.22) and (6.24) prove that

$$\begin{aligned}e^{ith_0} &= \Delta_0^{-1}\Delta_0(e^{ith_0}) = \Delta_0^{-1}\left(e^{itk(h_0,\Delta_0)}\right) = e^{ith(k(h_0,\Delta_0),\Delta_0^{-1})}, \text{ and}\\ e^{itk_0} &= \Delta_0\Delta_0^{-1}(e^{itk_0}) = \Delta_0\left(e^{ith(k_0,\Delta_0^{-1})}\right) = e^{itk(h(k_0,\Delta_0^{-1}),\Delta_0)},\end{aligned}$$

for all $t \in \mathbb{R}$, $h_0 \in M_{sa}$, $k_0 \in N_{sa}$, and hence

$$h(k(h_0, \Delta_0), \Delta_0^{-1}) = h_0, \text{ and } k(h(k_0, \Delta_0^{-1}), \Delta_0) = k_0,$$

for all $h_0 \in M_{sa}$, $k_0 \in N_{sa}$, that is, $k(\cdot, \Delta_0)$ and $h(\cdot, \Delta_0^{-1})$ are bijections with $h(\cdot, \Delta_0^{-1}) = k(\cdot, \Delta_0)^{-1}$.

Furthermore, given $h_0, h_0' \in M_{sa}$ and a real number t we have

$$\|\cdot\|\text{-}\lim_{t\to 0^+} \frac{e^{ith_0} - e^{ith_0'}}{t} = \|\cdot\|\text{-}\lim_{t\to 0^+} \frac{e^{ith_0} - \mathbf{1}_M}{t} - \frac{e^{ith_0'} - \mathbf{1}_M}{t} = ih_0 - ih_0'. \tag{6.25}$$

We deduce from (6.22) and the hypotheses on Δ that

$$\|e^{itk(h_0,\Delta_0)} - e^{itk(h_0',\Delta_0)}\|_{N} = \|\Delta_0(e^{ith_0}) - \Delta_0(e^{ith_0'})\|_{N} = \|e^{ith_0} - e^{ith_0'}\|_{M},$$

which by (6.25) leads to

$$\|k(h_0,\Delta_0) - k(h_0',\Delta_0)\|_{N} = \|h_0 - h_0'\|_{M},$$

witnessing that $k(\cdot,\Delta_0) : M_{sa} \to N_{sa}$ is a surjective isometry, and hence a surjective linear isometry by the Mazur–Ulam theorem. This proves the claim in (6.23).

Our next goal will consist in proving that

$$\Delta_0(U_{e^{ih_1}}(e^{ih_0})) = U_{\Delta_0(e^{ih_1})}(\Delta_0(e^{ih_0})), \text{ for all } h_1, h_0 \in M_{sa}. \tag{6.26}$$

Namely, fix $h_1 \in M_{sa}$ and consider the surjective isometry $\Delta_0^{h_1} := U_{\Delta_0(e^{ih_1})^*}\Delta_0 U_{e^{ih_1}} : \mathcal{U}^0(M) \to \mathcal{U}^0(N)$, which is well defined by Theorem 6.14. By (6.20) we deduce that

$$\begin{aligned}\Delta_0^{h_1}(\mathbf{1}_M) &= U_{\Delta_0(e^{ih_1})^*}\Delta_0 U_{e^{ih_1}}(\mathbf{1}_M) = U_{\Delta_0(e^{ih_1})^*}\Delta_0(e^{2ih_1}) = U_{\Delta_0(e^{ih_1})^*}(\Delta_0(e^{ih_1})^2)\\ &= U_{\Delta_0(e^{ih_1})^*}U_{\Delta_0(e^{ih_1})}(\mathbf{1}_M) = \mathbf{1}_N,\end{aligned}$$

which proves that the mapping $\Delta_0^{h_1} = U_{\Delta_0(e^{ih_1})^*}\Delta_0 U_{e^{ih_1}}$ is unital.

Let $k(\cdot,\Delta_0), k(\cdot,\Delta_0^{h_1}) : M_{sa} \to N_{sa}$ be the surjective linear isometries given by (6.22) and (6.23) for Δ_0 and $\Delta_0^{h_1}$, respectively.

Choose, by continuity, $\varepsilon > 0$ such that

$$\|e^{ih_1} - e^{-ith}\| < \frac{1}{2}, \ \forall t \in \mathbb{R}, \ h \in M_{sa} \text{ with } |t-1| < \varepsilon, \text{ and } \|h + h_1\| < \varepsilon.$$

Theorem 6.20(3), applied to Δ_0 and the unitaries $e^{ih_1}, e^{-ith} \in \mathcal{U}^0(M)$ with t in $\mathbb{R}$ and $h \in M_{sa}$ satisfying $\|h + h_1\| < \varepsilon$ and $|t-1| < \varepsilon$, proves that

$$\Delta_0(U_{e^{ih_1}}(e^{ith})) = U_{\Delta_0(e^{ih_1})}((\Delta_0(e^{-ith}))^*) = (6.21) = U_{\Delta_0(e^{ih_1})}(\Delta_0(e^{ith})).$$

This conclusion and the properties defining the mappings $k(\cdot,\Delta_0)$ and $k(\cdot,\Delta_0^{h_1})$ assure that

$$e^{itk(h,\Delta_0)} = e^{itk(h,\Delta_0^{h_1})}, \ \forall t \in \mathbb{R}, \ h \in M_{sa} \text{ with } |t-1| < \varepsilon, \text{ and } \|h + h_1\| < \varepsilon.$$

If we fix an arbitrary $h \in M_{sa}$ with $\|h + h_1\| < \varepsilon$ and we take a simple derivative at $t = 1$ in the above equality we obtain

$$k(h,\Delta_0) = k(h,\Delta_0^{h_1}), \text{ for all } h \in M_{sa} \text{ with } \|h + h_1\| < \varepsilon.$$

We conclude from the linearity of $k(\cdot,\Delta_0)$, $k(\cdot,\Delta_0^{h_1})$ that $k(\cdot,\Delta_0) = k(\cdot,\Delta_0^{h_1})$.

The conclusion in (6.26) now follows from the identity

$$\Delta_0(e^{ith_0}) = e^{itk(h_0,\Delta_0)} = e^{itk(h_0,\Delta_0^{h_1})} = \Delta_0^{h_1}(e^{ith_0}) = U_{\Delta_0(e^{ih_1})^*}\Delta_0 U_{e^{ih_1}}(e^{ith_0}),$$

for all real t and all $h_0 \in M_{sa}$ (cf. (6.22)).

We deal now with the first statement in the proposition. We observe that, since for each self-adjoint element $h \in M_{sa}$ we have $U_{e^{i\frac{h}{2}}}(\mathbf{1}_M)) = e^{ih}$, the desired identity is equivalent to

$$\Delta_0(U_{e^{ih_n}} \cdots U_{e^{ih_1}}(\mathbf{1}_M)) = U_{\Delta_0(e^{ih_n})} \cdots U_{\Delta_0(e^{ih_1})}(\mathbf{1}_N) \tag{6.27}$$

for every Δ_0, $n \in \mathbb{N}$, and $h_n, \ldots, h_1 \in M_{sa}$ as in the hypotheses.

The desired statement will be proved by induction on n. The case $n = 1$ follows from (6.27).

Suppose, by the induction hypothesis, that (6.27) holds for all natural $k \leq n$. Take an arbitrary unital surjective isometry $\Delta_1 : \mathcal{U}^0(M) \to \mathcal{U}^0(N)$ and h_{n+1}, h_n, $\ldots$, $h_1 \in M_{sa}$. Set $w_0 = e^{2ih_{n+1}}$ and $\widetilde{w}_0 = \Delta_1(e^{2ih_{n+1}}) \in \mathcal{U}^0(N)$. We consider the unital JB*-algebras given by the isotopes $M(w_0) = M_{(w_0^*)}$ and $N(\widetilde{w}_0) = M_{(\widetilde{w}_0^*)}$ with the corresponding Jordan products $\circ_{w_0}$ and $\circ_{\widetilde{w}_0}$ and involutions $*_{w_0}$ and $*_{\widetilde{w}_0}$, respectively. Since $w_0 \in \mathcal{U}^0(M)$ and $\widetilde{w}_0 \in \mathcal{U}^0(N)$, it follows from (6.6) that

$$\mathcal{U}^0(M(w_0)) = \mathcal{U}^0(M), \text{ and } \mathcal{U}^0(N(\widetilde{w}_0)) = \mathcal{U}^0(N),$$

and thus $\widetilde{\Delta}_1 : \mathcal{U}^0(M(w_0)) = \mathcal{U}^0(M) \to \mathcal{U}^0(N(\widetilde{w}_0)) = \mathcal{U}^0(N)$, $\widetilde{\Delta}_1(a) = \Delta_1(a)$ is a unital (i.e., $\Delta_1(w_0) = \widetilde{w}_0$) surjective isometry.

The mapping $U_{e^{ih_{n+1}}} : M \to M(w_0)$ is a unital surjective linear isometry, and hence a Jordan *-isomorphism between these two unital JB*-algebras (cf. [**115**, Theorem 6] or [**69**, Theorem 1.9] or Kaup's Banach–Stone in Theorem 4.3). Therefore $\hat{h}_k := U_{e^{ih_{n+1}}}(h_k) \in (M(w_0))_{sa}$ for all $1 \leq k \leq n$. By applying the induction hypothesis to $\widetilde{\Delta}_1 : \mathcal{U}^0(M(w_0)) \to \mathcal{U}^0(N(\widetilde{w}_0))$, and $\hat{h}_1, \ldots, \hat{h}_n$ we get

$$\begin{aligned} \Delta_1 U^{(w_0)}_{\exp_{(w_0)}(i\hat{h}_n)} \cdots U^{(w_0)}_{\exp_{(w_0)}(i\hat{h}_1)}(w_0) &= \widetilde{\Delta}_1 U^{(w_0)}_{\exp_{(w_0)}(i\hat{h}_n)} \cdots U^{(w_0)}_{\exp_{(w_0)}(i\hat{h}_1)}(w_0) \\ &= U^{(\widetilde{w}_0)}_{\widetilde{\Delta}_1(\exp_{(w_0)}(i\hat{h}_n))} \cdots U^{(\widetilde{w}_0)}_{\widetilde{\Delta}_1(\exp_{(w_0)}(i\hat{h}_1))}(\widetilde{w}_0), \end{aligned} \tag{6.28}$$

where, as in previous results, $\exp_{(w_0)}$ denotes the exponential in the unital JB*-algebra $M(w_0)$.

Let us carefully decode the identity in (6.28). Having in mind that the U-operator in the isotope $M(w_0) = M_{(w_0^*)}$ can be expressed in terms of the U-operator in M (cf. (2.3)) we derive that

$$U^{(w_0)}_{v_1}(v_2) = U_{v_1} U_{w_0^*}(v_2) = U_{v_1} U_{e^{-2ih_1}}(v_2). \tag{6.29}$$

Furthermore, since $\exp_{(w_0)}$ denotes the exponential in the unital JB*-algebra $M(w_0)$, it follows from the fact that $U_{e^{ih_{n+1}}} : M \to M(w_0)$ is a Jordan *-isomorphism that

$$\exp_{w_0}(i\hat{h}_k) = \exp_{w_0}(iU_{e^{ih_{n+1}}}(h_k)) = U_{e^{ih_{n+1}}}(e^{ih_k}), \text{ for all } 1 \leq k \leq n, \tag{6.30}$$

and by the fundamental identity (2.1) we derive that

$$\begin{aligned} U^{(w_0)}_{\exp_{w_0}(i\hat{h}_k)} &= U_{\exp_{w_0}(i\hat{h}_k)} U_{w_0^*} = U_{U_{e^{ih_{n+1}}}(e^{ih_k})} U_{w_0^*} \\ &= U_{e^{ih_{n+1}}} U_{e^{ih_k}} U_{e^{ih_{n+1}}} U_{e^{-2ih_{n+1}}} = U_{e^{ih_{n+1}}} U_{e^{ih_k}} U_{e^{-ih_{n+1}}}, \end{aligned} \tag{6.31}$$

(in the last step we applied that $e^{ih_{n+1}}$ and $e^{-2ih_{n+1}}$ operator commute).

Now, by combining (6.29) and (6.31), the left hand side of (6.28) writes in the form

$$\begin{aligned} &\Delta_1 U^{(w_0)}_{\exp_{(w_0)}(i\hat{h}_n)} \cdots U^{(w_0)}_{\exp_{(w_0)}(i\hat{h}_1)}(w_0) \\ &= \Delta_1 U_{e^{ih_{n+1}}} U_{e^{ih_n}} U_{e^{-ih_{n+1}}} \cdots U_{e^{ih_{n+1}}} U_{e^{ih_1}} U_{e^{-ih_{n+1}}}(e^{2ih_{n+1}}) \\ &= \Delta_1 U_{e^{ih_{n+1}}} U_{e^{ih_n}} U_{e^{ih_{n-1}}} \cdots U_{e^{ih_2}} U_{e^{ih_1}}(\mathbf{1}_M). \end{aligned} \tag{6.32}$$

Concerning the right hand side of (6.28) we observe that, by (6.30), for $1 \leq k \leq n$ we have

$$\begin{aligned}
U^{(\widetilde{w}_0)}_{\widetilde{\Delta}_1(\exp_{(w_0)}(i\hat{h}_k))} &= U^{(\widetilde{w}_0)}_{\Delta_1(\exp_{(w_0)}(i\hat{h}_k))} = U_{\Delta_1(U_{e^{ih_{n+1}}}(e^{ih_k}))}U_{\widetilde{w}_0^*} \\
&= (\text{by } (6.26)) = U_{U_{\Delta_1(e^{ih_{n+1}})}\Delta_1(e^{ih_k})}U_{\widetilde{w}_0^*} = (\text{by } (2.1)) \\
&= U_{\Delta_1(e^{ih_{n+1}})}U_{\Delta_1(e^{ih_k})}U_{\Delta_1(e^{ih_{n+1}})}U_{\Delta_1(e^{2ih_{n+1}})^*} \\
&= (\text{by } (6.20)) = U_{\Delta_1(e^{ih_{n+1}})}U_{\Delta_1(e^{ih_k})}U_{\Delta_1(e^{ih_{n+1}})}U_{(\Delta_1(e^{ih_{n+1}})^2)^*} \\
&= U_{\Delta_1(e^{ih_{n+1}})}U_{\Delta_1(e^{ih_k})}U_{\Delta_1(e^{ih_{n+1}})^*} \\
&= U_{\Delta_1(e^{ih_{n+1}})}U_{\Delta_1(e^{ih_k})}U^{-1}_{\Delta_1(e^{ih_{n+1}})}.
\end{aligned}$$

Therefore, the right hand side of (6.28) writes in the form

$$\begin{aligned}
&U^{(\widetilde{w}_0)}_{\widetilde{\Delta}_1(\exp_{(w_0)}(i\hat{h}_n))} \cdots U^{(\widetilde{w}_0)}_{\widetilde{\Delta}_1(\exp_{(w_0)}(i\hat{h}_1))}(\widetilde{w}_0) \\
&= U_{\Delta_1(e^{ih_{n+1}})}U_{\Delta_1(e^{ih_n})} \cdots U_{\Delta_1(e^{ih_1})}U^{-1}_{\Delta_1(e^{ih_{n+1}})}(\Delta_1(e^{2ih_{n+1}})) \\
&= (\text{by } (6.20)) = U_{\Delta_1(e^{ih_{n+1}})}U_{\Delta_1(e^{ih_n})} \cdots U_{\Delta_1(e^{ih_1})}U^{-1}_{\Delta_1(e^{ih_{n+1}})}(\Delta_1(e^{ih_{n+1}})^2) \\
&= U_{\Delta_1(e^{ih_{n+1}})}U_{\Delta_1(e^{ih_n})} \cdots U_{\Delta_1(e^{ih_1})}(\mathbf{1}_N).
\end{aligned} \tag{6.33}$$

Finally, (6.28), (6.32) and (6.33) give

$$\Delta_1 U_{e^{ih_{n+1}}}U_{e^{ih_n}} \cdots U_{e^{ih_2}}U_{e^{ih_1}}(\mathbf{1}_M) = U_{\Delta_1(e^{ih_{n+1}})}U_{\Delta_1(e^{ih_n})} \cdots U_{\Delta_1(e^{ih_1})}(\mathbf{1}_N),$$

which concludes the induction argument proving (6.27). □

The main technical results to understand the precise form of a surjective isometry between the principal components of the unitary sets in two unital JB*-algebras have been already established in Proposition 6.22. We can now proceed to present the desired description.

THEOREM 6.23 ([**31**, Theorem 3.4]). *Let $\Delta : \mathcal{U}^0(M) \to \mathcal{U}^0(N)$ be a surjective isometry between the principal components of two unital JB*-algebras. Then there exist self-adjoint elements $k_1, \ldots, k_n$ in N, a central projection p in N, and a Jordan *-isomorphism $\Phi : M \to N$ such that*

$$\Delta(u) = p \circ U_{e^{ik_n}} \cdots U_{e^{ik_1}}\Phi(u) + (\mathbf{1}_N - p) \circ (U_{e^{-ik_n}} \cdots U_{e^{-ik_1}}\Phi(u))^*,$$

for all $u \in \mathcal{U}^0(M)$. Moreover, the elements $k_1, \ldots, k_n \in N_{sa}$ can be replaced by any finite collection of self-adjoint elements $\tilde{k}_1, \ldots, \tilde{k}_m$ such that $\Delta(\mathbf{1}_M) = U_{e^{i\tilde{k}_m}} \cdots U_{e^{i\tilde{k}_1}}(\mathbf{1}_N)$. Under these hypotheses M and N are Jordan *-isomorphic, and there exists a surjective real-linear isometry (i.e., a real-linear triple isomorphism) from M onto N whose restriction to $\mathcal{U}^0(M)$ is Δ.

PROOF. Since $\Delta(\mathbf{1}_M)$ is a unitary in the principal component of $\mathcal{U}(N)$, by the algebraic description established in Theorem 6.14, there exist $k_1, \ldots, k_n \in N_{sa}$ such that $\Delta(\mathbf{1}_M) = U_{e^{ik_n}} \cdots U_{e^{ik_1}}(\mathbf{1}_N)$. Since the principal component is a quadratic subset (cf. Proposition 6.17), we deduce that the mapping $\Delta_0 = U_{e^{-ik_1}} \cdots U_{e^{-ik_n}}\Delta : \mathcal{U}^0(M) \to \mathcal{U}^0(N)$ is a surjective isometry. Clearly $\Delta_0(\mathbf{1}_M) = \mathbf{1}_N$.

Let $k(\cdot, \Delta_0) : M_{sa} \to N_{sa}$ be the surjective linear isometry given by Proposition 6.22 satisfying $\Delta_0(e^{ith}) = e^{itk(h,\Delta_0)}$, for all $h \in M_{sa}$.

The self-adjoint part of any JB*-algebra is a JB-algebra. Thus, $k(\cdot, \Delta_0) : M_{sa} \to N_{sa}$ is a surjective linear isometry between JB-algebras. Under these circumstances we know the existence of a central symmetry $k(\mathbf{1}_M, \Delta_0)$ in N_{sa} and a Jordan *-isomorphism $\Phi : M \to N$ such that

$$k(h, \Delta_0) = k(\mathbf{1}_M, \Delta_0) \circ \Phi(h), \tag{6.34}$$

for every $h \in M_{sa}$ (cf. [**69**, Theorem 1.4 and Corollary 1.11]).

Now, the arguments mix ideas from [**35, 63**] and [**31**]. Since $k(\mathbf{1}_M, \Delta_0)$ is a central symmetry, there exists a central projection p in N such that $k(\mathbf{1}_M, \Delta_0) = 2p - \mathbf{1}_N = p - (\mathbf{1}_N - p)$, with $p \perp (\mathbf{1}_N - p)$. It is not hard to check that for any $n > 0$ we have

$$(2p - \mathbf{1}_N)^n = (p - (\mathbf{1}_N - p))^n = p + (-1)^n(\mathbf{1}_N - p).$$

The mapping Δ_0 can be expressed in terms of p and Φ by some simple computations. Namely, given an arbitrary $h \in M_{sa}$, we have

$$\Delta_0(e^{ih}) = e^{ik(h,\Delta_0)} = e^{ik(\mathbf{1}_M,\Delta_0)\circ\Phi(h)} = e^{i(2p-\mathbf{1}_N)\circ\Phi(h)} = \sum_{n=0}^{\infty} \frac{(i(2p - \mathbf{1}_N) \circ \Phi(h))^n}{n!}.$$

We now make use of the properties of Φ as Jordan *-isomorphism and the expression of $(2p - \mathbf{1}_N)^n$ given above. It is worth noting that $k(\mathbf{1}_M, \Delta_0)$ (and hence p) is central, and so it operator commutes with any element in N. Therefore,

$$\begin{aligned}
\Delta_0(e^{ih}) &= \sum_{n=0}^{\infty} \frac{i^n(2p - \mathbf{1}_N)^n \circ \Phi(h)^n}{n!} = \sum_{n=0}^{\infty} \frac{i^n\,(p + (-1)^n(\mathbf{1}_N - p)) \circ \Phi(h^n)}{n!} \\
&= \sum_{n=0}^{\infty} \frac{i^n p \circ \Phi(h^n)}{n!} + \sum_{n=0}^{\infty} \frac{i^n(-1)^n(\mathbf{1}_N - p) \circ \Phi(h^n)}{n!} \\
&= p \circ \sum_{n=0}^{\infty} \frac{i^n\Phi(h^n)}{n!} + (\mathbf{1}_N - p) \circ \sum_{n=0}^{\infty} \frac{i^n(-1)^n\Phi(h^n)}{n!} \\
&= p \circ \Phi\left(\sum_{n=0}^{\infty} \frac{i^n h^n}{n!}\right) + (\mathbf{1}_N - p) \circ \Phi\left(\sum_{n=0}^{\infty} \frac{i^n(-1)^n h^n}{n!}\right) \\
&= p \circ \Phi(e^{ih}) + (\mathbf{1}_N - p) \circ \Phi(e^{-ih}) = p \circ \Phi(e^{ih}) + (\mathbf{1}_N - p) \circ \Phi(e^{ih})^*.
\end{aligned} \tag{6.35}$$

We now deal with a unitary element u in $\mathcal{U}^0(M)$. A new application of Theorem 6.14 assures the existence of a natural m and $h_1, \ldots, h_m \in M_{sa}$ such that $u = U_{e^{ih_m}} \cdots U_{e^{ih_1}}(\mathbf{1}_M)$. By combining Proposition 6.22 and (6.35) we arrive to

$$\begin{aligned}
\Delta_0(u) &= \Delta_0(U_{e^{ih_m}} \cdots U_{e^{ih_1}}(\mathbf{1}_M)) = U_{\Delta_0(e^{ih_m})} \cdots U_{\Delta_0(e^{ih_1})}(\Delta_0(\mathbf{1}_M)) \\
&= U_{\left(p\circ\Phi(e^{ih_m})+(\mathbf{1}_N-p)\circ\Phi(e^{ih_m})^*\right)} \cdots U_{\left(p\circ\Phi(e^{ih_1})+(\mathbf{1}_N-p)\circ\Phi(e^{ih_1})^*\right)}(\mathbf{1}_N)
\end{aligned}$$

Since p is a central projection in N, it is not hard to see that

$$U_{p\circ a+(\mathbf{1}_N-p)\circ b} = U_{p\circ a} + U_{(\mathbf{1}_N-p)\circ b} = p \circ U_a + (\mathbf{1}_N - p) \circ U_b,$$

for all $a, b \in N$. Since $p \perp \mathbf{1}_N - p$ and Φ is a Jordan *-isomorphism it follows that

$$\begin{aligned}
\Delta_0(u) &= p \circ \Phi(U_{e^{ih_m}} \cdots U_{e^{ih_1}}(\mathbf{1}_M)) + (\mathbf{1}_N - p) \circ \Phi(U_{e^{ih_m}} \cdots U_{e^{ih_1}}(\mathbf{1}_M))^* \\
&= p \circ \Phi(u) + (\mathbf{1}_N - p) \circ \Phi(u)^*.
\end{aligned}$$

Finally, by the definition of Δ_0, we arrive at

$$\Delta(u) = U_{e^{ik_n}} \dots U_{e^{ik_1}} \Delta_0(u) = U_{e^{ik_n}} \dots U_{e^{ik_1}} \left(p \circ \Phi(u) + (\mathbf{1}_N - p) \circ \Phi(u)^*\right),$$
$$= p \circ U_{e^{ik_n}} \dots U_{e^{ik_1}} \Phi(u) + (\mathbf{1}_N - p) \circ \left(U_{e^{-ik_n}} \dots U_{e^{-ik_1}} \Phi(u)\right)^*,$$

for all $u \in \mathcal{U}^0(M)$, which concludes the proof. □

REMARK 6.24. Let us observe that the conclusion in Theorem 6.23 is much finer under the extra assumption that Δ is unital, since in that case there exist a central projection p in N, and a Jordan *-isomorphism $\Phi : M \to N$ such that

$$\Delta(u) = p \circ \Phi(u) + (\mathbf{1}_N - p)\Phi(u)^* = p \circ \Phi(u) + (\mathbf{1}_N - p)\Phi(u^*),$$

for all $u \in \mathcal{U}^0(M)$.

The Anti-theorem 4.6 shows that the group of all surjective linear isometries on a unital JB*-algebra M does not always act transitively on $\mathcal{U}(M)$. The conclusion of Theorem 6.24 can be drastically simplified if we additionally assume that $\Delta(\mathbf{1}_M)$ can be mapped to $\mathbf{1}_N$ under a concrete surjective linear isometry. The next sufficient condition which is an easy consequence of Lemma 6.7(b).

LEMMA 6.25. *Let u be a unitary element in a unital JB*-algebra M. Suppose additionally that $\|\mathbf{1}_M - u\| < 2$, then there exists $v \in \mathcal{U}(N)$ satisfying $v^2 = u$. Consequently, $U_{v^*}(u) = \mathbf{1}_M$.*

PROOF. Lemma 6.7(b) implies the existence of a unitary ω in M such that $U_\omega(u) = \mathbf{1}_N$. Then $u = U_{\omega^*}U_\omega(u) = U_{\omega^*}(\mathbf{1}_M) = (\omega^*)^2$ (cf. Lemma 6.7(c)). The desired statement follows by just taking $v = \omega^*$. □

The next corollary goes in the line of [**35**, Theorem 3.6].

COROLLARY 6.26. *Let $\Delta : \mathcal{U}^0(M) \to \mathcal{U}^0(N)$ be a surjective isometry between the principal components of two unital JB*-algebras. Suppose there exists $v \in \mathcal{U}(N)$ such that $U_v(\Delta(\mathbf{1}_M)) = \mathbf{1}_N$. Then there exist a central projection p in N and a Jordan *-isomorphism $\Phi : M \to N$ such that*

$$\Delta(w) = p \circ U_{v^*}\Phi(w) + (\mathbf{1}_N - p) \circ (U_v\Phi(w))^*,$$

for all $w \in \mathcal{U}^0(M)$.

PROOF. The mapping $\tilde{\Delta} = U_v\Delta : \mathcal{U}^0(M) \to \mathcal{U}^0(N)$ is a surjective unital isometry. Theorem 6.23 (see also Remark 6.24) proves the existence of a Jordan *-isomorphism $\Phi : M \to N$ such that

$$\tilde{\Delta}(w) = U_v\Delta(w) = p \circ \Phi(w) + (\mathbf{1}_N - p)\Phi(w)^*,$$

for all $w \in \mathcal{U}^0(M)$. Since p is central we have

$$\Delta(w) = p \circ U_{v^*}\Phi(w) + (\mathbf{1}_N - p)U_{v^*}\Phi(w)^* = p \circ U_{v^*}\Phi(w) + (\mathbf{1}_N - p) \circ (U_v\Phi(w))^*,$$

for all $w \in \mathcal{U}^0(M)$. □

It is natural to ask what we can conclude in the case that we have a surjective isometry between two arbitrary connected components in two unital JB*-algebras. Let M be a unital JB*-algebra, let $\mathcal{U}^c(M)$ be a connected component of $\mathcal{U}(M)$, and let w be an element in $\mathcal{U}^c(M)$. We have already commented that $\mathcal{U}^c(M)$ is the principal component of the w-isotope $M(w) = M_{(w^*)}$ (cf. Remark 6.15). When particularized to the corresponding isotopes, the previous Theorem 6.23 can be re-stated as follows.

THEOREM 6.27 ([**31**, Theorem 3.6]). *Let M and N be unital JB*-algebras. Let $\Delta : \mathcal{U}^c(M) \to \mathcal{U}^c(N)$ be a surjective isometry between two connected components of $\mathcal{U}(M)$ and $\mathcal{U}(N)$, respectively. Fix $w_1 \in \mathcal{U}^c(M)$ and $w_2 \in \mathcal{U}^c(N)$. Then there exist $k_1, \ldots, k_n \in (N(w_2))_{sa}$, a central projection $p \in N(w_2)$ and a Jordan *-isomorphism $\Phi : M(w_1) \to N(w_2)$ such that*

$$\Delta(u) = p \circ_{w_2} U^{(w_2)}_{\exp_{(w_2)}(ik_n)} \cdots U^{(w_2)}_{\exp_{(w_2)}(ik_1)} \Phi(u)$$
$$+ (w_2 - p) \circ_{w_2} \left(U^{(w_2)}_{\exp_{(w_2)}(-ik_n)} \cdots U^{(w_2)}_{\exp_{(w_2)}(-ik_1)} \Phi(u) \right)^{*_{w_2}},$$

for all $u \in \mathcal{U}^c(M)$. In particular, the unital JB-algebras $M(w_1) = M_{(w_1^*)}$ and $N(w_2) = N_{(w_1^*)}$ are Jordan *-isomorphic, and there exists a surjective real-linear isometry (i.e., a real-linear triple isomorphism) from M onto N whose restriction to $\mathcal{U}^c(M)$ is Δ.*

Every unitary element of a unital JB*-algebra M need not be, in general, of the form e^{ih} for some $h \in M_{sa}$, the statement is not true even in the case of unital C*-algebras (cf. Remark 6.9 the discussion preceding Proposition 4.4.10 in [**74**]). However, in a von Neumann algebra, every unitary admits a logarithm (cf. [**74**, Theorem 5.2.5]). The same conclusion holds for JBW*-algebras.

LEMMA 6.28. *Let u be a unitary element in a JBW*-algebra M. Then there exists a self-adjoint element $h \in M$ such that $u = e^{ih}$. Therefore $\mathcal{U}(M) = \mathcal{U}^0(M) = \exp(iM_{sa})$.*

PROOF. Let $\mathcal{W}$ denote the JBW*-subalgebra of M generated by u, u^* and the unit element. Clearly $\mathcal{W}$ is an associative JBW*-algebra (cf. Lemma 2.1). Therefore $\mathcal{W}$ is a commutative von Neumann algebra. Theorem 5.2.5 in [**74**] implies the existence of an element $h \in \mathcal{W}_{sa} \subseteq M_{sa}$ such that $e^{ih} = u$ (in $\mathcal{W}$ and also in M). □

The next result is a consequence of Proposition 6.22, Theorem 6.23 and Lemma 6.28.

THEOREM 6.29 ([**35**, Theorem 3.9]). *Let $\Delta : \mathcal{U}(M) \to \mathcal{U}(N)$ be a surjective isometry, where M and N are two JBW*-algebras. Then there exist a self-adjoint element k_1 in N, a central projection $q \in N$, and a Jordan *-isomorphism $\Phi : M \to N$ such that*

$$\Delta(u) = q \circ U_{e^{ik_1}} (\Phi(u)) + (\mathbf{1}_N - q) \circ U_{e^{ik_1}} (\Phi(u)^*),$$

for all $u \in \mathcal{U}(M)$. If Δ is unital we can take $k_1 = 0$. Consequently, Δ admits a (unique) extension to a surjective real-linear isometry from M onto N.

PROOF. Lemma 6.28 implies that

$$\mathcal{U}(M) = \mathcal{U}^0(M) = \exp(iM_{sa}) \text{ and } \mathcal{U}(N) = \mathcal{U}^0(N) = \exp(iN_{sa}).$$

Therefore $\Delta(\mathbf{1}_M) = e^{2ik_1} = U_{e^{ik_1}}(\mathbf{1}_N)$ for some $k_1 \in N_{sa}$. Theorem 6.23 gives the desired conclusion. □

We can now obtain a stronger version of the previous Theorem 6.29.

THEOREM 6.30. *Let $\Delta : \mathcal{U}(M) \to \mathcal{U}(N)$ be a surjective isometry, where M and N are two unital JB*-algebras. Suppose additionally that M or N is a JBW*-algebra. Then M and N are JBW*-algebras and there exist a self-adjoint element k_1 in N, a central projection $q \in N$, and a Jordan *-isomorphism $\Phi : M \to N$ such that*

$$\Delta(u) = q \circ U_{e^{ik_1}} (\Phi(u)) + (\mathbf{1}_N - q) \circ U_{e^{ik_1}} (\Phi(u)^*),$$

for all $u \in \mathcal{U}(M)$. If Δ is unital we can take $k_1 = 0$. Consequently, Δ admits a (unique) extension to a surjective real-linear isometry from M onto N.

PROOF. There is no loss of generality in assuming that M is a JBW*-algebra. In such a case, Lemma 6.28 implies that $\mathcal{U}(M) = \mathcal{U}^0(M) = \exp(iM_{sa})$. It follows that $\mathcal{U}(N) = \Delta(\mathcal{U}(M)) = \Delta(\mathcal{U}^0(M)) = \Delta(\exp(iM_{sa}))$ is connected, and hence $\mathcal{U}(N) = \mathcal{U}^0(N)$. By Theorem 6.23 M and N are Jordan *-isomorphic, and thus N also is a JBW*-algebra (cf. the comments on the Dixmier–Ng theorem in page 288). The desired conclusion follows from Theorem 6.29. □

We can now rediscover a slightly improved version of a result proved by Hatori and Molnár in the setting of von Neumann algebras.

THEOREM 6.31 ([**63**, Corollary 3]). *Let A and B be two unital C*-algebras, and let $\Delta : \mathcal{U}(A) \to \mathcal{U}(B)$ be a surjective isometry. Suppose that A or B is a von Neumann algebra. Then there exist a central projection $q \in B$ and a Jordan *-isomorphism $\Phi : A \to B$ such that*

$$\Delta(u) = \Delta(\mathbf{1}_A) (q\Phi(u) + (\mathbf{1}_B - q)\Phi(u)^*),$$

for all $u \in \mathcal{U}(A)$. Consequently, Δ admits a (unique) extension to a surjective real-linear isometry from A onto B.

PROOF. The element $\Delta(\mathbf{1}_A)$ is a unitary in B and the mapping $\Delta(\mathbf{1}_A)^*\Delta : \mathcal{U}(A) \to \mathcal{U}(B)$ is a unital surjective isometry. By Theorem 6.30, there exist a central projection $q \in B$, and a Jordan *-isomorphism $\Phi : A \to B$ such that

$$\Delta(\mathbf{1}_A)^*\Delta(u) = q\Phi(u) + (\mathbf{1}_B - q)\Phi(u)^*$$

for all $u \in \mathcal{U}(A)$. The rest is clear. □

In the same way that Theorem 6.31 was obtained from Theorem 6.30, the next result, due to Hatori, follows from Theorem 6.23 or from Theorem 6.27.

THEOREM 6.32 ([**58**, Corollary 3.5]). *Let A and B be unital C*-algebras. Let $\Delta : \mathcal{U}^c(A) \to \mathcal{U}^c(B)$ be a surjective isometry between two connected components of $\mathcal{U}(A)$ and $\mathcal{U}(B)$, respectively. Fix $w_1 \in \mathcal{U}^c(A)$ and $w_2 \in \mathcal{U}^c(B)$ such that $\Delta(w_1) = w_2$, $\mathcal{U}^c(A) = w_1\mathcal{U}^0(A)$ and $\mathcal{U}^c(B) = w_2\mathcal{U}^0(B)$. Then there exists a central projection $p \in B$, a Jordan *-isomor-phism $\Phi : A \to B$, and a unitary $u = w_2(p\Phi(w_1^*) + (\mathbf{1}_B - p)\Phi(w_1^*)^*)$ such that*

$$\Delta(w) = u(p\Phi(w) + (\mathbf{1}_B - p)\Phi(w)^*),$$

for all $w \in \mathcal{U}^c(A) = w_1\mathcal{U}^0(A)$.

PROOF. We take part of the argument from [**58**, Corollary 3.5]. Consider the surjective isometry $\tilde{\Delta} : \mathcal{U}^0(A) \to \mathcal{U}^0(B)$ defined by $\tilde{\Delta}(a) := w_2^* \Delta(w_1 a)$. It follows from the definition that $\tilde{\Delta}$ is unital. By Theorem 6.23 (see also Remark 6.24) there exist a central projection p in B, and a Jordan *-isomorphism $\Psi : A \to B$ such that

$$\tilde{\Delta}(a) = p \circ \Psi(a) + (\mathbf{1}_N - p) \circ \Psi(a)^*,$$

for all $a \in \mathcal{U}^0(A)$. Therefore

$$\begin{aligned} \Delta(w) &= w_2 \left(p \circ \Psi(w_1^* w) + (\mathbf{1}_N - p) \circ \Psi(w_1^* w)^*\right), \text{ and} \\ \Delta(w) &= w_2 \left(p \circ \Psi(w_1^* w) + (\mathbf{1}_N - p) \circ \Psi(w_1^* w)^*\right), \end{aligned} \tag{6.36}$$

for all $w \in \mathcal{U}^c(A) = w_1 \mathcal{U}^0(A)$. Define a mapping $\Phi : A \to B$ by

$$\Phi(x) = p\Psi(w_1)\Psi(w_1^* x) + (\mathbf{1}_B - p)\Psi(w_1^* x)\Psi(w_1).$$

It is not hard to check that Φ is unital and $\Phi(w_1) = \Psi(w_1)$.

A surprising geometric argument based on Kadison's Theorem 4.1 (or in Kaup's Theorem 4.3) assures that Φ is a Jordan *-isomorphism. Namely, since orthogonality implies M-orthogonality

$$\begin{aligned} \|\Phi(x)\| &= \max\left\{\|p\Psi(w_1)\Psi(w_1^* x)\|, \|(\mathbf{1}_B - p)\Psi(w_1^* x)\Psi(w_1)\|\right\} \\ &= \max\left\{\|p\Psi(w_1^* x)\|, \|(\mathbf{1}_B - p)\Psi(w_1^* x)\|\right\} \\ &= \|p\Psi(w_1^* x) + (\mathbf{1}_B - p)\Psi(w_1^* x)\| = \|\Psi(w_1^* x)\| = \|w_1^* x\| = \|x\|. \end{aligned}$$

This proves that Φ is a (unital) surjective isometry, and thus Kadison's theorem implies that Φ is a Jordan *-isomorphism.

Finally, having in mind that p is central, (6.36), and the other properties, we get

$$\begin{aligned} &u(p\Phi(w) + (\mathbf{1}_B - p)\Phi(w)^*) = u(p\Psi(w_1)\Psi(w_1^* w) + (\mathbf{1}_B - p)\Psi(w_1)^*\Psi(w_1^* w)^*) \\ &= w_2(p\Phi(w_1^*) + (\mathbf{1}_B - p)\Phi(w_1^*)^*)(p\Phi(w_1)\Psi(w_1^* w) + (\mathbf{1}_B - p)\Phi(w_1)^*\Psi(w_1^* w)^*) \\ &= w_2\left(p \circ \Psi(w_1^* w) + (\mathbf{1}_N - p) \circ \Psi(w_1^* w)^*\right) = \Delta(w). \end{aligned}$$

□

The next corollary shows that the metric space given by the set of unitary elements in a unital JB*-algebra or by its principal component is a metric invariant.

COROLLARY 6.33 ([**35**, Corollary 3.8] and [**31**, Corollary 3.5]). *Let M and N be two unital JB*-algebras. Then M and N are isometrically isomorphic as (complex) Banach spaces if, and only if, they are isometrically isomorphic as real Banach spaces if, and only if, $\mathcal{U}(M)$ and $\mathcal{U}(N)$ are isometrically isomorphic as metric spaces. Furthermore, the following statements are also equivalent:*

(a) *M and N are Jordan *-isomorphic.*
(b) *There exists a surjective isometry $\Delta : \mathcal{U}(M) \to \mathcal{U}(N)$ satisfying $\Delta(\mathbf{1}_M) \in \mathcal{U}^0(N)$.*
(c) *There exists a surjective isometry $\Delta : \mathcal{U}^0(M) \to \mathcal{U}^0(N)$.*

PROOF. To see the first statement, we observe first that every surjective linear isometry from M onto N is a surjective real-linear isometry. Every surjective real-linear isometry $T : M \to N$ is a triple homomorphism (cf. Theorem 4.5). Therefore, T preserves triple products of the form $\{x, y, z\} = (x \circ y^*) \circ z + (z \circ y^*) \circ x - (x \circ z) \circ y^*$. In particular T maps unitaries in M to unitaries in N (see Lemma 4.4), and hence

$T(\mathcal{U}(M)) = \mathcal{U}(N)$. Therefore $\Delta = T|_{\mathcal{U}(M)} : \mathcal{U}(M) \to \mathcal{U}(N)$ is a surjective isometry. If there exists a surjective isometry $\Delta : \mathcal{U}(M) \to \mathcal{U}(N)$, the set $\Delta\left(\mathcal{U}^0(M)\right)$ must be a connected component (say $\mathcal{U}^c(N)$) of $\mathcal{U}(N)$ (cf. Corollary 6.8 and Remark 6.15). Since $\Delta|_{\mathcal{U}^0(M)} : \mathcal{U}^0(M) \to \mathcal{U}^c(N)$ is a surjective isometry, Theorem 6.27 proves that M and N are isometrically isomorphic as real Banach spaces.

We see next the equivalence of the three final statements. The implication $(a) \Rightarrow (b)$ is clear because every Jordan *-isomorphism is a unital linear isometry and maps unitaries to unitaries.

$(b) \Rightarrow (c)$ Suppose there exists a surjective isometry $\Delta : \mathcal{U}(M) \to \mathcal{U}(N)$ satisfying $\Delta(\mathbf{1}_M) \in \mathcal{U}^0(N)$. Since every surjective isometry maps connected components to connected components, $\Delta(\mathcal{U}^0(M))$ is a connected component of $\mathcal{U}(N)$ and contains an element in $\mathcal{U}^0(N)$, the equality $\Delta(\mathcal{U}^0(M)) = \mathcal{U}^0(N)$ holds, and thus $\Delta|_{\mathcal{U}^0(M)} : \mathcal{U}^0(M) \to \mathcal{U}^0(N)$ is a surjective isometry.

Finally, the implication $(c) \Rightarrow (a)$ is a consequence of Theorem 6.23. □

REMARK 6.34. If in Corollary 6.33 we additionally assume that $\mathcal{U}(M)$ or $\mathcal{U}(N)$ is connected the statements (a) to (c) are also equivalent to the existence of a surjective isometry $\Delta : \mathcal{U}(M) \to \mathcal{U}(N)$.

It is already pointed out by the Kadison's theorem (see Theorem 4.1) that two unital C*-algebras are isometrically isomorphic if and only if they are Jordan *-isomorphic. Since the set of all unitary elements in a unital C*-algebra is a multiplicative subgroup of the invertible elements, the conclusion in this setting is stronger (cf. [**62**, Corollary 2]).

COROLLARY 6.35. *Let A and B be two unital C*-algebras. Then the following statements are equivalent:*

(*a*) *A and B are isometrically isomorphic as (complex) Banach spaces.*
(*b*) *A and B are isometrically isomorphic as real Banach spaces.*
(*c*) *$\mathcal{U}(A)$ and $\mathcal{U}(B)$ are isometrically isomorphic as metric spaces.*
(*d*) *A and B are Jordan *-isomorphic.*
(*e*) *There exist connected components $\mathcal{U}^1(A) \subseteq \mathcal{U}(A)$ and $\mathcal{U}^2(B) \subseteq \mathcal{U}(B)$ and a surjective isometry $\Delta : \mathcal{U}^1(A) \to \mathcal{U}^2(B)$.*
(*f*) *There exists a surjective isometry $\Delta : \mathcal{U}^0(A) \to \mathcal{U}^0(B)$.*

PROOF. We only need to prove that $(e) \Rightarrow (d)$, the remaining implications have been established in Corollary 6.33. Suppose there exists a surjective isometry $\Delta : \mathcal{U}^1(A) \to \mathcal{U}^2(B)$, where $\mathcal{U}^1(A) \subseteq \mathcal{U}(A)$ and $\mathcal{U}^2(B) \subseteq \mathcal{U}(B)$ are connected components. Pick $w_1 \in \mathcal{U}^1(A)$ and $w_2 \in \mathcal{U}^2(B)$. It is easy to check that $L_{w_2^*}\Delta L_{w_1} : \mathcal{U}^0(A) \to \mathcal{U}^0(B)$, $u \mapsto w_2^*\Delta(w_1 u)$ is a unital surjective isometry. Then A and B are Jordan *-isomorphic by Corollary 6.33. □

We can now show that the conclusion in Corollary 6.33 is optimal. The properties of unital JB*-algebras differ from those in the strict subclass of unital C*-algebra.

REMARK 6.36. As we have seen in Anti-theorem 4.6, there exists a unital JB*-algebra M with unit $\mathbf{1}_M$ containing a unitary element w for which we cannot find a surjective linear isometry $T : M \to M$ mapping $\mathbf{1}_M$ to w, that is, the group of all surjective linear isometries on M –equivalently, triple automorphisms– on M is not transitive on $\mathcal{U}(M)$. Furthermore, the unital JB*-algebras M and $M(w)$ are not Jordan *-isomorphic.

Let $\mathcal{U}^c(M)$ denote the connected component of $\mathcal{U}(M)$ containing w. We claim that $\mathcal{U}^0(M)$ and $\mathcal{U}^c(M)$ are not isomorphic as metric spaces. Otherwise, Theorem 6.23 (or Theorem 6.27) implies that M and $M(w)$ are Jordan *-isomorphic, which is impossible.

6.3. Extendibility of surjective isometries between sets of unitary elements.

We have seen in Example 6.10 that a surjective isometry between the sets of unitary elements in two unital JB*-algebras is not always extendable to a surjective real-linear isometry between the algebras. Remark 6.36 shows an example of a unital JB*-algebra M containing a unitary element w_0 such that the principal component $\mathcal{U}^0(M)$ and the connected component $\mathcal{U}^c(M)$ containing w_0 are not isomorphic as metric spaces.

Another geometric property of JB*-triples which describes algebraically the extreme points of the closed unit ball of these Banach spaces has not been treated yet. Henceforth, let the symbol $\partial_e(B_X)$ stand for the set of all extreme points of the closed unit ball of a Banach space X. If E is a JB*-triple, the extreme points of its closed unit ball are precisely the complete tripotents in E, that is,

$$\partial_e(B_E) = \text{Trip}_{max}(E) = \{\text{complete tripotents in } E\} \tag{6.37}$$

(cf. [**78**, Proposition 3.5]).

A result by Youngson proves that a JB*-algebra is unital if and only if its closed unit ball has extreme points (see [**117**] or [**22**, Theorem 4.2.36]). It is also well known that for each unital JB*-algebra M we have $\mathcal{U}(M) \subset \partial_e(B_M)$ (and the containing can be, in general, strict). The set $\partial_e(B_M)$ seems to be an appropriate candidate to define a metric invariant for unital JB*-algebras. However, the answer under these weak conditions is not always positive, since we can define a counterexample via the same method employed in Example 6.10.

The lacking or scarcity of unitary elements in the setting of real C*-algebras makes it hopeless to conclude that every surjective isometry between sets of unitary elements in two unital real C*-algebras is extendable to a surjective linear isometry between the algebras. We present next a counterexample taken from [**33**, Remark 3.15].

EXAMPLE 6.37. Let us consider the real Banach space $A = \mathbb{R} \oplus^\infty \mathbb{R} = \ell^2_\infty(\mathbb{R})$. Note that A is actually a real von Neumann algebra with respect to the pointwise product in the usual sense (cf. [**84**]). The sets of all extreme points of B_A and the set of all unitary elements in A coincide, however, it is a relative small set formed by the following four points

$$\partial_e(B_A) = \mathcal{U}(A) = \{e_1 = (1,1), e_2 = (1,-1), e_3 = (-1,1), e_4 = (-1,-1)\}.$$

The distances between elements in $\partial_e(B_A)$ satisfy $\|e_i - e_j\| = 2$ for every $i \neq j$ in $\{1, \ldots, 4\}$. Therefore the mapping $\Delta : \partial_e(B_X) \to \partial_e(B_X)$

$$\Delta(e_1) = e_2, \; \Delta(e_2) = e_3, \; \Delta(e_3) = e_4, \; \text{and } \Delta(e_4) = e_1,$$

is a surjective isometry. We claim that Δ cannot be extended to a surjective linear isometry on A. Suppose, on the contrary, that $T : A \to A$ surjective real-linear isometry extending Δ. In such a case $e_4 = \Delta(e_3) = T(e_3) = T(-e_2)) = -T(e_2) = -\Delta(e_2) = -e_3$, which is impossible.

The natural question is whether there exist sufficient conditions to guarantee the extensibility of a surjective isometry between the sets of unitary elements.

A Jordan variant of the Russo–Dye theorem for unital JB*-algebras due to Wright and Youngson [**114**] was employed in [**31**] to find the following characterization of those surjective isometries between sets of unitary elements which admits an extension to a surjective real-linear isometry.

COROLLARY 6.38 ([**31**, Corollary 3.8]). *Let $\Delta : \mathcal{U}(M) \to \mathcal{U}(N)$ be a surjective isometry between the unitary sets of two unital JB*-algebras. Let $\{\mathcal{U}^0(M)\} \cup \{\mathcal{U}^j(M) : j \in \Lambda\}$ be the collection of all different connected components of $\mathcal{U}(M)$. For each $j \in \Lambda \cup \{0\}$, let $T_j : M \to N$ denote the surjective real-linear isometry extending the mapping $\Delta|_{\mathcal{U}^j(M)} : \mathcal{U}^j(M) \to \Delta(\mathcal{U}^j(M))$ whose existence is assured by Theorem 6.27. Then the following statements are equivalent:*

(*a*) *Δ admits an extension to a surjective (real-linear) isometry from M onto N.*

(*b*) *$T_{j_1} = T_{j_2}$ for all $j_1, j_2 \in \Lambda \cup \{0\}$.*

PROOF. (*a*) $\Rightarrow$ (*b*) Suppose there exists a surjective isometry $T : M \to N$ such that $T|_{\mathcal{U}(M)} = \Delta$. Fix $j \in \Lambda \cup \{0\}$, by hypothesis, $T(u) = \Delta(u) = T_j(u)$ for all $u \in \mathcal{U}^j(M)$.

Fix $u_0 \in \mathcal{U}^j(M)$. Let us consider the unital JB*-algebra given by the u_0-isotope $M(u_0) = M_{(u_0^*)}$. By the Wright–Youngson–Russo–Dye theorem (see [**114**, Corollary]), the closed unit ball of the Banach space $M = M(u_0)$ coincides with the closed convex hull of the set

$$\{\exp_{(u_0)}(ih) : h \in (M(u_0))_{sa}\}.$$

By applying Remark 6.15 we deduce that the latter set is contained in the principal component of $\mathcal{U}(M(u_0))$ which is precisely $\mathcal{U}^j(M)$. It follows from the real-linearity of T and T_j, combined with the fact that these two maps coincide on $\mathcal{U}^j(M)$, that $T = T_j$ on the closed unit ball of M, and hence on the whole M.

(*b*) $\Rightarrow$ (*a*) This implication is easy, it suffices to take $T = T_j$ for any $j \in \Lambda \cup \{0\}$. $\square$

We continue with the following identity principle, which is of independent interest, and has been taken from [**31**].

PROPOSITION 6.39 ([**31**, Proposition 3.9]). *Let $T, T_0 : M \to N$ be two surjective real-linear isometries between unital JB*-algebras. Let us assume that T_0 is a real-linear Jordan *-isomorphism, and suppose that $U_{T_0(u_0)} = U_{T(u_0)}$ for all $u_0 \in \mathcal{U}^0(M)$. Then there exists a central projection $p \in M$ such that $T|_{M \circ p \equiv M_2(p)} = T_0|_{M \circ p \equiv M_2(p)}$, and $T|_{M \circ (\mathbf{1}_M - p) \equiv M_2(\mathbf{1}_M - p)} = -T_0|_{M \circ (\mathbf{1}_M - p) \equiv M_2(\mathbf{1}_M - p)}$.*

PROOF. Assume first that T is unital and hence a real-linear Jordan *-isomorphism (cf. Theorem 4.5). It follows from the assumptions that given $h \in M_{sa}$ we have

$$U_{T_0(e^{ith})}(\mathbf{1}_N) = U_{T(e^{ith})}(\mathbf{1}_N),$$

for all $t \in \mathbb{R}$. Taking derivatives at $t = 0$ in both sides of the above identity we deduce that

$$2T_0(ih) = 2U_{T_0(ih),\mathbf{1}_N}(\mathbf{1}_N) = 2U_{T(ih),\mathbf{1}_N}(\mathbf{1}_N) = 2T(ih),$$

which proves that $T_0(ih) = T(ih)$ for all $h \in M_{sa}$ –recall that T and T_0 are merely real-linear maps–. Having in mind that T and T_0 are real-linear Jordan *-isomorphisms we deduce that

$$-T(h^2) = T(ih)^2 = T_0(ih)^2 = -T_0(h^2), \text{ for all } h \in M_{sa},$$

which proves that T and T_0 coincide on positive elements in M. Since every element in M_{sa} writes as the orthogonal sum of two positive elements in M, we conclude that T and T_0 also coincide on M_{sa}. A general element $x \in M$ can be written in the form $x = h + ik$ with $h, k \in M_{sa}$, and hence

$$T(x) = T(h + ik) = T(h) + T(ik) = T_0(h) + T_0(ik) = T_0(x).$$

We have therefore proved that $T = T_0$ in this case.

We consider next the general case. By hypothesis, $Id = U_{T_0(\mathbf{1}_M)} = U_{T(\mathbf{1}_M)}$, and hence $T(\mathbf{1}_M)^* = U_{T(\mathbf{1}_M)}(T(\mathbf{1}_M)^*) = T(\mathbf{1}_M)$, witnessing that $T(\mathbf{1}_M)$ is a symmetry in N, that is, there exist two orthogonal projections $q_1, q_2 \in N$ such that $T(\mathbf{1}_M) = q_1 - q_2$ and $q_1 + q_2 = \mathbf{1}_N$. Since $U_{q_1} + U_{q_2} - 2U_{q_1,q_2} = U_{q_1-q_2} = Id = U_{q_1+q_2} = U_{q_1} + U_{q_2} + 2U_{q_1,q_2}$, it can be easily deduced that $U_{q_1,q_2} = 0$ and hence $N = N_2(q_1) \oplus^\infty N_2(q_2)$, which implies that q_1 and q_2 are central projections.

Since T_0 is a real-linear Jordan *-isomorphism, the elements $p_1 = T_0^{-1}(q_1)$ and $p_2 = T_0^{-1}(q_2)$ are two orthogonal central projections in M with $p_1 + p_2 = \mathbf{1}_M$ and $M = M_2(p_1) \oplus^\infty M_2(p_2)$. We deduce from the above conclusions that every unitary u in M writes in the form $u = u_1 + u_2$ with $u_j \in M_2(p_j)$, and $\mathcal{U}^0(M)$ identifies with $\mathcal{U}^0(M_2(p_1)) + \mathcal{U}^0(M_2(p_2))$. It is clear from the hypotheses that T_0 is the orthogonal sum of the two real-linear Jordan *-isomorphisms $T_0|_{M_2(p_1)} : M_2(p_1) \to N_2(q_1)$ and $T_0|_{M_2(p_2)} : M_2(p_2) \to N_2(q_2)$. Moreover, since T is a surjective real-linear isometry (and hence a real-linear triple isomorphism, cf. Theorem 4.5) with $T(\mathbf{1}_M) = q_1 - q_2$. Therefore, the mapping $T : M \to N(q_1 - q_2)$ is a real-linear Jordan *-isomorphism, and T writes as the orthogonal sum of the real-linear Jordan *-isomorphisms $T|_{M_2(p_1)} : M_2(p_1) \to N_2(q_1)$ and $T|_{M_2(p_2)} : M_2(p_2) \to N_2(-q_2)$.

We shall next prove that the following pairs of (unital) maps $T_0|_{M_2(p_1)}, T|_{M_2(p_1)} : M_2(p_1) \to N_2(q_1)$ and $-T_0|_{M_2(p_2)}, T|_{M_2(p_2)} : M_2(p_2) \to N_2(-q_2)$ satisfy the hypothesis assumed for the pair T_0, T. To this end, let us take $u_j \in \mathcal{U}^0(M_2(p_j))$. The element $u = u_1 + u_2$ is a unitary in $\mathcal{U}^0(M)$, and by our hypotheses and the fact that $M = M_2(p_1) \oplus^\perp M_2(p_2)$ and $N = N_2(q_1) \oplus^\perp N_2(q_2)$ we obtain that

$$U_{T(u_1)} + U_{T(u_2)} = U_{T(u_1+u_2)} = U_{T_0(u_1+u_2)} = U_{T_0(u_1)} + U_{T_0(u_2)},$$

which implies that $U_{T_0(u_1)} = U_{T(u_1)}$ and $U_{T_0(u_2)} = U_{T(u_2)}$, because these maps have orthogonal ranges and supports.

We are therefore in a position to apply the conclusion in the first paragraph of this proof to the pairs $(T_0|_{M_2(p_1)}, T|_{M_2(p_1)})$ and $(-T_0|_{M_2(p_2)}, T|_{M_2(p_2)})$ to obtain that $T_0|_{M_2(p_1)} = T|_{M_2(p_1)}$ and $-T_0|_{M_2(p_2)} = T|_{M_2(p_2)}$. The desired conclusion follows by taking $p = p_1$. □

The extendibility of a surjective isometry Δ between the sets of unitary elements in two unital JB*-algebras M and N is not always an easy task. Actually, the extendibility is, in general, hopeless. It is natural to ask whether an additional preservation hypothesis can be added to guarantee a linear extension. The instability of the set of unitaries under Jordan products invites to discard the preservation of the Jordan product as extra hypothesis. In view of the properties of surjective real-linear isometries between JB*-algebras and the conclusion in Proposition 3.16, the natural hypothesis is the following:

$$\Delta\{u,v,u\} = \Delta U_u(v^*) = U_{\Delta(u)}\left(\Delta(v)^*\right) = \{\Delta(u),\Delta(v),\Delta(u)\}, \text{ for all } u,v \in \mathcal{U}(M).$$

This preservation hypothesis is enough under assuming a mild extra assumption.

PROPOSITION 6.40 ([**31**, Proposition 3.10]). *Let $\Delta : \mathcal{U}(M) \to \mathcal{U}(N)$ be a surjective isometry between the unitary sets of two unital JB*-algebras. Let $\{\mathcal{U}^0(M)\}\cup\{\mathcal{U}^j(M) : j \in \Lambda\}$ be the collection of all different connected components of $\mathcal{U}(M)$. Assume that for each $j \in \Lambda$ there exists $u_j \in \mathcal{U}(M)$ with $u_j^2 \in \mathcal{U}^j(M)$. Suppose additionally that*

$$\Delta\{u,v,u\} = \Delta U_u(v^*) = U_{\Delta(u)}\left(\Delta(v)^*\right) = \{\Delta(u),\Delta(v),\Delta(u)\}, \tag{6.38}$$

for all $u,v \in \mathcal{U}(M)$. Then Δ admits an extension to a surjective (real-linear) isometry from M onto N.

PROOF. Up to replacing the original Jordan product and involution on N by the one in the $\Delta(\mathbf{1}_M)$-isotope, we can assume that Δ is unital, and hence $\Delta(\mathcal{U}^0(M)) = \mathcal{U}^0(N)$. Let $T_j : M \to N$ denote the surjective real-linear isometry extending the mapping $\Delta|_{\mathcal{U}^j(M)} : \mathcal{U}^j(M) \to \Delta(\mathcal{U}^j(M))$ whose existence is assured by Theorem 6.27.

Fix $j,k \in \Lambda$ satisfying $\left(\mathcal{U}^j(M)\right)^* = \mathcal{U}^k(M)$. For each $u_j \in \mathcal{U}^j(M)$, it follows from (6.38) that

$$T_k(u_j^*) = \Delta(u_j^*) = \Delta U_{\mathbf{1}}(u_j^*) = U_{\mathbf{1}}(\Delta(u_j)^*) = T_j(u_j)^*,$$

which proves that $T_k(u_j^*) = T_j(u_j)^*$ for all $u_j \in \mathcal{U}^j(M)$. We fix $w_j \in \mathcal{U}^j(M)$. Since, by the Wright–Youngson–Russo–Dye theorem [**114**, Corollary], the convex hull of the set $\{\exp_{(w_j)}(ih) : h \in (M(w_j))_{sa}\}(\subset \mathcal{U}^j(M) = \mathcal{U}^0(M(w_j)))$ is norm dense in the closed unit ball of $M(w_j)$, it follows from the linearity and continuity of T_k and T_j that

$$T_k(x^*) = T_j(x)^* \text{ for all } x \in M. \tag{6.39}$$

Take now an arbitrary $u_0 \in \mathcal{U}^0(A)$. By (6.11) in Remark 6.18 we have $U_{u_0}(u_j^*) \in \mathcal{U}^k(M)$, and hence by the hypothesis (6.38) we get

$$\begin{aligned} U_{T_k(u_0)}(T_j(u_j)^*) &= U_{T_k(u_0)}(T_k(u_j^*)) = T_k U_{u_0}(u_j^*) = \Delta U_{u_0}(u_j^*) = U_{\Delta(u_0)}\left(\Delta(u_j^*)\right) \\ &= U_{T_0(u_0)}\left(T_k(u_j^*)\right) = (6.39) = U_{T_0(u_0)}\left(T_j(u_j)^*\right), \end{aligned}$$

for all $u_j \in \mathcal{U}^j(M)$. A new application of the Wright–Youngson–Russo–Dye theorem as in the previous paragraph –together with the surjectivity of T_j– implies that $U_{T_k(u_0)} = U_{T_0(u_0)}$ for all $u_0 \in \mathcal{U}^0(M)$. Proposition 6.39 assures the existence of a central projection $p_k \in M$ such that

$$T_k = T_0|_{M_2(p_k)} \oplus -T_0|_{M_2(\mathbf{1}_M - p_k)}. \tag{6.40}$$

Fix an arbitrary $l \in \Lambda$. By assumptions, there exists $u_l \in \mathcal{U}(M)$ with $u_l^2 \in \mathcal{U}^l(M)$. We can assume that $u_l \in \mathcal{U}^j(M)$ and $u_l^* \in \mathcal{U}^k(M)$ for some $j,k \in \Lambda$. By hypotheses,

$$T_l(u_l^2) = \Delta(u_l^2) = \Delta U_{u_l}(\mathbf{1}_M) = U_{\Delta(u_l)}\Delta(\mathbf{1}_M) = \Delta(u_l)^2 = T_j(u_l)^2. \tag{6.41}$$

By considering the central projections p_l and p_j given by (6.40), by orthogonality, we have

$$\begin{aligned} T_j(u_l)^2 &= (T_0(u_l \circ p_j) - T_0(u_l \circ (\mathbf{1}_M - p_j)))^2 \\ &= T_0(u_l^2 \circ p_j) + T_0(u_l^2 \circ (\mathbf{1}_M - p_j)) = T_0(u_l^2) \end{aligned}$$

and

$$T_l(u_l^2) = T_0(u_l^2 \circ p_l) - T_0(u_l^2 \circ (\mathbf{1}_M - p_l)),$$

which combined with (6.41) give $T_0(u_l^2 \circ (\mathbf{1}_M - p_l)) = 0$, and hence $u_l^2 \circ (\mathbf{1}_M - p_l) = 0$ (by the injectivity of T_0). Having in mind that p_l is a central projection, we have

$$0 = u_l^2 \circ (\mathbf{1}_M - p_l) = U_{u_l}(\mathbf{1}_M - p_l),$$

witnessing that $\mathbf{1}_M - p_l = 0$, and consequently, $T_l = T_0$. If we have in mind the arbitrariness of $l \in \Lambda$, the proof concludes by applying Corollary 6.38. □

The connections between the results in sections 3, 4 and 6 and the different metric invariants studied in each one of them are much deeper than one might expect. For example, the concrete description of the surjective isometries between sets of invertible elements in two unital JB*-algebras obtained in Theorem 4.8 can be also deduced from Theorem 3.8 and Theorem 6.23 (Remark 6.24). The argument relies on the following geometric characterization of unitary elements obtained in [**34**].

THEOREM 6.41 ([**34**, Theorem 3.8]). *Let u be an extreme point of the closed unit ball of a unital JB*-algebra M. Then u is a unitary if and only if the set*

$$\mathcal{M}_u := \{e \in \partial_e(B_M) : \|u \pm e\| \leq \sqrt{2}\}$$

contains an isolated point.

To see how to get the conclusion in Theorem 4.8, let $\Delta : \mathfrak{M} \to \mathfrak{N}$ be a surjective isometry in the conditions of the just quoted theorem. Having in mind that every JB*-algebra is semisimple, Theorem 3.8 proves the existence of a surjective real-linear isometry $T_0 : M \to N$ whose restriction to $\mathfrak{M}$ is Δ. Clearly T_0 maps $\partial_e(\mathcal{B}_M)$ onto $\partial_e(\mathcal{B}_N)$. Moreover, Theorem 6.41 assures that $T_0(\mathcal{U}(M)) = \mathcal{U}(N)$, and hence $T_0|_{\mathcal{U}(M)} : \mathcal{U}(M) \to \mathcal{U}(N)$ is a surjective isometry. In particular, $u = T_0(\mathbf{1}_M)$ is a unitary in N.

Let $\mathcal{U}(N_{(u^*)}) = \mathcal{U}(N(u))$ stand for the set of unitaries in the u-isotope $N(u)$ of N. As we observed in Lemma 4.4 $\mathcal{U}(N(u)) = \mathcal{U}(N)$. We can therefore conclude that $T_0|_{\mathcal{U}(M)} : \mathcal{U}(M) \to \mathcal{U}(N(u))$ is a unital surjective isometry. By applying Theorem 6.23 we deduce the existence of a central projection p in M and an isometric Jordan *-isomorphism $J : M \to N(u)$ such that

$$T_0(w) = J(p \circ w) + J((\mathbf{1}_M - p) \circ w^*), \tag{6.42}$$

for all $w \in \exp(iM_{sa}) \subseteq M_{\mathbf{1}}^{-1} \subseteq \mathfrak{M}$. A new application of the Wright–Youngson–Russo–Dye theorem for unital JB*-algebras (see [**114**, Corollary 2.4] or [**22**, Corollary 3.4.7]) asserts that the closed unit ball of M coincides with the closed convex-hull of the set $\exp(iM_{sa})$. Since the maps T_0 and $x \mapsto J(p\circ x + (\mathbf{1}_M - p)\circ x^*)$ are real-linear and continuous, we deduce from (6.42) that $T_0(x) = J(p\circ x) + J((\mathbf{1}_M - p)\circ x^*)$, for all $x \in M$, and consequently

$$\Delta(a) = T_0(a) - J(p \circ a) + J((\mathbf{1}_M - p) \circ a^*),$$

for all $a \in \mathfrak{M}$. The rest follows from the arguments given in the final part of the proof of Theorem 4.8.

Acknowledgments

The author would like to express his gratitude to the reviewers for their exceptional help to improve the final presentation of the manuscript, and their patience in revising text and results.

References

[1] E. M. Alfsen, *Compact convex sets and boundary integrals*, Ergebnisse der Mathematik und ihrer Grenzgebiete [Results in Mathematics and Related Areas], Band 57, Springer-Verlag, New York-Heidelberg, 1971. MR445271

[2] E. M. Alfsen and F. W. Shultz, *Geometry of state spaces of operator algebras*, Mathematics: Theory & Applications, Birkhäuser Boston, Inc., Boston, MA, 2003, DOI 10.1007/978-1-4612-0019-2. MR1947002

[3] K. Alvermann, *Real normed Jordan Banach algebras with an involution*, Arch. Math. (Basel) **47** (1986), no. 2, 135–150, DOI 10.1007/BF01193684. MR859264

[4] E. Andruchow, G. Corach, and D. Stojanoff, *Geometrical significance of Löwner-Heinz inequality*, Proc. Amer. Math. Soc. **128** (2000), no. 4, 1031–1037, DOI 10.1090/S0002-9939-99-05085-6. MR1636922

[5] W. Arveson, *A short course on spectral theory*, Graduate Texts in Mathematics, vol. 209, Springer-Verlag, New York, 2002, DOI 10.1007/b97227. MR1865513

[6] L. Asimow and A. J. Ellis, *Convexity theory and its applications in functional analysis*, London Mathematical Society Monographs, vol. 16, Academic Press, Inc. [Harcourt Brace Jovanovich, Publishers], London-New York, 1980. MR623459

[7] B. Aupetit, *Spectral characterization of the radical in Banach and Jordan-Banach algebras*, Math. Proc. Cambridge Philos. Soc. **114** (1993), no. 1, 31–35, DOI 10.1017/S0305004100071371. MR1219911

[8] B. Aupetit, *Spectral characterization of the socle in Jordan-Banach algebras*, Math. Proc. Cambridge Philos. Soc. **117** (1995), no. 3, 479–489, DOI 10.1017/S030500410007331X. MR1317491

[9] T. J. Barton and Y. Friedman, *Bounded derivations of* JB*-*triples*, Quart. J. Math. Oxford Ser. (2) **41** (1990), no. 163, 255–268, DOI 10.1093/qmath/41.3.255. MR1067482

[10] T. Barton and R. M. Timoney, *Weak*-continuity of Jordan triple products and its applications*, Math. Scand. **59** (1986), no. 2, 177–191, DOI 10.7146/math.scand.a-12160. MR884654

[11] C. J. K. Batty and D. W. Robinson, *Positive one-parameter semigroups on ordered Banach spaces*, Acta Appl. Math. **2** (1984), no. 3-4, 221–296, DOI 10.1007/BF02280855. MR753696

[12] J. Becerra-Guerrero, M. Cueto-Avellaneda, F. J. Fernández-Polo, and A. M. Peralta, *On the extension of isometries between the unit spheres of a JBW*-triple and a Banach space*, J. Inst. Math. Jussieu **20** (2021), no. 1, 277–303, DOI 10.1017/S1474748019000173. MR4205783

[13] J. Becerra Guerrero and A. M. Peralta, *Subdifferentiability of the norm and the Banach-Stone theorem for real and complex JB*-triples*, Manuscripta Math. **114** (2004), no. 4, 503–516, DOI 10.1007/s00229-004-0477-6. MR2081949

[14] R. P. Boas Jr., *Entire functions*, Academic Press, Inc., New York, 1954. MR68627

[15] F. F. Bonsall and J. Duncan, *Complete normed algebras*, Ergebnisse der Mathematik und ihrer Grenzgebiete [Results in Mathematics and Related Areas], Band 80, Springer-Verlag, New York-Heidelberg, 1973. MR423029

[16] O. Bratteli and D. W. Robinson, *Operator algebras and quantum statistical mechanics. 1: C^*- and W^*-algebras, symmetry groups, decomposition of states*, 2nd ed., Texts and Monographs in Physics, Springer-Verlag, New York, 1987, DOI 10.1007/978-3-662-02520-8. MR887100

[17] R. Braun, W. Kaup, and H. Upmeier, *A holomorphic characterization of Jordan C^*-algebras*, Math. Z. **161** (1978), no. 3, 277–290, DOI 10.1007/BF01214510. MR493373

[18] M. Burgos, F. J. Fernández-Polo, J. J. Garcés, J. Martínez Moreno, and A. M. Peralta, *Orthogonality preservers in C^*-algebras,* JB**-algebras and* JB**-triples*, J. Math. Anal. Appl. **348** (2008), no. 1, 220–233, DOI 10.1016/j.jmaa.2008.07.020. MR2449340

[19] M. Burgos, J. J. Garcés, and A. M. Peralta, *Automatic continuity of biorthogonality preservers between weakly compact JB^*-triples and atomic JBW^*-triples*, Studia Math. **204** (2011), no. 2, 97–121, DOI 10.4064/sm204-2-1. MR2805534

[20] M. Burgos, J. J. Garcés, and A. M. Peralta, *Automatic continuity of biorthogonality preservers between compact C^*-algebras and von Neumann algebras*, J. Math. Anal. Appl. **376** (2011), no. 1, 221–230, DOI 10.1016/j.jmaa.2010.11.007. MR2745402

[21] M. Burgos, E. A. Kaidi, A. Morales Campoy, A. M. Peralta, and M. Ramírez, *Von Neumann regularity and quadratic conorms in JB^*-triples and C^*-algebras*, Acta Math. Sin. (Engl. Ser.) **24** (2008), no. 2, 185–200, DOI 10.1007/s10114-007-0985-x. MR2383347

[22] M. Cabrera García and Á. Rodríguez Palacios, *Non-associative normed algebras. Vol. 1: The Vidav-Palmer and Gelfand-Naimark theorems*, Encyclopedia of Mathematics and its Applications, vol. 154, Cambridge University Press, Cambridge, 2014, DOI 10.1017/CBO9781107337763. MR3242640

[23] M. Cabrera García and Á. Rodríguez Palacios, *Non-associative normed algebras. Vol. 2: Representation theory and the Zel'manov approach*, Encyclopedia of Mathematics and its Applications, vol. 167, Cambridge University Press, Cambridge, 2018, DOI 10.1017/9781107337817. MR3791798

[24] A. Chigogidze, A. Karasev, K. Kawamura, and V. Valov, *On C^*-algebras with the approximate n-th root property*, Bull. Austral. Math. Soc. **72** (2005), no. 2, 197–212, DOI 10.1017/S0004972700035012. MR2183403

[25] J. B. Conway, *A course in functional analysis*, 2nd ed., Graduate Texts in Mathematics, vol. 96, Springer-Verlag, New York, 1990. MR1070713

[26] G. Corach, *Operator inequalities, geodesics and interpolation*, Functional analysis and operator theory (Warsaw, 1992), Banach Center Publ., vol. 30, Polish Acad. Sci. Inst. Math., Warsaw, 1994, pp. 101–115. MR1285601

[27] G. Corach, H. Porta, and L. Recht, *A geometric interpretation of Segal's inequality* $\|e^{X+Y}\| \leq \|e^{X/2}e^{Y}e^{X/2}\|$, Proc. Amer. Math. Soc. **115** (1992), no. 1, 229–231, DOI 10.2307/2159590. MR1075945

[28] G. Corach, H. Porta, and L. Recht, *The geometry of the space of selfadjoint invertible elements in a C^*-algebra*, Integral Equations Operator Theory **16** (1993), no. 3, 333–359, DOI 10.1007/BF01204226. MR1209304

[29] G. Corach, H. Porta, and L. Recht, *Geodesics and operator means in the space of positive operators*, Internat. J. Math. **4** (1993), no. 2, 193–202, DOI 10.1142/S0129167X9300011X. MR1217380

[30] G. Corach, H. Porta, and L. Recht, *Convexity of the geodesic distance on spaces of positive operators*, Illinois J. Math. **38** (1994), no. 1, 87–94. MR1245836

[31] M. Cueto-Avellaneda, Y. Enami, D. Hirota, T. Miura, and A. M. Peralta, *Surjective isometries between unitary sets of unital* JB**-algebras*, Linear Algebra Appl. **643** (2022), 39–79, DOI 10.1016/j.laa.2022.02.003. MR4384686

[32] M. Cueto-Avellaneda and A. M. Peralta, *On the Mazur-Ulam property for the space of Hilbert-space-valued continuous functions*, J. Math. Anal. Appl. **479** (2019), no. 1, 875–902, DOI 10.1016/j.jmaa.2019.06.056. MR3987064

[33] M. Cueto-Avellaneda and A. M. Peralta, *The Mazur-Ulam property for commutative von Neumann algebras*, Linear Multilinear Algebra **68** (2020), no. 2, 337–362, DOI 10.1080/03081087.2018.1505823. MR4048060

[34] M. Cueto-Avellaneda and A. M. Peralta, *Metric characterisation of unitaries in* JB**-algebras*, Mediterr. J. Math. **17** (2020), no. 4, Paper No. 124, 21, DOI 10.1007/s00009-020-01556-w. MR4119511

[35] M. Cueto-Avellaneda and A. M. Peralta, *Can one identify two unital JB^*-algebras by the metric spaces determined by their sets of unitaries?*, Linear Multilinear Algebra **70** (2022), no. 22, 7702–7727, DOI 10.1080/03081087.2021.2003745. MR4571052

[36] T. Dang, *Real isometries between* JB**-triples*, Proc. Amer. Math. Soc. **114** (1992), no. 4, 971–980, DOI 10.2307/2159615. MR1056677

[37] T. Dang, Y. Friedman, and B. Russo, *Affine geometric proofs of the Banach Stone theorems of Kadison and Kaup*, Proceedings of the Seventh Great Plains Operator Theory Seminar (Lawrence, KS, 1987), Rocky Mountain J. Math. **20** (1990), no. 2, 409–428, DOI 10.1216/rmjm/1181073116. MR1065839

[38] M. M. Deza and E. Deza, *Encyclopedia of distances*, 4th ed., Springer, Berlin, 2016, DOI 10.1007/978-3-662-52844-0. MR3559482

[39] S. Dineen, *The second dual of a JB* triple system*, Complex analysis, functional analysis and approximation theory (Campinas, 1984), North-Holland Math. Stud., vol. 125, North-Holland, Amsterdam, 1986, pp. 67–69. MR893410

[40] C. M. Edwards, *On Jordan W^*-algebras*, Bull. Sci. Math. **104** (1980), no. 2, 393–403.

[41] C. M. Edwards and G. T. Rüttimann, *Compact tripotents in bi-dual* JB*-*triples*, Math. Proc. Cambridge Philos. Soc. **120** (1996), no. 1, 155–173, DOI 10.1017/S0305004100074740. MR1373355

[42] G. M. Escolano, A. M. Peralta, and A. R. Villena, *Lie–Trotter formulae in Jordan-Banach algebras with applications to the study of spectral-valued multiplicative functionals*, Result. Math. **79** (2024), no. 1, Paper No. 17.

[43] A. Fernández López, E. García Rus, E. Sánchez Campos, and M. Siles Molina, *Strong regularity and generalized inverses in Jordan systems*, Comm. Algebra **20** (1992), no. 7, 1917–1936, DOI 10.1080/00927879208824440. MR1167082

[44] F. J. Fernández-Polo, J. Martínez Moreno, and A. M. Peralta, *Geometric characterization of tripotents in real and complex* JB*-*triples*, J. Math. Anal. Appl. **295** (2004), no. 2, 435–443, DOI 10.1016/j.jmaa.2004.02.027. MR2072023

[45] F. J. Fernández-Polo, J. Martínez, and A. M. Peralta, *Surjective isometries between real JB*-triples*, Math. Proc. Cambridge Philos. Soc. **137** (2004), no. 3, 703–723, DOI 10.1017/S0305004104007893. MR2103925

[46] R. J. Fleming and J. E. Jamison, *Isometries on Banach spaces: function spaces*, Chapman & Hall/CRC Monographs and Surveys in Pure and Applied Mathematics, vol. 129, Chapman & Hall/CRC, Boca Raton, FL, 2003. MR1957004

[47] R. J. Fleming and J. E. Jamison, *Isometries on Banach spaces. Vol. 2: Vector-valued function spaces*, Chapman & Hall/CRC Monographs and Surveys in Pure and Applied Mathematics, vol. 138, Chapman & Hall/CRC, Boca Raton, FL, 2008. MR2361284

[48] Y. Friedman and B. Russo, *Structure of the predual of a JBW^*-triple*, J. Reine Angew. Math. **356** (1985), 67–89, DOI 10.1515/crll.1985.356.67. MR779376

[49] Y. Friedman and B. Russo, *The Gel′fand-Naĭmark theorem for* JB*-*triples*, Duke Math. J. **53** (1986), no. 1, 139–148, DOI 10.1215/S0012-7094-86-05308-1. MR835800

[50] J. J. Garcés and A. M. Peralta, *One-parameter groups of orthogonality preservers on* C*-*algebras*, Banach J. Math. Anal. **15** (2021), no. 1, Paper No. 24, 23, DOI 10.1007/s43037-020-00100-y. MR4181733

[51] J. J. Garcés and A. M. Peralta, *One-parameter semigroups of orthogonality preservers of* JB*-*algebras*, Adv. Oper. Theory **6** (2021), no. 2, Paper No. 43, 25, DOI 10.1007/s43036-021-00135-4. MR4234256

[52] J. Glück and M. R. Weber, *Almost interior points in ordered Banach spaces and the long-term behaviour of strongly positive operator semigroups*, Studia Math. **254** (2020), no. 3, 237–263, DOI 10.4064/sm190111-18-10. MR4118668

[53] K. R. Goodearl, *Notes on real and complex C^*-algebras*, Shiva Mathematics Series, vol. 5, Shiva Publishing Ltd., Nantwich, 1982. MR677280

[54] H. Hanche-Olsen, E. Størmer, *Jordan operator algebras*, Pitman, London, 1984.

[55] L. A. Harris, *Bounded symmetric homogeneous domains in infinite dimensional spaces*, Proceedings on Infinite Dimensional Holomorphy (Internat. Conf., Univ. Kentucky, Lexington, Ky., 1973), Lecture Notes in Math., Vol. 364, Springer, Berlin-New York, 1974, pp. 13–40. MR407330

[56] R. Harte, *The exponential spectrum in Banach algebras*, Proc. Amer. Math. Soc. **58** (1976), 114–118, DOI 10.2307/2041369. MR407603

[57] O. Hatori, *Isometries between groups of invertible elements in Banach algebras*, Studia Math. **194** (2009), no. 3, 293–304, DOI 10.4064/sm194-3-5. MR2539557

[58] O. Hatori, *Isometries of the unitary groups in C^*-algebras*, Studia Math. **221** (2014), no. 1, 61–86, DOI 10.4064/sm221-1-4. MR3194062

[59] O. Hatori, *Algebraic properties of isometries between groups of invertible elements in Banach algebras*, J. Math. Anal. Appl. **376** (2011), no. 1, 84–93, DOI 10.1016/j.jmaa.2010.11.027. MR2745390

[60] O. Hatori and T. Miura, *On a characterization of the maximal ideal spaces of commutative C^*-algebras in which every element is the square of another*, Proc. Amer. Math. Soc. **128** (2000), no. 4, 1185–1189, DOI 10.1090/S0002-9939-99-05454-4. MR1690991

[61] O. Hatori, G. Hirasawa, T. Miura, and L. Molnár, *Isometries and maps compatible with inverted Jordan triple products on groups*, Tokyo J. Math. **35** (2012), no. 2, 385–410, DOI 10.3836/tjm/1358951327. MR3058715

[62] O. Hatori and L. Molnár, *Isometries of the unitary group*, Proc. Amer. Math. Soc. **140** (2012), no. 6, 2127–2140, DOI 10.1090/S0002-9939-2011-11216-4. MR2888199

[63] O. Hatori and L. Molnár, *Isometries of the unitary groups and Thompson isometries of the spaces of invertible positive elements in C^*-algebras*, J. Math. Anal. Appl. **409** (2014), no. 1, 158–167, DOI 10.1016/j.jmaa.2013.06.065. MR3095026

[64] O. Hatori and K. Watanabe, *Isometries between groups of invertible elements in C^*-algebras*, Studia Math. **209** (2012), no. 2, 103–106, DOI 10.4064/sm209-2-1. MR2943837

[65] L. Hogben and K. McCrimmon, *Maximal modular inner ideals and the Jacobson radical of a Jordan algebra*, J. Algebra **68** (1981), no. 1, 155–169, DOI 10.1016/0021-8693(81)90291-X. MR604300

[66] T. Ho, J. Martinez-Moreno, A. M. Peralta, and B. Russo, *Derivations on real and complex JB *-triples*, J. London Math. Soc. (2) **65** (2002), no. 1, 85–102, DOI 10.1112/S002461070100271X. MR1875137

[67] S.-t. Hu, *Homotopy theory*, Pure and Applied Mathematics, Vol. VIII, Academic Press, New York-London, 1959. MR106454

[68] J. M. Isidro, *Holomorphic automorphisms of the unit balls of Hilbert C^*-modules*, Glasg. Math. J. **45** (2003), no. 2, 249–262, DOI 10.1017/S0017089503001216. MR1997703

[69] J. M. Isidro and Á. Rodríguez-Palacios, *Isometries of* JB*-algebras*, Manuscripta Math. **86** (1995), no. 3, 337–348, DOI 10.1007/BF02567998. MR1323796

[70] N. Jacobson, *Structure theory of Jordan algebras*, University of Arkansas Lecture Notes in Mathematics, vol. 5, University of Arkansas, Fayetteville, AR, 1981. MR634508

[71] F. B. Jamjoom, A. A. Siddiqui, H. M. Tahlawi, and A. M. Peralta, *Approximation and convex decomposition by extremals and the λ-function in* JBW**-triples*, Q. J. Math. **66** (2015), no. 2, 583–603, DOI 10.1093/qmath/hau036. MR3356839

[72] P. Jordan, J. von Neumann, and E. Wigner, *On an algebraic generalization of the quantum mechanical formalism*, Ann. of Math. (2) **35** (1934), no. 1, 29–64, DOI 10.2307/1968117. MR1503141

[73] R. V. Kadison, *Isometries of operator algebras*, Ann. of Math. (2) **54** (1951), 325–338, DOI 10.2307/1969534. MR43392

[74] R. V. Kadison and J. R. Ringrose, *Fundamentals of the theory of operator algebras. Vol. I: Elementary theory*, Pure and Applied Mathematics, vol. 100, Academic Press, Inc. [Harcourt Brace Jovanovich, Publishers], New York, 1983. MR719020

[75] O. F. K. Kalenda and A. M. Peralta, *Extension of isometries from the unit sphere of a rank-2 Cartan factor*, Anal. Math. Phys. **11** (2021), no. 1, Paper No. 15, 25, DOI 10.1007/s13324-020-00448-2. MR4194426

[76] W. Kaup, *A Riemann mapping theorem for bounded symmetric domains in complex Banach spaces*, Math. Z. **183** (1983), no. 4, 503–529, DOI 10.1007/BF01173928. MR710768

[77] W. Kaup, *On spectral and singular values in* JB**-triples*, Proc. Roy. Irish Acad. Sect. A **96** (1996), no. 1, 95–103. MR1644656

[78] W. Kaup and H. Upmeier, *Jordan algebras and symmetric Siegel domains in Banach spaces*, Math. Z. **157** (1977), no. 2, 179–200, DOI 10.1007/BF01215150. MR492414

[79] S. Kowalski and Z. Słodkowski, *A characterization of multiplicative linear functionals in Banach algebras*, Studia Math. **67** (1980), no. 3, 215–223, DOI 10.4064/sm-67-3-215-223. MR592387

[80] M. Kusuda, *A simple proof for* JB**-triple structures in Hilbert C^*-modules*, Sci. Math. Jpn. **77** (2014), no. 1, 139–142. MR3330094

[81] B. Lemmens and R. Nussbaum, *Nonlinear Perron-Frobenius theory*, Cambridge Tracts in Mathematics, vol. 189, Cambridge University Press, Cambridge, 2012, DOI 10.1017/CBO9781139026079. MR2953648

[82] B. Lemmens, M. Roelands, and M. Wortel, *Isometries of infinite dimensional Hilbert geometries*, J. Topol. Anal. **10** (2018), no. 4, 941–959, DOI 10.1142/S1793525318500255. MR3881047

[83] B. Lemmens, M. Roelands, and M. Wortel, *Hilbert and Thompson isometries on cones in JB-algebras*, Math. Z. **292** (2019), no. 3-4, 1511–1547, DOI 10.1007/s00209-018-2144-8. MR3980302

[84] B. Li, *Real operator algebras*, World Scientific Publishing Co., Inc., River Edge, NJ, 2003, DOI 10.1142/9789812795182. MR1995682

[85] L. Li, A. M. Peralta, L. Wang, and Y.-S. Wang, *Weak-2-local isometries on uniform algebras and Lipschitz algebras*, Publ. Mat. **63** (2019), no. 1, 241–264, DOI 10.5565/PUBLMAT6311908. MR3908793

[86] O. Loos, *On the set of invertible elements in Banach Jordan algebras*, Results Math. **29** (1996), no. 1-2, 111–114, DOI 10.1007/BF03322210. MR1377684

[87] P. Mankiewicz, *On extension of isometries in normed linear spaces* (English, with Russian summary), Bull. Acad. Polon. Sci. Sér. Sci. Math. Astronom. Phys. **20** (1972), 367–371. MR312214

[88] J. Martínez and A. M. Peralta, *Separate weak*-continuity of the triple product in dual real* JB*-*triples*, Math. Z. **234** (2000), no. 4, 635–646, DOI 10.1007/s002090050002. MR1778403

[89] H. Matui, M. Nagisa, and M. Yamamoto, *Square root closed C^*-algebras*, Kyushu J. Math. **62** (2008), no. 1, 107–122, DOI 10.2206/kyushujm.62.107. MR2413786

[90] K. McCrimmon, *A general theory of Jordan rings*, Proc. Nat. Acad. Sci. U.S.A. **56** (1966), 1072–1079, DOI 10.1073/pnas.56.4.1072. MR202783

[91] K. McCrimmon, *The radical of a Jordan algebra*, Proc. Nat. Acad. Sci. U.S.A. **62** (1969), 671–678, DOI 10.1073/pnas.62.3.671. MR268238

[92] K. McCrimmon, *A characterization of the radical of a Jordan algebra*, J. Algebra **18** (1971), 103–111, DOI 10.1016/0021-8693(71)90129-3. MR277583

[93] L. Molnár, *Thompson isometries of the space of invertible positive operators*, Proc. Amer. Math. Soc. **137** (2009), no. 11, 3849–3859, DOI 10.1090/S0002-9939-09-09963-8. MR2529894

[94] M. Mori and N. Ozawa, *Mankiewicz's theorem and the Mazur-Ulam property for* C*-*algebras*, Studia Math. **250** (2020), no. 3, 265–281, DOI 10.4064/sm180727-14-11. MR4034747

[95] R. Mortini and M. von Renteln, *Ideals in the Wiener algebra W^+*, J. Austral. Math. Soc. Ser. A **46** (1989), no. 2, 220–228. MR973315

[96] T. Nieminen, *On ultrapseudocompact and related spaces*, Ann. Acad. Sci. Fenn. Ser. A I Math. **3** (1977), no. 2, 185–205. MR644158

[97] T. Oikhberg, A. M. Peralta, and D. Puglisi, *Automatic continuity of orthogonality or disjointness preserving bijections*, Rev. Mat. Complut. **26** (2013), no. 1, 57–88, DOI 10.1007/s13163-011-0089-0. MR3016618

[98] T. Oikhberg and A. M. Peralta, *Automatic continuity of orthogonality preservers on a non-commutative $L^p(\tau)$ space*, J. Funct. Anal. **264** (2013), no. 8, 1848–1872, DOI 10.1016/j.jfa.2013.01.019. MR3022728

[99] T. Oikhberg, A. M. Peralta, and M. Ramírez, *Automatic continuity of M-norms on C^*-algebras*, J. Math. Anal. Appl. **381** (2011), no. 2, 799–811, DOI 10.1016/j.jmaa.2011.03.066. MR2802115

[100] T. W. Palmer, *Banach algebras and the general theory of *-algebras. Vol. I: Algebras and Banach algebras*, Encyclopedia of Mathematics and its Applications, vol. 49, Cambridge University Press, Cambridge, 1994, DOI 10.1017/CBO9781107325777. MR1270014

[101] A. L. T. Paterson and A. M. Sinclair, *Characterisation of isometries between C^*-algebras*, J. London Math. Soc. (2) **5** (1972), 755–761, DOI 10.1112/jlms/s2-5.4.755. MR324431

[102] G. K. Pedersen, *C^*-algebras and their automorphism groups*, London Mathematical Society Monographs, vol. 14, Academic Press, Inc. [Harcourt Brace Jovanovich, Publishers], London-New York, 1979. MR548006

[103] A. M. Peralta, *On the axiomatic definition of real* JB*-*triples*, Math. Nachr. **256** (2003), 100–106, DOI 10.1002/mana.200310073. MR1989381

[104] A. M. Peralta, *Surjective isometries between sets of invertible elements in unital Jordan-Banach algebras*, J. Math. Anal. Appl. **502** (2021), no. 2, Paper No. 125284, 21, DOI 10.1016/j.jmaa.2021.125284. MR4254558

[105] A. M. Peralta, *On the extension of surjective isometries whose domain is the unit sphere of a space of compact operators*, Filomat **36** (2022), no. 9, 3075–3090. MR4520880

[106] A. M. Peralta and B. Russo, *Automatic continuity of derivations on* C*-*algebras and* JB*-*triples*, J. Algebra **399** (2014), 960–977, DOI 10.1016/j.jalgebra.2013.10.017. MR3144622

[107] A. M. Peralta and R. Švarc, *A strengthened Kadison's transitivity theorem for unital JB*-algebras with applications to the Mazur–Ulam property*, arXiv:2301.00895, 2023.

[108] C. E. Rickart, *General theory of Banach algebras*, The University Series in Higher Mathematics, D. Van Nostrand Co., Inc., Princeton, N.J.-Toronto-London-New York, 1960. MR115101

[109] M. Roelands and M. Wortel, *Hilbert isometries and maximal deviation preserving maps on JB-algebras*, Adv. Math. **352** (2019), 836–861, DOI 10.1016/j.aim.2019.06.027. MR3975028

[110] M. Takesaki, *Theory of operator algebras. I*, Encyclopaedia of Mathematical Sciences, Springer-Verlag, Berlin, 2003.

[111] J. Väisälä, A proof of the Mazur-Ulam theorem, *Amer. Math. Monthly* **110** (2003), 633–635.

[112] A. Vogt, *Maps which preserve equality of distance*, Studia Math. **45** (1973), 43–48, DOI 10.4064/sm-45-1-43-48. MR333676

[113] J. D. M. Wright, *Jordan C^*-algebras*, Michigan Math. J. **24** (1977), no. 3, 291–302. MR487478

[114] J. D. M. Wright and M. A. Youngson, *A Russo-Dye theorem for Jordan C^*-algebras*, Functional analysis: surveys and recent results (Proc. Conf., Paderborn, 1976), North-Holland Math. Stud., Vol. 27, North-Holland, Amsterdam-New York-Oxford, 1977, pp. 279–282. MR487472

[115] J. D. M. Wright and M. A. Youngson, *On isometries of Jordan algebras*, J. London Math. Soc. (2) **17** (1978), no. 2, 339–344, DOI 10.1112/jlms/s2-17.2.339. MR482212

[116] M. A. Youngson, *A Vidav theorem for Banach Jordan algebras*, Math. Proc. Cambridge Philos. Soc. **84** (1978), no. 2, 263–272, DOI 10.1017/S0305004100055092. MR493372

[117] M. A. Youngson, *Nonunital Banach Jordan algebras and C^*-triple systems*, Proc. Edinburgh Math. Soc. (2) **24** (1981), no. 1, 19–29, DOI 10.1017/S0013091500003965. MR621449

[118] W. Żelazko, *Banach algebras*, Elsevier Publishing Co., Amsterdam-London-New York; PWN—Polish Scientific Publishers, Warsaw, 1973. Translated from the Polish by Marcin E. Kuczma. MR448079

Instituto de Matemáticas de la Universidad de Granada (IMAG). Departamento de Análisis Matemático, Facultad de Ciencias, Universidad de Granada, 18071 Granada, Spain.

Email address: aperalta@ugr.es

Selected Published Titles in This Series

106 **Gilles Godefroy, Mohammad Sal Moslehian, and Juan Benigno Seoane-Sepúlveda, Editors,** Geometry of Banach Spaces and Related Fields, 2024

105 **A. Connes, C. Consani, B. I. Dundas, M. Khalkhali, and H. Moscovici, Editors,** Cyclic Cohomology at 40: Achievements and Future Prospects, 2023

104 **A. Kechris, N. Makarov, D. Ramakrishnan, and X. Zhu, Editors,** Nine Mathematical Challenges, 2021

103 **Sergey Novikov, Igor Krichever, Oleg Ogievetsky, and Senya Shlosman, Editors,** Integrability, Quantization, and Geometry, 2021

102 **David T. Gay and Weiwei Wu, Editors,** Breadth in Contemporary Topology, 2019

101 **Avraham Aizenbud, Dmitry Gourevitch, David Kazhdan, and Erez M. Lapid, Editors,** Representations of Reductive Groups, 2019

100 **Chiu-Chu Melissa Liu and Motohico Mulase, Editors,** Topological Recursion and its Influence in Analysis, Geometry, and Topology, 2018

99 **Vicente Muñoz, Ivan Smith, and Richard P. Thomas, Editors,** Modern Geometry, 2018

98 **Amir-Kian Kashani-Poor, Ruben Minasian, Nikita Nekrasov, and Boris Pioline, Editors,** String-Math 2016, 2018

97 **Tommaso de Fernex, Brendan Hassett, Mircea Mustaţă, Martin Olsson, Mihnea Popa, and Richard Thomas, Editors,** Algebraic Geometry: Salt Lake City 2015 (Parts 1 and 2), 2018

96 **Si Li, Bong H. Lian, Wei Song, and Shing-Tung Yau, Editors,** String-Math 2015, 2017

95 **Izzet Coskun, Tommaso de Fernex, and Angela Gibney, Editors,** Surveys on Recent Developments in Algebraic Geometry, 2017

94 **Mahir Bilen Can, Editor,** Algebraic Groups: Structure and Actions, 2017

93 **Vincent Bouchard, Charles Doran, Stefan Méndez-Diez, and Callum Quigley, Editors,** String-Math 2014, 2016

92 **Kailash C. Misra, Daniel K. Nakano, and Brian J. Parshall, Editors,** Lie Algebras, Lie Superalgebras, Vertex Algebras and Related Topics, 2016

91 **V. Sidoravicius and S. Smirnov, Editors,** Probability and Statistical Physics in St. Petersburg, 2016

90 **Ron Donagi, Sheldon Katz, Albrecht Klemm, and David R. Morrison, Editors,** String-Math 2012, 2015

89 **D. Dolgopyat, Y. Pesin, M. Pollicott, and L. Stoyanov, Editors,** Hyperbolic Dynamics, Fluctuations and Large Deviations, 2015

88 **Ron Donagi, Michael R. Douglas, Ljudmila Kamenova, and Martin Rocek, Editors,** String-Math 2013, 2014

87 **Helge Holden, Barry Simon, and Gerald Teschl, Editors,** Spectral Analysis, Differential Equations and Mathematical Physics: A Festschrift in Honor of Fritz Gesztesy's 60th Birthday, 2013

86 **Kailash C. Misra, Daniel K. Nakano, and Brian J. Parshall, Editors,** Recent Developments in Lie Algebras, Groups and Representation Theory, 2012

85 **Jonathan Block, Jacques Distler, Ron Donagi, and Eric Sharpe, Editors,** String-Math 2011, 2012

84 **Alex H. Barnett, Carolyn S. Gordon, Peter A. Perry, and Alejandro Uribe, Editors,** Spectral Geometry, 2012

83 **Hisham Sati and Urs Schreiber, Editors,** Mathematical Foundations of Quantum Field Theory and Perturbative String Theory, 2011

For a complete list of titles in this series, visit the AMS Bookstore at **www.ams.org/bookstore/pspumseries/**.